Finance for Engineers

T0189554

F.K. Crundwell

Finance for Engineers

Evaluation and Funding of Capital Projects

 Springer

F.K. Crundwell, PhD
CM Solutions
Wimbledon
London SW19
United Kingdom

ISBN 978-1-84996-708-2   e-ISBN 978-1-84800-033-9

DOI 10.1007/978-1-84800-033-9

British Library Cataloguing in Publication Data
Crundwell, F. K.
   Finance for engineers : evaluation and funding of capital projects
   1. Capital investments 2. Project management - Finance
   3. Engineering - Finance
   I. Title
   620'.00681

Cover design: eStudio Calamar S.L., Girona, Spain

Printed on acid-free paper

9 8 7 6 5 4 3 2 1

springer.com

For Natascha and Nicholas

Preface

Investment decisions are critical part of a company's success, and capital budgeting is a central topic of financial management. However, in contrast to other strategic decisions in business, decisions on capital projects are decentralized, from corporate to divisions, divisions to operations, from operations to departments. Engineers at various levels within the organization who make a capital proposal are contributing to the strategic success of the organization. This book will assist engineers to contribute strategically to the organization, so that they can contribute to the success of their company and can make a success of their careers.

The purpose of this book is to provide engineers and managers with a working knowledge of the financial evaluation and the funding of capital projects. Four main topics concerning capital investment are covered, namely, the *context* of these decisions, the assessment of their *returns*, the assessment of their *risks*, and their funding and *financing*. Engineers have developed the field of *engineering economics*, and financial managers have developed the discipline of *capital budgeting*, each with its own approach and slightly different terminology. This book borrows from both these disciplines in order to provide engineers with the best possible knowledge about capital projects.

An important aspect of the approach adopted here is to provide context to the techniques, methods and concepts of the economic assessment of engineering decisions. An engineer who knows her subject is able to communicate her ideas to people of different professions and with different training. An essential part of effective communication is to understand the other person's paradigm, to know their basic assumptions. As a result, it is important to understand the basics of decision making, of financial management, and accounting. It is also important to understand the terminology used by different professions. Part I of the book examines these topics, which forms the introduction to the economic assessment of an engineering project.

A project is assessed on its economic merits, its profitability. Part II of the book covers the techniques and methods of capital budgeting and engineering economics. In-depth case studies demonstrate the application of these techniques to practical situations.

The assessment of the economic merits of a project usually refers to the returns expected from the project. The cornerstones of finance are risk and return, and it is essential to address the risk of the project in a discussion of its merits. The risk is the chance of the expected returns not materializing. The types of risk, such as stand-alone risk and portfolio risk, for a capital project are examined in Part III of this book.

It is also important to understand the financing of a project, the sources of finance, and the structure of financial arrangements such as project finance and public-private partnerships. Part IV of the book examines both the internal and external sources of funding and the structuring of the financial and legal relationships that enable the project to be built.

Intended Readers

This book is designed for use both as an undergraduate textbook for engineers and as a reference for practising engineers. The material assumes little prior knowledge of engineering, accounting or financial management. It is intended for use as a text in undergraduate courses on either engineering economics or financial management. It may also be used in an engineering design course where the financial implications of design are under discussion.

This book has been designed to accommodate all branches of engineering. Examples and case studies have been drawn from as wide a range of the engineering disciplines as possible. For example, there are case studies on, amongst others, hydroelectric power, pumping stations, toll roads, mining, processing technologies, and petroleum production. In addition, these examples cover different geographies, such as the US, the UK, Australia and Brazil, so that this book is relevant to the experience of engineers across the World. Another important example of this is that the difference in the accounting terminology used in the US and the UK for the preparation of financial statements is examined, and a translation table is provided as an Appendix.

Educational Package

The organization of the themes and the major topics of the book is shown in the diagram. A number of different courses can be derived from this book. An Introductory course in Engineering Economics might include a topic from the Part I as an introduction, and then focus on Part II, in which the evaluation of capital projects is discussed in detail. A more advanced course in Engineering Economics could select Chaps. 1 and 4 from Part I, move through Part II and end with Part IV. A course with the title of this book, Finance for Engineers, might cover the material in Part I,

Part II, Chapter 11 from Part III, and then end with Part IV. An advanced course in Financial Management for Engineers might cover all the topics.

Since this book is intended as a textbook and a reference, each chapter includes numerous worked examples and case studies. Each chapter also includes review questions and tutorial exercises. The review questions are designed to emphasize

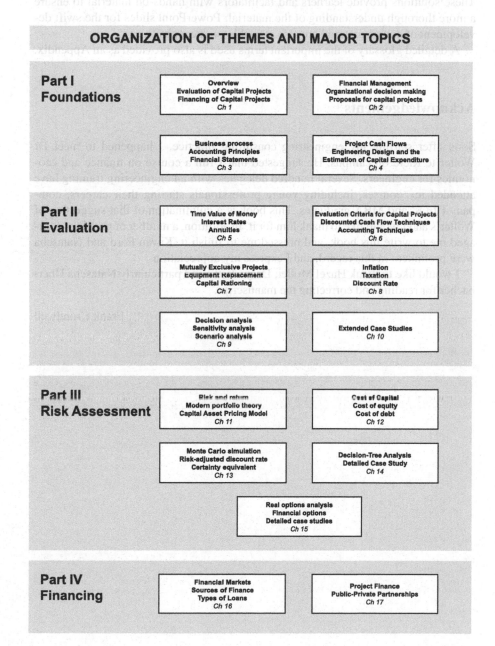

ORGANIZATION OF THEMES AND MAJOR TOPICS

Part I
Foundations

| Overview
Evaluation of Capital Projects
Financing of Capital Projects
Ch 1 | Financial Management
Organizational decision making
Proposals for capital projects
Ch 2 |

| Business process
Accounting Principles
Financial Statements
Ch 3 | Project Cash Flows
Engineering Design and the
Estimation of Capital Expenditure
Ch 4 |

Part II
Evaluation

| Time Value of Money
Interest Rates
Annuities
Ch 5 | Evaluation Criteria for Capital Projects
Discounted Cash Flow Techniques
Accounting Techniques
Ch 6 |

| Mutually Exclusive Projects
Equipment Replacement
Capital Rationing
Ch 7 | Inflation
Taxation
Discount Rate
Ch 8 |

| Decision analysis
Sensitivity analysis
Scenario analysis
Ch 9 | Extended Case Studies
Ch 10 |

Part III
Risk Assessment

| Risk and return
Modern portfolio theory
Capital Asset Pricing Model
Ch 11 | Cost of Capital
Cost of equity
Cost of debt
Ch 12 |

| Monte Carlo simulation
Risk-adjusted discount rate
Certainty equivalent
Ch 13 | Decision-Tree Analysis
Detailed Case Study
Ch 14 |

| Real options analysis
Financial options
Detailed case studies
Ch 15 |

Part IV
Financing

| Financial Markets
Sources of Finance
Types of Loans
Ch 16 | Project Finance
Public-Private Partnerships
Ch 17 |

the concepts that are covered in the chapter, while the exercises allow the reader to practise the application of the techniques discussed in the chapter. Worked solutions to almost all of the exercises are included in an Appendix.

Spreadsheet solutions of all the questions and the case studies presented in the book can be obtained from the website (www.springer.com/978-1-84800-032-2). These solutions provide learners and facilitators with hands-on material to ensure a more thorough understanding of the material. PowerPoint slides for the swift development of course material are also available on the website.

A detailed glossary of the important terms used is also provided as an Appendix.

Acknowledgements

Soon after starting my engineering consulting practice, I happened to meet Dr Wolter te Riele on campus. He suggested that I run a course on finance and economics for engineers. Several hundred delegates with an engineering training have attended my courses, including young professionals starting their careers, company CEOs, and MBA graduates. This book is a culmination of that suggestion of Wolter's and I would like to thank him for it. In addition, a number of people encouraged me to write this book, and pressed me to finish it. Kevan Ford and Natascha were prominent in this regard, and I express my appreciation.

I would like to thank Hazel Moller, Deryn Petty and particularly Natascha Uberbacher for reading and correcting the manuscript.

Frank Crundwell

Contents

List of Abbreviations

AOC	Annual Operating Charge
APR	Annual Percentage Rate
B/C	Benefit Cost Ratio
BOOT	Build, Own, Operate Transfer
BTO	Build, Transfer, Operate
CEV	Certainty Equivalent Value
CF	Cash Flow
CPI	Consumer Price Index
CR	Capital Recovery
DDB	Double Declining Balance
EAC	Equivalent Annual Charge
EPC	Engineering, Procurement and Construction
EPCM	Engineering, Procurement and Construction Management
ESL	Economic Service Life
FV	Future Value
FVA	Future Value of an Annuity
FVIFA	Future Value Interest Factor of an Annuity
GAAP	Generally Accepted Accounting Practices
GDP	Gross Domestic Product
ICO	Initial Capital Outlay
IPO	Initial Public Offering
LIBOR	London Interbank Offer Rate
MACRS	Modified Accelerated Cost Recovery System
MARR	Minimum Attractive Rate of Return
NPV	Net Present Value
OTC	Over The Counter
PV	Present Value
QDB	Quarter Declining Balance
RADR	Risk-Adjusted Discount Rate
ROA	Return On Assets
ROE	Return On Equity
SEC	Securities Exchange Commission
SGA	Sales, General and Administrative
SL	Straight Line Depreciation
SOYD	Sum Of Years Digits Depreciation
TV	Terminal Value
VaR	Value At Risk
WACC	Weighted Average Cost of Capital

Part I
Foundations

The engineering design and business assessment of capital projects are intertwined with another. The business and the engineering aspects of investing in a capital project form the foundations for financial planning and capital budgeting. The business context includes the roles and responsibilities of the financial management of a company, the sources and use of funds, and the decision-making functions within a company. An understanding of accounting, particularly of the financial statements, is important in determining the effects of capital investments on the financial health of the company. The purpose of an engineering design is twofold: to enable the project to be built and to estimate the cost to build it. In this way, the engineering design plays a central role within a company by determining the financial requirements of a project and by ensuring its successful implementation.

The foundation for the evaluation of capital projects is an understanding of the components of the cash flows of a project. Chapter 1 provides an overview of evaluation and financing activities and an introduction to the main themes and topics of the other chapters. Particularly important in Chapter 1 are the concepts of project financials and of free cash flow. After this overview, the foundation topics are examined. Chapter 2 is a discussion of the context within a company for decision-making concerning capital projects. Chapter 3 presents the financial statements of a company. The development of the free cash flows for the project is discussed in Chapter 4. A full appreciation of the components of the cash flows often requires an understanding of the financial statements of a company and an understanding of engineering design. The integration of the engineering design with the decision-making process within a company is discussed in Chapters 2 and 4.

Chapter 1
An Overview of the Evaluation and Financing of Capital Projects

1.1 Evaluation and Funding of Projects

The activities of a company can be seen as a collection or a succession of projects. These projects involve the acquisition of assets and the operational use of these assets for production. For example, an oil company will acquire lease rights for an area of land, explore for oil, drill wells, and produce oil from the wells for a number of years. This may be the only activity, or one of many of the activities, that constitute the company. As such, the project is the basic unit of activity for a company.

Projects range from those that are large relative to the size of the company to those that are relatively small. Usually, the company engages in lots of smaller projects, such as the replacement of ageing equipment, and fewer large projects, such as the construction of a new production facility or the building of a high-rise office block. The decision-making process for projects is usually decentralized, so that the evaluation and approval of a project is performed at a level of authority appropriate to the size of the project.

The evaluation of a project is a multidisciplinary task, and is not solely the domain of any one of the professions or of management. It involves elements of engineering economics, capital budgeting, financial management, and strategic planning. It is performed by a variety of members of staff throughout the organization who may have different professional approaches. Many companies standardize the procedures for the evaluation of projects. However, these methods differ from company to company, indicating that there is no consensus between different companies on the best methods to adopt. There are also slight differences in terminology and notation within the different disciplines. All these different approaches have been used in this book in order to provide students and practitioners with a thorough and clear understanding of evaluation procedures. An integrated approach allows employees and management at the different levels and functions to have a common basis for communication and decision-making where analysis and decision-making occurs.

The funding of large capital projects, either from the internal resources of the company or by raising additional finance from external sources, is the field of corporate finance and investment banking. However, because engineers are integrally involved in capital projects, it is important for them to understand the basic mechanisms of finance for these large projects. This includes a clear understanding of return and risk from the viewpoint of both the company's investors and the financial markets.

The aim of this chapter is to provide an overview of the evaluation and the funding of projects. The evaluation of a project is comprised of forecasting the project's financials and evaluating these financials based on decision criteria. The financials for a project are discussed in the next section, and after that the evaluation of these project financials are discussed.

1.2 Project Financials

In order to assess the merits of a project, the engineer or manager must determine the project's ability to consume and produce cash for each year of the project's life. In other words, they must determine the project's cash flow. The project financials represent the forecast of the project's cash flow for each year of the life of the project. Broadly, the components of the cash flow are the capital expenditure, the revenue generated, the costs incurred in generating the revenue, the taxes paid and the working capital requirements. A forecast of these values and their timing is needed to calculate the cash flow for each year of the project's life and to prepare the project financials.

There are five basic inputs to the project financials. Each component is discussed separately as follows:

1. *Revenue.* The revenue is the money that flows to the project because of the project's activities. Consider the following simple example. The project aims to sell 100,000 items per year at $1.50 per item. The total revenue is $150,000 per year. The projections of revenue may be based on marketing data and forecasts or on contractual agreements.

2. *Costs.* These are the operating and overhead costs of the project. Operating costs are those costs that are incurred in the direct manufacture of the items. They include the costs of purchasing the raw materials, the energy, and the labour required for manufacturing the company's products. These costs vary with the amount of production; as a result, they are sometimes called variable costs. Overhead costs are those that are not operating costs, such as those arising from administration and from selling and marketing. Overhead costs generally do not vary with the amount of production. If this is the case, they are also referred to as fixed costs.

3. *Taxes and royalties.* The taxes are the charges made by the government, such as income tax and capital gains tax. Royalties may be charged by the government for the use of a natural resource, such as in mining or oil production.

4. *Capital expenditure.* The sum of money required to develop and install a manufacturing facility is the capital expenditure. The capital expenditure is also referred to as the fixed capital in order to distinguish it from working capital.

5. *Working capital.* Working capital is the net amount of money required for stock, debtors and creditors.

These inputs are used to calculate the cash flow and the free cash flow. The cash flow in a year is equal to the revenues less the costs less the taxes and royalties. The free cash flow is the cash flow less the capital expenditure and the working capital requirement. The development of the financials for a project, including the calculation of the cash flow and the free cash flow, is illustrated in the following example. The example is fictitious but is based on a real case.

Example 1.1: Project financials for the Santa Clara Hydroelectric Power Scheme.

A number of power outages have occurred recently in Northern Brazil due to an anticipated increase in demand in the region. The local government is soliciting bids to produce power from a new hydroelectric dam to be built on unoccupied state property. A power supply company has been approached to bid in a consortium with two other companies. These other two companies are consumers of electricity, and wish to secure supply for their own long-term needs.

The company's projects department has studied the Santa Clara project. The hydroelectric power scheme (HEPS) will deliver 460 MW of power, and will sell the electricity at $40/MWhr. It will be built at a total cost of $481 million. The costs to build the hydroelectric scheme are scheduled to occur over the next seven years, as shown in Table 1.1. The project will begin to generate electricity in the fifth year, before the dam is completely built. As the project nears completion in the sixth and seventh year, the production of electricity increases, as shown in Table 1.1. The costs to operate the facility are $4.4 million per year from the year that electricity generation begins. The state wishes to encourage private investment in infrastructure. It has provided a zero rating for income tax for the life of the project to act as an incentive for private investment.

It has cost the company about $4 million in management time, engineering costs and legal fees to date. Is this a good project for the company? Should the company join the consortium?

Solution:

Before these questions can be answered, it is necessary to determine the project's ability to consume and generate cash. This is done by preparing the project financials for the project.

Table 1.1 Projected profiles for the capital expenditure and power supplied for the Santa Clara HEPS

Year	Capital expenditure US$ M	Available power MW
1	10	0
2	75	0
3	78	0
4	114	0
5	141	314
6	60	448
7	4	460
Total	482	

Revenue

The revenue generated by the project is determined by the units of good or services sold and the price per unit received.

The amount of power generated at Santa Clara is expected to be 460 MW, and the received price is expected to be \$40/MWhr. If the hydroelectric power plant operates for the entire year without stop, and all the electricity generated is purchased by customers, then the revenue each year of full production is \$40/MWhr × 365 days/year × 24 hr/day × 460 MW = \$161.2 million. The revenue for the first ten years of the Santa Clara project is shown in Table 1.2.

Costs

The operating costs are the costs incurred in the production of the product. These consist, for example, of raw materials, labour, utilities and operating supplies. The operating costs for Santa Clara are estimated to be \$4.4 million per annum. The operating costs for the first ten years of the Santa Clara operations are shown in Table 1.2.

Taxes and Royalties

The Santa Clara HEPS has the feature that profits will not be taxed due to the unique circumstances of this project. Typically, tax is usually charged as a proportion of the taxable income. The income, also referred to as profit or earnings, is the revenue less expenses and depreciation. Income tax, depreciation and other accounting concepts are discussed in more detail in the Chapter 1.3.

Although income tax is not charged, a royalty is charged by the regional government for the use of the land at a rate of 2% of revenue. This is reflected in the project financials in the line labelled "Taxes and royalties."

Cash Flow

The cash flow for the project represents the actual cash consumed or generated by operational activities. The cash generated by the project is the revenue less the operating costs less the taxes and royalty to the regional government for the use of the land. The anticipated cash flow for the first ten years of the Santa Clara operations is shown in Table 1.2.

Capital Expenditure

The fixed-capital investment for Santa Clara is the amount of money that must be invested in the construction and installation of the entire power scheme. Land has to be purchased from the state, a dam has to be built, and the turbines and power lines have to be installed before power can be delivered and revenue produced.

Table 1.2 Projected revenue, operating costs and cash flow for the Santa Clara HEPS

Year	1	2	3	4	5	6	7	8	9	10
Revenue	0.0	0.0	0.0	0.0	109.6	156.3	161.2	161.2	161.2	161.2
Costs	0.0	0.0	0.0	0.0	4.4	4.4	4.4	4.4	4.4	4.4
Taxes and royalties	0.0	0.0	0.0	0.0	2.2	3.1	3.2	3.2	3.2	3.2
Cash flow from operations	0.0	0.0	0.0	0.0	103.0	148.8	153.6	153.6	153.6	153.6

Table 1.3 Projected free cash flow for the Santa Clara HEPS

Year	1	2	3	4	5	6	7	8	9	10
Cash flow	0.0	0.0	0.0	0.0	103.0	148.8	153.6	153.6	153.6	153.6
Capital investment	10.0	74.9	77.8	114.2	140.6	60.3	3.8	0.0	0.0	0.0
Free cash flow	−10.0	−74.9	−77.8	−114.2	−37.6	88.6	149.8	153.6	153.6	153.6

Working Capital

The working capital for a manufacturing facility is the net amount of money invested in raw materials in stock, finished goods in stock, monies owed to suppliers, employees and taxes, and monies owed by customers. The working capital investment for Santa Clara is negligible: power cannot be stored, and there is only a small delay in the payment for power supplied to bulk consumers.

Free Cash Flow

The free cash flow of the project is the cash flow generated by operational activities of the project less the cash consumed due to investment in fixed and working capital. It represents the net cash available. (The terms introduced up to now are similar to those found in accountancy. Free cash flow is a financial management and corporate finance term, rather than an accounting term.) The free cash flows generated by the Santa Clara project for the first ten years of operations are shown in Table 1.3.

Project Financials

The project financials represent the forecast of the cash flow and free cash flow for the project over the life of the project. The project financials for the Santa Clara HEPS are summarized in Table 1.4. An examination of the values in Table 1.4 shows that the free cash flow is negative for the first five years and positive after that. This means that the project consumes cash for the first five years, and after this, the project generates cash for the owners.

Table 1.4 Project financials for the Santa Clara HEPS

Year	1	2	3	4	5	6	7	8	9	10
Revenue	0.0	0.0	0.0	0.0	109.6	156.3	161.2	161.2	161.2	161.2
Costs	0.0	0.0	0.0	0.0	4.4	4.4	4.4	4.4	4.4	4.4
Taxes and royalties	0.0	0.0	0.0	0.0	2.2	3.1	3.2	3.2	3.2	3.2
Cash flow	0.0	0.0	0.0	0.0	103.0	148.8	153.6	153.6	153.6	153.6
Capital expenditure	10.0	74.9	77.8	114.2	140.6	60.3	3.8	0.0	0.0	0.0
Free cash flow	−10.0	−74.9	−77.8	−114.2	−37.6	88.6	149.8	153.6	153.6	153.6

1.3 Evaluating the Project Based on Free Cash Flow

The evaluation of a project is concerned with determining the merits of the project. This evaluation is an assessment of the soundness of the business, of whether the project is economically favourable. Some of the factors that are important to assess are the following: that there is a strategic fit of the project within the company's current business; that there are opportunities for the products of the project in the market; that the project is economically viable; and that the project is technically feasible. The outcome of the evaluation should be an assessment not only of the intrinsic value of the project proposal but also the suitability of the project within the context of the company.

The economic viability of the project is assessed on its free cash flows. The free cash flow is the amount of money that the business generates or consumes. If it generates money, it has a positive free cash flow. Money generated by the business is available to the owners of the business. It can be distributed to the owners as dividends or it can be re-invested in the business. If the business consumes money, the free cash flow is negative. This represents a shortfall in cash, and the managers need to make up this shortfall either by getting money from the owners or by getting it from lenders who loan the money to the company.

The project financials for the Santa Clara HEPS (given in Table 1.4) indicate that the project will consume cash for the first five years, while the dam and hydroelectric scheme are being constructed. This is because the free cash flow is negative in these years. Only after that will it begin to produce cash. The owners invest the capital and have to wait for the project to make returns. This represents a risk to the owners. The owners of the business must be rewarded for taking on this risk. The returns must therefore be sufficient for investors to risk their money. As will be discussed in later chapters, the value of a project depends on the amount, timing and risk of the free cash flows generated by the project.

The free cash flows are used to calculate measures of the economic attractiveness of the project. These measures can be used as decision criteria to determine whether a project should be recommended for approval. There are a number of measures of the economic attractiveness. Some of these are the payback period, the return on investment, the net present value and the internal rate of return. These criteria, and others, are discussed in detail in Chapter 6.

One useful way of analysing the free cash flow is to determine the cumulative free cash flow, which is the sum of the free cash flows for all the prior years. The cumulative free cash flow represents the total cash position of the project at that point. Initially, a typical project shows a negative cash position due to the construction and set-up costs. As the project begins to generate and accumulate cash, the cumulative cash position will gradually become more positive. The time taken for the cumulative free cash flow to reach zero is known as the payback period.

In the following example, the calculation of the cumulative cash position and the payback period are demonstrated for the Santa Clara project.

Table 1.5 Cumulative free cash flow for the Santa Clara HEPS

Year	0	1	2	3	4	5	6	7	8	9	10
Free cash flow (FCF)	−10.0	−74.9	−77.8	−114.2	−37.6	88.6	149.8	153.6	153.6	153.6	153.6
Cumulative FCF	−10.0	−84.9	−162.8	−277.0	−314.6	−226.0	−76.2	77.4	230.9	384.5	538.0

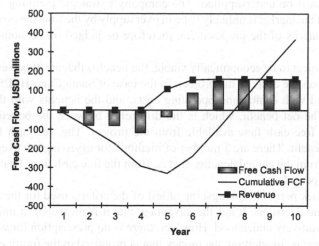

Figure 1.1 The projected free cash flow and cumulative free cash flow generated by the Santa Clara Hydroelectric Scheme

Example 1.2: The payback period for the Santa Clara HEPS project.

This example is a continuation of Example 1.1. Determine the cumulative cash position and payback period for the Santa Clara HEPS.

Solution:

The cumulative cash position is determined by summing the free cash flows for all the previous years. This is equivalent to the running total of the free cash flows. The cumulative free cash flow for the Santa Clara operation is given in Table 1.5.

The free cash flow and the cumulative free cash flow for each year are shown in Figure 1.1. The point at which the cumulative cash position changes from negative to positive represents the time taken for the project to "pay back" the entire investment in the project. At this point, the project has generated as much cash as it initially consumed. The payback period represents the length of time the investor in the project must wait to retrieve his or her investment from the project. In other words, this is the amount of time until the investor "breaks even."

The payback period for the Santa Clara Hydroelectric Scheme is close to seven years. Is this a good opportunity? Is it good, considering that the life of the Santa Clara Hydroelectric Scheme is expected be 35 years?

Is it is worth building the hydroelectric scheme? Implicit in this question is the alternative to do nothing. In the absence of any evidence to the contrary, it is assumed that these consequences will not negatively impact on the company's current business and its owners. Thus, the decision is a "yes/no" decision; either the company invests, or it does not. Many investment decisions are of this type, whereas decisions regarding the replacement of equipment are of the "either/or" type. For example, if the turbines need to be replaced because they are at the end of their lives, this represents an "either/or" decision; they are replaced now or they are replaced later.

The project is clearly a strategic fit with the company's current business, that is, the generation and supply of electricity to customers. Since there are power outages, the market must be undersupplied. The company's strategic planning department confirms that the market is unlikely to be in oversupply by the time the project comes online. The merits of the proposal can therefore be judged on economic grounds alone.

For the project to be economically viable, the benefits that are expected to accrue are greater than the costs of the project. In the case of Santa Clara HEPS, the costs are associated with capital and operating costs, and the benefits with the revenue generated. The net benefit, which is the difference between the benefits and the costs, is the free cash flow available from the project. The free cash flow is the annual net benefit. There are a number of methods for assessing the overall benefit of a project from the annual benefits, that is, from the free cash flow. One of these is the payback period.

The payback period is amongst the oldest of the criteria used for the analysis of the investment decisions. It has the advantage that it is measured in units of time, which are intuitively understood. However, there is no prescription for what a good payback period is. In addition, the money that is promised in the future is less valuable than cash that is actually in hand now. This concept, called the time value of money, is discussed in further detail in Chapters 5 and 6. The payback period method does not account for the time value of money.

For Santa Clara, the length of the project is much longer than the payback period. Based on this, the Santa Clara project appears to be a good investment in the long-term. The payback period is one amongst many other investment criteria that can assist management and investors in deciding whether a project is suitable for investment.

Another useful method of analysis of the project's financials is the return on investment. The return on investment is a measure of the project's ability to generate funds for its investors. For the Santa Clara project, the return on investment can be defined as the ratio of the annual income to the investment costs. The annual income at full capacity is \$161.2 million, while the total investment cost is \$481.6 million. The return on investment is therefore equal to $33.5\% = 161.2/481.6 = 0.335$. This return can be compared with that of other business opportunities to determine if this is the most profitable opportunity available to the company for the use of its resources.

The project financials, determined by establishing the free cash flow for each year of the life of the project, and the decision criterion, such as the payback period, are the foundations of engineering economics and capital budgeting. The statement of the project financials is not the same as the financial statements prepared by accountants (even though some of the line items have similar names to those used in accountancy, such as revenue). The accounting statements establish the profit of the company and summarize the financing of the company. The project financials forecast the project's ability to generate free cash flow. The relationship between the two is explored in Chapter 3. The details of determining the components of the project financials are discussed in Chapter 4.

Three important principles for the assessment of capital projects are briefly discussed in the following three sections. These principles are as follows: (i) the assessment is based on cash flow rather than on profit; (ii) the investment decision is assessed independently of the financing of the project; and (iii) the project is assessed on future cash flows without regard for past expenses.

1.4 Profit (or Earnings) is Not Required for the Assessment of the Project

The profit, or earnings, is the difference between the revenue (or income) and the expenses (or costs) including taxes and other items such as depreciation. The project financials for the Santa Clara HEPS did not mention the profit. The concept of profit is not relevant to the evaluation of the project. This is because profit includes non-cash items, such as depreciation and amortization. The topics of profits, depreciation and amortization will be treated in some detail in Chapter 3. The economic evaluation of the project is based on the cash flow of the project that represents actual monetary transactions.

1.5 Investment Decision Precedes the Financing Decision

The analysis of the Santa Clara project included no reference to the mechanisms or cost of financing the project. For example, if the project were to be financed by a loan, there would be interest charges that impact on the costs. These costs were deliberately ignored for the following reason. There are two decisions that need to be made: Firstly, is this an economically worthwhile project? Secondly, if it is, how is it to be funded? The decisions are separated in order to clarify and simplify the decision-making process. The project is evaluated as if the project is completely owned by the company and entirely financed from the company's own resources. This is referred to as the "entity basis." The investment decision is made on this basis. If the project is evaluated with the interest for loans included, it is referred to as the "equity basis." The reconciliation of the two approaches is discussed in Chapter 12.

Only once it has been determined that it is a profitable project and worth pursuing, is the source of funding considered. The separation of the two decisions helps to make the decision-making process more transparent, especially since funds are not raised specifically for a project. Rather, the funding requirements are lumped together and funds are raised for the total requirements. This is referred to as the pooling of funds within a company. In addition, the two decisions, that of the investment decision and that of the financing decision, are functionally separate in most organisations. Separating the decisions separates the roles and responsibilities of each of these different departments. Thus, evaluating the project on the entity basis is not done for theoretical reasons, but for reasons of practical management.

1.6 Past Expenses are Excluded from the Project Financials

In developing the Santa Clara opportunity, the company incurred significant costs mainly for legal fees and engineering studies. However, in preparing the project financials, these costs were ignored. This was deliberate, since past events have no bearing on the economic viability of the opportunity at the present. "Gambler's Ruin" is a well-known problem in probability studies, where there is a finite and significant probability that a gambler who tries to recover past losses will loose everything. In an economic analysis, past expenses, referred to as *sunk costs*, are irrelevant to the decision now. The evaluation is focused on the future.

1.7 Assessment of the Risk of the Project

The forecast of the project financials included no mention of the chance that these forecasts may not be achieved. These projections implicitly assumed that they were known with certainty. In reality, a large number of sources of uncertainty can influence the outcome. Although the forecasts of the project financials are made with the best possible information and techniques, they are uncertain. On the other hand, a project is not risky if the project financials can be predicted with certainty. Risk is the chance of an undesirable outcome caused by the unpredictability of the future.

The source of risk for a project is the variation or change in the values of the constituents of the project financials. For example, if the revenues change from year to year in a fashion that is different from its prediction in the project financials, then this is a source of risk. Some of the risks that are relevant are the business risk, the investment risk, and the financing risk. The variability in the revenues and costs of the project is the source of business risk. The change or variation in the estimate of the initial capital expenditure to build and install the project is the source of investment risk. The financing risk refers to the variation in factors that affect the financing of the project, such as changes in the interest rate. Together, the business, investment and financing risks are referred to as the financial risks of the project.

Risk is important in the assessment of a project for two reasons. Firstly, the value of the project may differ from what it was forecast to be. This type of risk represents a loss to the owner of the project. It is the *stand-alone risk* of the project. Secondly, the risk of the project contributes to the risk of the company and to the risk of the investors in the company. The risk of the project to an investor who has a diversified portfolio of investments is referred to as the *market risk*.

The topic of risk, which is the subject of Chapters 11 through 15, is more complex than it might appear at first. For example, we know that risks of one project influence the risks of another. Risks can be combined or separated. The risk of a portfolio of assets is not the average of the risks of individual assets. The combined risk of two projects may be lower than either one. This means that diversifying investments lowers the overall risk. That is why there is the saying: "Do not keep all your eggs in one basket."

The methods for the analysis of financial risk and the formation of the optimal portfolio of assets are the subjects of modern portfolio theory. The discussion of the topic of risk, including modern portfolio theory, is continued more detail in Chapter 11.

1.8 Financing of a Capital Project

There are two major financing issues associated with large capital projects. These are the following: (i) how is the project to be financed during construction? and (ii) how is the project to be financing permanently after construction? The separation into these two phases is important, since during construction there is the additional risk of the project not being completed on time or within budget.

The funding during construction depends on the owner's contractual arrangements with engineering contractors and equipment vendors, that is, it depends on how the project is "procured." The owner of a project can procure a capital project in a number of ways, each of which carries different risks. There are three main models: owner managed, cost-reimbursable and lump sum turnkey. The way in which the project is procured or delivered influences the financing during construction of the project.

If the project is owner managed, the owner prepares the engineering design package that defines and specifies the project and the equipment required. The owner then procures all of the equipment and constructs the project at the owner's cost. In other words, the owner finances the construction of the project directly from its own source of funds, that is, the equity and debt of the company. The prime advantage of this method is that it allows the owner to select the engineering company or construction company that is best suited to each stage of the project. The owner is the active manager of the project and maintains control and overall responsibility.

The cost-reimbursable contract is only slightly different to an owner-managed project from a financing point of view. In this type of contract, the owner appoints a contractor, who undertakes the engineering design, procures the equipment, and manages the construction until completion, on behalf of the owner. The contractor integrates all the aspects of the project. The owner finances the project directly from its own resources. Although the owner delegates responsibility to the contractor, most of the project risk remains with the owner. The owner pays the contractor based on the costs incurred by the contractor and fees charged by the contractor.

The owner may develop a specification and issue tenders to engineering contractors for the delivery of the project at a particular date for a fixed, or "lump sum," price. These contracts are called lump sum or lump-sum turnkey. The risks of and responsibilities for the construction of the project are transferred to the contractor. The financing of the construction of the project is transferred to the contractor, who maybe required to arrange a construction loan. The engineering contractors would also be required to provide guarantees of performance. On completion, the construction loan terminates and the owner provides permanent finance. On the other hand, the owner may agree to pay the agreed price according to a construction schedule, in which case the owner may request the contractor to provide a performance bond that the contractor would forfeit in the case that the project is not completed.

Permanent financing is the funding of the project after construction. In some cases, there may be no difference between the construction finance and the permanent finance because the owner funds all requirements. Either way, once the project has been constructed, the permanent financing can be arranged on the basis of the general standing of the owner, that is, from the owner's own resources, or as "project finance."

The owner, who is generally a company, can raise finance to fund the capital project from one of two sources: equity, or debt. The company sells shares or stock in the company to investors to raise equity capital. These shares represent part ownership in the company and the investor bares the risk of ownership. On the other hand, the company can raise funding from lenders who be willing to loan money to the company because of the creditworthiness of the company. Such loans that are raised by the company, usually for extended periods, are referred to as debt financing or the debt capital of the company. The lenders may require security or collateral, and the assets of the company are often used as collateral.

An alternative form of permanent financing for a project is to secure the finance against the anticipated cash flows of the project. This means that the lenders are not relying on the company's past performance in assessing the loan, not do they require the company's assets as collateral. Instead, the lenders look at the profitability of the project and its ability to repay the debt. In this case, the project is separated legally and financially from the original owner, who is now called a sponsor, and the finance is structured to suit the needs of the project. This form of finance, called project finance, has found application in large infrastructure and industrial projects in both the private and public sectors.

The sources of finance and the financing of projects are discussed in further detail in Chapters 16 and 17.

1.9 Summary

A capital project is the investment in equipment, plant, machinery and property that is employed in productive work. Examples of capital projects are the construction of a factory, the replacement of equipment, and the building of a toll road. This chapter has provided an overview of the assessment of capital projects. The assessment is based on a variety of evaluation criteria, such as the strategic fit of the project within the organisation's core business, the market opportunities for the project's products, the economic viability of the project, the technical feasibility of the project, and the ability of the company to implement the project.

The project financials is the forecast of the free cash flow for each year of the life of the project. This forecast is used to assess the economic viability of the project. The free cash flow is calculated from five main variables: the revenues, the expenses, the taxes and royalties, the capital expenditure and the working capital requirements. The revenue is the in-flow of cash to the project from the sale of the project's products or services. The expenses are the costs incurred in the operations. They include the operating costs, which are directly associated with production, and the overheads, which are related to the administrative and the marketing functions of the business. The taxes and royalties are the charges made by government, such as income tax and mining licences. The capital expenditure is the cost of the project's investment in the equipment, plant and property required to manufacture the project's products. The working capital requirement is the amount of money that is invested in inventory and creditors. The free cash flow is revenue less expenses, taxes, capital expenditure and working capital.

The project financials indicate the ability of the project to generate or consume cash on a year-by-year basis. An overall assessment of the value or the attractiveness of the project can be made by determining criteria such as the payback period, the return on investment, the net present value and the internal rate of return. There are many more criteria that can be used than these four. The first two criteria are calculated directly from the values of the project financials, while the last two require the application of the time value of money, a concept that is discussed in detail in Chapter 5.

The project financials are prepared with three principles in mind. Firstly, the project financials represent future cash flows; past expenses are irrelevant to the assessment of the project. Secondly, the project financials are based on cash flows rather than on accounting concepts such as profit that include non-cash items. Thirdly, the project is evaluated as if it is owned and funded entirely by the company (or the investor), called the entity basis. The entity basis separates the investment and financing decisions.

Once the recommendation is adopted, then the activity of determining how the project is to be funded commences. For example, will the project be funded entirely by the owner, or will the owner raise financing in the capital markets? The types and structure of the financing will be discussed in Chapter 16.

Other steps can be included in the assessment of a project. Examples of these are an assessment of the assumptions made, an assessment of the possible outcomes and an assessment of the uncertainty or risk of the project.

The above description of the assessment of the project is based on the returns of the project to the owner of the project. The risk of the project from the owner's viewpoint is that it does not meet the expectations given in the project financials.

The source of this risk is the difference between the outcome and the forecast for any of the variables that make up the project financials. Another way of viewing risk is as the variability in the values of these variables with time. If they change over a wide range, they are risky. If they vary within a narrow range, they are predictable, and hence less risky. Financial risk is the overall project risk, comprised of business, investment and financing risks. Business risk is the variation in the revenue and expenses of the project. Investment risk is the variation in the cost of the capital equipment for the project. Financing risk is the variation in the factors such as interest rate that affect the financing of the business.

1.10 Looking Ahead

The next chapter provides the business context for decision-making on capital projects. The chapter discusses the goals of the financial management of a business, the structuring of the financial function within the business, the theory of business decision-making and the practice of gaining approval for capital projects.

1.11 Review Questions

1.   Provide definitions for the following terms:
      (i)   Cash flow
      (ii)   Assets
      (iii)   Liabilities
      (iv)   Capital
      (v)   Fixed capital
      (vi)   Working capital
      (vii)   Revenue
      (viii)   Taxes
      (ix)   Royalties
      (x)   Stock
      (xi)   Inventory
2.   What are capital projects?
3.   What are the two types of decisions that are commonly found?
4.   What are sunk costs?
5.   Why do the project financials ignore the past expenses?
6.   What are the components of the project financials?
7.   What are three basic principles in evaluating a project?
8.   What does the payback period mean?
9.   Name three other criteria for the assessment of the economics of the project other than the payback period.
10.   What are the financial risks for a project?

11.   How does stand-alone risk differ from market risk?
12.   What does "raising capital" mean?

1.12 Exercises

1.   A capital project under consideration has the following cash flow profile:

Year	0	1	2	3	4	5
Free cash flow	−100	35	40	45	50	55

Determine the payback period.

2.   A project consists of an investment of $100 million in capital for the construction of the manufacturing operations. Each year after the completion of the construction, the project will sell product worth $70 million. The costs of manufacturing this product are expected to by 60% of the value of the sales. The taxes are expected to be 30% of taxable income. No deductions, such as those for depreciation or capital allowance, are permitted for the determination of the taxation charge. The working capital is 15% of the fixed capital. The working capital charge is incurred in the first year of operation and is recovered in the fifth year. Determine the free cash flow for the first five years of operation.

3.   A new capital project has the following cash flow profile:

Year	Cash flow
0	−10,000
1	−5,000
2–9	3,000
10–14	2,000
15	3,000

Determine the payback period.

4.   An international manufacturer of hard disk drives is considering three locations for its new plant. The location conditions and distance from the markets affect the cash flows for the project for the different locations. Determine the preferred location, if the cash flow profiles for the three different locations are those given in the table below:

Year	Philippines	Malaysia	Thailand
0	−30,000	−40,000	−38,000
1	−10,000	−6,000	−7,000
2	9,000	8,000	5,000
3–10	7,800	8,000	7,500

5. The manufacturer of power tools for the home market is considering the intro-
    duction of a new product, a laser-measuring device to replace tape measures.
    The plant is expected to cost $10 million and generate a net income (also
    called net profit) of $3 million each year for five years. The company has de-
    cided that the appropriate measure of the project's performance is the return
    on investment. Unfortunately, there is no single definition of the return on in-
    vestment, which is also called the accounting rate of return. Determine the
    rate of return if the following possible definitions are adopted:

    (i)   Return on investment $= \dfrac{\text{Annual income (profit)}}{\text{Investment}}$

    (ii)   Return on investment $= \dfrac{\text{Annual income (profit)}}{\text{Investment}/2}$

    (iii) Return on investment $= \dfrac{\text{Total income (profit)-Investment}}{(\text{Investment}/2)(\text{Project life})}$

6. Johnson Engineering is considering building a plant. The plant will cost $8
    million, and will generate an after tax income (profit) of $750,000. If the in-
    come is expected to grow at 5% pa, determine the payback period.

7. An engineer is considering the replacement of equipment in the plant. The
    equipment that is currently in the plant cost $500,000 three years ago, and is
    currently valued in the books at $350,000. It was expected to last ten years.
    The equipment that she is considering for the replacement is technically and
    economically superior. It will cost $700,000. The company will be able to
    sell the incumbent equipment for $100,000. Is the original price or the current
    book value relevant in her assessment of whether to replace the equipment?

8. A project will generate $10,000 each year. What is the maximum amount that
    can be invested if the company requires a payback period of 2.5 years? How
    would your answer change if the income were to increase at 10% per year?

9. Distil the assessment of the Santa Clara HEPS into a number of basic steps.
    Generalize these steps to those required for any decision-making problem.

10. Research the problem known as gambler's ruin in probability theory and de-
    scribe the analogies with the investment in capital projects by companies.

Chapter 2
The Theory and Practice
of Decision-making Concerning Capital Projects

2.1 Introduction

The purpose of the assessment of projects is to determine if the project justifies investment. There are usually many projects that the organization assesses and approves each year. The available money, or capital resources, must be divided between the different projects. In other words, the capital resources must be allocated to the projects. The activities for the assessment of projects and the allocation of capital are known variously as cost-benefit analysis, engineering economics, and capital budgeting, terms that arise from the different professions of economics, engineering and finance. They are all concerned with the allocation of resources to projects in a company in the most cost effective and profitable manner. The success of all enterprises, both private and public, depends on how well the enterprise chooses projects and allocates resources to its projects. As a result, these are critically important activities for an enterprise.

The objective of this chapter is to present the context within a company for decision-making concerning capital projects. This context concerns the history of these activities, the organization of the finance function within a company, the framework for decision-making in general, and the practice of decision-making for capital projects. An important part of this context is the understanding of the aims and objectives of the professions that impact on the investment in capital projects. The classification of projects and the practice of promoting a project are also discussed. The history of cost-benefit analysis, engineering economics and capital budgeting is described in the next section.

2.2 Cost-benefit Analysis, Engineering Economics and Capital Budgeting

(i) Cost-benefit analysis

Cost-benefit analysis is the practice of assessing the desirability of projects from the perspective of an economist. It takes both a long-term and a wide view. It is long-term in the sense of examining the effects, implications and repercussions both in the short and long-term. It takes a wide view in the sense of examining the effects of the project on different peoples, industries and regions. It is a broad treatment, and has more in common with economics than with business and commerce. It draws on a range of subdisciplines within economics, such as resource economics and public finance, to create a coherent view of the project.

Cost-benefit analysis has a long history, particularly in France, where the engineer Dupuit published a paper in 1844 on the utility of public works, a groundbreaking contribution to the field of economics. In the US, the early practice of cost-benefit analysis was closely associated with control of navigation. It came about as an administrative function, little related to economics. The River and Harbor Act of 1902 required the Army Corps of Engineers to report on the desirability of their river and harbour projects by accounting for their benefits to commerce and their associated cost. The terms of reference were broadened in the 1930s. The Flood Control Act of 1936 authorised projects by the Army Corps of Engineers "if the benefits to whomsoever they may accrue are in excess of the estimated costs." It should be pointed out that the practice of cost-benefit analysis in these cases was not only to justify projects, but also to determine who should pay for them. This practice spread to other US government agencies, and in the 1950s, the general principles of cost-benefit analysis, including the considerations of welfare economics, were codified. As a result of this developmental history, the application of cost-benefit analysis was and still is clearly situated in the domain of public works.

Cost-benefit analysis is primarily an economic analysis. Although market prices are a starting point for the determination of the benefits and the costs, they imperfectly represent the interests of various parties and stakeholders in the project. Market prices may be distorted by political intervention, taxes, subsidies, incentives, lack of competition, price control and other factors. In an economic cost-benefit analysis, prices are adjusted towards their efficiency prices, those that would be achieved in a perfect market as a result of the best allocation of resources due to supply and demand. This is in contrast with financial or commercial analysis, which considers the flow of cash at market price.

While the engineers of the Corps des Ingenieurs des Ponts et Chausses in France, most notably Dupuit, contributed significantly to the early development of economics itself, endeavours in the US focused on the combination of engineering design with economic choice.

(ii)   Engineering economics

The involvement of engineers in the evaluation of the benefits and costs of projects led to the development of the field of *engineering economics*. Arthur M. Wellington, the pioneer of the field, published "The Economic Theory of the Location of Railways" in 1877 that included the notion of "the judicious use of capital" in the engineering design process. Wellington was concerned with the most cost effective location of railways, and the application of economic choices to other engineering design questions, such as the grade and the gauge of the railway. For example, he demonstrated that light rail was false economy using his analysis techniques.

Wellington was the first to describe the application of present value techniques for the allocation of capital within a company. Present value techniques, also called discounted cash flow techniques, account for the time value of money. The concept of the time value of money has been known since the early 1800s. Today it is regarded as necessary to account for the time value of money in the evaluation of projects that last for more that a few years. Wellington wrote for engineers. Early adopters of the discounted cash flow techniques that incorporated Wellington's ideas of the time value of money were the engineers of mining companies and public utility industries. Beginning in the 1920s, the new field of engineering economics influenced the way in which AT&T made capital allocations to possible projects. These ideas of the use of time value of money in the assessment of the economics of a project slowly diffused through engineers, reaching the petroleum industry by the 1950s.

Engineering economics is more restricted than cost-benefit analysis, limiting itself to the application of a number of techniques for economic choice rather than engaging in a full economic analysis as would be expected in cost-benefit analysis. The analysis is performed from one point of view, rather than representing the views of all the stakeholders. Engineering economics has more in common with investment analysis than with economics.

The relatively slow rate of adoption of techniques incorporating the time value of money between the 1920s and the 1960s is due, in part, to Du Pont's development of the concept of return on investment, which was a development of equal importance to the time value of money for the analysis of the performance of capital investments, and their development of a system for the allocation of capital resources to projects. This system is known as capital budgeting, and is discussed next.

(iii)   Capital budgeting

In contrast with cost-benefit analysis, which takes a broad view, the assessment of projects within a company takes a narrower view. The view is narrowed by mainly considering the benefits or costs to the company, not to society. This is not to say that the companies do or should act irresponsibly; it is merely that the company's viewpoint is that the main beneficiary of its investments should be itself. The activities within a company of assessing long-term opportunities and allocating capital to them are known as *capital budgeting* within the field of financial management. It addresses essentially the same questions as engineering economics. Examples of capital budgeting decisions are the acquisition of land, buildings, equipment, and vehicles, amongst others, for productive use by the company.

The history of capital budgeting runs parallel to that of engineering economics. In 1903 the E.I. du Pont de Nemours Powder Company merged the functions of scores of smaller, specialized firms, and centralized the manufacturing and distribution process. Pierre du Pont wished to measure both the profitability of the current operations and the attractiveness of future investments in terms of the capital invested in each venture or department. He and his accounting staff developed the concept of return on investment accompanied by a system for allocating capital resources to projects. The return on investment combined the profit margin with the asset turnover. This concept, and the system of reporting that it created, significantly decreased the cost of managing a complex company like Du Pont.

Du Pont owned 23% of General Motors, a company also created from the merger of many smaller firms. When faced with financial and administration problems in 1920, General Motors consulted Pierre du Pont. The capital-budgeting system GM constructed allowed operational managers to decide how best to employ their division's resources, while at the same time maintaining central control by senior executives.

Although the techniques used at Du Pont and General Motors, and subsequently adopted by other industries, were highly effective, they were designed for the projection of the next year's performance. At the time, they did not forecast performance or evaluate projects on a long-term basis; as a result, they did not need to incorporate Wellington's ideas of the time value of money into their assessment of the project or capital investment. However, these concepts were gradually incorporated into the capital-budgeting framework. Indeed, today they are the dominant measures of economic viability used by companies to assess capital projects.

Investment in projects is extremely important to a company. Such investment represents a fairly large commitment of the company's resources; they are invested for a long time, and the operating needs of the company will be driven by these investment choices. Since the company may be viewed as the sum of its investments, the future development of the company depends on the efficient choice of these investments. Capital budgeting is a strategic activity, central to the business objectives and business plan of the company.

Cost-benefit analysis, engineering economics and capital budgeting offer methods and techniques for the choice of investments in both the government and private sectors. The analysis of government projects is closer to economic cost-benefit analysis than it is to capital budgeting, because government has a broader remit than does a company. Many of the infrastructure projects that government invests in, such as dams, power schemes, and highways, are typically very long lived. However, both government and company decisions are championed and made by people, not necessarily in the best interests of the public or the company, and are subject to the politics of the organization. In a company, it is the duty of the executive management to set the direction and the strategy of the company clearly, so that the most efficient choices can be made. The understanding of different viewpoints and the organization and the business context in which investment decisions are made is important, and is discussed in the following few sections.

2.3 Perspectives for the Assessment of Projects

Various views of investment decisions other than those of engineers and economists are important in both the public and private sectors. Within the private sector, there are two divergent viewpoints that are important to bear in mind when the results of any analysis are communicated. These viewpoints are those of financial accounting and financial management.

Financial accounting has a particular perspective. For these professionals, the measure of financial success is based on liquidity in the short-term and on profitability in the long-term. Liquidity means that sufficient cash is available for the business and that it is not bound in assets that are difficult to sell, that is, are illiquid. A profitable business is one in which the revenues exceed the costs. The liquidity requirement translates into measures of success based on how quickly the investment can recoup its costs, as encapsulated in the payback period discussed in Chapter 1. The profitability measure translates to how productive the investment is at generating cash flows or profit. This notion is encapsulated in the return on investment as a measure and decision criterion. These are both accounting measures made with accounting data.

Financial management has a worldview different to that of financial accounting. The aim, and hence the measure of financial success, of financial management is to create and maximize shareholder wealth. Shareholder wealth is not seen as being driven by profitability; rather it is seen as a function of returns of cash to the investor in the medium and long-term. The goal of maximizing shareholder value leads to measures for the evaluation of investments and projects that maximize the future cash flows for the allocated capital. These notions are encapsulated in a number of decision criteria such as net present value and internal rate of return, that account for both the cost of the funds required to pursue the opportunity and the time value of money. The details of these criteria are discussed in Chapter 6.

Both of these perspectives are of value. Generally, the perspective that is required by engineers, scientists and managers is one that is closer to financial management than it is to either economics or accounting, and this theme dominates the treatment followed in this text. However, it is essential to understand the accounting view, since the engineers, scientists and managers must be able to communicate effectively with other professions. Since the perspective required of engineers is closer to financial management, it is instructive to examine the components of value for a shareholder or an investor.

2.4 Enhancing Value for Investors

Investors will choose to invest in a business based on what they expect the business to return to them. In evaluating an investment opportunity, investors will compare the expected returns with those from the many other choices that they have for the investment of their money. Once the investors have placed their money in the business, the onus is on the management to ensure that the expected returns materialize. The overall aim of the company's management, including the company's engineers and scientists, is therefore to maximize the value of the investment made by investor for the benefit of investors and other stakeholders in the company.

There are three main drivers for enhancing the value of a company. These are (i) through the superior use of *finance*; (ii) through superior *organization*; and (iii) through superior *strategy*. The first driver has three elements: the control of costs, the use of capital, and the raising of capital. The second driver refers to the management of processes, people, performance and talent. The third driver refers to the superior anticipation of the future, of trends and events that may affect the business, and the superior deals that place the company in a better position than competitors. These factors, and the company's performance in managing them, influence the value of the company.

Engineers and other technical professionals contribute significantly to all of the main drivers of value. The focus of most engineers and scientists is on enhancing value through the optimum use of resources, the control of costs and the management of processes. Some may find themselves working in research and development on strategic projects that anticipate, or take a leading role in shaping, future trends. Others have personal leadership qualities, and rise to become part of the leadership of the organisation.

The value of a company is most easily measured by the generation of cash by the company. Value is measured by cash flows. More particularly, value depends on the amounts of cash flow, the anticipated timing of future cash flow, and the risk of these anticipated cash flows not materializing. If the company is listed on a public exchange, the value of the company is reflected directly in the share price. The same factors that affect value for an investor in a public company, that is, the amount, timing and risk of the expected cash flows, affect the value for investors in a private company.

Clearly, a major task of the engineers and scientists employed in any organization is to ensure that capital resources are allocated optimally to the benefit of all the organization's stakeholders. The business context in which these decisions take place is sketched in the next section.

2.5 Business Context

In order to pursue a project or an investment opportunity, the organization needs funds. The sources of the funds, the stewardship of the funds within the organization and the interaction between financing and investment constitute the context that impacts on the evaluation of capital projects, the decisions concerning which projects are pursued and those which are not. These topics are discussed next.

2.5.1 Financial Stewardship Within a Business

The two major functions of the finance department in a company or corporation are the treasury and the controller. This is illustrated schematically in Figure 2.1. The treasury is responsible for managing the cash and investments of cash in the financial

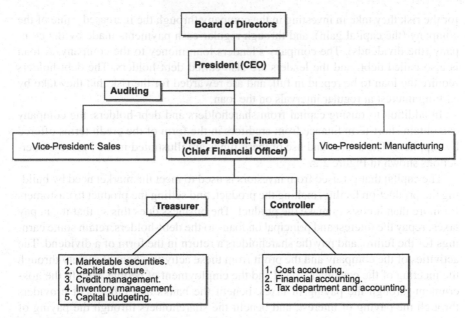

Figure 2.1 Functional roles of the finance in a corporation

markets (marketable securities) and managing the company's debt. The controller is responsible for functions like cost accounting, financial accounting, and tax. The chief financial officer (CFO) is responsible for both of these functions.

Another function within the domain of the CFO is broadly called the financial management of a company. Financial management aims to maximize value for the shareholders while managing the risk profile of the company. The opportunities for the financial manager to create value for the company are in arranging the company's financing efficiently and in using these finances effectively. As a result, the type of decisions required of financial managers can be classified as either financing decisions or investment decisions. The raising and allocating of funds is discussed next.

2.5.2 Sources and Use of Funds

The aim of a company is to meet a market need. Purchasing the company's products will satisfy the customer's needs. In order to provide this product or service, the company embarks on a project that requires a production facility to manufacture the product and a distribution network to bring the product or service to market. The company requires capital for the acquisition of the assets to do this.

The company raises the capital it requires from two main sources: shareholders and debt-holders. Shareholders are part owners of the company. They are rewarded for the risk they take in investing in the company through the increased value of the company (the capital gain), and through regular cash payments made by the company (the dividends). The company's lenders loan money to the company. A loan is also called debt, and the lenders are also called debt-holders. The debt-holders require the loan to be repaid in full, and are rewarded for the risk that they take by earning interest at regular intervals on the loan.

In addition to raising capital from shareholders and debt-holders, the company can obtain short-term finance from creditors in the form of the credit terms offered by suppliers. The source and use of these funds are illustrated in the business interactions shown in Figure 2.2.

The capital that is raised from investors is used to meet the market need by building the production facility, making the product, and selling the product to customers for more than it costs to make the product. The business does this so that it can pay taxes, repay the interest and principal on loans to the debt-holders, retain some earnings for the future, and pay the shareholders a return in the form of a dividend. The activities of the company and the profit from these activities benefit society through the meeting of the customer's need and the employment of people, benefit the government through the paying of taxes, benefit the banks and other debt providers through the paying of interest, and benefit the shareholders through the paying of dividends and the increased value of the company.

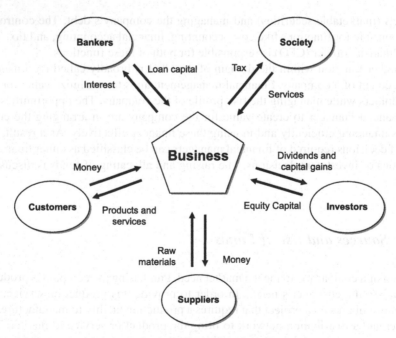

Figure 2.2 Interactions between a business, its markets (suppliers and customers), funders (investors and bankers) and society represented by the government

The source and use of funds is clearly seen in the balance sheet of a company. The balance sheet will be discussed more comprehensively in Chapter 3. It is a statement of the assets, liabilities and owners' equity of a company. The assets are the items that the company owns, the liabilities are the amounts of money that company owes to others, and the owner's equity is the contribution by the shareholders to the value of the company. The source of funds is represented by the liabilities and the owner's equity, and the use of funds is represented by the assets. Thus, the balance sheet reflects the interactions between the investment and financing activities of the company.

2.5.3 Investment and Financing Decisions Within the Business

Another way of viewing the balance sheet of a company is as a representation of the investment and financing decisions made by the company.

Investment decisions concern the acquisition of operating assets or financial assets. Operating assets are items such as machinery, vehicles, property, inventory and buildings. A distinction is usually drawn between fixed assets, such as production equipment, and working capital, which is the net amount of money required for stock, inventory, debtors and creditors. The company may own shares in other companies, or it may loan money to other companies. These investments are in financial assets. In other words, the company may acquire financial assets by investing in other businesses or by investing in financial instruments sold in the financial markets.

Financing, or the financing decision in a company, is the function of determining the most suitable financing arrangement or structure to fund the company's opportunities. The financing decisions are mostly concerned with the following three issues: (i) how much debt can the company afford to have (called the capital structure); (ii) how much credit the company can afford to provide to its clients (called the credit policy); and (iii) how much of the company's profits should be retained by the company (by not paying all the profits to the shareholders) to have sufficient resources for all anticipated needs and future investments (called the dividend policy). In later chapters, it will be shown that it is advantageous to have as much debt as possible, because the returns to shareholders increase with increasing debt. However, increasing debt also increases the risk for the company. In structuring the finances for the company, the financial manager makes a trade-off between return and risk. Similar trade-offs are involved in considering the other two issues.

From a strategic viewpoint, the financial manager must ensure that the company has sufficient resources to meet its goals. The financial manager must forecast what the resource requirements are, and determine if additional resources are required. If insufficient, the financial manager must raise the additional capital that matches the requirement from either shareholders or debt-holders. The funding requirements for the company are pooled or lumped together; finance is not generally arranged for a particular project. The financial manager determines the total requirements for the company, and determines the best method for meeting these requirements. The sources of funds are discussed in more detail in Chapter 16.

The investment and financing decisions represent the core of what a company does. It raises the funds to invest in projects. It raises capital, which are the financing activities, and it allocates that capital, which are the investment activities. The analysis, planning and evaluation of these opportunities are intimately tied to the company's strategic objectives.

The functions of the investment and financing activities are usually separated in corporations. The functional teams working on investment proposals are usually not those working on financing structure. In other words, they occur in different departments. This separation is also in timing. As mentioned in Chapter 1, the first decision is whether it is a good opportunity for the company in the context of all the other opportunities available to the company, and the second decision is how best to access the funds to implement the investment proposal. The separation of these decisions is useful for a number of reasons: (i) it clarifies what decision is required; (ii) it allows opportunities to be evaluated on a common basis; and (iii) it clarifies the roles and responsibilities of different functional units within the organization or company.

2.5.4 Evaluation of Investment Opportunities

The decision to make an investment in an asset (operating or financial) is taken on the basis of a business evaluation. The evaluation has the objective of building a business case for the investment proposal. It requires an overall perspective of the business, of the company's positioning on strategy, marketing, and production. It requires an understanding of the company's risks and returns and it must integrate aspects of tax, commercial agreements, and possible liabilities. It requires knowledge of the project, through the construction, commissioning ramp-up and production stages. In addition to analytical skills, business evaluation involves judgement, experience and wisdom.

There are essentially four parts to the evaluation of an investment opportunity. These are the strategic evaluation, the economic evaluation, the technical evaluation and the financial evaluation. Not all of these are relevant to all organizations or to all investments.

(i) Strategic evaluation

The strategic evaluation considers key factors for the success of the project, such as the company's ability to penetrate the market and the structure of competition in the industry. The markets that most companies operate in are competitive, resulting in a drive for efficiency and effectiveness. The company must be able to understand the dynamics of the industry and harness this knowledge profitably. Strategic evaluation encompasses an overall knowledge of the company's current and future activities, including the company's anticipated projects.

The company does not only exist in the context of its market. Beyond its markets, the company exists within a system of law, society and politics. Knowledge of, and an anticipation of the changes in, the external factors, such as public opinion and the regulatory environment, that may impact on the business's ability to execute its business plan, is essential for the strategy of a business to be successful.

(ii)   Financial evaluation

The financial evaluation is the function within the business evaluation that examines all the available information from a financial viewpoint. The merits of the investment are examined on the basis of the investment costs and the cash flows that will be generated from the investment. It includes the synthesis and financial quantification of the company's knowledge of the key factors for success, and an assessment of the risks to the company.

(iii)   Technical evaluation

The technical evaluation is usually a staged process that occurs with the design of the equipment for the operation. Within the various engineering disciplines and industries there are differing names for the stages, but generally they consist of concept, pre-feasibility, feasibility and final design. The staged approach to project design and evaluation is discussed further in Chapter 4.

(iv)   Economic evaluation

The aim of an economic evaluation is to assess the costs and benefits of the project to all stakeholders in the project. This is a much broader view than the financial evaluation mentioned earlier. Another important difference is that in the opinion of the economists performing the analysis, market values and prices may not be a true reflection of the costs and benefits to all the stakeholders. In this case, the economic analysis may include adjustments to account for these anomalies.

Once the project has been assessed, a decision needs to be taken on whether to invest in it or not. Prior to discussing the practice of investment decisions for capital projects, it is worth discussing the theory and practice of decision-making in general within a business.

2.6 Business Decision-making

The process of arriving at a decision in an organization is a complex interaction of factors such as personal psychology, group dynamics, context, access to information and self-interest. These decisions are either programmed or not programmed. Programmed decisions are those in which a policy, a standard operating procedure, a rule or some precedence exists. Creativity and analytical thought is not required. Most decisions, even at executive level, are of this kind. In contrast, non-programmed decisions are ill structured and elusive – there is no standard way to

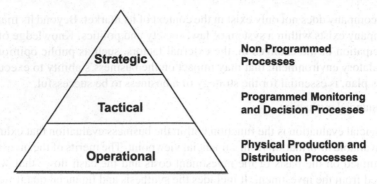

Figure 2.3 Three layers of decision processes found within organisations and corporations

answer them. Very few decisions within an organization are of the non-programmed variety.

Within any organization or corporation, there is a hierarchy of decisions that is taken, illustrated in Figure 2.3. At the bottom of the hierarchy is the mass of production-related decisions that are largely programmed, the middle layer is concerned with monitoring first layer processes and making largely programmed decisions, and at the top of the hierarchy is a layer of non-programmed decisions relating to redesigning or changing parameters for first layer processes.

The investment and financing decisions are at the top of the hierarchy, and contain significant strategic content. They are long-term, have significant cost and risk, and set the course for the business for some years to come. Programmed, operational decisions have less strategic content. They are the day-to-day decisions that have impact in the shorter-term, are largely based on past experience, and generally do not clearly or directly impact on the direction of the company.

Decisions can be reversible or irreversible. If the money invested can be easily recovered without cost, the decision is reversible. For example, consider an investor who purchases an item in a market. If, after a short time, a better opportunity arises and she decides to sell the item and receives almost the price paid originally, then the decision is reversible. If, on the other hand, the investor cannot sell the item, or can only dispose of it for a small fraction of its cost, then the decision is irreversible. Irreversible decisions consume resources and reduce flexibility. If all an investor's resources are invested in assets that cannot be easily sold, then the investor does not have the flexibility to engage in more profitable ventures that might arise, or to liquidate their assets to meet expenses if economic conditions deteriorate. The decision to invest in a capital project is usually an irreversible decision. The costs of buildings, facilities and plants that are constructed for a company cannot be easily recovered.

The function of the financial and strategic evaluation of business opportunities by engineers and scientists is to maximize value for shareholders. Tactics that may be adopted in the short-term, such as increasing sales, decreasing costs or maintaining market position, are subordinate to the long-term goal of maximizing the value of the company. However, in practice, it is often easier to implement short-term tactics and the results are seen more quickly. Since managers are largely focused on their individual career paths, not necessarily on creating shareholder wealth, programmed decisions with mostly short-term, tactical contact are favoured. Thus, it is often easier to get a decision to replace ageing equipment than to get one to increase production capacity by de-bottlenecking, and easier to get a decision to increase production capacity by de-bottlenecking than to get one to build a new production facility.

In evaluating a decision, a distinction should be made between the decision process, and the outcome of the decision. It was not a good decision because the outcome was good, or vice versa. It was a good decision because it was considered, transparent, inclusive and rational. It was a good decision if the assumptions in both the problem formulation and the solution process were interrogated and found to be valid. It was a good decision if information was collected thoroughly, and was presented in an unfiltered and unbiased fashion.

2.7 A Framework for Decision-making

Decision-making can be examined from two different viewpoints: the *normative* view and the *descriptive* view. The normative view suggests what ought to take place, that is, how the ideal decision maker would set about the task. The descriptive view is how it actually happens. A prime criticism of the normative process is that there isn't time to make a full analysis of the alternatives to arrive at the best option. Besides, people usually aren't interested in the best option, they will settle for the first option that meets the minimum requirements, a process known as "satisficing," meaning both to satisfy and to sacrifice.

The focus of our attention is on providing the methods and the means to make successful capital investment decisions. These decisions are not made on the spur of the moment. They usually involve the outlay of significant resources and consequently they more closely represent the normative process. For this reason, the discussion of the decision-making process that follows is based on the normative view.

2.7.1 Steps in the Decision-making Process

The decision-making process can be broken down into different steps. One formulation that might be useful is shown in Figure 2.4. In this formulation, the decision-making process occurs in three stages: frame, evaluate and decide. Each of these stages is discussed below.

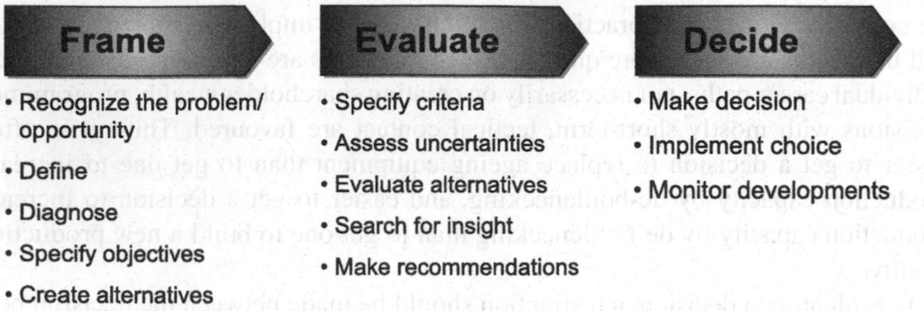

Frame	Evaluate	Decide
• Recognize the problem/ opportunity	• Specify criteria	• Make decision
• Define	• Assess uncertainties	• Implement choice
• Diagnose	• Evaluate alternatives	• Monitor developments
• Specify objectives	• Search for insight	
• Create alternatives	• Make recommendations	

Figure 2.4 The decision-making process

2.7.2 Frame: the Decision Context and Possible Alternatives

Management has to solve problems of a wide-ranging nature. For example, the pump station of a major water supply utility is not performing to specification, requiring maintenance or replacement. Should management repair or replace the equipment? Another example is the following. The capacity of the manufacturing facility will be exceeded in three years at the current rate of increase in production. Since it takes at least that long to authorize the approvals and to build a new facility, should management expand the current operations or should they build a new factory?

The decision frame is the context of the decision. The particular problem or opportunity usually comes with an "in-built" context. However, simply accepting this context limits the possibilities for a creative and better solution.

The object of the framing stage is to define thoroughly the problem or opportunity, and to specify criteria and objectives for the decision that is required. A technique that can assist in clarifying the decision frame is to ask simple questions, such as the following: (i) what is the problem? (ii) what question are we trying to address? (iii) what decisions might we have to make? (iv) what are the uncertainties? (v) what is the important background information? and (vi) what questions are we not trying to answer?

Another technique that is useful in framing a decision is to ask stakeholders what the top issues are with regards to the opportunity. These issues and their associated information can be categorized into facts, uncertainties and decisions. In this context, a fact is known information, an uncertainty is a quantity or variable about which the decision maker or the company has no control, and a decision is a choice that the decision maker can control. The category of uncertainties is used in building a model of the possible outcomes. A decision hierarchy of three different groups can be used to segregate the decisions group: the policy decisions are those that have already been made; the strategic decisions are those that are needed to solve the problem; and the tactical decisions are those that will be made later. The decision hierarchy is useful in determining and maintaining focus during the decision steps (Figure 2.5). The decision hierarchy is discussed in Chapters 9 and 14.

Other well-known management techniques may also be useful in this regard. SWOT analysis, in which the strengths, weaknesses, opportunities and threats are examined, and PEST analysis, in which the political, environmental, social and

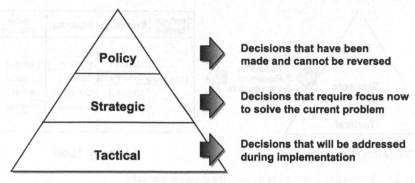

Figure 2.5 The decision hierarchy

technical factors are discussed, can both be useful. Scenario descriptions, in which a number of different "states of the world" are envisaged, can be useful to indicate some of the issues at hand.

Some alternative solutions to the problem will generally emerge during a "framing session," in which decision makers, project sponsors and other stakeholders meet to determine the problem, the boundaries of the problems and other elements of the decision frame. Additional alternatives can be generated using creative techniques, such as brainstorming and blockbusting. Brainstorming is the creation of ideas and alternatives, while blockbusting is used to remove mental "road-blocks."

One method for seeking out viable alternatives is to generate a strategy table. The strategic decisions from the decision hierarchy are used to generate alternatives using the strategy table. For example, consider the situation in which an opportunity existed to lease a new oil property. Should the company go ahead? Decisions that are required now concern the exploration, the development and the infrastructure alternatives. Tactical decisions are those that are required later, such as, the choice of engineering contractor. The strategic decisions become the column headings for the strategy table, and the various options for each strategic decision are listed under each heading. These options are then linked together in a coherent set of choices, called a theme. The theme should have an objective and a rationale that links the various choices together in an identifiable solution strategy. These themes are viable alternatives for the solution of the problem as a whole.

An example of a strategy table is shown in Figure 2.6. There are three elements to the strategic decisions for this problem, which are the headings for columns of the strategy table. Options for each of these strategic decisions are listed in each of the columns. A strategy is the set of choices from each column that is coherent and coordinated, and is linked by a theme or rationale that drives that particular set of choices. The strategic theme shown in Figure 2.6 is the "low cost" strategy, which links the choice of the Alpha Field as the exploration target with the Platform as the means of production and the Pipeline option for the infrastructure choice. The use of the strategy table to create alternative solutions is illustrated in Chapters 9 and 14.

The framing stage of the decision process determines the problem and the boundaries of the problem, that is, what has already been decided, what needs to be decided and what is permissible in moving forward. The framing stage should deliver

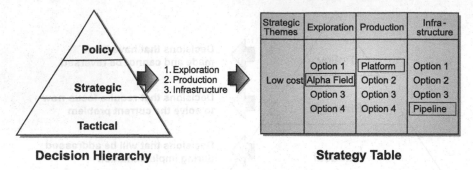

Decision Hierarchy   **Strategy Table**

Figure 2.6 Constructing a strategy table from the decision hierarchy

a clear understanding of the problem, its limits, the decision objectives and several themed strategies that are viable possible solutions. The next step of the decision process is the analysis of these alternatives to determine the most attractive alternative.

2.7.3 Evaluate: the Assessment of Alternatives Based on Criteria

The second step is the evaluation of alternatives against a set of requirements or criteria. Three things are required to perform this step: (i) alternatives, which have been generated in the decision framing step; (ii) a set of criteria; and (iii) a method of evaluation. The set of criteria in general decision-making can be broad. However, in investment decisions, the main criterion is usually that the company should create wealth for its shareholders, that is, the project seeking approval must be economically viable or profitable. The senior management of a company usually specify a measure of profitability that must be used, like the payback period and return on investment discussed in Chapter 1. The method of evaluation involves both the methods for the calculation of the decision criteria and a method for assessing the most attractive alternative. A number of methods are discussed in Chapter 6.

A number of tools can assist in the evaluation stage. Influence diagrams are one these. They link the uncertainties and choices that were delineated in the framing step to the decision criteria. The influence diagrams are used to determine that all the factors that influence the decision have been included in the evaluation, and they are used to build a model of the decision criteria. They create a visual understanding of the interactions between the factors that influence the decision criteria. Influence diagrams and their application to capital project decisions are discussed in detail in Chapter 9.

Decision trees link the outcomes of events and decisions to the value of the decision criteria. Choices and events create different outcomes and these are represented as different paths on the decision tree. The values of the final outcomes are used to determine the optimal choice or set of choices. Decision tree analysis is presented in further detail in Chapter 14.

Other useful techniques in the evaluation stage are sensitivity analysis and scenario analysis. Sensitivity analysis is the determination of how much the decision criterion changes in response to variations in input values. For example, say profit was the decision measure, and two of the uncertainties are the level of market penetration and the production cost per unit. The sensitivity of the first uncertainty is the amount of change of the profit to a change in market share, and the sensitivity of the second uncertainty is the amount of change of the profit to a change in production cost. Scenario analysis is the description of several scenarios, and the examination of the impact of the outcomes expressed in the different scenarios on the decision criteria. The application of scenario and sensitivity analysis to the assessment of capital projects is discussed in further detail in Chapter 9.

2.7.4 Decide: the Act of Decision-making

The final step is the act of choosing the preferred alternative and implementing the decision. This may take a variety of forms depending on the stage of development of the project and the decision that has been made. For example, there are a number of decision stages in the approval of a capital project, each one of which will require the three decision-making steps discussed. The engineering and approval processes are generally aligned, as shown in Figure 2.7. Once approval has been granted for the next stage, the engineering, projects and business teams implement that next phase.

The project moves through the stages outlined in Figure 2.7. Finally, the engineering and projects team enter the stage of detailed engineering. Detailed engineering is followed by the implementation phase, which involves procurement, construction and commissioning. Once the project has been implemented, a post audit is usually conducted to assess the effectiveness of the decision-making.

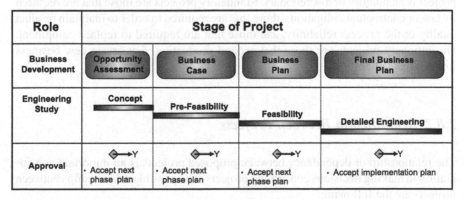

Role	Stage of Project			
Business Development	Opportunity Assessment	Business Case	Business Plan	Final Business Plan
Engineering Study	Concept	Pre-Feasibility	Feasibility	Detailed Engineering
Approval	◇→Y · Accept next phase plan	◇→Y · Accept next phase plan	◇→Y · Accept next phase plan	◇→Y · Accept implementation plan

Figure 2.7 An overview of the business development, engineering design and approval phases for a project

2.8 The Practice of Decision-making for Capital Projects

Capital projects involve significant capital resources, the benefits of which will be experienced over many years. As such, there is a sense in which the present is sacrificed for the future. In the practice of making decisions about whether to invest in a project, companies will classify projects and determine the relationship between projects. These topics are discussed before discussing the decision authority for capital projects.

2.8.1 Classification of Capital Projects

Most organizations specify guidelines that assist managers in their motivation of a particular proposal. The guidelines may include a classification system for investment proposals that is appropriate to the company. The decisions concerning capital investment may be classified according to a variety of different measures depending on what is considered important about the project and the decision to invest in it. For example, the relationship between projects under consideration may be important. Projects may be independent of one another, or, they may be mutually exclusive, which means that only one of those under consideration can be chosen. They may also be classified on basis of the method of evaluation. For example, they may be classified into projects in which little or no detailed analysis is required, such as projects that are required by law, and those where detailed analysis is both possible and necessary, such as new equipment. A further classification might be made based on project risk. Low risk projects might be those such as equipment replacement decisions, normal risk projects may be those such as the expansion of the production capacity of the manufacturing operations, while high risk projects may be those such as the development of a new line of business.

The most practical distinction between projects is made based on whether the project is mandatory or discretionary. Mandatory projects are those that are required by law or contractual obligations, those that are required in order to maintain product quality or the process reliability, and those that are required to replace equipment. Discretionary projects are those that expand production, that create new business lines or that cut costs.

2.8.2 Relationship Between Projects

The relationship or dependency between proposed projects is an important consideration in making decisions concerning projects. The possible relationships between projects are the following:

(i)   Mutually exclusive. Only one of the proposals may be accepted, so that the acceptance of one proposal precludes the other possibilities.
(ii)   Independent. More than one proposal can be accepted, and each proposal has no influence on the others.
(iii)   Complementary. The acceptance of one proposal enhances the prospects of another proposal.
(iv)   Contingent. The acceptance of one proposal is dependent on the prior acceptance of another project.

These relationships between the projects influence the decisions concerning them, and should be acknowledged clearly in the decision process.

2.8.3 Decision Authority for Capital Projects

It has been argued that investment decisions are a critical part of a company's success. However, in contrast to other strategic decisions, decisions on capital projects are decentralized, from corporate to divisions, divisions to operations, from operations to departments. Central control is exercised by setting budget limits for capital expenditure, hence the name capital budgeting. The board of directors sets approval limits on capital expenditure for the chief executive officer and the senior executives. The senior management then cascades these limits down to middle management positions along with other objectives, roles and responsibilities. Thus, the site of the decision-making authority depends on the size of the proposal. An example of the decision-making authority for different sized projects within a company is illustrated in Table 2.1. As shown in Figure 2.7, most projects are executed in stages. The early stages are exploratory, and even though it may eventually be a large proposal requiring the approval of the board of directors, it starts off as a divisional project requiring divisional approval for an initial engineering study.

The aims of the company and the purpose and role of the management of the company have been examined in broad terms in previous sections of this chapter. Two case studies of the practice of decision-making for the allocation of capital are described next. These examples are typical, generic descriptions of actual practice in a company. The first concerns a small project, and the other, a large project.

Table 2.1 An illustration of the investment authority for projects of different sizes

	Project size	Authority
Very small	Less than $100,000	Plant
Small	$100,000 to $1 million	Division
Medium	$1 million to $10 million	Corporate Investment Committee
Large	Over $10 million	CEO and Board of Directors

2.8.4 Case Study: Small Project

Rob Sinclair is the member of the company's divisional development team. The development team's primary responsibility is to investigate improvements to the operations. These can consist of a variety of projects, from those for improved efficiency through to those for cost cutting. One of Rob's projects concerns the processing of the raw material to improve the overall cost position of the division. In discussion with the plant engineer, Rob has identified that the mill, a major consumer of energy in the plant, could benefit from advanced control techniques. Rob estimates that advanced control techniques might make savings in the electricity costs and might improve product quality. Rob obtains quotations from the two vendors who supply advanced controls systems for mills, and estimates the benefits.

Armed with a rough "business case," Rob needs to get the approval of the operations manager. The operations manager is the king of the plant; he can veto or block any initiative on the plant. Reasons for his decisions are not for questioning. If other managers seek to challenge him, he replies that their proposals are a risk to "his" operation. The operations manager views technological change with suspicion; however, competitors have implemented the same control system at their operations without disruption. Much to Rob's surprise, both the operations manager and Rob's superior give approval to prepare the engineering design and the investment proposal.

Rob contacts the two vendors, each of whom provide an engineering design and an estimate of the capital and operating costs. Rob determines the benefits to the operations, estimating the electricity savings based on the savings achieved by the competitors that they reported at the local branch meeting of the engineering professional institute. Rob decides that although the improved quality from the mill is important, it is not possible for him to quantify this, so he excludes it from his estimate of cost savings.

The capital cost is higher than Rob originally anticipated. He was hoping that he could keep the project decision at plant level, but this is not possible. As a result, the decision will be made at divisional level. He believes that it is more difficult to get projects approved at plant and divisional level than at corporate level. However, significantly more work is required to prepare a proposal for approval at corporate level, so he does not want it to be large enough that it needs the approval from the corporate investment committee. He has observed that higher-level managers in the company are more interested in projects that improve chances for the growth of the business, while the lower level managers are interested in projects that improve the efficiency of the manufacturing process. Cost-cutting measures have the best chance of approval at plant level.

Rob's calculations indicate that the payback period is 2 years, which is longer than those approved by the division in the previous year. Part of the cost of the project is the lost production while the new system is being installed and commissioned. Rob gets the vendors to redesign an implementation programme that results in minimal disruption to the production from the plant in an attempt to decrease the payback period. The vendors' work results in reduced disruption, so that the payback period is reduced to 1.6 years.

Rob also computes several measures of economic viability based on discounted cash flow (DCF) techniques, which account for the time value of money. The company has several defined procedures for doing this. For this type of project, he must assume that the project has a ten-year life, that all prices are escalated at 6% for inflation, and that the calculations are based on after-tax cash flows. Rob assumes that the reduction in electricity consumption will be constant over the period. Rob does not perform an assessment of the financial risk. Although he calculated the more sophisticated DCF measures required by the company, the project is discussed solely in terms of its payback period at plant level.

Rob prepares a request for funds. It is a two-page document with a technical description and comments on the benefits, including the possible effect the control system may have on product quality. Rob sends it to the plant manager, who has an overview of all the requests for capital funds sent from the plant to the division. The plant manager is not particularly impressed with the project, but he has respect for Rob because of his success in previous projects. The plant manager sends it to the plant engineer for comment, before submitting it, along with other requests, to the divisional headquarters. The company has an annual capital budgeting and planning cycle during which the budgets for capital equipment are set for the year ahead. The plant manager knows that he has not exceeded the budgeted amount for the plant.

The person at the divisional headquarters who has the most influence on the decision is a vice president in charge of operations. He has prior knowledge of all of the requests for capital from the scheduled meetings with the plant manager. The company policy is to decide on capital expenditure based on the internal rate of return (IRR), one of the DCF measures that Rob has calculated. He notes that the IRR for this project is above the rate required by the company for acceptance, called the "hurdle rate." He trusts that Rob has calculated these values according to the company's procedures, and he trusts the review process undertaken at the plant.

The division has a fixed budget for capital expenses, which is split between mandatory and discretionary projects. The funding for mandatory projects across all the division's plants has increased, so that there is a significant shortfall of funds available for the discretionary projects. The division could make a request for additional funds from corporate, but the divisional president views this unfavourably, because he believes it will create a perception at corporate that he cannot control his budget or his operations.

If all the projects were ranked according to IRR, and only those with the highest IRR selected, Rob's project would not be selected. However, the vice president believes that Rob has been conservative in his estimates so he personally backs it at the meeting to review capital projects. He also argues that it is necessary to begin to examine advanced control systems for all their plants, and that this is an excellent way to do that with minimal cost and disruption to production. The project is approved.

2.8.5 Case Study: Large Project

Sarah Charlton is the divisional technology manager responsible for three of the division's nine operations. The other six operations are upstream of the three plants for which she is responsible. Demand for the division's products has increased over a number of years. Sarah believes that one of the plants within her remit will become capacity constrained within the next three years. The plant is not core to the division's primary business; however, it produces commodity products, and because most of the production costs are assigned to the core products, the plant ranks as one of the producers with the lowest costs in the industry.

She is concerned; no one in the divisional planning department or strategy department has mentioned the impending capacity constraint. In her opinion, this is a major oversight, because it takes about three years to approve a large project and begin construction. She raises this concern with the vice president in charge of operations. He asks her to prepare a proposal with a budget to perform an initial estimate of the capital costs, called an order-of-magnitude estimate. This proposal is to be submitted to the divisional steering committee, which is a subcommittee of the divisional executive committee.

Although there are few similar operations in the world, and none that have been constructed in the last twenty years, Sarah estimates that the capital cost will be in the region of $500 million if a new operation is built, and $350 million if a significant amount of the current equipment can be used. She obtains this estimate by escalating the capital costs of the current plant in order to account for inflation from the date when it was built. Generally, the cost of obtaining an initial capital estimate (order of magnitude estimate) costs between 0.1 and 0.5% of the estimate itself. Sarah's budget request for an engineering firm to perform the order-of-magnitude estimate is about $2.5 million. Her proposal to the steering committee is based on two factors: firstly, the strategic need for an entirely new plant, and, secondly, an analysis of the project financials and economics.

There has been talk in the past to sell this operation because it is non-core business, so the economics must be attractive and the business case compelling to justify the construction of an entirely new operation. The methods for the analysis of the project financials are specified by corporate: all decisions are based on net present value, a discounted cash flow technique, using a discount rate derived from the weighted average cost of capital plus two percent for country risk. She has to value the operations based on a twenty-year life for the plant. She estimates three different scenarios based on different capital requirements. While all three scenarios are positive, they are only just positive.

At the next monthly meeting of the steering committee, she gets approval for the initial phase of the order-of-magnitude cost study.

Sarah and her technology manager at the facility recognize the opportunity to install modern systems and possibly new process technology. Her choice of the engineering contractor for this initial cost estimate is based on her assessment of their technical ability at the conceptual level of design. She is acutely aware of the need to keep the capital as low as possible, and instructs the engineering company to use as much of the existing equipment as possible.

While the capital budget for the division is larger than $100 million a year, proposals such as this do not occur often, perhaps once every four years across the entire division. As a result, they are studied in more detail and more departments are involved than for the smaller projects where the economic analysis is standardized and the decision-making is decentralized throughout the division. Two other departments at divisional headquarters will have input into the decisions as the project moves forward. These are the Projects Department and the Strategic Planning Department.

Sarah assesses that she needs to build alliances within the Strategic Planning Department and the Projects Department, both of whom will assess the project as it moves through the early engineering approval stages. The Projects Department is primarily a project management department, responsible for managing engineering projects. The Projects Department usually prepares the investment proposals for major capital expenditure for approval by the Corporate Investment Committee. The Strategic Planning Department is responsible for production forecasting and the overall scheduling of these production plans within the division. Any planned capital expenditure must also fit into their forecasts of the division's activities. At a later stage, Sarah's project will also be assessed by the division's Corporate Finance Department.

Due to some superb engineering by Sarah's technology team and the engineering company, the estimate for the capital cost of the new operation is $250 million. The error limits on this estimate are +/- 30%. Sarah reports the findings to the steering committee. The project is assigned to the Projects Department, which is responsible for guiding it through the engineering design and capital approval process. A project manager is assigned to the project. Sarah's role now is as technical leader on the project. In addition, Sarah has secured for herself the role of producing the project financials and the financial evaluation of the project.

The Projects Department convene a framing meeting. A wide range of departments within the division is represented at the meeting, which is held to clarify the decision frame for the project. Following the framing meeting, the project moves through two further stages of engineering design: the pre-feasibility and feasibility stages. At the end of each stage, approval is sought from the steering committee for the funds to engage in the next stage. At the end of the feasibility stage, before the basic engineering work is approved, the project is assessed and examined from both technical and economic viewpoints. The project financials that Sarah has produced and refined throughout the various stages of design are scrutinised by the division's corporate finance department. The design is reviewed by a team of experts chosen from within the division and from external consultants.

A project of this size needs corporate approval. The project is placed on the agenda for the Corporate Investment Committee and onto the agenda for the meeting of the corporate Board of Directors. This establishes a deadline for the completion of the feasibility stage of the engineering and the financial evaluation of the project. Prior to the meeting of the Corporate Investment Committee, the corporate

Technical Department, the corporate Finance Department and the Strategic Planning Department need to assess and approve the project.

An investment proposal is prepared in accordance with the prescribed format and topics required by the Corporate Investment Committee. Because Sarah has maintained close contact with the Projects Department, she and the project manager are the prime authors of the proposal. She does this work in addition to her regular duties as technology manager. Sarah's project eventually is approved by all the departments, the Corporate Investment Committee and, finally, by the Board of Directors.

2.9 Summary

An overview of an investment in a capital project by an organization was given in Chapter 1. This chapter has provided an overview of the business context of the decision to invest in projects. The history of the field has roots in economics, engineering, financial management and accounting. As such, it is a truly interdisciplinary activity.

The financial evaluation of projects within a company is known more commonly within the fields of financial management and corporate finance as capital budgeting and within the engineering community as engineering economy. Since the process of allocating funds to the investment in capital projects is a managerial function that is at the heart of the company's future, it is often assumed that the decision-making process is a financial function that lies with the chief financial officer (CFO). In reality, many of the decisions are evaluated and made by technical and engineering staff several layers deeper in the hierarchy than the CFO. Capital budgeting is a decentralized function, governed managerially by setting approval limits. Capital budgeting is also part of long range planning and it must be integrated into the company's business plan and strategy.

The goal of the financial management of a company is to increase the value of the company for the owners of the company. The owners have tasked the management, through the Board of Directors, to run the company so that the owner's value increases. The company must be run profitably on both a year-by-year basis, and in the longer-term. The benefits of a profitable company flow to society as a whole through the payment of tax, and to other stakeholders such as the suppliers and customers, employees and owners. For the company to run a profitable operation in the longer-term, the management must continuously approve projects that they anticipate will create value for the shareholders. Management must make the right decisions about which projects to approve for a profitable future.

A decision is a good decision if process for making it is transparent, inclusive, rational and unbiased. There are a number of descriptions of the decision-making process, and all of them have similar elements. The three-step process is described. The first step, called "frame," is the framing of the decision. It consists of the recognition for the need for a decision, the definition of the decision required, its boundaries and permitted solutions, the definition of the decision objectives, and the creation of alternatives. The second step, called "evaluate," is the assessment of the alternatives. It consists of the specification of the decision criteria, the assessment of the alternatives, the generation of further insight into the decision problem, and the recommendation of the preferred choice. The third step, called "decide," is the actual decision. It consists of making the decision, implementing the preferred choice, monitoring developments and evaluating the performance.

The project proposals for the investment in fixed capital are usually prepared in accordance with procedures specified by the company. There are no globally accepted procedures, and each company has different requirements for the preparation of proposals. A practical distinction between projects that is often useful can be on the basis of whether the project is mandatory or discretionary. Mandatory projects are those that must be implemented. Reasons for these projects may be that they are required because of contracts or regulations, or, more commonly, they are required to maintain product quality and process reliability. Discretionary projects are those that are not mandatory. Examples are process improvements, expansions, new products or cost-cutting exercises.

The decision authority for the approval of projects is cascaded down through an organization. The site of the decision depends on the size of the project and the levels of authority within the organization. The approval process for a larger project is usually staged. The engineering design and approval stages are coordinated so that the correct approvals are obtained with the correct engineering information.

The engineers and scientists in an organization must be able to communicate effectively with one another about financial issues and with the personnel from the accounting and finance departments. This requirement to communicate effectively becomes a prime skill as the career of the engineer or scientist develops.

2.10 Looking Ahead

The key concepts of accounting are discussed in the next chapter. These concepts are formulated as the three financial statements: the balance sheet, the income statement and the cash flow statement. The construction of these financial statements from business transactions is discussed. The relationship between the financial statements and with project financials is explained. Finally, the measures of the performance of a business are presented.

2.11 Review Questions

1. What is the difference between cost-benefit analysis, engineering economics, and capital budgeting?

2. What does financial success mean for financial accounting and financial management?
3. Why is it important for engineers and scientists to understand the viewpoints of financial accountants and financial managements?
4. What are the main drivers of value in a company?
5. What is the aim of the company's management?
6. What affects the value of a company?
7. What are the sources of funds for a company?
8. How do the activities of a company benefit society?
9. What are the roles of financial management within a company?
10. What are the main types of decisions made by the financial managers of a company?
11. Outline the types of evaluation that a project decision can be based on.
12. What are the basic stages of a decision?
13. What are programmed decisions?
14. How are projects classified?
15. Describe the decision-making process for a small and a large project in a company.

2.12 Exercises

1. In the description of both small and large projects, there seemed to be a shortage of capital for investment in projects. In fact, the name "capital budgeting," for the approval process and techniques for assessment of projects used in the field of financial management suggests a shortage of funds that must be budgeted. Surely if a project is a winner, the company should raise the extra financing to fund it, and thereby grow the company. This limitation of funds is called capital rationing. Discuss reasons why there might be a rationing of funds within a company.

Chapter 3
Financial Statements

3.1 Introduction

A high-level understanding of accounting is required by most engineers and scientists within a company in order to communicate with the company's managers and accounting staff. Knowledge of the concepts and principles is also necessary in order to determine the relationship between the project cash flows, as discussed in Chapter 1 for the Santa Clara Hydroelectric Power Scheme, and their effect on the financial statements. Although this is usually the role of the corporate finance department of a company, an understanding of the issues important to the financial manager will enhance communication between the different functions in the company.

This chapter discusses the major accounting principles and the terminology required in order to interpret the financial accounts for a company. This terminology is also required in order to understand the economic viability for a particular engineering project, and the impact the project may have on the company. The discussion of accounting is restricted to the transactions that constitute the financial statements; a more detailed study of the mechanics of accounting is left to the interested reader.

3.2 Business Process and the Dual Nature of Transactions

A company begins life with neither assets nor liabilities. It owns nothing and nobody owes it anything. The company raises funds, called share capital, from its owners for the purpose of entering the field of business in which the company wishes to engage. For example, the company raises $200,000 from an entrepreneur in order to start a factory to produce gas nozzles. The $200,000 is known as share capital or equity capital, and the entrepreneur has shares in the company. If the entrepreneur is the only investor, he owns all the shares and is the owner of the company.

45

Further capital can be raised in the form of loans, bonds, and mortgages. For example, say the nozzle company required a further $200,000 in order to start production. The company may borrow this capital from a bank in the form of a long-term loan. This type of capital is called debt capital.

The company invests the total capital in building the production facility and getting production started. There are different components to this investment: (i) the company acquires buildings and equipment that are used in the manufacture of the product; (ii) it acquires some equipment that is not used in the manufacturing process, but is required for the business; and (iii) it purchases stock of the raw materials that are used in making the product. Items (i) and (ii) are called fixed capital items, while item (iii) is called working capital. So, the company uses the capital to invest in fixed capital and working capital in order to begin business.

3.2.1 The Dual Nature of Transactions

Each of these initial transactions has a dual nature to it. For example, the company raised initial equity of $200,000 from the owner. The owner provided cash in exchange for a share of ownership of the company. This means that after this transaction the company has $200,000 of cash assets, and the owner's portion of the business, which is called equity or owner's equity, exactly matches these cash assets. This is illustrated in Figure 3.1a.

The company raised a further amount of $200,000 in the form of a loan from the bank. This transaction increases the cash assets, and increases the amount of money the company owes to others, called liabilities. This is shown in Figure 3.1b.

In the next transaction, the company uses some of this cash that it has obtained from the owner and the bank to purchase equipment. This decreases the cash but adds a physical asset of the same value as the cash spent to the list of assets. This transaction is shown in Figure 3.1c.

These three transactions each have a dual nature to them. If cash comes into the business from an owner, it increases the owner's "claim" to the assets of the business by that amount. If equipment is purchased for cash, the amount of cash decreases and the value of equipment that the company owns, called fixed assets, increases. The dual nature of transactions leads to the important idea of the accounting equation, which is a balance between what the company owns and what the company is owed.

The company has raised both equity from its owners and debt from it lenders and has purchased assets. In other words, the company has assets that it owns, and it has liabilities that it owes to its lenders. The difference between these two quantities is called the shareholders' equity, which is "owned by" or "due to" the shareholders. The relationship between these variables is given by the following equation:

$$Assets - Liabilities = Equity \qquad (3.1)$$

The assets of the company are items of value owned by the company or items of value in which the company has an interest (called a beneficial interest). Liabilities

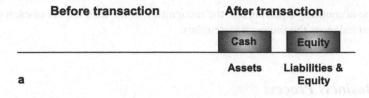

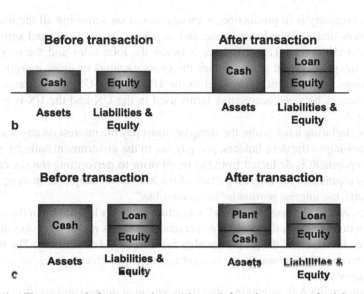

Figure 3.1 a The company's first transaction: raising cash from the owners to fund the company's business plan. **b** The second transaction: raising more cash from the bank to fund the company's business plan. **c** The third transaction: purchasing of manufacturing equipment to execute the company's business plan.

are items that the company owes to others, usually debts in some form or another, while equity is the amount that is deemed to be due to the shareholders or owners of the company. The accounting equation can be rearranged to express this idea in the following form:

$$Assets = Liabilities + Equity \qquad (3.2)$$

The accounting equation in this form expresses the idea that all the assets of a company are either owed to lenders or "belong" to the owners. More formally, the assets have a claim against them, either in form of liabilities, or in the form of the claim by the owners. If the business of the company was stopped today, and all the assets sold, the proceeds from the sale would be distributed to the company's lenders (and creditors) and the company's owners. The term on the right-hand side of Equation 3.2 can be called the "claims" against the value of the company's assets.

It is the accounting equation and the recognition of the dual aspect of each trans-action that balances the financial statements.

3.2.2 Business Process

Once the company is in production, it spends money on acquiring all the materials and services that are needed to make and distribute its products, and sells these products to customers. The difference between the total sales and the total costs is called the gross profit, or sometimes the gross earnings or gross margin. Profit and earnings are different terms used in the UK and the US for the same thing. The differences between accounting terms used in the UK and the US is given in Appendix A.

Before declaring a net profit, the company must pay the interest on any loans that are outstanding to the debt-holders, and pay tax to the government authorities. The interest repayment is deducted from the profit prior to performing the tax calcula-tion. This means that the interest portion of the loan is not subject to income tax; in other words, the interest portion is "tax deductible."

The tax is paid on "taxable income" according to the tax legislation of the country or jurisdiction in which the company operates. The tax is normally proportional to the taxable income, and the proportionality factor is called the tax rate. The tax rate varies from country to country, and in most countries has changed significantly over the last thirty years.

The amount of profit remaining after the payment of interest and taxation is called the net profit, or the "profit after tax." Part of the profit can be paid to shareholders as a dividend, and part is retained to add to the company's asset base, so that the company has further resources to pursue its business activities. Retaining a portion of the profits allows the business to maintain the operations at the same level of profitability, and, if sufficiently profitable, to grow the operations. This creates value for the owners, which it was argued in Chapter 2, is the aim of the management of a business.

Profit is a feedback loop. If the company is meeting customer needs, controlling costs, and has an appropriate mix of funding from owners and lenders, the profit should be sufficient to maintain operations. If the company is doing these excel-lently, there should be sufficient funds to expand operations. As a result, profit is a prime measure of the effectiveness of a business and its management.

The business needs to keep track of all the transactions that make up each of these steps so that management can make the correct operational, financial and strategic decisions, that the correct amount of tax is paid, and that the shareholders and other interested parties, such as potential customers and suppliers, can assess the well-being of the company. The collection and collation of these records into a form that is useful is the field of accounting. The financial records of a business, and the principles and conventions of accounting are discussed in the next couple of sections. After this brief background, the financial statements are examined.

3.3 Financial Records

Businesses keep records of their transactions in order to manage their operations. These records assist in tracking the movement of goods and money. The records of transactions are consolidated into the company's financial statements, which provide a snapshot of the state of the business at a point in time and of the performance of the business since the last financial statements were prepared.

A wide range of interested parties uses the financial statements of the business. The business' management, the owners, the bankers, the potential investors, the tax authorities, the employees, the trade unions and the pension funds are examples of some of the parties who may be interested in the company's financial statements.

The directors of the company have a duty to ensure that the appropriate accounting records are maintained, prepared and submitted. This duty is a legal responsibility that cannot be delegated to anyone else, for example, to the financial director. All directors are jointly liable for the financial records. The directors of a company are required to produce an annual report, which consists of the following reports:

(i)   Directors' report
(ii) Auditor's report
(iii) Financial statements

The financial statements must consist of a balance sheet and an income statement. The income statement is also known as a "profit and loss account," and in this chapter the terms are used interchangeably. This difference in terminology arises from differences in terms used in the UK and the USA. As mentioned before, a table of equivalent terms for the US and the UK is given in Appendix A. Often a cash flow statement is included in the financial statements. Other statements, such as a statement of change in equity may be included.

The financial statements must be a fair reflection of the state of the company. In order to prepare these financial statements so that they can be assessed by the various parties interested in the business and so that the financial statements of different businesses can be compared on a similar basis, accounting has created and adopted a number of principles and conventions. The conventions that are important for a high-level understanding of the financial statements are discussed in the next section.

3.4 Accounting Principles and Conventions

The financial statements of a company are prepared according to a set of rules, precedents and guiding principles. They are formulated as the "Generally Accepted Accounting Principles" (GAAP). The accounting profession in each country has a different variation of the accounting principles. For example, companies within the US are currently required to report according to US GAAP, which is the variation of GAAP adopted by the accounting profession in the USA. Recently there has been a movement to harmonize the accounting standards across different countries with the formulation of the "International Financial Reporting Standards" (IFRS). The European Union and many other countries have adapted their standards to comply with IFRS, or have adopted IFRS completely. Since 2005, the accounts of all public companies in the EU have been reported according to IFRS.

Although it is not necessary to have a detailed understanding of the GAAP or the IFRS rules in order to understand the financial statements, it is necessary to understand some of the basic principles and conventions in accounting. Of the conventions of accounting, there are three that impact on the understanding of accounting for the purposes of the chapters that follow. These are: (i) business entity; (ii) accrual accounting; and (iii) historical cost.

3.4.1 Business Entity

The business and its owners are treated separately for the purposes of accounting. This differs from the legal treatment, where there is no separation between the business and its owners in some types of business.

The three main forms of legal structures of a business are a sole proprietorship, a partnership, and a corporation or limited company. Legally, the owner of a proprietorship is personally responsible for the debts of the business. The owners of a partnership are jointly and severally responsible, meaning that creditors can attempt to recover bad debts from all the owners, or each one separately, in their personal capacities. A limited company, or corporation, is limited in the sense that the most that the owners can lose is their equity investment in the company. Creditors cannot attempt to recover bad debts from the personal assets of the owners of a company. Personal assets are not at risk for owners of limited companies. Directors of limited companies or corporations have additional fiduciary duties that may not provide them with the same protection as owners or shareholders. Irrespective of the legal structure, from one another the accounts for the business and for the owners of the business are separate.

3.4.2 Accrual Accounting

In cash accounting, a sale is recognised when the cash comes in and a purchase when cash departs. Cash accounting is easy to conceptualise. Imagine a local hardware store. If the owner sells an item, cash comes into the till. If the owner needs to buy stock or pay rent, cash comes out of the till. A business like this does not generally sell to customers on credit. The result is that the profitability of the business can be measured by how full the cash register is. Because everything is done on a cash basis, there is no difference between cash and profit.

Accrual accounting, on the other hand, is designed to allow for more complex transactions, like credit sales. It recognises a sale when it occurs, that is, when the sale documents are signed, not when the customer finally pays the cash in settlement. This means that there is a separation between the sales event and the cash settlement event. Although more complex than cash accounting, it allows for the more accurate reflection of transactions like credit sales and credit purchases. The amount of creditors and debtors is critical for a company. Tools for their accurate measurement and control are essential. Accrual accounting provides the mechanism for this.

If the transaction occurs before the cash transfer, the transaction is accrued, and if the transaction is recognised after the cash transfer, it is deferred. For example, accrued revenue refers to revenue that is recognized before the cash is received for the delivery of the product. Deferred taxation is the recognition of taxes owed but not yet paid.

Because of the difference in timing between the recognition of a transaction and cash receipt or payment, there is the need to match the expenses to the correct period in which the revenue is recognised. This allows for the correct calculation of the profit that a company has made in a particular period. For example, if a retailer sells an item at the end of the accounting period, but does not pay the supplier for the item until some time in the next accounting period (because the company simply delays payment to the supplier), the cost of the item must be allocated to the previous period, that is, the period in which the sale occurred.

The situation is slightly different for the purchase of assets whose useful life is longer than a year. In this case, the cost of the asset is not allocated to the year in which it was purchased. If the cost of the asset were allocated in total to the year of purchase, it would have the effect of making the company less profitable or even unprofitable in that year and then more profitable in other years when no assets are purchased. This will undoubtedly have the detrimental and undesirable effect of encouraging management to not purchase new equipment or to not replace ageing equipment. Rather than allocating the entire cost of the asset to the year in which it was purchased, the cost of the asset is spread out over its expected economic life. The cost of the asset is allocated to the income statement over a number of years in order to correctly reflect the profits and the performance of the company. This procedure is called depreciation, and is discussed in further detail later.

3.4.3 Historical Cost

The figures in the accounts represent historical values. The accounts represent the cost of acquiring the assets or goods at the date of acquisition. They do not represent current values of the assets or goods, either as market values or as replacement values. The value of an asset in the accounts, that is, its "book value," is unrelated to its value either within the business (replacement value) or to its value to others (market value). Although this seems unrealistic, reporting that is based on historic costs removes the requirement for valuation based on subjective opinion, which could result in manipulation of the financial position. This convention also removes the need to adjust the figures for inflation, which would considerably complicate accounting and the interpretation of the accounts.

This convention is supported by the "going concern" convention, which holds that the business will continue operations for the foreseeable future. In other words, there is no expectation that the business will need to sell off it assets and realise values different from those based on historic costs represented in the financial statements.

The historical costs principle is also in accord with the conservative position adopted by accounting. For example, if losses are probable, and can be reasonably estimated, they are recorded. However, if gains occur, these are only recorded when they occur.

In the next section, the composition of the three main financial statements for a company is discussed.

3.5 Financial Statements

The three main financial statements that are prepared for each reporting period are the income statement, the balance sheet and the cash flow statement. These financial statements are balances describing the profit, the assets and the cash position of a company. The income statement and the cash flow statement are balances for the accounting period under consideration, and the balance sheet is a balance over the life of the company's existence. This is reflected in the language used: the income statement will state "for the year ending," while the balance sheet will state "as at the year ending." This concept is shown in Figure 3.2.

Prior to discussing the construction of the financial statements, the business process discussed in Section 3.2 is examined in more detail.

3.5.1 Basic Transactions in a Business

The business process that was discussed in Section 3.2 can be represented as a flow diagram that illustrates the interaction between capital, operations and profit

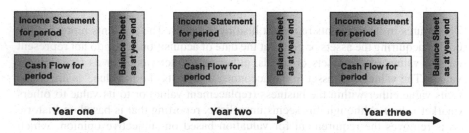

Figure 3.2 The three main financial statements and their relationship with the reporting period

in a business. This flow diagram is shown in Figure 3.3 for the reporting period under consideration.

The beginning of the diagram is at the "capital reservoir," shown as point (1) on Figure 3.3. The capital reservoir is a source or sink of capital. The flow of cash coming into the capital reservoir at point (1) represents cash raised as capital, either as equity capital from shareholders, or as debt capital from debt-holders, such as banks. This capital is used to invest in fixed and working assets (2).

The assets, liabilities and owner's equity are balanced by the accounting equation presented in Section 3.2, so that the value of the company's assets is equal to the company's liabilities plus the owners' equity.

The assets are used in the operations, shown as point (3) on Figure 3.3, in the production process of the company. Costs are incurred for the purchase of raw materials, the hiring of labour, purchase of energy and similar expenses. The products or services of the business are sold for more than the costs that were incurred, which generates profit for the project. The gross profit is the difference between the sales, given by s_t, and the costs, given by c_t, for the operations:

$$Gross\ profit_t = s_t - c_t \tag{3.3}$$

The subscript t on each of these variables represents the period under consideration.

The taxation system in most countries allows the deduction from the profit of an amount that is proportional to the fixed capital investment. This deduction reduces the taxable income and encourages, but does not prescribe, investment in fixed capital. This deduction from the taxable income is called depreciation, or capital allowance, or wear-and-tear allowance. Depreciation is not a cost or an expense, and it does not reflect real expenses for the replacement of equipment. It is an allowance that reduces the amount of tax paid. The concept of depreciation, and its calculation, will be discussed in more detail later in this chapter. The operating profit is calculated by deducting depreciation, given by d_t, from the gross profit.

$$Operating\ profit_t = s_t - c_t - d_t \tag{3.4}$$

This is also called profit before interest and taxation (PBIT) or earnings before interest and taxation (EBIT). It is shown as point (4) on Figure 3.3.

Any interest payable on loans (i_t) is deducted from the operating profit, so that the profit after interest before tax, shown as point (5) on Figure 3.3, is given by the following equation:

$$Profit\ after\ interest\ before\ tax_t = Gross\ profit_t - d_t - i_t = s_t - c_t - d_t - i_t \tag{3.5}$$

Income tax is paid to the government at a rate T. Income tax is not paid if the taxable income is negative, that is, if there is a loss. If the company makes a profit, the income tax paid is given by the following equation:

$$Income\ tax = T(s_t - c_t - d_t - i_t) \tag{3.6}$$

The profit after tax, shown at point (6) on Figure 3.3, is given by the following equation:

$$Profit\ after\ tax = (1 - T)(s_t - c_t - d_t - i_t) \qquad (3.7)$$

This is the accounting profit, net income or earnings for the business for the period under consideration.

It is worth pointing out that the profit is not related to cash, since it includes depreciation, which is a non-cash allowance that can be deducted from the profit in order to calculate the tax. In order to calculate the cash that is generated in the period, the depreciation must be added to the profit after tax. This is expressed as follows:

$$Cash\ generated = (1 - T)(s_t - c_t - d_t - i_t) + d_t \qquad (3.8)$$

The cash generated, shown as point (7) on Figure 3.3, is added to the capital reservoir (1) where the diagram began. The company can use the resources in the capital reservoir to pay the shareholders a dividend, to repay the principal on any outstanding loans, or to invest in other business opportunities.

The cycle repeats itself for each reporting period.

This flow diagram will be used to construct the financial statements.

The interaction of capital and profit shown in Figure 3.3 is the basis for the development of the balance sheet and the income statement. The reporting "domain" for both the income statement and the balance sheet is shown in Figure 3.4.

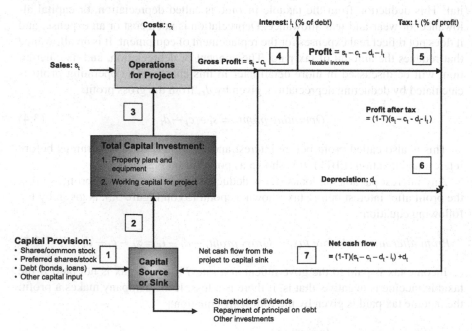

Figure 3.3 Flow diagram representing the investment and operating transactions of the company

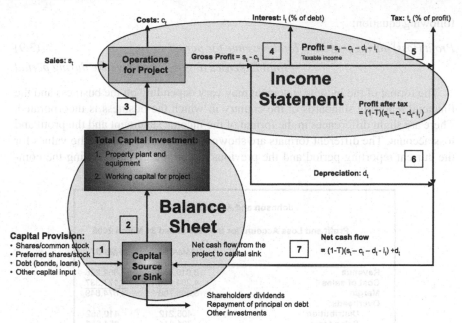

Figure 3.4 The areas of interest to the income statement and balance sheet in reporting the company's transactions

3.5.2 Income Statement

The purpose of a business, as discussed in Chapter 2, is to create value by making a profit. The income statement, also known as the "profit and loss" account in the UK, reports how much profit a business has made during a particular period of operation, known as the reporting period.

In order to determine the profit, the revenues and the expenses must be identified. The revenues, the result of a cash sale or a credit sale, may be generated in different ways depending on the nature of the business. For a manufacturer, this is the amount received for the sale of the manufactured goods; for an engineering consultancy, this is the fee received for the services; for a bank, this is the interest received on loans.

The expenses represent the out-flows incurred in generating the revenues. In the same way in which revenue can be a result of cash or credit sales, expenses may result in direct cash out-flows or in increases in the liabilities. Expenses are the costs of purchasing the raw materials that are transformed into the company's products, as well as the salaries and wages, advertising, rent, transportation, insurance, telephone and many others costs of doing business.

The profit is the difference between the total revenue (for the period) and the total expenses (for the period). If this is negative, that is, the expenses exceed the revenues, then a loss is incurred. Thus, the profit or the loss is expressed by the following equation:

Profit (loss) during period $=$ *Total Revenue for period*   (3.9)

− Total Expenses in generating revenue during period

The format of the income statement may vary depending on the business and the financial reporting standards of the country in which the business is incorporated. There are slight differences in the format of the income statement and the profit and loss account. The different formats are shown in Figure 3.5. Usually the values for the current reporting period and the previous period are given, enabling the com-

Johnson and Associates		
Profit and Loss Account for the year ended 31 March 2008		
	31-Mar08	*31-Mar07*
Revenue	8,010,402	7,852,236
Cost of sales	4,204,696	4,177,387
Margin	3,805,706	3,674,849
Overheads:		
Distribution	406,212	410,852
Sales force	701,844	654,655
Marketing	496,450	465,267
R&D	320,404	194,705
Finance	198,680	194,352
Administration	282,301	289,514
Depreciation	281,005	275,652
Total overheads	2,686,896	2,484,997
Profit before interest and taxation	1,118,810	1,189,852
Interest payable	47,269	65,281
Taxation	82,489	98,665
Net Profit	989,052	1,025,906

Johnson and Associates		
Income Statement for the year ended 31 March 2008		
	31-Mar08	*31-Mar07*
Sales	8,010,402	7,852,236
Cost of goods sold	4,204,696	4,177,387
Gross profit	3,805,706	3,674,849
Selling, General and Administrative		
Distribution	406,212	410,852
Sales force	701,844	654,655
Marketing	496,450	465,267
R&D	320,404	194,705
Administration	282,301	289,514
Finance	198,680	194,352
Total	2,405,891	2,209,345
Depreciation	281,005	275,652
Operating Profit (EBIT)	1,118,810	1,189,852
Interest payable	47,269	65,281
Taxation	82,489	98,665
Net Income	989,052	1,025,906

Figure 3.5 Different reporting formats for the profit and loss account and the income statement

parison of the performance of the company in the current period with that in the previous period.

The expenses are separated in more useful categories than lumping them all together as implied in Equation 3.9. These different categories will be discussed in the order that they are usually found in the income statement. The first section of the income statement is concerned with calculating the gross profit (also called gross margin). The Revenue, or Sales, is the first item in the income statement, and the Cost of Sales, or the Cost of Goods Sold, follows this item. The costs of goods sold are the direct costs incurred in the delivery of the product or the service to the customer. The difference between the Revenue and the Cost of Goods Sold is the gross margin or gross profit.

The next section of the income statement is concerned with the Overheads or the Selling, General and Administrative category. These are the expenses other than the direct costs of producing the company's products or services that are incurred in the running of the business. They include the distribution costs of the product, the costs of the marketing and the sales departments, finance and administration costs, and research and development costs.

The final section involves the deduction of interest, depreciation (if it hasn't been included in the overheads) and taxation. The figure arrived at is the net profit, or "bottom line," which indicates how much value the company has created during the period.

The income statement can be derived from Figure 3.3 and reported in the same order in which it was discussed. Thus beginning with the operations, following the flow diagram around to net profit, the income statement for the company is given in Table 3.1.

This discussion has provided an overview of the general principles of the income statement. The details of the line items are now discussed.

(i)   Sales or revenue

The revenue or sales is the total invoiced to customers for the goods or services provided. If, as is common, value-added tax (VAT) or general sales tax (GST) is

Table 3.1 Construction of the income statement

Income statement	Value for reporting period
Sales	s_t
Cost of goods sold	c_t
Gross profit	$s_t - c_t$
Selling, general and administrative	sga_t
Depreciation	d_t
Operating profit (EBIT)	$s_t - c_t - sga_t - d_t$
Interest Payable	i_t
Taxation	$T(s_t - c_t - sga_t - i_t - d_t)$
Net income (profit after interest and taxation)	$(1 - T)(s_t - c_t - sga_t - i_t - d_t)$

charged on goods and services, the sales quoted in the income statement excludes the VAT or GST charge.

(ii)   Cost of sales or cost of goods sold

The cost of sales is the sum of the "direct" costs in producing the goods or services, that is, all the production related expenses. For a manufacturing company, this includes the purchase of raw materials, the employment of labour used directly in the manufacturing process, and the other direct costs used in production, such as the cost energy, rental of the factory premises, and so forth. If there was no production, the direct production costs would be zero. This can be used as a practical test for determining if the costs are costs of goods sold or not.

The cost of goods sold is often explained by the following relationship that constitutes a balance on the inventory:

$$COGS = (Beginning\ Inventory) + (Purchases\ of\ Inventory) - (Ending\ Inventory)$$

$$(3.10)$$

However, this ignores the production labour, and other production-related expenses, like the need for energy and production premises. The inventory balance is augmented by these additional factors to give the full cost of sales:

$$COGS = (Beginning\ Inventory) + (Purchases\ of\ Inventory) - (Ending\ Inventory)$$
$$+ (Cost\ of\ Production\ Labour) + (Other\ Direct\ Costs) \qquad (3.11)$$

(iii)   Overheads or selling, general and administrative expenses

The overheads, also called the selling, general and administrative expenses (SGA), are the expenses incurred that are not directly associated with the production activities, but are necessary in the running of the business. In the example shown in Figure 3.5, these expenses are broken down in the categories by department. (Note that depreciation is included in overheads in the profit and loss account but is not part of SGA expenses in the income statement).

(iv)   Interest

Interest is the charge for borrowing money and is deducted from the profit before calculating the amount of tax owed. The interest reduces the tax charged.

(v)   Depreciation

The full cost of an asset is not borne in the year of purchase on the income statement; rather the cost of the asset is spread out over the life of the asset. This spreading out over a number of years of the cost of the asset is called depreciation. More formally, it is the allocation of the asset's cost over its economic life. The total depreciation for an asset must add up to the original cost of the asset. Depreciation is the amount deducted from the profit to account for the purchase of fixed assets, such as equipment.

Depreciation also reduces the amount of tax paid, as shown for the taxation line in Table 3.1. If the depreciation is spread over fewer years, the depreciation amount is increased, and the tax paid over those years is reduced. The depreciation need not be spread evenly, so that there may be a greater reduction in tax in the earlier years of the asset's life than in the later years. The amount of depreciation that is allowed for the determination of taxes is prescribed by the taxation system of the country. In the US, there is a system called MACRS (Modified Accelerated Cost Recovery System), while in the UK it is fixed proportion of the book value of the asset, which is a more straightforward system.

Depreciation is such a commonly misunderstood topic that it requires a more full discussion than presented above. Because it has implications for both the income statement and the balance sheet, a more detailed discussion is delayed until after the discussion of the balance sheet. However, it is important to acknowledge three aspects relating to depreciation: (i) that the costs of purchasing assets are not represented directly on the income statement; (ii) that they are spread out over a number of years as depreciation; and (iii) that depreciation reduces the tax charge.

(vi)   Taxation

This is the corporate income tax charged. In some countries, this is a flat rate, while in others it is a sliding scale. For example, in the UK, the corporate tax rate is a flat rate of 30% of taxable income, while federal income tax rate in the USA is according to a sliding scale given in a tax table. An example of a tax table is given in Table 3.2. The corporate tax rate in different countries varies widely, as shown in Table 3.3. Countries such as the USA and Canada also have a state or provincial tax that is charged in addition to the federal or national tax.

Generally, losses that are incurred may be carried forward to the following year, which means that these losses can be used to reduce the taxable income in the following year by the amount of the loss in the previous year.

(vii)   Net income or net profit

The net income or net profit is the final line item (hence "bottom line") on the income statement, and represents a measure of the performance or productive effort of the company for the year. The larger the profit, the better the performance.

Table 3.2 Example of a tax table

Taxable income over	Not over	Tax rate
0	50,000	15%
50,000	75,000	25%
75,000	100,000	34%
100,000	335,000	39%
335,000	10,000,000	34%
10,000,000	15,000,000	35%
15,000,000	18,333,333	38%
18,333,333		35%

Table 3.3 Corporate Tax Rate (combined national and subnational) for OECD countries for 2000 and 2006 (OECD, 2006)

Country	Corporate Tax Rate in 2000	Corporate Tax Rate in 2006
Japan	40.9	39.5
United States	39.4	39.3
Germany	52.0	38.9
Canada	44.6	36.1
France	37.8	35.0
Spain	35.0	35.0
Belgium	40.2	34.0
Italy	37.0	33.0
New Zealand	33.0	33.0
Greece	40.0	32.0
Netherlands	35.0	31.5
Luxembourg	37.5	30.4
Mexico	35.0	30.0
Australia	34.0	30.0
Turkey	33.0	30.0
United Kingdom	30.0	30.0
Denmark	32.0	28.0
Norway	28.0	28.0
Sweden	28.0	28.0
Portugal	35.2	27.5
Korea	30.8	27.5
Czech Republic	31.0	26.0
Finland	29.0	26.0
Austria	34.0	25.0
Switzerland	24.9	21.3
Poland	30.0	19.0
Slovak Republic	29.0	19.0
Iceland	30.0	18.0
Hungary	18.0	16.0
Ireland	24.0	12.5
OECD Average	33.6	28.7

Note: Small changes are attributable to changes in provincial or state (subnational) rates.

3.5.3 Balance Sheet

The income statement measures performance of a company over the reporting period; the balance sheet reports the value of the company at a point in time. It details the forms in which the wealth of the company is held, and how much is held in each form. The balance sheet is an implementation of the accounting equation discussed in Section 3.2.1, in which assets of the business are balanced by the claims against the business. It was shown in that section that the claims against a business are of two types, the liabilities of the business, and the owners' equity. Before we examine the format of the balance sheet, we need to examine each of these terms.

(i)   Assets

An asset is a resource that has monetary value and is owned by the business. The value of an asset can arise through its use, or through its hire or sale. An obsolete piece of equipment that can be sold for scrap is an asset, while that which cannot be sold is not an asset. The owner must have an exclusive right to or control of the asset and the transaction that resulted in this right must have already been concluded.

Assets are normally divided into two categories: fixed assets and current assets. Fixed assets, or non-current assets, are tangible items that are acquired for the purpose of being used to generate revenue (rather than for resale). Fixed assets are the equipment used by the business in generation of the company's products; they are not consumed quickly during their use. Typical fixed assets are items like land, buildings, plant and equipment, fixtures and fittings, and vehicles. Most companies will have an authorisation procedure for the acquisition of fixed assets. Although definitions will differ from one organisation to another, this procedure is commonly known as capital expenditure.

The value of a fixed asset represented on the balance sheet is the "book value," that is, the original cost of the asset less the accumulated depreciation.

The current assets are those items that can or will be turned into cash relatively quickly, that is, in less than a year. For example, the company purchases stock, and sells the product to the customer on credit. These two transactions involve three items that are current assets: cash; stocks or inventory; and debtors. The stock is purchased with cash and the product sold on credit increases the debtors. (Debtors still owe the company for the products they have purchased from the company. Debtors are also called accounts receivable.)

(ii)　Liabilities

Liabilities represent claims that others, except the owner of the business, have against the business. These claims usually arise from lending money to the business, or supplying goods to the business on credit terms. Once a liability has been incurred, it remains an obligation of the business until it is settled.

The liabilities are divided into current liabilities and long-term liabilities. Long-term liabilities are not due for more than a year, whereas current liabilities are due within a year. Long-term liabilities are items such as loans, while current liabilities are items such as creditors and the bank overdraft. Creditors, to whom the company owes money for the supply of stock, are sometimes called accounts payable.

(iii)　Equity

The equity in a business is the amount that is owed to the owners or shareholders of the business. The equity arises from two main mechanisms: (i) the capital that was raised from shareholders and owners to start the business; and (ii) the profit earned by the business and not paid out to the shareholders. The investment by the owners of the business is called the share capital or the common stock on the balance sheet. The profits that are retained by the company (that is, not paid to the shareholders) are called the reserves or retained earnings. Thus the retained earnings or profits are the sum of all profits made by the company during its history less all the dividends paid to the shareholders during the company's history.

The relationship between these categories is given by the accounting equation, which is expressed as follows:

$$Assets = Liabilities + Equity \qquad (3.12)$$

Johnson and Associates
Balance Sheet as at 31 March 2008

	31-Mar-08	31-Mar-07		31-Mar-08	31-Mar-07
Fixed assets	3,820,046	4,034,274	**Shareholders funds**		
Book value	6,245,742	6,178,965	Capital	10,000	10,000
Accumulated depreciation	2,425,696	2,144,691	Profit reserves	2,854,612	2,544,792
Net book value	3,820,046	4,034,274	Revaluation reserve	812,000	812,000
			Total Shareholders funds	3,676,612	3,366,792
Current assets					
Stocks	612,550	589,365	**Long-term liabilities**		
Debtors	3,254,522	2,740,005	Loan	1,700,000	1,700,000
Cash	440,548	112,450	Long-term creditors	75,442	88,354
Total Current Assets	4,307,620	3,441,820	Total long-term liabilities	1,775,442	1,788,354
Investments	200,000	200,000	**Current liabilities**		
			Creditors	2,875,612	2,520,948
Total Investment	8,327,666	7,676,094	Total finance employed	8,327,666	7,676,094

Figure 3.6 Balance sheet for Johnson and Associates in the horizontal format

Since the assets are comprised of fixed assets and current assets and the liabilities consist of long-term liabilities and current liabilities, the accounting equation can be expanded to give the following equation:

$$(Fixed\ Assets) + (Current\ Assets) = (Long\ term\ Liabilities)$$
$$+ (Current\ Liabilities) + Equity \quad (3.13)$$

There are two formats of the balance sheet: the horizontal format and the vertical format. Equation 3.13 represents the format of the balance sheet in the "horizontal" layout. A simple rearrangement represents the format for the "vertical" layout:

$$(Fixed\ Assets) + (Current\ Assets) - (Current\ Liabilities)$$
$$= (Long\ term\ Liabilities) + Equity \quad (3.14)$$

The difference between current assets and current liabilities is called the working capital.

The horizontal and vertical formats are shown in Figures 3.6 and 3.7.

3.5.4 Cash Flow Statement

The income statement provides a view of the business based on the revenue and expenses, not the cash receipts and payments. Because accrual accounting is used and because profit includes non-cash items such as depreciation, there may not be a one-to-one relationship between profit and cash. It is possible for a profitable company to go out of business due to a shortage of cash; indeed, a shortage of cash is the most common cause of business failure. A shortage of cash in a business is referred to as commercial insolvency.

The cash flow statement is a formal statement of the cash payments and receipts for the reporting period. The cash flow statement, shown in Figure 3.8, represents

```
┌─────────────────────────────────────────────────────────────────┐
│                     Johnson and Associates                       │
│                                                                   │
│                  Balance Sheet as at 31 March 2008                │
│                                                                   │
│                              31-Mar08          31-Mar07           │
│                                                                   │
│   Fixed assets                                                    │
│   Book value                 6,245,742         6,178,965          │
│   Accumulated depreciation   2,425,696         2,144,691          │
│   Net book value             3,820,046         4,034,274          │
│                                                                   │
│   Investments                 200,000           200,000          │
│                                                                   │
│   Current assets             4,307,620         3,441,820          │
│   Stocks                       612,550           589,365          │
│   Debtors                    3,254,522         2,740,005          │
│   Cash                         440,548           112,450          │
│                                                                   │
│   Current liabilities        2,875,612         2,520,948          │
│   Creditors                  2,875,612         2,520,948          │
│                                                                   │
│   Net working capital        1,432,008           920,872          │
│                                                                   │
│   Total net assets           5,452,054         5,155,146          │
│                                                                   │
│   Shareholders funds                                              │
│   Capital                       10,000            10,000          │
│   Profit reserves            2,854,612         2,544,792          │
│   Revaluation reserve          812,000           812,000          │
│   Total Shareholders funds   3,676,612         3,366,792          │
│                                                                   │
│   Long-term liabilities                                           │
│   Loan                       1,700,000         1,700,000          │
│   Creditors due after more than one year  75,442     88,354       │
│   Total long-term liabilities  1,775,442       1,788,354          │
│                                                                   │
│   Total capital employed     5,452,054         5,155,146          │
└─────────────────────────────────────────────────────────────────┘
```

Figure 3.7 Balance sheet for Johnson and Associates in the vertical format

the cash generated or consumed in the operating, investment and financing activities of the business.

(i)   Operating activities

The operating cash flow represents the cash received from customers and cash paid to suppliers and employees. There are two methods for determining the cash flow from operating activities: the direct method and the indirect method. The direct method involves separating all the cash transactions into those that involve operating activities and those that do not. The indirect method determines the cash generated by calculation from profit (net income) obtained from the income statement. The indirect method is the method most adopted in practice. The operating cash flow is calculated by adjusting the net income for two things: the non-cash items, such as depreciation, and the change in working capital.

The flow diagram presented in Figure 3.3, and the associated income statement represented in Table 3.1 can be used to illustrate the method of determine the cash flow from operating activities. The net income is given by the following expression:

$$Net\ Income = (1 - T)(s_t - c_t - sga_t - i_t - d_t) \qquad (3.15)$$

Johnson and Associates		
Cash Flow Statement for year ended 31 March 2008		
	31-Mar-08	*31-Mar-07*
OPERATING ACTIVITIES		
Income from operating activities	1,399,815	1,465,504
Adjusted for increase in working capital	183,038	161,684
Taxation paid	82,489	98,665
Net cash provided by operating activities	1,134,288	1,205,155
INVESTMENT ACTIVITIES		
Cash used to acquire fixed tangible assets	66,777	50,000
FINANCING ACTIVITIES		
Decrease in long term liabilities	12,912	16,095
Interest paid	47,269	65,281
Dividends paid	679232	984328
Net cash used by financing activities	739,413	1,065,704
NET CASH AND EQUIVALENTS PROVIDED	328,098	89,450
Cash at beginning of year	112,450	23,000
Cash at end of year	440,548	112,450
Net change in cash position	328,098	89,450

Figure 3.8 Cash flow statement for Johnson and Associates

Since depreciation is a non-cash item, the depreciation must be added to net income in order to calculate the cash flow from the business activities. This calculation is encapsulated in this expression:

$$Cash\ Flow\ from\ Income\ Statement = (1 - T)(s_t - c_t - sga_t - i_t - d_t) + d_t \quad (3.16)$$

A change in the inventory or the amount of debtors and creditors also changes the cash position of the company. Since these changes are a result of operations, they are assigned to the operating activities category of the cash flow statement. The working capital is the current assets less the current liabilities. The change in the working capital is the working capital for the current period less the working capital for the previous period. It should be possible to determine the change in working capital from the values presented in the balance sheet.

The cash flow from operating activities is calculated by adding the change in working capital over the period to the cash flow of Equation 3.16. Thus, the cash flow is given by the following expression:

$$Cash\ Flow\ from\ Operating\ Activities = (1 - T)(s_t - c_t - sga_t - i_t - d_t)$$
$$+ d_t - \Delta(WC)_t \quad (3.17)$$

The symbol $\Delta(WC)_t$ represents change in working capital for the reporting period.

(ii)   Investment activities

Investing activities report the acquisition or disposal of long-term assets. The change between the current period and the previous period in the fixed assets on the balance sheet represents the change in the cash position as a result of investment activities by the company. This is given by the following expression:

$$Cash\ Flow\ from\ Investment\ Activities\ =\ A_t \qquad (3.18)$$

The symbol A_t represents investment in fixed or non-current assets during the period.

(iii)   Financing activities

The company can raise funds for its activities from a combination of equity and loans. The financing activities of the cash flow statement reflect these activities of the company for the reporting period. The results for these two activities can be derived from the change in the long and short-term loans over the reporting period, and the change in equity capital (not retained earnings).

The payment of a dividend to the shareholders by the company will reduce the cash position and is represented in this section. Therefore the cash flow as a result of financing activities is given by the following expression:

$$Cash\ Flow\ from\ Financing\ Activities\ =\ L_t + E_t - Div_t \qquad (3.19)$$

The symbols L_t, E_t, and Div_t represent the loans raised, the equity capital raised, and the dividend paid in the period, respectively.

(iv)   Net cash flow and cash equivalents

The net cash flow for the reporting period is the sum of the cash flows for the operating, investment and financing activities of the company. This is expressed in the following equation:

$$Net\ Cash\ Flow\ =\ CF(Operating\ Activities) + CF(Investment\ Activities)$$
$$+\ CF(Financing\ Activities) \qquad (3.20)$$

The net cash flow can be expressed in terms of the symbols for each of the components of these cash flows. This is given in the following equation:

$$Net\ Cash\ Flow\ =(1 - T)(s_t - c_t - sga_t - i_t - d_t) + d_t - \Delta(WC)_t - A_t$$
$$+ L_t + E_t - Div_t \qquad (3.21)$$

Equation 3.21 represents a comprehensive view of the major factors that affect the cash position of a company.

The final section on the cash flow statement summarises the cash position of the company.

3.6 Depreciation

In the discussion of the income statement earlier it was mentioned that profit is a measure of the performance or productivity of a company. If the company were performing consistently, it would be expected that the profit would be consistent. However, if the company buys an expensive piece of equipment that is expected to last for a number of years, and this cost were charged to the income statement in the year of purchase, the profit would be significantly lower in the year of purchase. The readers of the income statement may interpret the lower profit as being a result of poor performance on the part of management. If this were the case, management would be encouraged not to spend on new opportunities or on replacing ageing equipment. In order to overcome this, the cost of the equipment is not charged all at once to the income statement, but apportioned over its economic life, and these portions are charged to the income statement each year.

The division of the company's investments in fixed capital into portions that are charged to the income statement over a number of years is called depreciation. The difference between the original cost of the equipment and all the depreciation accumulated to date is called the "book value."

For example, a company buys a computer at the beginning of the first year for $10,000. The computer has an economic life of 5 years. The company calculates depreciation on the basis that the portion for each year is the same. This is called the straight-line basis. As a result, the annual depreciation charge is equal to $10,000/5 = $2,000. The book value of the computer at the end of each year is given in Table 3.4.

For this example, assume that the gross profit of the company is constant, at $20,000. The company decides to sell the computer in the fourth year for $4,300. The profit on the sale of the computer is the proceeds from the sale less the book value, that is $4,300 - $2,000 = $2,300. The effect of the purchase and sale of the computer on the income statement at the end of each year is shown in Table 3.5.

Without depreciation, the purchase of the computer would have reduced the profit before tax in year 1 to $10,000, and in year 2 it would have returned to $20,000. It is this drop in profit, and the subsequent rise, that depreciation smoothes out. If there was no depreciation and management is measured, even partly, on the profit achieved, there would be a strong incentive for management not to make investments in fixed capital, and this would be detrimental to the longevity of the company.

Table 3.4 Depreciation schedule for computer

Year	0	1	2	3	4	5
Cost	10,000					
Depreciation		2,000	2,000	2,000	2,000	2,000
Book value		8,000	6,000	4,000	2,000	0

Table 3.5 Calculation of profit

Year	1	2	3	4	5
Gross profit	20,000	20,000	20,000	20,000	20,000
Depreciation	2,000	2,000	2,000	2,000	0
Profit from sale of computer				2,300	
Profit before tax	18,000	18,000	18,000	20,300	20,000

Table 3.6 Cash position as a result of the purchase and sale of the computer

Year	1	2	3	4	5
Cost	(10,000)				
Sale				4,300	

The cash position is different from that of the profit represented on the income statement. The cash flows resulting from the sale and purchase of the computer are represented by Table 3.6. In the first year there is a cash outflow of $10,000 due to the purchase of the computer, and in the fourth year there is a cash inflow of $4,300 due to the sale of the computer.

The value of the assets reflected on the balance sheet is not their market value, but their book value. The depreciation of the assets means that the book value is decreasing with time, and this is reflected on the balance sheet.

To sum up so far, depreciation affects the profits on the income statement and the value of the assets on the balance sheet. Depreciation acts to smooth the profit from year to year, resulting in a fair reflection of the productive effort of the company on the income statement.

Depreciation affects the cash flows of the company indirectly. Because the depreciation charge to the income statement is deducted from the profit before the calculation of income tax, depreciation lowers the tax charged, resulting in higher cash flown. This can be more clearly explained by examining the expression for cash flow from income, given by the following equation:

$$Cash\ Flow\ from\ Income\ Statement = (1-T)(s_t - c_t - sga_t - i_t - d_t) + d_t \tag{3.22}$$

To refresh, the term T represents the tax rate and the term d_t represents the depreciation charged in the reporting period. Collecting the terms that involve depreciation, this equation can be rewritten in the following form:

$$Cash\ Flow\ from\ Income\ Statement = (1-T)(s_t - c_t - sga_t - i_t) + Td_t \tag{3.23}$$

This expression demonstrates that as the depreciation increases, so does the cash flow. The income tax charged is given by the expression:

$$Income\ Tax = T(s_t - c_t - sga_t - i_t) - Td_t \tag{3.24}$$

A comparison of the expression for the cash flow from income with the income tax charged indicates that the increase in cash flow due to an increase in depreciation is a result of the decrease in taxation.

When an asset reaches the end of its useful life, its book value is zero. It is removed from the register of assets of the company, or "written off." If, during the asset's life, the asset is damaged, and the disposal value is less than the book value, it is said to be impaired. The circumstances leading to impairment must be unusual and unexpected. For example, if a vehicle is involved in a collision, and it is not possible to repair it, it is impaired to the extent of the difference between its book value and its scrap value. The accident is an unexpected event that resulted in the loss in value.

If an asset is worth more than its book value, the directors may elect to revalue the asset. If the asset value increases, the claims against the company on the balance sheet must also increase. This is reflected in the owners' equity section of the balance sheet under a heading such as "revaluation reserves." The company's directors may not distribute the revaluation reserves to the shareholders. For this reason, the revaluation reserves may also be called "non-distributable reserves" on the balance sheet.

There are a number of misconceptions concerning depreciation. It is not a capital replacement fund, it is not a "wear and tear" allowance, and it does not represent the obsolescence or impairment of the asset. Depreciation allocates the cost of an asset to the income statement over the economic life of the asset, and simultaneously reduces the value of the asset on the balance sheet. It is a calculation that has an indirect effect on the cash flows by reducing the income tax. It is unrelated to the market value or the productivity of the asset.

There are a number of ways of calculating depreciation and these are largely discretionary. However, because the depreciation affects the amount of tax involved, the tax authorities have specified the depreciation method for the purposes of determining taxable income, and hence the corporation tax. While it is important to understand the concept of depreciation and its implications in the books of account, it is the effect of depreciation on the amount of tax paid that is of primary concern to engineers and scientists.

The tax authorities specify the method of calculation and the economic life of classes of assets, and this method must be used to determine the taxable income. In the US, the method is the modified accelerated cost recovery system (MACRS), while in other countries the method may be the declining-balance method, or some other method. In an attempt to separate the issue of depreciation for the purpose of smoothing the accounting profit and for the purpose of determining corporation tax, the tax authorities in some countries do not refer to it as depreciation. For example, in the UK the depreciation that can be charged to the profit as a result of capital investments for tax purposes is called *capital allowance*, not depreciation. This emphasises the difference between depreciation for the company's books and depreciation for the calculation of tax. While the use of the term capital allowance provides clarity, the terminology used in tax systems of other countries, terms such as "wear and tear allowance," adds to the confusion and to the misunderstanding of the concept of depreciation.

The main methods of calculation of depreciation are:

(i)   Straight-line depreciation
(ii)   Declining balance
(iii) Sum-of-years digits
(iv)   MACRS

Because the methods of calculating the depreciation or capital allowance directly affect the after-tax cash flows to a project by affecting the amount of tax paid, it is necessary to understand the details of these methods. However, this discussion is delayed until Chapter 8. For the purposes of the discussion before Chapter 8, depreciation is calculated on a straight-line basis. This means that the capital cost of the equipment is spread evenly over the useful life of the equipment. Thus, if the tax authorities deem that the useful life of a vehicle is 5 years, the depreciation charged to the income statement in each year is the cost of the vehicle divided by five.

3.7 The Interaction Between the Financial Statements

The three financial statements represent three views of the company: the balance sheet represents the assets and liabilities of the company, that is, the value of the company; the income statement represents the revenue, costs and profit of the company, that is, the productive effort of the company; and the cash flow statement represents the net flow of cash into and out of the company, that is, the cash position of the company.

Changes to one of the statements will affect the others. Although the discussion on depreciation was instructive in its own right, it also illustrated some of the interactions between the financial statements. Changes to depreciation affect the income statement, the balance sheet and the cash flow statement.

In this section, the links between the financial statements will be examined in order to more fully understand the accounting relationships.

The balance sheet represents the fundamental accounting equation:

$$Assets = Liabilities + Equity \qquad (3.25)$$

or, using the symbols A, L and E to represent Assets, Liabilities, and Equity, respectively, Equation 3.25 can be rewritten in the following form:

$$A = L + E \qquad (3.26)$$

The assets and liabilities can both be divided into categories of current and non-current. Using the symbols CA and NCA to represent the current assets and the non-current assets, respectively, and CL and NCL to represent current liabilities and non-current liabilities, respectively, the accounting equation can be written as follows:

$$CA + NCA = CL + NCL + E \qquad (3.27)$$

The non-current assets are the fixed assets, and the non-current liabilities are the long-term liabilities. The current assets consist of cash (C), inventory or stock (S) and debtors (D). Substitution of these values into the accounting equation yields the following expression:

$$C + S + D + NCA = CL + NCL + E \qquad (3.28)$$

Rearrangement of this expression yields an expression that links the cash position of the company with the other values on the balance sheet. This is given as follows:

$$C = CL + NCL + E - S - D - NCA \tag{3.29}$$

This equation is the link between the cash flow statement and the balance sheet. In order to raise more cash, the company can either increase the current liabilities, the non-current liabilities or the equity, or decrease the stock or debtors. Increasing the current liabilities can be done by delaying payment to suppliers or by getting a short-term loan, such as an overdraft from the bank. Increasing the non-current liabilities can be achieved by increasing the long-term loans. Increasing the owners' equity can be achieved by getting the owners to provide more capital for the business or by reducing the dividend payment to the owners.

The income statement and the balance sheet are linked primarily through the income retained in the business. This was clearly shown in Figure 3.4. The owners' equity is comprised of share capital, accumulated earnings (from all the previous years) and the earnings retained in the current reporting period. If the symbols SC, AE, and RE are used to represent the share capital, the accumulated earnings and the earnings retained, respectively, the accounting equation can be written in the following form:

$$A = L + SC + AE + RE \tag{3.30}$$

The retained earnings for a particular year are directly proportional to the net income or profit of the business in that year. The accounting equation can be written as follows:

$$A = L + SC + AE + \alpha_t(NI_t) \tag{3.31}$$

The symbol NI_t represents net income, and α_t represents the proportion of the net income that is retained by the business. Both of these symbols represent values for the current reporting period. If none of the income is retained, α_t is equal to zero; on the other hand, if all the income in a particular year is retained, α_t is equal to one. This equation indicates that the higher the net income and the higher the value of α_t, the higher the value of the assets, which, in general, translates into a value of the business.

The components of the net income, NI_t, are described by the following relationship:

$$NI_t = (1 - T)(s_t - c_t - sga_t - i_t - d_t) \tag{3.32}$$

The substitution of this expression into the accounting equation links the components of the income statement with the balance sheet, as shown in the following equation:

$$A = L + SC + AE + \alpha_t(1 - T)(s_t - c_t - sga_t - i_t - d_t) + d_t \tag{3.33}$$

This equation succinctly summarizes the idea that increasing the sales and decreasing the costs, overheads and interest payments increases the asset value of the company.

As was argued in the Section 3.6, depreciation affects both the profits and the value of the assets. The value of the assets is decreased by the depreciation charge in the reporting period. The asset value is the asset value at the beginning of the period, A_{t-1}, plus the assets acquired in the period, AA_t, less the depreciation for the period, d_t. Therefore, increasing the depreciation increases the profits and the cash flow, but decreases the book value of the company.

The relationship between the financial statements has been examined in this section. In the next section, the relationship between the financial statements and the project financials are explored.

3.8 Relationship Between the Financial Statements and the Project Cash Flows

The cash flows for a project, called the project financials, were discussed in Chapter 1. The aim of the project financials is to determine the free cash flows, not the profit. It is clear that the project cash flows, as presented in that chapter, do not follow the same format as the financial statements. Because they seem to be a mix between the cash flow statement and the income statement, a reader may be tempted to conclude that a mistake was made. However, no mistake has been made.

Unlike the financial statements, there is no prescribed or standardized format for the project cash flows. The format adopted by most practitioners is similar to that shown in Chapter 1. It is similar in layout to the income statement. However, the line items include items from the cash flow statement, such as investments that are not represented in the income statement, and exclude items from both, such as the interest paid.

The guidelines for the layout and format adopted for the project cash flows are that they must be easy to construct, to read and to interrogate. The final result of the project cash flow schedule is to determine the free cash flow required or generated by the project. This is different from the financial statements because there are regulatory requirements set by the accounting profession and by various bodies of government that govern the form and content of the financial statements.

The financial projections of the project can also be reported in the form of the financial statements. This is generally not done, because it is unnecessary for the evaluation of the fitness of the project. It is important to understand how the financial statements and the project financials interact in order to determine the impact of the project on the company's overall performance and to assess the business risks that this project may bring to the company.

The best way to examine the relationship between the company financials and the project cash flows is to build a model of the company and the project using the equations for the financial statements given in the sections above. We will tackle this as a case study.

3.9 Case Study: Santa Anna Hydroelectric Power Scheme

Santa Anna Hydroelectric Power Scheme (HEPS) is a project similar to the examples concerning the Santa Clara HEPS studied in Chapter 1. The main difference is that the construction, and hence the investment, all occurs in the first year. A new company, Santa Anna HEPS Company Limited has been launched to pursue this opportunity.

The total cost of the project is $400 million. In order to fund this project, $200 million in equity has been obtained from the company's owners and the company has borrowed $250 million from a syndication of investment banks. The capital cost is to be depreciated on a straight-line basis over 35 years. This means that each year the assets are depreciated by $400/35 = \$11.4$ million. The book value of the investment for each can be determined in the form of a depreciation schedule, shown in Table 3.7.

The project will generate electricity, and hence revenue, from year 2. The revenue in year 2 is 68% of the final revenue of $161.2 million per year, while that in year 3 is 97% of the final amount. After that, it is expected that the revenue will be $161.2 million per year.

The direct costs of producing the electricity are estimated to be 5% of the revenue, while indirect costs for the administration of the operations is expected to be about $10 million per year. Both local and federal authorities charge tax; the aggregate tax rate is 35%. From this information the project financials can be constructed.

Table 3.7 Investment, depreciation and book value for the Santa Anna HEPS project

Investment and depreciation	1	2	3	4	5	6	7	8	9	10
Investments	400.0	0.0	0.0	0.0	0.0	0.0	0.0	0.0	0.0	0.0
Depreciation	11.4	11.4	11.4	11.4	11.4	11.4	11.4	11.4	11.4	11.4
Cumulative depreciation	11.4	22.8	34.2	45.6	57.0	68.4	79.8	91.2	102.6	114.0
Book value	388.6	377.2	365.8	354.4	343.0	331.6	320.2	308.8	297.4	286.0

3.9.1 Project Financials

The project financials for the first six years of the project's life are shown in Table 3.8.

The first three lines for each year represent the revenue, the costs of goods sold and the administration expenses, respectively. The figures in each line are calculated directly from the information supplied above.

The tax calculation requires an assessment of the taxable income, and since depreciation is allowed, this will reduce the taxable income. In this case, the taxable

Table 3.8 Project financials for Santa Anna HEPS for the first six years

Year	1	2	3	4	5	6
Revenue	0.0	109.6	156.3	161.2	161.2	161.2
Cost of goods sold	0.0	5.5	7.8	8.1	8.1	8.1
Administration	10.0	10.0	10.0	10.0	10.0	10.0
Tax	0.0	29.0	44.5	46.1	46.1	46.1
Depreciation	11.4	11.4	11.4	11.4	11.4	11.4
Profit before tax	(21.4)	82.7	127.1	131.7	131.7	131.7
Cash flow	(10.0)	65.2	94.0	97.0	97.0	97.0
Investment	400.0	0.0	0.0	0.0	0.0	0.0
Free cash flow	(410.0)	65.2	94.0	97.0	97.0	97.0
Cumulative free cash flow	(410.0)	(344.8)	(250.8)	(153.8)	(56.8)	40.3

income, or profit before tax, is calculated from the following equation:

$$Profit\ before\ tax = Revenue - COGS - Administration\ expenses - Depreciation \tag{3.34}$$

This is the same as that presented in Table 3.1 for the income statement, except that the project financials are presented on an entity basis with no debt, so there is no interest deduction from the taxable income.

The depreciation is obtained directly from the depreciation schedule given in Table 3.7. The amount of tax paid is given by the tax rate multiplied by the taxable income. For year one, there is no taxable income because the project makes an operating loss. In year 2, the taxable income is calculated from the following equation:

$$Profit\ before\ tax = 109.6 - 5.5 - 10 - 11.4 = 82.7$$

The tax is then given by:

$$Tax\ payable = 35\%(82.7) = 28.95$$

Notice that the loss that was made in the first year was not used to reduce the tax paid in the second. This was done for the sake of simplicity. In many tax jurisdictions, losses can be carried forward to reduce profits made in subsequent years.

The cash flow from the project is given by the following expression:

$$Cash\ flow = Revenue - COGS - Administration\ expenses - Tax\ charged \tag{3.35}$$

For the first year the cash flow is an out-flow of $10 million, while in the second year the cash flow is an in-flow of $65.2 million.

The free cash flow is the cash flow less the costs of investments, which is the capital cost of the HEPS and amounts to $400 million in the first year. In the first year the free cash flow is given by an out-flow of $410 million. In the second year, there are no investment costs, and the free cash flow is the same as the cash flow.

The next step is to prepare the income statement.

3.9.2 Income Statement

All the information used in the preparation of the project financials is used to prepare the income statements for each year. The additional information required for the income statement for the Santa Anna Company is the interest charged on the loan of $250 million. The aggregate interest for this long-term loan is 13%. The income statement for Santa Anna HEPS, Limited for the first six years of operation is shown in Table 3.9.

For each year, the income statement is prepared using the equations given in Table 3.1. In the first year, there is no revenue or direct operating costs. The cost of the administration and the depreciation results in negative earnings before interest and tax (EBIT) of $21.4 million. The interest charge on the loan of $250 million is $13\%(250) = \$32.5$ million. The taxable income is therefore minus $53.9 million.

In the second year, the company produces and sells electricity. The costs and earnings are calculated in the same manner as before, resulting in an EBIT of $82.7 million. The interest charged is $32.5 million, so the taxable income is $50.2 million. Tax is charged at 35%, resulting in a net income (or profit) of $32.6 million. Some $4.9 million is distributed to the shareholders, while the company retains the remainder.

Table 3.9 Income statement for the Santa Anna HEPS Company, Limited

Year	1	2	3	4	5	6
Revenue	0.0	109.6	156.3	161.2	161.2	161.2
COGS	0.0	5.5	7.8	8.1	8.1	8.1
Operating income	0.0	104.1	148.5	153.1	153.1	153.1
Administration	10.0	10.0	10.0	10.0	10.0	10.0
Depreciation	11.4	11.4	11.4	11.4	11.4	11.4
Earnings (EBIT)	(21.4)	82.7	127.1	131.7	131.7	131.7
Interest paid	32.5	32.5	32.5	32.5	32.5	32.5
Income before tax	(53.9)	50.2	94.6	99.2	99.2	99.2
Tax	0.0	17.6	33.1	34.7	34.7	34.7
Net income	(53.9)	32.6	61.5	64.5	64.5	64.5
Dividends	0.0	4.9	9.2	9.7	9.7	9.7
Retained earnings	(53.9)	27.8	52.3	54.8	54.8	54.8

Notice that the tax paid is different to that calculated in preparing the project financials. This is because the project financials are prepared as if the project was funded entirely from equity resources; in other words, there is no interest in the calculation of the project financials. This is because the project financials are prepared on the entity basis.

3.9.3 Cash Flow Statement

The cash flow statement for the company can now be prepared for each year. The cash flow statement reports on the movement of cash between operating activities, investment activities and financing activities. The cash flow statement for the first six years of the Santa Anna HEPS, Limited is shown in Table 3.10.

The income from operating activities is the operating income less the administrative expenses (or equivalently, the EBIT plus the depreciation). In the case study, there were no changes in working capital, and the taxation paid is obtained from the income statement. Therefore, in the first year, the operating activities generate a negative cash flow of $10 million. In the second year, the income from operating activities is $94.1 million, which is obtained from the operating income of $104.1 million less the administration expenses of $10 million. The taxation paid in the sec-

Table 3.10 Cash flow statement for Santa Anna HEPS, Limited

Year	1	2	3	4	5	6
Operating activities						
Income from operating activities	(10.0)	94.1	138.5	143.1	143.1	143.1
Adjusted for increase in working capital	0.0	0.0	0.0	0.0	0.0	0.0
Taxation paid	0.0	(17.6)	(33.1)	(34.7)	(34.7)	(34.7)
Net cash provided by operating activities	(10.0)	76.5	105.4	108.4	108.4	108.4
Investment activities						
Cash used to acquire fixed tangible assets	400.0	0.0	0.0	0.0	0.0	0.0
Financing activities						
Equity capital raised	200.0	0.0	0.0	0.0	0.0	0.0
Increase in long-term liabilities	250.0	0.0	0.0	0.0	0.0	0.0
Increase in short-term liabilities	0.0	0.0	0.0	0.0	0.0	0.0
Interest paid	(32.5)	(32.5)	(32.5)	(32.5)	(32.5)	(32.5)
Dividends paid	0.0	(4.9)	(9.2)	(9.7)	(9.7)	(9.7)
Net cash provided by financing activities	417.5	(37.4)	(41.7)	(42.2)	(42.2)	(42.2)
Net cash and cash equivalents		7.5	39.1	63.7	66.2	66.2	66.2
Cash at beginning of year	0.0	7.5	46.6	110.3	176.5	242.8
Cash at end of year	7.5	46.6	110.3	176.5	242.8	309.0
Net change in cash position	7.5	39.1	63.7	66.2	66.2	66.2

ond year is $17.6 million, which results in a net cash flow due to operating activities of $76.5 million.

The investment activities of the company are the investment in the project itself, amounting to $400 million. This company only makes this investment in the first year, and makes no other investments.

The financing activities are the raising of equity and debt capital, the payment of interest to the debt-holders and the payment of dividends to the shareholders. In the first year the company raises a loan of $250 million and equity of $200 million. It also pays $32.5 million in interest. In the second year, no further monies are raised, but the company pays interest to the debt-holder and a dividend to the shareholders.

The net cash flow is the sum of the cash flows for the operating activities, the investment activities and the financing activities. During the first year, the company's activities have generated a cash flow of $7.5 million, while these activities have generated a cash flow of $39.1 million during the second year.

The company starts off at the beginning of the first year with no cash in the bank. At the end of the first year the company has $7.5 million in cash, while at the end of the second year, this figure amounts to $46.6 million ($= 7.5 + 39.1$).

3.9.4 Balance Sheet

The balance sheet is the accounting equation in tabular format. The balance sheet for the first six years of the operation of Santa Anna HEPS Limited is shown in Table 3.11. The company begins existence with no assets, liabilities or equity. During the course of the first year, the company raises $200 million in equity from its shareholders and $250 million from debt-holders. The cash raised from this capital raising exercise is invested in fixed assets.

The assets were purchased at a cost of $400 million; by the end of the year, their book value, obtained from the depreciation schedule given in Table 3.7, is $388.6 million, the purchase price less the depreciation.

The only current asset of the company is cash. The amount of cash that the company has on hand is obtained from the cash flow statement. At the end of the first year this is $7.5 million, and at the end of the second year it is $46.6 million. Therefore, the total assets for the company amounts to $396.1 million and $423.6 million at the end of the first and the second year, respectively.

The company raised $250 million in debt to fund its activities, and since it does not pay back any of the principal on this debt in the first six years, this value is reflected on the balance sheet for each year.

The company raised equity of $200 million from the owners, and this value is reflected as the share capital of the company. As was shown in the discussion of the income statement, the company retains some of its income each year to fund its ongoing activities. (If a loss is made, all of its losses are "retained.") This cumulative amount of retained earnings is added to the share capital to reflect the owner's equity in the company.

Table 3.11 Balance sheet for the Santa Anna HEPS, Limited

Year	1	2	3	4	5	6
Assets						
Net fixed assets	388.6	377.2	365.8	354.4	343.0	331.6
Total current assets	7.5	46.6	110.3	176.5	242.8	309.0
Cash	7.5	46.6	110.3	176.5	242.8	309.0
Total assets	396.1	423.9	476.1	531.0	585.8	640.6
Liabilities						
Short-term liabilities	0.0	0.0	0.0	0.0	0.0	0.0
Long-term liabilities	250.0	250.0	250.0	250.0	250.0	250.0
Total liabilities	250.0	250.0	250.0	250.0	250.0	250.0
Equity						
Investors equity	200.0	200.0	200.0	200.0	200.0	200.0
Retained earnings	(53.9)	(26.1)	26.1	81.0	135.8	190.6
Total equity	146.1	173.9	226.1	281.0	335.8	390.6
Total liabilities and equity	396.1	423.9	476.1	531.0	585.8	640.6

In the first year, the company made a loss of $53.9 million, and this whole amount reduces the owner's equity. In the second year, the company was profitable and retained $27.8 million of the income. This is added to the retained income from the year before to reflect a total retained income of $-$26.1$ million ($= 27.8 - 53.9$).

Because of the dual nature of transactions, that is, if an asset is transferred, payment is also transferred in the opposite direction, the accounting equation holds, and the value of the total assets is equal to the total liabilities plus owner's equity. This is true for each of the years shown in Table 3.11.

This case study has demonstrated the interaction between the three financial statements and the project financials.

3.10 Examining the Business Risks

The discussion of the financial statements is sufficiently detailed to be used to develop a model of the business. The calculations can be performed manually or more easily on a spreadsheet. The model can be used to determine the effect of some of the key decisions that the financial manager or CFO may make, and the influence these might have on the viability of the company. These decisions are the amount of capital required, the ratio of debt to equity, the dividend policy, and the allocation of capital to different projects. These factors are examined in the following case study, which covers business risks.

3.11 Case Study: Apex Foods

Apex Pumps and Valves Limited is a manufacturer of pumps and valves. Apex wishes to set up a new business that manufactures and sells pumps and valves for use in the food and beverage industry. This business will operate as a separate company, called Apex Foods. Apex cannot afford to completely fund the business and has approached a variety of lenders to provide debt for the new venture. The CFO is concerned about the working capital requirements, given that it pays on a 30-day basis, but its customers may pay on a 90-day basis. Because there is probably going to be significantly more debt in the new business than Apex is accustomed to, the CFO would like to investigate the effect of different amounts of debt in the business, and the possible effects of the interest rate that Apex may be able to obtain from its lenders. In addition, the CFO would like to know the effect of the credit terms that it offers clients, called credit management, on the finance requirements of the business.

The financial statements are constructed based on the assumption that the accounts payable, accounts receivable and the inventory can be estimated from the following expressions:

$$Accounts\ Payable = (Annual\ COGS)(Creditor\ Days)/365 \qquad (3.36)$$

$$Account\ Receivable = (Annual\ Revenue)(Debtors\ Days)/365 \qquad (3.37)$$

$$Inventory = (Annual\ Revenue)(Inventory\ days)/365 \qquad (3.38)$$

The base case financial statements have been prepared in the same manner as those in the previous case study (Section 3.9), except for the addition of the estimation of working capital items expressed in Equations 3.36 to 3.38 above. The projected financial statements are given in Tables 3.12 to 3.14.

The investment in the business can be obtained from the balance sheet. This indicates that Apex, the owners of Apex Foods, intends to invest $30,000 of its own money as an equity investment, and borrow a further $30,000 in long-term loans. The amount of debt throughout the five-year period remains unchanged, which means that Apex Foods does not intend to repay the principal on the loan. A quick look at the income statement shows that Apex Foods does intend to meet the interest payments on this loan. The cash flow statement reveals that the money obtained from these two sources is spent on an investment in fixed capital that amounts to a total of $52,500 in the first year.

The income statement shows that Apex Foods expects to earn revenues amounting to $30,000 in the first year, and amounting to $60,000 for each subsequent year. The costs of producing the products are just over half of the revenue, while there is a small fee for administration, selling and other overhead expenses of $2,000.

Table 3.12 Forecast of the income statement for Apex Foods for five years (amounts in thousands)

Year	1	2	3	4	5
Revenue	30.0	60.0	60.0	60.0	60.0
COGS	20.0	35.0	35.0	35.0	35.0
Operating income	10.0	25.0	25.0	25.0	25.0
Administration	2.0	2.0	2.0	2.0	2.0
Depreciation	10.0	10.5	11.0	11.5	12.0
EBIT	(2.0)	12.5	12.0	11.5	11.0
Interest paid	3.9	3.9	3.9	3.9	3.9
Profit before tax	(5.9)	8.6	8.1	7.6	7.1
Tax	0.0	2.5	2.3	2.2	2.1
Net profit	(5.9)	6.1	5.8	5.4	5.0
Dividends	0.0	3.1	5.5	5.1	4.8
Percentage retained	90%	90%	20%	5%	5%
Retained earnings	(5.9)	5.5	1.2	0.3	0.3

There is a depreciation charge of $10,000 in the first year, and this increases over the period of this projection. This indicates that there is further capital investment during the period, otherwise the depreciation would remain constant or decrease. A quick check of the cash flow statement indicates that an additional capital investment of $2,500 is made each year. Presumably this is to replace equipment. Interest is paid on the outstanding loan at a rate of 13% ($= 3,900/30,000$), and tax is paid at a rate of 29% ($= 2,500/8,600$). The projected income statements indicate that Apex Foods expects to make a profit for the entire period. Dividends are retained at a rate of 90% for the first two years, 20% for the third year and at 5% for the subsequent years.

The projected income statements indicate that the company expects to make a loss in the first year and to be profitable for the subsequent years.

The forecast of the balance sheet provides some further detail of the anticipated activities of Apex Foods. Of the $30,000 invoiced in the first year, $7,500 in payments is outstanding at the end of the year (accounts receivable). There is an inventory of $4,900 of manufactured but unsold products at the end of the year. These figures double in the second year, and remain constant after that. Throughout the period, Apex Foods will have a positive cash position.

The gross fixed assets on the balance sheet represent the total investments made by Apex Foods in fixed capital such as equipment. This figure rises throughout the projected period, suggesting that the company is expecting to make some replacements to the capital equipment. The accumulated depreciation is the total amount of depreciation since the company's beginning that has been charged to the income statement. The difference between the gross fixed assets and the accumulated depreciation is the book value.

Table 3.13 Forecast of the balance sheet for Apex Foods for five years (amounts in thousands)

Year	1	2	3	4	5
Assets					
Total current assets	13.3	28.0	37.7	46.9	56.7
Cash	0.8	3.1	12.8	22.1	31.8
Accounts receivable	7.5	15.0	15.0	15.0	15.0
Inventory	4.9	9.9	9.9	9.9	9.9
Net fixed assets	42.5	34.5	26.0	17.0	7.5
Gross fixed assets	52.5	55.0	57.5	60.0	62.5
Less accumulated Depreciation	10.0	20.5	31.5	43.0	55.0
Total assets	55.8	62.5	63.7	63.9	64.2
Liabilities					
Short-term liabilities	1.7	2.9	2.9	2.9	2.9
Accounts payable	1.7	2.9	2.9	2.9	2.9
Long-term liabilities	30.0	30.0	30.0	30.0	30.0
Loans	30.0	30.0	30.0	30.0	30.0
Total liabilities	31.7	32.9	32.9	32.9	32.9
Equity					
Investors equity	30.0	30.0	30.0	30.0	30.0
Retained earnings	(5.9)	(0.4)	0.7	1.0	1.3
Total equity	24.1	29.6	30.7	31.0	31.3
Total liabilities and equity	55.8	62.5	63.7	63.9	64.2

The liabilities of the company consist of those suppliers that the company has not yet paid (accounts payable) and the long-term bank loan. The equity of the company consists of the original amount of $30,000 invested by the owner and the retained earnings.

The cash provided by the operating activities in the first year amounts to $8,000, which is obtained from the income statement as revenue, less cost of goods sold, less administration. The cash tied up in working capital is $10,800, which is made up of $7,500 in accounts receivable and $4,900 in inventory less $1,700 in accounts payable. With the tax paid during the year, the net cash provided by operating activities is projected to be negative $2,800 in the first year.

In the first year, the company raised $30,000 in equity from the owners, and $30,000 in debt from the bank. Interest is paid on the outstanding loan and dividends are paid to the owners. An amount of $52,500 is invested in operating equipment for producing the pumps and valves that the company intends to sell. Overall, the company expects to have a cash balance of $800 at the end of the first year. This may be skating on thin ice though.

Table 3.14 Forecast of the cash flow statement for Apex Foods for five years (amounts in thousands)

Year	1	2	3	4	5
Operating activities					
Income from operating activities	8.0	23.0	23.0	23.0	23.0
Adjusted for increase in working capital	(10.8)	(11.2)	0.0	0.0	0.0
Taxation paid	0.0	(2.5)	(2.3)	(2.2)	(2.1)
Net cash provided by operating activities	(2.8)	9.3	20.7	20.8	20.9
Investment activities					
Cash used to acquire fixed tangible assets	(52.5)	(2.5)	(2.5)	(2.5)	(2.5)
Financing activities					
Equity capital raised	30.0	0.0	0.0	0.0	0.0
Increase in long-term liabilities	30.0	0.0	0.0	0.0	0.0
Increase in short-term liabilities	0.0	0.0	0.0	0.0	0.0
Interest paid	(3.9)	(3.9)	(3.9)	(3.9)	(3.9)
Dividends paid	0.0	(0.6)	(4.6)	(5.1)	(4.8)
Net cash provided by financing activities	56.1	(4.5)	(8.5)	(9.0)	(8.7)
Net cash and cash equivalents	0.8	2.3	9.7	9.3	9.8
Cash at beginning of year	0.0	0.8	3.1	12.8	22.1
Cash at end of year	0.8	3.1	12.8	22.1	31.8
Net change in cash position	0.8	2.3	9.7	9.3	9.8

After having met all its obligations in the first year, the company still has money in the bank in spite of making a loss. By the end of the second year the company has money in the bank and is declaring a profit.

The changes to the base case are examined from the point of view of their effect on the cash position of the business. This is the second last line of the cash flow statement with the label of "Cash at end of year." While profit is important, "cash is king," to quote a cliché.

The effects of changes to the inputs to this scenario are examined. Firstly, the effect of changes in the interest rate is shown in Figure 3.9. These results indicate that if the interest rate increases to 20% the company will need more capital in the first year, but a mild increase from 13 to 15% will not affect the need for more cash.

The effect of the amount of debt as a function of the total capital on the projected cash flows for Apex Foods is shown in Figure 3.10. Debt above 50% results in a negative cash position, that is, the company is bankrupt or insolvent. While the company can easily afford to repay the interest on the debt after the third year, it cannot afford higher debt repayments in the first two years. This limits the amount of debt in these few years. A number of different financing strategies may be developed to tide the company through this period. It is worth mentioning here that accounting insolvency means that the liabilities of the company exceed the assets, while commercial insolvency means that the company has run out of cash.

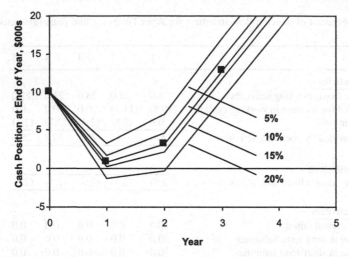

Figure 3.9 The effect of the interest charged on the cash flow for Apex Foods. The points represent the base case assumptions

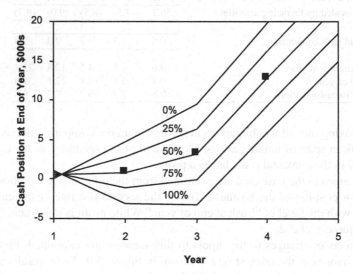

Figure 3.10 The effect of the percentage of debt on the cash flow for Apex Foods

The effect of the number of days receivable is shown in Figure 3.11. A reduction in the number of days receivable from 90 days to 30 days will reduce the working capital requirements from $21,900 to $11,900. This is a significant reduction in the light of the total capital of the company of $60,000. This is the basis for the simplest of all financial strategies: "collect early, pay late" to quote another cliché.

While spreadsheet projections can provide a lot of numbers relatively quickly, it is more important to gain insight and formulate a financial strategy. Insight can be gained from determining the effect of various parameters on the financial position of the company. Further insight can be gained by varying the projections for different

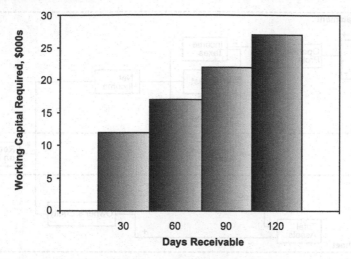

Figure 3.11 The effect of the days receivable on the working capital requirements for Apex Foods

economic conditions, that is, to create different scenarios and examine the performance under them. A third way of gaining insight is to examine the consequences of decisions on the performance, for example, changing the dividends paid, the amount of debt carried, the days payable, amongst others.

Another way to develop insight and understanding is through a thorough analysis of the financial statement using financial ratios and financial trees (also called Du Pont diagrams). The use of these techniques is discussed in the next section.

3.12 Ratio Analysis and Financial Trees

There are two main methods of analysing financial statements. The first is trend analysis and the other is ratio analysis.

In trend analysis, a value from the financial statements, or a ratio of these values is plotted over time to discern whether any trends exist in the performance of the company.

In ratio analysis, the company's performance is analysed against its own historical performance, and those of its peers with respect to a number of ratios between combinations of values from the financial statements. For example, the return on assets is the ratio of operating profit (income, EBIT) to the total assets, and this ratio is compared with that of the company's previous performance, and with that of companies in the same industry. From this analysis, the investors and the management can reach conclusions on the operating performance of the company compared with its past and with its peers.

The relationship between the various ratios and the financial statements is often best visualized in the form of a financial tree, such as that shown in Figure 3.12. Two ratios are shown together with their relationship between the income statement

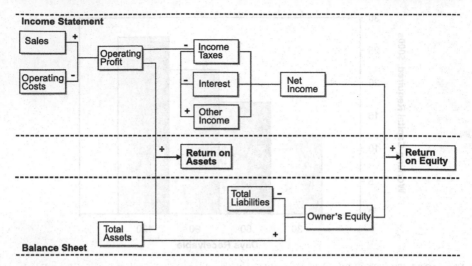

Figure 3.12 A financial tree showing the relationship of the return on assets and the return on equity to the income statement and balance sheet

and the balance sheet. These two ratios are the return on assets and the return on equity.

The return on assets (ROA) is the ratio of the operating income to the total assets, while the return on equity (ROE) is the ratio of the net income to the owner's equity. These two ratios are measures of the financial performance of the company. The first, ROA, is a measure of the performance of the company's operations since it is a ratio of the operating profit to the assets in operations. The second, ROE, is a measure of the performance of the company from the owner's or investor's point of view since it is a ratio of the net income to the owner's share of the company.

These two measures are important to different sections of the company's management. As a measure of operating performance, the ROA is of paramount importance to the general manager or chief operating officer. Often, this is the point of measurement of the performance for the divisions of a company. Another way of expressing this is to say that operations or divisions must be concerned up to the EBIT level, while the chief executive and other senior executives are concerned with the financing, tax, and equity issues that constitute the inputs to ROE.

Another way of viewing these two ratios is through the Du Pont equation. The role of the Du Pont Powder Company in the formalization of a comprehensive capital budgeting process in the early twentieth century was mentioned in Chapter 2. Pierre du Pont and his executives strove to measure the efficiency with which assets were utilized. They developed the concept of return on investment, ROI, which was stated in the following form:

$$ROI = \frac{Earnings}{Sales} \frac{Sales}{Investment} \tag{3.39}$$

The earnings for the Du Pont equation is the same as the net income or profit, and the investment is the expenditure on fixed capital. Therefore, it is clear that the return on assets mentioned earlier is the same as Du Pont's return on investment. The return on investment allowed Du Pont to measure the performance of each division and department within the company as a function of its profit margin and its asset turnover. The ranking of the company's operations and future investments based on the return on investment significantly lowered the cost of managing a complex and diverse company.

The return on assets is the same measure as Du Pont's return on investment and is given in the following formula:

$$ROA = \frac{Net\ Income}{Sales} \frac{Sales}{Total\ Assets} \tag{3.40}$$

The interpretation of the return on assets is given in the following expression:

$$ROA = (Profit\ Margin)(Asset\ Turnover) \tag{3.41}$$

The company can improve performance by increasing the profit margin or by increasing the asset turnover, which is the sales divided by fixed investment in assets. Increasing the asset turnover is often referred to as "sweating the assets" or, more formally, the capital utilization.

The return on equity is sometimes referred to as the extended Du Pont equation. It can be expressed as follows:

$$ROE = \frac{Net\ Income}{Sales} \frac{Sales}{Total\ Assets} \frac{Total\ Assets}{Equity} \tag{3.42}$$

The third factor on the right hand side is referred to as the equity multiplier, and expresses the asset value of the company as a function of the equity owner's value. The equity multiplier is a measure of the amount of debt that is used in the business. This can be shown from the following algebra:

$$Equity\ Multiplier = \frac{A}{E} = \frac{A}{A-L} = \frac{1}{1-L/A} = \frac{1}{1-Debt\ Ratio} \tag{3.43}$$

A, L and E represent the assets, liabilities and equity, respectively. Thus, the higher the ratio of debt to assets, the higher the equity multiplier.

The interpretation of the extended Du Pont equation reveals the critical components in managing a successful business:

$$ROE = (Profit\ Margin)(Total\ Asset\ Turnover)(Equity\ Multiplier) \tag{3.44}$$

This equation implies that the company can improve its performance, as viewed by the owners, by increasing the profit margin, increasing the asset turnover and increasing the equity multiplier. Increasing the equity multiplier is achieved by increasing use of debt. Too much debt, however, will have a negative impact on the net income and hence the profit margin. Consequently, the return on equity encourages the judicious use of debt.

3.13 Summary

The basics of the financial statements were examined in this chapter. There are three financial statements: the balance sheet, the income statement and the cash flow statement. The balance sheet represents the funding and financial position of the business, the income statement represents the performance of the business and the cash flow statement represents the health of the business.

The financial statements are prepared according to accounting principles and conventions. Two of those that impact on the general understanding of the financial statements are accrual accounting and the historical cost conventions. Accrual accounting recognizes transactions when a firm commitment has been made, rather than when the payment is made. This allows for the accurate recording of credit transactions. The financial accounts are prepared on the basis of historical costs, rather than current market values. They represent a true reflection of the transactions at the time at which they occurred. This particularly affects assets, whose book value may have no bearing on its market value or its operational value.

The dual nature of transactions was considered, which lead to an expression known as the accounting equation. The accounting equation is the basis of the balance sheet, and states that the value of the assets of a company is equal to the liabilities plus the owner's equity.

The financial statements are prepared for a reporting period. The balance sheet, which represents conserved quantities because of the dual nature of transactions, is the balance from the beginning of the company's life to the reporting date. The income statement and the cash flow statement do not represent conserved quantities, and are therefore balances between two points in time, that is, from the beginning of the reporting period to the end of the reporting period.

The business process was considered in the form of a flow diagram with a feedback loop. The construction of the income statement from the business process was discussed. The construction of the balance sheet from the accounting equation was presented, and the derivation of the cash flow statement from the other two statements was discussed. The interaction between the financial statements was derived.

The ratio method for analysing the performance of a company from its reported financial statements was examined. The Du Pont diagram showed how interrelation between the income statement and the balance sheet could be analysed to obtain useful measures of the performance of the company at the level of operations and owner's interest.

3.14 Review Questions

1.   What must be provided in the annual report of a company?
2.   What are the financial statements?

3.   What is another word for profit that is used in the income statement?

4.   What is the income statement known as in the UK?

5.   What is owner's equity?

6.   Who regulates the accounting profession?

7.   Does accounting treat a sole proprietorship the same was as it is treated in law?

8.   Why is accrual accounting used?

9.   Is the market value of the assets of a company the same as those represented on the balance sheet?

10.   Can the reporting period for the preparation of the financial statements be three months?

11.   What is a credit sale?

12.   Are the values of the assets in the company's accounts increased for inflation?

13.   What are the basic transactions in a business?

14.   How is EBIT determined?

15.   Is the principle amount for a loan tax deductible?

16.   What is the difference between SGA and overheads?

17.   Why is the Profit and Loss Account different from the Income Statement?

18.   How is corporate tax calculated?

19.   What is the "bottom line?" What is the "top line?"

20.   Is "Accounts Payable" a current asset?

21.   What is the "book value" of an asset?

22.   Is "Accounts Receivable" and "Debtors" the same thing?

23.   What are the categories of the Cash Flow Statement?

24.   What is working capital?

25.   What are the two formats for the Balance Sheet?

26.   Is depreciation a fund for the replacement of capital?

27.   What is the book value of an asset at the end of its useful life?

28.   If the company is short of cash, how can this be rectified?

29.   How do retained earnings affect the cash position of the company?

30.   What is the Du Pont diagram?

31.   What is return on equity?

32.   What is limited in a limited company?

33.   Who is responsible for the debts of a partnership and a limited company?

3.15 Exercises

1. An engineering business is expected to have revenues of $1 million each year. The costs of producing the manufactured items, consisting of labour and raw materials, amounts to $600,000. The overheads are $50,000, the interest on debt is $50,000 and the tax rate is 35%.

    (i)   What is the gross profit?
    (ii) What is the EBIT (PBIT)?
    (iii) How much tax is paid?
    (iv) What is the net income (profit)?

2. Over a period of a year, a business engages in the following transactions:

    (i)   Borrows $600,000 from the bank
    (ii) Purchases a lathe for $50,000
    (iii) Purchases a delivery vehicle for $15,000
    (iv) Purchases stock for $200,000
    (v)   Sells $100,000 of product, which has not been paid for by the end of the year

    Construct a balance sheet representing these transactions.

3. An engineering consultancy has the following income and expenses during a particular reporting period:

Item	Amount, $
Vehicle expenses	1,200
Rent and rates payable	4,000
Depreciation of office equipment	1,500
Sales	100,000
Goods purchased	67,000
Lights and water	900
Telephone	450
Internet	300
Insurance	800
Interest on business loan	500
Salaries and wages	11,000

    (i) Determine the profit for this business by preparing an income statement.
    (ii) Explain the meaning of sales for an engineering consultancy.

4. An engineering consultancy employs five civil engineers who charge their services at $1000/day. There are 260 working days in a year. The engineers get 20 days leave a year, and only half of the remaining time can be charged to clients. The gross profit margin is 65%. The consultancy employs a number of administrative staff and have other expenses related to overheads. The overheads are 50% of the gross margin. The consultancy recently purchased furni-

ture and office equipment for $100,000, which is depreciated on a straight-line basis over five years. The company paid for this with a loan, which carries an interest rate of 9%. Tax is charged at 35%. Forecast the income statement over the next six years if the consultancy adds a new engineer to their staff each year.

5. Construct a forecast of the balance sheet and the cash flow statement for the engineering consultancy in Question 5. Determine the amount of equity investment the owners of the consultancy need to provide for the business to be viable.

6. A company has an EBIT of $3 million, and has interest expenses of $500,000 and depreciation of $300,000. If the tax rate is 35%, and the company retains 40% of the profit, determine the dividends paid to the shareholders, and the cash flow from these activities during the period.

7. Consider the financial statements for Apex Foods given earlier.

  (i) Prepare a forecast of the company's income statement for each year of the project.

  (ii) Prepare a spreadsheet of the company's balance sheet for each year of the project. Allow the user to enter different values for the amount of equity and debt raised, the interest rate, and the dividend rate paid to shareholders.

  (iii) Prepare a spreadsheet of the company's cash flow statement for each year of the project. Allow the user to enter different values for the amount of equity and debt raised, the interest rate, and the dividend rate paid to shareholders.

  (iv) Prepare a spreadsheet of the project financials for the project.

8. Compare the return on assets and the return on equity for the Santa Anna HEPS with Apex Foods. Which business is more attractive?

9. Construct a Du Pont diagram for the Apex Food's performance in the fifth year of operation. What are the values for the return on assets and return on equity?

10. An engineer starts a business by paying $100,000 in equity capital. During the course of the first year, the business made sales of $89,000, incurred variable costs of $35,000 and fixed costs of $10,000. The business also loaned $75,000 from a bank, incurring interest of $2,000 during the period. The funds were used to purchase a machine for $120,000 and to fund working capital of $30,000, which consisted of $10,000 in accounts receivable, and $20,000 in inventory. The machine is depreciated on a straight-line basis over ten years, and the tax rate is 40%. Establish the income statement, cash flow statement and balance sheet for the business at the end of the first year.

11. A common method of valuing a business is to use Price/Earnings ratios, or P/E multiples. Determine the value of Apex Pumps at the end of the fifth year is the P/E multiple is 15.

12. The following items are relate account balances for Johnson Engineering as at year end 5 April:

Item	Amount
Equipment	82,000
Creditors	12,000
Bank overdraft	16,000
Product inventory	20,000
Debtors	11,000
Buildings and land	92,000
Capital at previous year end	190,000
Cash	12,000
Vehicles	5,000
Office equipment	9,000
Profit for year	25,000
Dividend payment	12,000

Establish the balance sheet for the company.

Chapter 4
Cash Flows for a Project

4.1 Introduction

It was argued in Chapter 1 that value of the project was determined by the amounts, timing and risk associated with the cash flows for the project. The case study of the Santa Clara Hydroelectric Power Scheme in Chapter 1 was introduced to illustrate the construction of these cash flows to obtain the financials for a project. The case study of the Santa Anna HEPS Company was introduced in Chapter 3 to illustrate the interaction between the project financials and the financial statements. In this chapter, the estimation of the cash flows for a project is examined in more detail.

4.2 Determining the Cash Flows for a Project

The forecast of the cash flow and free cash flow for a project has the following five basic items:

1. Estimation of the capital costs
2. Estimation of the operating costs or expenses
3. Forecasts of the sales or revenue
4. Calculation of the direct taxes and royalties
5. Estimation of the working capital

An overview of each of these items was presented in Chapter 1. Each of them will be discussed in further detail in this chapter.

The level of detail required for each of these items is dependent on the accuracy of the estimate required. In addition, the level of detail and the accuracy of each item must be consistent with one another. It is common to build an estimate in stages, adding further detail and refining the accuracy at each stage. This staging of the cost estimates is mirrored in the staging of the engineering design.

The estimation of these five items for a project is a forecasting process. It must not be confused with the cost allocation performed by cost (or management) accountants, which is a retrospective function aimed at providing management with tools for the control of operations.

Before discussing methods for determining each of the cash flows, it is useful to outline the phases or stages of the engineering design process and the approval procedure within a company.

4.3 Overview of the Stages of Engineering Design and Construction

The design and construction of a capital investment project usually proceeds through a number of phases. These can be classified as follows:

(i) Design

The design process will depend on the type of project. The scope may be defined in terms of the following: either greenfields site (new) or brownfields site (where utilities and infrastructure may exist); the battery limits (an assigned geographic boundary to specifically identify a portion of the plant or equipment); and the available infrastructure and offsites. Generally, the design phase will proceed in stages during which the level of detail is enhanced and the accuracy of the capital cost estimate is improved as the project matures. For example, in a major chemical engineering project, design usually proceeds through the stages of conceptual, basic and detailed design. The deliverables for the conceptual and basic stages may include flow sheets, basic control strategies, equipment lists, and equipment specification sheets.

(ii) Detailed engineering

Detailed engineering is the phase of producing the specifications for the project that are sufficiently detailed for procurement and construction. Items that may be included, depending on the project, are: major equipment list, including sizing calculations and materials of construction specification; piping and instrumentation diagrams, including control loop specifications; motor lists; piping specifications; plot plans, and general arrangement drawings; hazard and operability analysis; planning schedules.

(iii) Procurement

Procurement is the process of obtaining quotations on enquiry and bids on tender from vendors and service providers, and the placing of the orders for equipment and construction packages.

(iv) Construction

The buildings, infrastructure and facilities are erected and the equipment installed.

(v)   Commissioning

The commissioning phase concerns the start-up of the operation. It may include the ramp-up to full capacity. The standard operating procedures for the operations are produced during this stage.

4.4 Approval Procedure

The approval for a capital project occurs in stages. Each of these approval stages represents the approval to embark on the next stage of the engineering design. As a result, the design phase itself occurs in a number of stages. Usually, the design phase and the approval process are matched, so that the design does not continue to the next stage unless approval has been provided. Thus, an approval step, in which a decision is taken to continue to the next stage or not, separates each stage of the design process. The approval process is sometimes referred to as a "gate," and the design work the "stage." The interaction between the technical development and the decision approval process is shown in Figure 4.1.

The nomenclature for the design stages and the estimates of the capital costs are used slightly differently from company to company and from industry to industry. In spite of this, they do convey a sense of the status of the design in the overall process and the accuracy of the current estimate. For example, a chemical engineering design might go through the stages of concept, basic and detailed design, while a mining project might go through stages of exploration, pre-feasibility and bankable feasibility.

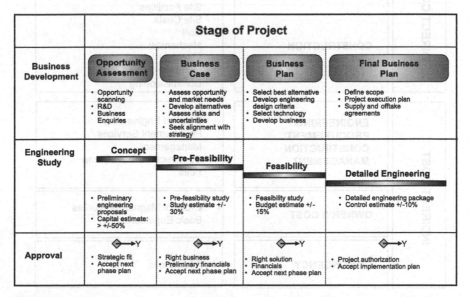

Figure 4.1 An overview of the progress of a project through the different stages of engineering design

The decision to proceed from one of the design stages to the next usually requires formal approval. Major capital expenses will require the approval of the company's board of directors. The technical and financial aspects of the project may include peer review, either within the organization or from independent reviewers.

4.5 Estimation of the Capital Costs

A cost estimate is the forecast of the probable costs of a project of a given scope at a particular location at a particular time. The components of the capital cost, and methods for their estimation, are discussed in the following sections.

4.5.1 Components of Capital Cost Estimates

The capital cost estimate consists of two main categories: direct fixed costs and indirect fixed costs. This is illustrated in Figure 4.2. Direct fixed costs consist of all the production equipment, the facilities and the utilities. The cost of site preparation, such as, cutting and filling, structures, service and office buildings, roads, rail tracks and boundary fencing is included in the fixed capital cost category.

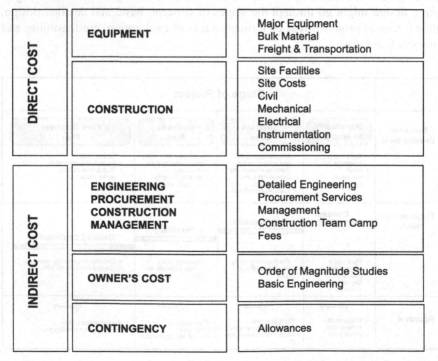

Figure 4.2 Elements of the capital cost estimate for a project managed by the owner

There are three elements to the indirect costs. These are engineering, procurement, and construction management costs, owner's costs and contingency.

There are two main categories of contract that the owner can enter into with the contractor for the engineering, procurement, construction and construction management of a project: lump sum and cost-reimbursable.

In a lump-sum contract, the contractor agrees to provide all the work specified in the scope for a fixed price. For example, the contract might be for the delivery of a crystallization plant. The contractor agrees to provide the plant with a certain capacity and performance on a certain date for a fixed price. The project's owner is not exposed to the risk that the capital cost may escalated between the time the project is approved and the time it is handed over to the owner after construction.

The lump-sum price includes cost of the equipment and the cost to the contractor for the construction and installation of the operations. It might include the costs of any outstanding design work that is required. It also includes the contractor's overheads, profits and a provision for the risk of budget and schedule overruns. Lump-sum contracts are generally awarded on the basis of a competitive tender and often lead to the lowest direct cost.

In a cost-reimbursable contract, the owner directly purchases all the equipment, and pays for the construction. The engineering contractor is paid for their costs incurred, usually fees charged on an hourly basis. Cost-reimbursable contracts are common when the project's definition is unclear, so that the contractor cannot properly estimate the costs involved. These contracts provide the owner with greater flexibility to specify changes during the process. In this case the contractor is usually appointed earlier on in the design process than that for a lump-sum contract and is more intimately involved in the design process.

The main difference between these contracts is in the risk assumed by the two parties. In a lump-sum contract, the engineering contractor assumes the majority of the risk. In a cost-reimbursable contract, the owner assumes most of the risk. More detail and definition of these contracts and the variations on them can be obtained from the International Federation of Consulting Engineers, (www.fidic.org) or the Institution of Civil Engineers (www.ice.org.uk). The International Federation of Consulting Engineers is also known as Fédération Internationale Des Ingénieurs-Conseils (FIDIC).

Commonly, an engineering, procurement, and construction (EPC) cost implies a lump-sum contract, while an engineering, procurement, and construction management (EPCM) cost implies a cost-reimbursable contract.

The owners' costs for the early design and project definition and for the owners' team during the remainder of the design and construction process are typically not charged to the capital of a project; instead the owner treats them as an expense of, for example, the projects department, and the costs flow through the project owner's accounts.

The contingency cost is a provision for unforeseeable events or elements of cost. The contingency accounts for elements of cost that may arise from inadequacies in estimating methods, and other factors within the project scope, such as unproven technologies. Changes in project scope, commonly called "scope creep," change the basis of the estimate and should be dealt with in a transparent manner rather than as a contingency. Contingency should not account for events such as price escalation, work stoppages and disasters.

When an idea for a new project is first mooted, a quick appraisal of costs is required. If it passes muster, further engineering and cost studies will be conducted to define more accurately the estimate of the capital costs. The accuracy of the estimates of costs depends on the maturity of the project, the viewpoint adopted and the end use of the estimate. Two common systems of classifications of estimates are those based on the end use, and those based on the level of accuracy. There is significant overlap between these two conventions. If the approval procedures and the engineering design are completely aligned, as shown in Figure 4.1, then there is no difference between them. These two conventions are discussed next.

4.5.2 Classification of Capital Cost Estimates Based on their End Use

Three common types of estimates, that is, design, bid and control estimates, are based on the end use of the estimate.

(i)   Design estimates

This is the cost estimate that the design team determines on behalf of the project owner. The design estimate is typically staged into at least three, and possibly more stages in a major capital project; each of these stages adds detail to the engineering design and accuracy to the cost estimate. These estimates guide the owner's decision to proceed with the investment or to abandon or delay the project.

(ii)   Bid estimates

This is the estimate provided by a contractor or vendor. It is often submitted for competitive bidding in response to a request for tenders or proposals. The scope depends on the tender request, but could include construction and supervision. The bid estimate is used by the contractor to secure the job, and by the owner's team to prepare their estimates. The bid reflects the contractor's desire to secure the tender, and bids can vary widely between different suppliers.

(iii)   Control estimates

The owner and the construction company use the control estimate as a basis for cost control during construction. This estimate may also be called the control budget or budget estimate. For a contractor, the bid estimate is used for planning and control.

Other classifications of the costs are based on the level of accuracy of the estimate.

4.5.3 Classification of Capital Cost Estimates Based on their Level of Accuracy

The level of detail and consequently the level of accuracy of the estimate required are dependent on its end use. In practice, however, the level of detail represents a trade-off between time and cost. Improvements in both the design and the capital cost estimate require the allocation of further skilled human resources, which are typically scarce and expensive. Importantly, acquiring more accuracy takes time, which may be in shorter supply than either money or skilled people.

The typical cost of obtaining the estimate as a fraction of the total design cost is illustrated in Figure 4.3. The level of accuracy is expressed in terms of confidence limits. In statistical terms, these limits express the range about the expected value in which the outcome can occur with a certain probability. The confidence limits are usually those at a probability of 90%. From the data represented in Figure 4.3 it may cost about 1% of the final capital cost to obtain an estimate that has an accuracy level of $\pm 15\%$. The accuracy level means that there is a 90% probability that the final cost will be between $e - 0.15e$ and $e + 0.15e$, where e is the estimate of the capital cost. In other words, if the capital cost of the plant is \$350,000 at an accuracy level of $\pm 15\%$, there is a 90% probability that the cost is between \$297,500 and \$402,500.

Cost estimation is not an exact science. In the same way, the naming conventions are not uniform through all the fields of engineering and project management. A number of cost engineering associations have suggested classifications of capital estimates. For example, the Association for the Advancement of Cost Engineers International (AACE) has suggested a classification of the capital cost estimates that

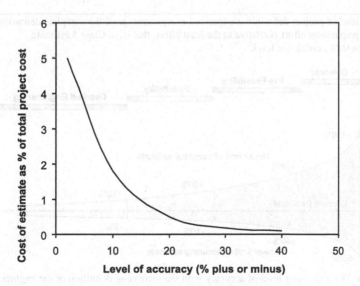

Figure 4.3 Cost of the estimate as a percentage of the total capital cost

is shown in Table 4.1. The level of accuracy of the estimate within each design stage is shown schematically in Figure 4.4, although different terminology is used to that used in the Table 4.1. Since there is no one standard terminology, it is important to be able to translate between the different systems.

Table 4.1 Cost estimation classes defined by the Association for the Advancement of Cost Engineers International

Estimate class	Level of project definition[+]	End usage of estimate	Expected or probable accuracy range[+++]	Preparation effort[++]	Method of preparation of estimate
Class 1	50–100%	Control or bid/tender	Low: –3 to – 10% High: +3 to +15%	5 to 100	Detailed estimating data by trade with firm quantities
Class 2	30–70%	Control or bid/tender	Low: –5 to –15% High: –5 to +20%	4 to 20	Detailed estimating data by trade with detailed quantities
Class 3	10–40%	Budget, authorisation or control	Low: –10 to –20% High: +10 to +30%	3 to 10	Semi-detailed unit costs with assembly level line items by trade, historical relationship factors
Class 4	1–15%	Study or feasibility	Low: –15 to –30% High: +20 to +50%	2 to 4	Equipment factored, historical relationship factors, broad unit cost data
Class 5	0–2%	Concept screening	Low: –20 to –50% High: +30 to +100%	1	Capacity factored, judgment, analogy, historical comparisons, gross unit cost

[+]   The level of project definition is expressed as a percentage of the complete definition.
[++]   The preparation effort is relative to the least effort, that is, to Class 5 estimate.
[+++]   At the 90% confidence level.

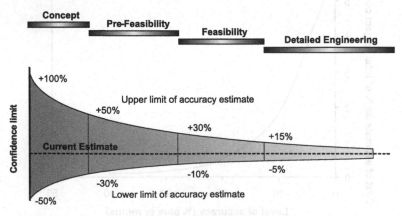

Figure 4.4 The increasing level of accuracy with the increasing definition of the engineering design

The increase in the level of detail results in the increase in the accuracy of the capital cost estimate, as shown in Table 4.2.

The methods typically used in the preparation of the estimate of the direct fixed costs are shown in Table 4.1. In the earlier stages, estimates are based on factors and unit costs. Both these methods are discussed later. In the latter stages, quotations from suppliers and tendered bids are used.

(i)   Conceptual estimate (order of magnitude estimate)

This an approximate forecast of the capital cost. The easiest method of achieving an order of magnitude cost is to scale the cost for a similar operation to the same production rate and apply escalation factors to account for the increase in capital equipment costs. It is not realistic to apply confidence limits to this type of estimate.

Table 4.2 Cost estimation level guide for the estimate classes

Estimate	Class 1	Class 2	Class 3	Class 4	Class 5
Level of capital cost estimate	Conceptual	Feasibility	Budget	Control	Revised
Confidence limits	> +25% to −15%	+25% to −15%	+15% to −5%	+10% to −10%	+5% to −5%
Site:					
Plant capacity	Assumed	Preliminary	Finalised	Finalised	Finalised
Visits by team	Not essential	Recommended	Essential	Essential	Essential
Infrastructure requirements	Assumed	Recommended	Essential	Finalised	Finalised
Process:					
Flow diagrams	Assumed	Preliminary	Optimum	Finalised	Finalised
Laboratory tests	If available	Recommended	Essential	Essential	Finalised
Pilot plant testing	Not needed	Recommended	Recommended	Essential	Finalised
Energy and material balances	Not essential	Preliminary	Optimum	Finalised	Finalised
Project facility definition:					
Plant layouts	Conceptual	Preliminary	Probable	Finalised	Finalised
Equipment selection	Notional	Preliminary	Optimum	Finalised	Finalised
General arrangements: mechanical	None	Preliminary	Recommended	Finalised	Complete
General arrangements: structural	None	Preliminary	Recommended	Finalised	Complete
General arrangements: civil	None	Preliminary	Recommended	Finalised	Complete
General arrangements: infrastructure	None	Preliminary	Recommended	Finalised	Complete
Piping drawings	None	None	Preliminary	Line	Mostly complete
Electrical drawings	None	None	Preliminary	Line	Mostly complete
Instrumentation drawings	None	None	Preliminary	Line	Mostly complete
Specifications	None	Preliminary	Recommended	Complete	Complete
Overall draughting:					
% completion	Minimal	5%	10–15%	30–40%	80%

(ii)   Study estimate (pre-feasibility, preliminary estimate)

Some definition has been added to the project from the conceptual stage. The duty, or the operating capacity or the size has been defined in broad terms, the main items of equipment are known and the location of the operation is known. For chemical engineering type projects, a mass and energy balance of the preliminary flowsheet has been calculated. The cost of the major equipment items is estimated by reference to the costs of similar items adjusted for the difference in size and to account for the change in cost since the reference item was installed. The costs of minor items and of other installations, such as electrical, instrumentation, piping, and structural work can be estimated based on factors expressed as a percentage of major equipment. An estimate obtained in this manner can have an error limit of ±30%.

(iii)   Budget estimate (feasibility, funding, scope estimate, bankable feasibility)

As is shown in Table 4.2, the site is specified, the process engineering is largely complete, and the site layout, civils and general arrangements are mostly complete. The necessary pilot plant work has been done; even a trial mine has been completed if a mining project. The capital cost is estimated from individual items, groups of equipment, structures and buildings. The instrumentation can usually be specified from the piping and instrumentation diagrams, and the control system can be costed from the number of I/O (input/output) loops required. Electrical costs can be determined, for example, from the electrical motor lists and the installed power required. An estimate obtained from this level of detail can be within an error limit of ±20% or even ±10%.

The budget estimate is frequently used as the basis for the securing of funds, either internally or externally. For this reason it is sometimes called the bankable feasibility study. This estimate forms the basis for the decision to proceed with the investment. As result, it will also detail a schedule for the completion of the work and a cost forecast for the completion of the design.

(iv)   Control estimate (definitive estimate)

The vast majority of the design information is known. Costs are determined by quotations from suppliers or bids from vendors and contractors. The control estimate will be within less than ±10%. The control estimate is used for cost control during the construction and commissioning phases. The project manager will monitor all costs and authorize funds against this estimate.

(v)   Detailed estimate

At this stage all design and draughting work has been essentially completed. The costs can be estimated from bills of materials and unit costs, from quotations by vendors for purchased equipment and contractors' costs. An estimate with an accuracy of less than ±5% can be achieved.

4.6 Estimation Techniques for Capital Costs

There are two main techniques that are used for estimating capital costs. These are factored estimation techniques and unit cost techniques. Factored techniques are generally used in the earlier stages of design when bills of quantities are not yet available or the vendors' quotations have not yet been obtained. Unit cost techniques are used when a bill of quantities is available, which is generally when the design is at a defined stage.

4.6.1 Factored Estimate Techniques

If the price of a similar item is known, either from historical data or for a similar item of a different size, the cost or design engineer can update the cost for the current situation. Thus different types of factoring account for a change in time, that is, updating a cost from the past to the present, while others account for a change in capacity, that is, correcting a cost of item of a particular size or capacity to another size or capacity. Another factoring method estimates the cost of an item based on the cost of a reference item. For example, the costs of civil, structural and other items can be estimated as a factor of the major equipment costs.

Each of these different types of factored methods is discussed next.

(i)   Cost index

A cost index is the ratio of an item's cost today to its cost in the past. The consumer price index (CPI) is the most well-known of these types of indices.

The use of the cost index is expressed in the formula:

$$C_t = C_0 \frac{I_t}{I_0} \tag{4.1}$$

where C_t and I_t represent the cost and the index today, respectively, and C_0 and I_0 represent the cost and the index at some other time t_0.

There are a number of published cost indexes for overall operations and various individual groups within an operation. For example, the journal Chemical Engineering publishes the Chemical Engineering Plant Cost Index (CEPCI), and there is the Marshall and Swift equipment cost index. These cost indices represent the price inflation for various categories of capital equipment.

The application of the cost index is illustrated in the following two examples.

Example 4.1: Cost index.

A fluorspar plant cost \$15 million in the year 2000. The plant cost index for operations similar to this one was 1,179 in the year 2000, and it is 2,162 today. What is the current cost of the plant today?

Solution:

Cost today = \$15 million (2162/1179) = \$27.5 million.

Example 4.2: Calculation of cost index.

A sulphuric acid plant cost $12 million in 2001 and a plant of similar capacity and design today costs $19 million. Determine a cost index for acid plants today.

Solution:

Use 2001 as the reference year, and assign it an index of 100, then the index for today is given by:

Index $= 100(19/12) = 158.3$

(ii)   Cost-capacity factors

The cost-capacity relationship corrects for capacity differences. The commonly used relationship is a power-law function, and because the exponent is often found to be close to 0.67, it referred to as the "rule of two-thirds." The functional form for this relationship is given by the following equation:

$$C_2 = C_1 \left(\frac{Q_2}{Q_1} \right)^n \tag{4.2}$$

where C_1 is the cost at capacity Q_1 and C_2 is the cost at capacity Q_2. The exponent n is commonly between 0 and 1. For n less than one, economies of scale exist, while if n is greater than one, the equipment costs increase at a rate greater than the increase in capacity.

Exponents are available for individual pieces of equipment, or whole plants. Sources are Chemical Engineers' Handbook, Plant Design and Economics for Engineers, amongst others. Examples of the values of the exponential factor n for various types of equipment are in Table 4.3.

The application of the cost-capacity factor method is illustrated in the following example.

Example 4.3: Cost-capacity factors.

A fluorspar plant cost $15 million for a 50,000 t per year facility. What would the cost be for a 120,000 t per year facility if the "rule of two thirds" were applicable?

Solution:

Cost for 120,000 tpa plant $= \$15$ million $(120,000/50,000)^{0.67} = \$27$ million.

(iii)   Equipment factors

A method that relates the total plant cost to the cost of the major equipment is commonly used. This factor method, sometimes called the Lang method after the originator of the idea, has the functional form given by the following equation:

Table 4.3 Capacity factors for different equipment

Equipment	Range	Factor n
Tanks	1–100m^3	0.67
Reactor	1–200m^3	0.74
Compressor	150–1500 kW	0.32

Table 4.4 Equipment factors for plants

Type plant	Factor, f
Solid process	3.1
Solid-fluid process	3.6
Fluid process	4.7

$$C_T = fC_E \qquad (4.3)$$

where C_T and C_E are the total costs and the major equipment costs, respectively, and f is a factor dependent on the type of operation. Values of factors for different types of facilities are given in Table 4.4.

Table 4.5 Example of equipment factors for a process plant

Category	Factor
Equipment	100
Erection	
Simple proprietary equipment	10
Some site fitting or fabrication	11
Extensive site fitting or fabrication	12
Structural	
Mainly ground level access	8
Some above ground level access required	10
Significant access structure or platforming	12
Civils	
For equipment, minor structures	10
For equipment, structures, some acid proofing	15
For equipment, extensive structures/pits/acid proofing	20
Buildings	
Mainly personnel use	6
Small provision for workshops	10
Larger workshops/warehouses	16
Piping	
Relatively simple, mainly small pore piping	60
More complex, medium bore piping	80
Complex, significant recycle	100
Electrical	
Mainly small, ancillary drives (*e.g.*, conveyors)	12
Machine main drives (pumps, compressors, crushers)	16
Major anticipated electrical *vs.* equipment cost	20
Instrumentation	
Simple control system	15
Moderately complex control system	23
Complex control system	30
Protective cover	
Mostly painting	4
Some insulation	6
Heavily insulated surfaces	10

Table 4.6 Equipment factors for the plant

Item	Factor
Plant and equipment	100
Erection	10
Structural	10
Civils	20
Buildings	6
Piping	80
Electrical	16
Instruments	15
Protective cover	6
Subtotal (direct fixed cost)	263
Site development (add 5%)	13
EPCM (add 20%)	53
Offsites (add 15%)	39
Subtotal	368
Contingency (add 20%)	74
Grand total	442

Refinements to this method allow for different categories of the total capital cost to be estimated from the equipment costs. In this case, the factor f is composed of components arising from elements of the design and construction, such as structural, civils, buildings, piping, electrical, instrumentation, amongst others. In general, the functional form for the decomposition of the factor f into its component parts is given by the following expression:

$$f = 1 + \sum_{i=1}^{n} f_i \qquad (4.4)$$

where f_i represents the component factors. For example, Table 4.5 provides an illustration of typical factors for a process plant.

The application of the equipment factor method is illustrated in the following example.

Example 4.4: Equipment factors.

The major equipment is estimated to cost in the region of $20 million for a moderately complex process. Determine an estimate, using the Lang factors for a process operation given in Table 4.5, to determine the total capital cost.

Solution:

The estimation of the factors is given in the Table 4.6 above.

From the results given in the Table 4.6, the total capital cost is estimated to be $88.4 million (= 442/100 × $20 million). Note EPCM can vary from 10 to 30% of the fixed capital; for large projects the EPCM costs are closer to 10%, for smaller projects they are higher. The contingency accounts for the poor quality of the estimate, amongst others. For a factored approach, a contingency of 10 to 25% is common.

4.6.2 Unit Costs Techniques

If a project is sufficiently specified, a bill of quantities can be estimated or determined. The unit cost approach uses the bill of quantities and the cost per unit of each of these items to determine the total cost. This can be expressed mathematically as the following equation:

$$C = \sum_i c_i q_i \qquad (4.5)$$

where C is the total cost, and c_i and q_i are the cost per unit and the quantity of units required for each element in the bill of quantities.

The unit cost may not be simply specified. For example, a piece of equipment may require unpacking from the delivery crates, transport to the site and installation. Another example may be the building of a foundation, which requires formwork, rebars and concrete. In cases like these the unit cost approach is expanded to include the additional factors. If the unit cost consists of materials, equipment and labour, then the total cost may be calculated from the following equation:

$$C = \sum_i (m_i + e_i + w_i L_i)\, q_i \qquad (4.6)$$

where m_i and e_i refer to the unit cost of materials and equipment, respectively, while the term $w_i L_i$ refers to the unit cost of the labour. The symbol L_i represents the amount of labour required per unit of q_i, and w_i represents the wage associated with the labour task.

The following example illustrates the application of the unit cost method to estimate the total cost.

Example 4.5: Unit costs.

A contractor needs to determine the costs for the concrete foundations for a building in order to submit a bid. The estimated quantities required are given in the Table 4.7.

Determine the cost estimate.

Solution:

The unit costs for each item are given in Table 4.8. The total is calculated using Equation 4.6. The bid total for the contractor is $78,191.

Table 4.7 Unit costs

Item	Quantity
Formwork	1,120 m^2
Rebars	1,800 kg
Concrete	400 m^3

Table 4.8 Calculation of costs based on unit costs

Item	Quantity	Unit material cost	Unit equipment cost	Wage rate	Labour required	Total
Formwork	1,120 m^2	$4.2/m^2	$8.3/m^2	$20/hr	1.5hr/m^2	$47,600
Rebars	1,800 kg	$0.8/kg	$1.1/kg	$20/hr	0.06hr/kg	$ 5,791
Concrete	400 m^3	$5/m^3	$45/m^3	$20/hr	0.6hr/m^3	$24,800
Grand total						$78,191

4.7 Estimation of the Total Operating Costs

Once the project has been built, it must be operated over the course of its lifetime. The total operating costs of the project are all the costs of the manufacturing process, the administration of the operation and the distribution and marketing of the product. The manufacturing costs are also known as production costs or operating costs. The composition of the total operating costs for a typical manufacturing facility is shown in Figure 4.5.

The operating costs can be calculated on the basis of an annual cost or a cost based on a unit of production. The preferred method is the former, mainly because

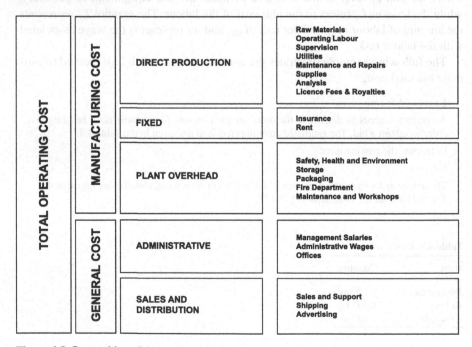

Figure 4.5 Composition of the total operating costs

the results are to be used in a cash flow analysis that is usually presented on an annual basis.

The components of each of the cost items are discussed next.

4.7.1 Direct Production or Manufacturing Costs

All the costs associated with the manufacturing process, such as the direct production costs and the plant, are regarded as manufacturing costs.

(i)   Raw materials

The raw material requirements are determined from mass balances or production schedules. The calculation of the required annual quantities must account for all losses and inefficiencies. These costs should include all transport and delivery to the production site. The delivery quantities should be determined as a trade-off between lower costs associated with bulk quantities and the cost of on-site storage.

(ii)   Labour

The costs of labour can vary widely for a project, ranging from a major new greenfields site to a small addition to the current manufacturing process, and from an integrated and automated chemical plant to a labour intensive manufacturing process. Labour costs are highly dependent on the location of the plant.

In the initial stages of cost estimation, the labour component can be estimated from either the owner's experience or from published information on similar operations. In order to adjust for capacity differences, an expression similar to the cost-capacity relationship used in Equation 4.2 can be used with the value of the factor n to be about 0.25.

The labour requirement can also be estimated from the amount of work that must be done. This can be constructed from the bottom up by determining the labour requirement for each manufacturing step or unit of operation. The labour requirement for each manufacturing step can be determined from equipment suppliers and vendors, and the company's experience.

For example, the equipment supplier indicated that two operators are required for a continuous filtration operation per shift, making the total complement six (three shifts per day, allowances for weekends and vacation leave). The filter operators report to a foreman who has a number of operational steps under his or her control. The foreman reports to a plant superintendent. The supervisory positions would not be required on a shift basis. The labour complement is determined by constructing an organisation chart for the operation in this manner.

(iii) Utilities

The utilities are the energy, cooling and air requirements for the operations. Examples are steam, cooling water, compressed air, instrument air, refrigeration, fuels, electricity, water and waste disposal. The cost of the main utilities can be estimated from local costs for their supply or generation and the level of consumption required.

(iv)   Maintenance and repairs

The annual maintenance costs can be expressed as a fraction of the capital cost. In the process industries, the total annual maintenance costs for material and labour ranges between 4 and 12% of the fixed capital cost. Of course, this is dependent on the type and scale of the operation, the processing conditions, the maturity of the technology and the type of equipment.

(v)   Operating supplies (consumables)

There are numerous supplies required to keep an operation functioning. These items range widely, from lubricants to packaging materials, and from personal protection equipment to stationery.

4.7.2 Fixed Manufacturing Costs

The fixed manufacturing costs do not vary with the amount of production. Examples of these are insurance and rentals.

(i)   Insurance

The insurance charges depend on the type of operation and the extent of insurance cover. Typical annual charges are about 1% of the fixed capital cost.

(ii)   Rentals

Rentals are charges for rented land and property, and will be dependent on the area.

4.7.3 Plant Overheads

The overhead costs involve essential services, such as fire protection, safety, warehouses, and cleaning, amongst others, that are not dependent on the level of production. These services allow the plant to operate efficiently as a unit.

4.7.4 General and Administrative

The main contributors to this cost category are the administrative functions and the sales and marketing functions. These costs are also a fixed cost in the sense that they are not dependent on the level of production.

(i)   Administrative costs.

These are the costs for managers, engineers not directly involved in operations (projects department), and accountants, and their associated office costs, such as office supplies and communications (internet and telephone).

(ii)   Sales and distribution

These are the expenses associated with getting the product to market. They include the salaries, wages, travelling expenses and commissions for the sales staff, shipping expenses, and technical sales service. The costs vary widely, depending on the location of the operation, the location of the market, the need for advertising and the type of product.

4.7.5 Royalties and Production Payment

There are two types of royalties that should be considered in the evaluation of projects. These are the charges for the use of intellectual property or know-how, and those levied by the owner of rights, such as the rights for the extraction of natural resources. The second type is normally levied by a government as an additional form of taxation, and is discussed in the section on taxes.

The agreement between two parties to allow the one to use the intellectual property of the other on a royalty basis is generally known as a licence agreement. The agreement will specify the term, the territory, the rights and calculation of payment of the licence. Typically, royalties are paid as a proportion of the annual revenue generated or on the production rate. The royalty rates vary widely, depending on the industry and the product.

In the calculation of the cash flow for the project, the royalties for the use of intellectual property or know-how are included in the operating expenses. The level of detail in these calculations should match the level of detail required for the whole analysis.

4.8 Forecasts of the Sales or Revenue

The forecasts of sales revenue should always be derived from an analysis of the market needs or requirements. The driving factors in each industry will be different. For example, the competitive process in one commodity industry may be focused on price, while in another commodity industry it may be focused on share of the market. Factors like these must be taken into account in the development of a business or marketing plan.

The sales revenue can be forecast based on the levels of production, market penetration or based on actual sales or off-take agreements. In a commodity industry, it is usual to expect that the entire production can usually be sold. However, the unknown factor is the price that can be received. In the early stages of the design process, a price forecasting method might be used while later on revenue predictions may be based on secured contracts with customers.

The forecasting of sales is a major effort in its own right, as important as the engineering design. Depending on the industry and the particular dynamics of the industry, the new operation may face stiff competition from established players. These actions must be anticipated and taken into account in the forecast.

For large capital projects, long-term contracts for the sale of the products may be a prerequisite for obtaining funding for the project. The sale of the products or the use of the facilities can be made under a number of different types of arrangements. Some of these sales or product off-take agreements are the following: (i) take-or-pay contract; (ii) take-if-offered contract; (iii) hell-or-high-water contract; (iv) toll-treatment contract; (v) throughput contract; and (vi) cost-of-service contract. The main features of each of these agreements are as follows:

(i)   Take-or-pay contract. The purchaser is obliged to pay for the output of the project whether or not the purchaser takes delivery. If the purchaser does not take delivery, payment is still required. These payments are credited to future delivery. If the project is unable to deliver the service or produce the product, the purchaser does not have to pay. For example, a chemical manufacturer enters a take-or-pay offtake contract with a purchaser who agrees to pay for the entire product from the facility whether or not he accepts delivery. This provides the funders of the chemical manufacturer with security that the revenue projections will be met.

(ii) Take-if-offered contract. The purchaser is obliged to accept and pay for a delivery of the project's output if the project delivers it. Thus, whatever the project can produce, the purchaser must accept. If the project fails to deliver, the purchaser has no obligation.

(iii) Hell-or-high-water contract. In this type of contract, the purchaser must pay, whether the project delivers or not.

(iv) Toll-treatment contract. In this type of arrangement, the project charges fees for the processing of the materials, which continue to be owned by the supplier. The toll fees must meet the operating and financing costs of the project. For example, a gold refinery accepts gold ingots with a gold content of 90%, and delivers gold back to the supplier at 99.99% purity. The charges are levied in accordance with the costs of operation and financing. Penalties may be imposed for material that is regarded as problematic to the refinery. For example, in copper refining, penalties are charged for elements such as bismuth and arsenic, which are regarded as "poisons."

(v)   Throughput agreement. A throughput agreement has some conceptual similarities to a toll-treatment agreement. The purchaser of the service agrees to purchase sufficient amounts of the service so that all the operating costs and financing costs are met. For example, a gas producer will agree to ship enough gas through a pipeline project to meet its operating and funding expenses.

(vi) Cost-of-service agreement. In this type of contract, the purchasers meet their share of the full costs of providing for the service in return for the proportionate share of the project's output. For example, a consortium of electricity consumers that together build and operate a hydroelectric scheme might enter a cost-of-service agreement. Each of the parties is obliged to pay for its share of the electricity, whether it uses it or not, and whether it is supplied or not. This type of agreement provides debt funders with a high level of security.

4.9 Calculation of the Direct Taxes and Royalties

The aim of a company is to address a market need, and thereby make a profit. Society benefits from the company's activities from the company addressing the market need, and from taxation on the company's taxable profits, which the government uses for the general benefit of all.

Taxation has a direct effect on the cash flow for a project and must therefore be included in the financials for the project. Another category of charges that also affects the cash flow to a project is that of royalties. Royalties may be charged either by the owner of intellectual property that the company uses in order to make its products, or by a government for the extraction of a natural resource. In this section the royalties that are charged by government are discussed. The royalties for the use of intellectual property were discussed in Section 4.7.5. Taxes and royalties are discussed separately.

4.9.1 Corporate Tax

The tax on companies and corporations clearly affects the cash flow to a project, and, hence, affects the evaluation of investment opportunities. As a result, the project financials are prepared on an after-tax basis.

In spite of the complexity of the taxation system, the details can be calculated precisely. However, the level detail of the calculations that are included in the calculation of the taxes for the evaluation of the project must match the level of detail used to obtain the other cash flows. The major effects and variables of the taxation system must be included; minor or second order effects can be neglected. The test is that of materiality: does its inclusion significantly alter the cash flow of the project? If the answer is yes, then its inclusion is justified.

(i)   Income tax

The general principles of taxation are clear and straightforward. Companies are assessed for income tax annually on their profits at a corporation rate. In some countries, like the USA, corporate tax is charged on a sliding scale, while in the others, such as the UK, it is charged at a flat rate. In addition, the taxation rate for corporations has changed widely over the last thirty years. Tax is charged not on the gross profits (income less costs) but on the "taxable profits," "taxable income" or "assessed income." There are two deductions from the pre-tax profits that affect the taxable income. These are allowances and interest expense.

The interest payable on debt is deductible from the profit in determining the tax charge. It essentially has the effect of treating interest as an expense. In order to separate the investment and the financing decision, the evaluation of the project is made on the basis of funding the project from the company's own resources, known as the entity basis, or "100% owner-funded basis." As a result of this choice, interest is not charged as an expense in determining the tax due by the project. The effect of the funding of the project, and hence the effect of interest on the taxable amount, is included in the cost of capital against which the attractiveness of the project is to be assessed. The concept of the entity basis is discussed in Chapter 12 and the cost of capital is discussed in Chapters 8 and 12.

(ii)   Timing of tax payments

In Chapter 3, it was assumed that when tax is due, it is paid immediately. In many countries, the taxation system allows for payment about six months later. This allows for time to prepare the books and determine the tax liability. This delay in the payment of tax can be incorporated in the project financials, but usually at a later stage when the additional effort is justified. Other taxation systems may require the payment of a provisional tax every six months, which is adjusted in the final assessment six months after the close of books. This is the same as paying the tax when it is due.

(iii)   Capital allowances and depreciation

The taxation system grants allowances that are deducted from the pre-tax profits to determine the taxable profit. These allowances may be called tax deductions in some taxation systems. The main tax allowance applicable in the assessment of projects is the depreciation, also known as the capital allowance. In some other countries, the depreciation for tax purposes (capital allowance) might be called a "wear and tear" allowance. The concept of depreciation was discussed in Chapter 3 and the main methods of calculating the depreciation are presented in Chapter 8.

In some countries, the costs of research and development can be deducted in the year of purchase.

(iv)   Tax credits

Tax credits are not the same as deductions, which are subtracted from the project's profit. Tax credits, if applicable, are deducted from the tax amount. Examples of tax credits in the US are the investment credit, the research credit and the alternative fuels credit. In the UK, the investment credit has been withdrawn.

4.9.2 Capital Gains Tax

A distinction between the proceeds earned from the sale of capital items and those from the sale of goods or services is made by the taxation systems of most countries.

A capital gain is realized if a capital item is sold for more than it originally cost. There may be a separate taxation rate for capital gains; different taxation systems treat capital gains differently. It is difficult to forecast a capital gain because most assets loose value with use and over time. An exception may be for assets that are known to appreciate, such as the value of land.

For assets that have been depreciated, the profit realized from the sale of an asset above its book value is not a capital gain. If it is sold at a price higher than the book value, this "book" profit is taxed as ordinary income. Sometimes the difference between the sale price and the book value is called depreciation recovery or depreciation recapture.

The sale of an asset below the book value causes a loss; this capital loss usually cannot be used to offset ordinary income, it can only offset other capital gains. These factors are usually not important in the forecasts of the project cash flow, except in the case of replacement studies, where the equipment to be replaced may incur a gain or loss on sale.

4.9.3 Royalties

In some cases the government charges royalties for certain activities, for example, for the extraction of natural resources. These royalties can be for mining, oil and gas recovery, or even leases for the right to explore for natural resources. These royalties, where applicable, must be accounted for in the project financials.

4.10 Working Capital

Working capital, as discussed in Chapter 3, is the funding required by the project or company in order to get it going, and to sustain its operations. Working capital includes inventories, accounts receivable (debtors) and accounts payable (creditors).

The working capital is defined as:

$$Working\ capital = Current\ assets - Current\ liabilities \qquad (4.7)$$

$$Working\ capital = Accounts\ receivable + Inventory + Accounts\ payable \quad (4.8)$$

As a rule of thumb, the working capital is about 15% of the fixed capital. The working capital requirements can be estimated from an estimate of the creditor days and debtor days that the business expects. Similarly, the inventories of raw materials, work-in-progress, and finished product stock can be estimated in the same manner. For example, for accounts payable, it is usually expected that the company will meet its payment obligations within four weeks. Therefore, the accounts payable can be estimated from the following equation:

$$Accounts\ payable = (Annual\ COGS)(Creditor\ days)/365 \qquad (4.9)$$

Similarly, for the other items of working capital:

$$Account\ receivable = (Annual\ revenue)(Debtors\ days)/365 \qquad (4.10)$$

$$Raw\ materials\ inventory = (Raw\ materials\ purchased\ in\ a\ year) \qquad (4.11)$$
$$(Raw\ materials\ inventory\ days)/365$$

$$Product\ inventory = (Annual\ revenue)(Product\ inventory\ days)/365$$
$$(4.12)$$

$$Maintenance\ inventory = (Annual\ maintenance\ requirements) \qquad (4.13)$$
$$(Maintenance\ inventory\ days)/365$$

The following guidelines can be valuable in estimating the parameters for these equations:

(i)   Creditor days. Usually 4 weeks is the recommended value.
(ii)   Debtor days. Locally payments are usually made within 4 weeks. However, if the sale is to an offshore buyer, there may be a significant delay between the sales event and the payment event. In these cases, three to four months is not uncommon.
(iii)   Raw materials. Usually four weeks. If the materials are imported, this may be longer, say eight weeks.
(iv)   Products. The product inventory accounts for both finished goods and work-in-progress. An average estimate is four weeks.
(v)   Maintenance spares. Usually a company holds about six weeks of spares.

The following example illustrates the estimation of the working capital requirements.

Example 4.6: Determination of the working capital requirement.

The annual revenue for an operation is estimated to be $10 million, and the cost of goods sold is forecast to be $8 million. The creditor days are expected to be 30 days, while the debtor days are expected to be about 60 days. The product inventory days are expected to be 40 days.

Determine the working capital requirement.

Solution:

$$Working\ capital = Accounts\ receivable + Inventory + Accounts\ payable$$
$$= 10(60/365) + 10(40/365) - 8(30/365)$$
$$= \$2.08\ million.$$

4.11 Case Study: Order of Magnitude Estimate of the Capital Cost of a Plant

The cost of an oil refinery with a capacity of 200,000 barrels per day was constructed in Liverpool six years ago. The cost was £100 million. Estimate the cost of a refinery with a capacity of 300,000 barrels per day in Cardiff. The average inflation rate was 7% over the six years, the cost-capacity exponent for oil refineries is 0.65, and the location index for Liverpool is 0.95, while that for Cardiff is 1.02.

Solution:

The value of the capital cost of the Liverpool plant would be equal to $100(1 + 0.07)^6 = £150$ million today.

The cost-capacity exponent for oil refineries is 0.65. This means that the cost for the larger plant would be equal to $150(300/200)^{0.65} = £195$ million.

There is a difference in the costs for construction between the two sites. This different in the costs between the two locations is accounted for in a location index. Since Liverpool has a location index of 0.95 Cardiff's is 1.02, the cost of the plant in Cardiff would be equal to $195(1.02/0.95) = £209$ million.

The order of magnitude estimate for the plant is £209 million.

4.12 Case Study: Factored Estimate of the Capital Cost of a Plant

The major equipment items for a chemical plant have been established. They are a furnace, a reactor, a distillation column, and a tank farm. The costs for these items are given in the Table 4.9.

Estimate the fixed capital investment and the working capital requirement for a 300,000 t/yr plant.

Solution:

The process plant is mainly a fluid-processing plant. The Lang factors for this type of plant are given in the Table 4.10.

The total equipment costs are for the two capacities are $7,500,000 and $25,000,000 respectively. The costs for the total fixed capital is therefore equal to 7.5(483/100) and 25(483/100) for the two different capacities. If linear interpolation is used

Table 4.9 Major equipment costs

Equipment	Equipment cost ($000s)	
	100,000 t/yr	400,000 t/yr
Furnace	3,000	10,000
Reactor	2,000	6,000
Distillation column	1.500	5,000
Tank farm	1,000	4,000

Table 4.10 Estimation of the capital expenses

Item	Factor
Direct costs	
Purchased equipment – delivered	100
Purchased equipment – installed	47
Instrumentation and control	18
Piping	66
Electrical	11
Buildings	18
Yard	10
Service facilities	70
Land	6
Total direct costs	346
Indirect costs	
Engineering	33
Construction	41
Contractor's fee (5%)	21
Contingency (10%)	42
Fixed capital investment	483

to obtain the value at a capacity of 300,000 t/yr, the capital costs are estimated to be equal to $92,500,000.

The working capital is often between 10 and 20% of the fixed capital investment. If an average of 15% is used, the working capital cost is estimated to be $13,900,000.

4.13 Summary

There are five components to the estimation of the project financials. They are as follows:

1. Estimation of the capital costs
2. Estimation of the total operating costs
3. Forecasts of the sales revenue
4. Calculation of the direct taxes and royalties
5. Estimate of the working capital

The estimation of the capital costs for a project is part of the engineering design process. The design and the cost estimation of major capital projects are staged activities in which further detail and greater accuracy are obtained at each stage. The cost estimates for the capital costs are classified according to either their end-use or their level of accuracy.

The estimate may be a design, bid or control estimate, depending on its end use. Design estimates are usually prepared by the owner to estimate the costs prior to soliciting bids. A bid estimate is prepared by a contractor in order to submit a response to a tender for work. The owner and the construction company use the control estimate to control costs during construction and commissioning. There are two main ways in which a contractor can be engaged: either on a lump-sum basis or a cost reimbursable basis. If the contractor submits a bid estimate on a lump-sum basis, and he wins the tender, this is the total that the owner will pay the contractor, and the bid estimate is the contractor's control estimate. If the contact is on the basis of a cost reimbursable contract, the owner and the contractor will agree to a particular estimate that forms the budget for the control of costs during construction.

The estimate could also be classified as a conceptual, pre-feasibility, feasibility or basic engineering estimate, depending on its level of accuracy. The nomenclature for this system of classification is inconsistent, although cost engineering associations have attempted to introduce some rigor into the system. The cost of preparing an estimate can be a significant amount compared with the final capital cost. These estimates are primarily used to assess whether a project is to be approved or not. As a result, the design process for producing the estimate and the approval process for decision-making are aligned in most major corporations.

The capital cost is determined from a specification of the equipment and materials required and their construction or installation costs. This is achieved from the engineering design, which establishes the equipment list, the layout and the general arrangement of the equipment. The most accurate method of determining the cost is to obtain quotations from vendors, suppliers and contractors. This is an enormous amount of work. As a result, two techniques are used to estimate the overall costs without having to expend the full level of effort in the earlier stages of the design and approval process while still maintaining some level of validity. These are the factor techniques and unit cost techniques.

There are three types of factor techniques: cost index, cost-quantity and equipment factor techniques. The cost index techniques scale the costs from historical costs to today's costs based on an index that has been determined from cost data. The cost-quantity equation scales the equipment or entire operation for different duties or capacities. The equipment factor technique estimates the cost of the overall operation as factors of the cost estimate of the major items.

The operating costs are determined from the manufacturing costs and the general costs of operations. The manufacturing costs are the direct production costs, the fixed costs and the plant overhead. The general costs are the costs of the administrative functions, sales and distribution functions. These categories can be divided further into line items such as raw materials costs, operating labour costs, *etc.* The materials costs are determined from the production schedule or the mass and energy balances, while labour and supervision costs can be established from an organisational chart.

The sales for the project can be estimated by a variety of means, depending on the product and the industry. For example, copper from a mine is unlikely to remain unsold. The unknown factor is not the level of sales, but the price that will be received for the copper. On the other hand, the sales of a chiral oxyrane, which is a group of chemicals used in pharmaceutical manufacture, can be as specific as that for a particular drug for only one client. In this case, the sales forecast requires an in-depth knowledge of the market, the relationships in the market and an ability to assess the future demands.

Income taxes are charged on the "taxable income" at the tax rate for that level of income. There may be some incentives provided by the authorities that reduce the amount of tax paid, and these must be included in the calculation. Other forms of tax are capital gains tax that may be applicable. Royalties are commonly charged by the state for the use of state property. An example is the lease right for oil, or mining royalties for the extraction of a metal. These royalties, like income taxes, depend on the country. The difference between royalties and taxes is that taxes are charged on the taxable income, whereas royalties are charged on the revenue from the operation.

The working capital is the amount of money required for inventory, stores, spares and creditors. These costs can be estimated from the hold-up of material or money in these items.

As a checklist, the elements that are generally required for all projects are the following:

1. Capital costs
    a. Equipment
        • Purchased equipment
        • Installation, including insulation and painting
        • Instrumentation and control
        • Piping
        • Electrical
        • Spare parts
    b. Buildings
    c. Facilities and yard
    d. Land
    e. Services
    f. Engineering contractor and associated costs
    g. Contingency
2. Operating costs
    a. Direct operating costs
        • Raw materials
        • Operating labour
        • Supervision
        • Power and utilities
        • Maintenance
        • Supplies

- Analytical
- Licence fees

b. Fixed production costs
- Insurance
- Rent

c. Plant overheads
- Safety, health and environment
- Storage
- Packaging
- Disposal
- Fire
- Maintenance and workshops

d. Administrative costs
- Management salaries
- Administrative salaries
- Offices

e. Sales and distribution expenses
- Sales and support staff salaries
- Shipping and distribution expenses
- Advertising and marketing expenses

3. Forecasts of the sales revenue

4. Direct taxes and royalties

- Income tax: national and local
- Tax credits
- Depreciation
- Royalties and property taxes

5. Working capital

- Accounts receivable
- Accounts payable
- Product inventory
- Raw materials inventory
- Maintenance inventory

4.14 Looking Ahead

The next chapter introduces the concept of the time value of money, a key idea in developing comprehensive criteria for the assessment of a project.

4.15 Review Questions

1. What are the different ways of classifying the estimate of the capital costs?
2. Discuss the different types of engineering contracts.
3. At what confidence level are cost estimates usually established?
4. Compare and contrast three different methods for the estimation of capital costs.
5. Is the feasibility estimate the same as the bid estimate?
6. What is the most accurate way of establishing the capital costs?
7. What are the elements of the operating costs?
8. How are the operating costs established?
9. What are the plant overheads?
10. How is tax charged?
11. On what basis are royalties usually charged?
12. What is working capital?
13. What are the elements of the project financials?
14. Are interest charges included in the project financials?
15. What are the different methods for the estimation of capital costs?

4.16 Exercises

1. Mr Darrel, a manager with a company developing solutions for an environmental problem known as acid-mine drainage, has been asked by the board of directors to prepare a "class 3 estimate" of the capital costs. Mr Darrel wants to know, what is a "class 3 estimate?"

2. A chemical reactor cost $3 million five years ago. What is the cost today if the plant equipment index was 211 then and is 567 now?

3. The price for steam boilers of a particular capacity has increased at a rate 4.7% pa. What is the cost of the boiler if it cost $200,000 five years ago?

4. The price of a 300 kW compressor is $2,000. Estimate the price of a 1,500kW compressor if the rule of two-thirds applies.

5. The Lang factors for an ammonia plant are given in the table below:

    (i) What is a Lang factor?
    (ii) How are they used?
    (iii) Describe the meaning of each of the items in the table.
    (iv) If the major equipment costs are $4 million, determine the cost of the plant if these Lang factors apply.

Item	Factor
Plant and equipment	100
Erection	10
Structural	10
Civils	20
Buildings	6
Piping	80
Electrical	16
Instruments	15
Protective cover	6
Subtotal (direct fixed cost)	263
Site development (add 5%)	13
EPCM (add 20%)	53
Offsites (add 15%)	39
Subtotal	368
Contingency (add 20%)	74
Grand total	442

6. The research department has developed a new electronic device. The parts and their costs are as follows:

Item	Quantity	Unit price, $
12-k resistor	2	0.11
20-k resistor	3	0.15
150-k potentiometer	1	0.98
Diode	4	0.43
Switch	1	0.66
Capacitor	2	0.67
Transistor	1	1.53

The labour costs for manufacture and quality control are $25/hr. The machine shop requires 6 hr to process 100 units, and assembly and inspection require 5 and 2 hr, respectively for 100 units. The overhead costs for manufacturing are equal to the labour costs. Determine the cost of the device.

7. A heat exchanger costs $20,000 and is depreciated on a straight-line basis over ten years. Construct the depreciation schedule showing the depreciation charge and the book value over the ten-year period.

8. A manufacturing facility will cost $10 million in fixed capital. The working capital is 25% of the fixed capital, and the fixed capital is depreciated over ten years. The revenues are $3 million, the expenses $1 million and the tax rate is 35%.

   (i) Determine the project financials for the first ten years of operation.

        (ii)   Determine the payback period for this operation.

        (iii)   Determine the return on investment.

9.   Plot the price of stainless steel tanks between 1 m^3 and 1000 m^3 if the rule of two thirds applies. The price of a 10 m^3 tank is $20,000.

10.   If the cost of a tank is priced in terms of weight of steel, determine the cost of a 100 m^3 tank with a diameter of 4 m and a wall thickness of 7.5 mm. The cost of steel is $20,000 per ton.

11.   The total capital investment for a chemical plant that produces epoxide is $10 million. The plant produces 3.12 million kg of the epoxide annually. Working capital is 15% of fixed capital. The selling price of the epoxide is $0.93/kg, and raw materials cost $0.21/kg. The utilities for the plant amount to $0.13/kg, and the packaging and delivery amounts to 4.5% of the total cost. The tax rate is 35%.

        (i)   Determine the costs of production.

        (ii)   Determine the revenues.

        (iii)   Determine the taxes.

        (iv)   Determine the free cash flow.

        (v)   Construct the project financials for the first five years of the plant's life.

Part II
Evaluation of Capital Projects

Capital projects have lives that extend over several years, sometimes over decades. However, the value of a cash flow that is anticipated in a few years does not have the equivalent value today. This is because of inflation and risk, which in turn means that investors prefer to get their money back from the investment sooner rather than later. It is important to account for this loss in value of money in the assessment of the project. In addition, if money is borrowed interest is charged. Accounting for these factors is referred to as the time value of money, which is the subject of Chapter 5.

Several techniques have been devised for the assessment of the economic viability or profitability of capital investments. Those techniques that incorporate the concepts of the time value of money are generally referred to as discounted cash flow techniques. There is no one preferred method of analysis, and usually several are calculated. Both accounting and discounted cash flow techniques are examined in Chapter 6. The application of these techniques to a variety of different decision-making situations is discussed in Chapters 7 and 8.

The assessment of a capital project occurs in the broader context of the company and its business context. The context for the decision on whether to invest in the project establishes some of the questions that the analysis of the project's free cash flows should answer. The techniques of decision analysis, such as decision hierarchies and strategy tables, compliment scenario analysis and sensitivity analysis in providing insight into the decision. The application of these techniques to decisions concerning capital projects is discussed in Chapter 9.

A number of case studies are discussed in Chapter 10. These case studies represent a broad range of projects. These case studies are fictionalised versions of real situations.

Chapter 5
Time Value of Money

5.1 Introduction

The value of an investment or project depends on the amount of cash flows expected from the project, the timing of these cash flows and the risk associated with each of them. Methods for the calculation and estimation of the amount and the timing of the cash flow expected from the investment were presented in Chapter 4. In this chapter, the effect of the timing of the cash flows is explored.

It is a truism that "time is money," but the measure is different for different people. This is reflected in their economic choices. A student who has low earning power takes a bus to Paris, while an executive with high earning power prefers to fly, getting there in half the time. A retiree takes hours shopping for bargains, while the plastic surgeon takes short vacations at expensive resorts. People from the first world are seen by people from developing countries to be wasteful with material goods and economical with time, while first world inhabitants see the citizens of developing countries as casual with time, or "laid back," and economical with their cars, furniture and other material possessions.

In general, people prefer a dollar today than a dollar in a year's time. Why is this true? It is true because the owner of the dollar today has the use of that dollar for a whole year longer than if she were to wait a year to receive it. That dollar has value to its owner because it can be used. For example, a dollar received today can be invested and as a result will be worth more tomorrow. There may be some risk associated with the use of the dollar; however, a risk-averse owner may loan the dollar to the government, which is the most secure investment, and still generate earnings on that dollar. On the other hand, if the owner does not invest it, but simply holds it, it may have a lower purchasing power due to erosion by inflation.

There are three reasons the value of money decreases with time: *inflation*, *risk*, and preference for *liquidity*. Inflation refers to the loss of purchasing power of money with time. Risk refers to variation in the outcome from the expected or anticipated value; that is, risk refers to the chance of events not happening as anticipated. Liquidity is the ability of the owner to sell the asset easily. For example, it is harder to find buyers for shares that trade infrequently, that is, they are illiquid.

In this chapter, the effect of timing on the value of the cash flows that are expected from a project is discussed. This is material that forms the foundation of the decision-making criteria that are developed in the next chapter. The time value of money is the most important concept in engineering economics and forms one of the cornerstones of finance. Most of the examples in this chapter use interest-bearing investments, similar to a bank account, in which the interest rate is fixed, to illustrate the concept or technique. This is done only for the sake of simplicity of the presentation. Of course, in reality there are all sorts of investments, many of which do not have these features. The treatment of other types of investments, particularly capital investments where equipment and plant is purchased and put to work, is considered in the next chapter.

5.2 Interest and Interest Rates

Interest is the price paid for the use of money. Borrowers pay lenders interest for being able to use the money now. It can be thought of as the rental payment for the use of the money for a particular time.

The interest payments are usually expressed as a percentage of the total amount borrowed and as a rate (per unit time). So if the interest charged for the use of $100 for a year is $9, then the interest rate is 9% per annum. The interest is expressed as a percentage to allow easy comparison between loans of different sizes. Like any rental, the charge for the use of the money depends on the period, and this needs to be stated for clarity. If the time period is not stated, the assumption is that it is the rate per year. For example, the statement that "the interest rate has increased" means that the interest rate per annum has increased. Interest is also affected by multiple periods, in which interest may be earned on the interest that was earned in previous periods. This is called compounding, and will be discussed later.

5.3 Effect of Timing on the Value of Money

Time is money, and generally (economically rational) people prefer cash now rather than cash in the future. Can this be quantified? Consider the following case in which the financial officer has to make a decision.

The financial officer for the company wishes to invest an amount of money for ten years. He has identified two options that seem to be attractive: (i) An electrical utility is offering a security, known as a zero-coupon bond, for sale. The terms of the security are that investors will pay $2,337.57 today to purchase the security, and the utility will pay the owner of the security $10,000 in ten year's time. The owner of the security will receive nothing until then. (ii) The government is offering a similar security to that of the utility company, except that this security will pay $500 each year for the duration of the security and in the last year will pay the full $10,000 plus the $500. The government is selling this security, known as a coupon bond, for $4,787.76.

(It is important to remember that a choice is always available: an investor, either an entrepreneur, or a company or a financial investor, always has the alternative of putting his or her money in the bank.)

The timing of the cash flows for each of these options is different, and the financial officer, as the decision maker, needs tools to enable him to make these decisions. In the next sections, two concepts, the future value and the present value of cash flows, are developed, and these concepts will be put to work to uncover criteria that will enable this decision to be made rationally. Briefly, the method relies on expressing any cash flow that is expected in the future in terms of its value today. Then all options are compared on a common basis, that is, on the basis of the present value of each option. This is schematically shown in Figure 5.1. Criteria other than that of comparing on the basis of present value can also be used. A number of decision-making criteria will be discussed in the next chapter.

Before discussing the calculations required to determine present value of a cash flow in the future, the next section discusses the future value of cash flow that is received now.

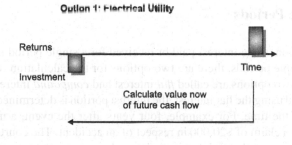

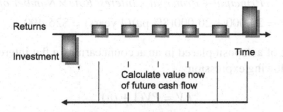

Figure 5.1 A schematic of one of the methods of comparing investment options with different cash flow profiles. These determine the equivalent value for all future cash flows in today's terms, and then compare the options based on their equivalent value today

5.4 Future Value

Cash held today will earn interest with time if invested in an interest paying account such as a bank account. As a result, the value of the investment will grow. For example, if $100 is invested for a year at an annual interest of 10%, the value of the investment will be $110 at the end of the year. This is equal to the value of the principal ($100), plus the value of the interest that is received, $10 (= 10%×100). This can be expressed in the following form:

$Value\ after\ 1\ year = Principal + (Principal)(Interest\ Rate)(Number\ of\ Periods)$

$Value\ after\ 1\ year = 100 + (100)(10/100\ per\ year)(1\ year) = 110$

It is common to refer to the value at some date in the future as the *future value*, and the value at present as the *present value*. In this example, the present value is $100, and the future value after one year is $110 at an interest rate of 10%. That is,

$$FV(after\ one\ year) = PV + PVi = PV(1+i) \qquad (5.1)$$

where FV is the future value, PV is the present value or the principal, and i is the interest rate.

5.5 Multiple Periods

The determination of the interest paid or received for a single period is straightforward. For multiple periods, there are two options for the calculation of the interest portion. These two options are called *flat* interest and *compound* interest.

In the method using the flat interest, the interest portion is determined in a manner proportional to the time. For example, four years after the event, a motorist wins a court case for a claim of $20,000 in respect of an accident. The courts determined that the motorist is entitled to interest on the settlement at a flat rate of 8% pa. The total settlement amount is calculated as follows:

$Settlement\ = Principal + Interest\ Earned$

$Settlement\ = Principal + Principal \times Interest\ Rate \times Number\ of\ Periods$

$Settlement\ = 20,000 + 20,000(8\%\ pa)(4\ years) = \$26,400\ .$

The future value of a deposit placed in an account earning a flat interest is therefore given by the following expression:

$$FV = PV(1+in) \qquad (5.2)$$

where FV is the future value, PV is the present value, i is the interest rate, and n is the number of periods.

If the principal amount of $10,000 were invested for a period less than a year, say 5 months, at a flat interest rate of 5%, then the value after five months is calculated as follows:

$$Value = Principal + Principal \, (5/12 \text{ years})(5\% \text{ per year})$$
$$Value = 10,000 + 10,000(5/12)(0.05) = 10,000 + 208.33 = \$10,208.33$$

Sometimes the *flat* interest is called *simple* interest. This is, however, confusing, because simple interest sometimes refers to the *effective* interest, which is synonymous with *annual percentage rate*. In order to not confuse the two terms, the term flat interest is used. The two terms "effective rate" and "annual percentage rate" will be discussed shortly. The term flat interest is used to prevent confusion.

The second option for calculating the interest over multiple periods is called compound interest. In this method it is imagined that the investor withdrew her money from the bank at the end of each period and re-invested both the principal and the interest earned for the next period. For example, if the investor placed $100 in the bank for year at an interest rate of 5% pa, the value at the end of the first year is calculated as follows:

$$Value \ at \ end \ of \ year \ 1 = Principal + Interest \ Earned$$
$$Value \ at \ end \ of \ year \ 1 = 100 + 100(5\% \text{ pa})(1 \text{ year}) = \$105$$

Now the entire amount is re-invested at the same rate, so the value at the end of year 2 is calculated as follows:

$$Value \ at \ end \ of \ year \ 2 = 105 + 105(5\% \text{ pa})(1 \text{ year}) = \$110.25$$

This contrasts strongly with the calculation based on flat interest, where there was no interest earned in the second period on the interest earned in the first period. In the case of flat interest, the amount at the end of year 2 would have been equal to $110. The additional $0.25 is the interest earned on interest. The quarter dollar may not seem significant, but consider the additional earnings if the principal was $1,000,000.

If the amount were to be re-invested for a third year, the value of the amount in the bank account at the end of the third year can be calculated as follows:

$$Value \ at \ end \ of \ year \ 3 = 110.25 + 110.25(5\% \text{ pa})(1 \text{ year}) = \$115.50$$

It can be easily verified that the future value of a principal amount, *PV*, at an interest rate of i per period after n periods is given by the following equation:

$$FV = PV(1+i)^n \tag{5.3}$$

The process of determining the future value is called *compounding*, and the opposite process, that of determining the present value from the future value, is called *discounting*. Prior to discussing each of these operations in turn, it is necessary to discuss the interest rate in more detail.

5.6 Types of Interest Rates

In the previous section, it was shown that interest could be charged either as flat interest or as compound interest. Up until now the examples have been limited to compounding once a year and quoting the interest rate on an annual basis. However, neither of these two factors needs to be fixed, or even the same. For example, the interest rate can be quoted on a monthly basis, and the compounding period can be daily. Two concepts and their associated terminology are important in understanding interest rates. These are the concepts of nominal rate and effective rate, which are discussed in the next two sections.

5.6.1 Nominal and Period Interest Rate

The nominal rate is the quoted rate without the effect of compounding. For example, if the bank quotes an interest rate of 18% pa on debit balances on credit cards, how will the interest charge be calculated?

The quoted rate, called the nominal rate, is an annual rate that does not include the effect of compounding. The outstanding balances are determined monthly, which means that the compounding period is monthly. The monthly rate, called the period rate, can be simply obtained from the following expression:

$$Period\ Rate\ =(Nominal\ Rate\ per\ annum)$$
$$/(Number\ of\ Compounding\ Periods\ per\ year) \qquad (5.4)$$

In this case, the period rate is equal to 1.5% per month (= 18% pa/12 months per year).

Then, assume that there was a constant balance of $1 in the account. The total amount in the account after one year would be equal to $1(1 + 0.015)^{12} = \$1.1956$, since the account earns 1.5% each month, which is then paid into the account so that interest is also earned on the interest in the next month. This result is obviously higher than a flat rate of 18%, where the account would be worth $1.18 at the end of the year.

The difference between flat interest and compound interest is the result of compounding twelve times during the year. This means that the compound interest formula in terms of the nominal rate is written in the following form:

$$FV = PV\left(1 + \frac{i}{m}\right)^{mn} \qquad (5.5)$$

where i is the nominal interest rate per period compounded m times per period over n periods.

For example, suppose we wish to determine the future value of an amount of $100 deposited in an account for two years that earns 7% pa compounded monthly.

This means that there are 2 periods of a year with compounding 12 times per year. The value of the deposit at the end of the two-year period is calculated as follows:

$$FV = 100 \left(1 + \frac{0.07}{12}\right)^{24} = \$114.98$$

The term *nominal* is equivalent to the terms *annual percentage rate*, or APR. The interest rate of what is actually earned or paid is called the effective rate.

5.6.2 Effective Rate

The effective rate is the rate with compounding once per period that gives the same amount of interest as a nominal rate compounded a number of times per period. This definition is used to derive a formula for the effective rate.

Consider an investment that earns a nominal interest rate i per annum compounded m times per year. The value of such an investment at the end of a year is given by:

$$FV = P(1 + i/m)^m$$

where P is the principal and m is the number of compounding times per period. The effective rate, e, is the interest rate that is earned if the compounding period is only once a year:

$$FV = P(1 + e)$$

Since the future value is the same for both equations (by the definition of e), the relationship between the effective rate and the periodic rate is obtained by equating the expressions:

$$P(1 + e) = P(1 + i/m)^m$$

Solving for e gives the following expression for the effective rate:

$$e = \left(1 + \frac{i}{m}\right)^m - 1 \qquad (5.6)$$

For example, if a credit card company quoted an annual percentage rate of 18% per annum compounded monthly, then the effective rate is given by

$$e = (1 + 0.18/12)^{12} - 1 = 19.56\%$$

This value coincides with the value of the debit balance discussed earlier in Section 5.6.1. The calculation of the effective rate is illustrated in the following example.

Example 5.1: Annual percentage rate and effective rate.

An interest rate of 12% APR is charged on outstanding amounts on a bank loan. For this particular loan, interest is charged on a daily basis. What is the actual rate or effective rate that the lender is paying?

Solution:

APR is the quoted rate, or as was discussed in the previous section, it is the nominal rate. The effective rate is then given by Equation 5.6 and is calculated as follows:

$$e = (1 + 0.12/365)^{365} - 1 = 12.75\%$$

The concepts of nominal, period and effective rates are important in understanding the manner in which interest is charged. In the next section, the compounding of interest is discussed.

5.7 Compounding

Compounding is the process of determining the future value of a deposit made today, and represents the growth in value with time due to interest earned. The effect of single and multiple payments at regular intervals are examined.

5.7.1 Single Payments: Growth

The value in the future of a single payment into an interest-bearing investment is determined by the compound interest formula given in Equation 5.3. For example, suppose $100 was deposited at 9% for ten years. The cash flows at various periods can be visualized using a timeline. The timeline for this example, which is shown in Figure 5.2, is straightforward: a single deposit is made at the beginning of the first year, that is, at time 0. No other deposits are made, and the deposit earns interest at 9% per annum while in the account.

The future value, given the interest rate and the number of years, is calculated as follows:

$$Future\ Value\ At\ Year\ 10 = 100(1 + 0.09)^{10} = 236.74\ .$$

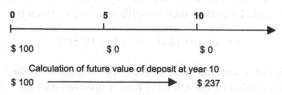

Calculation of future value of deposit at year 10

Figure 5.2 Timeline representing the deposit of $100 for 10 years

Table 5.1 Contributions of flat interest and compound interest to the total value of an interest-bearing investment at an interest rate of 9% pa

Year	Opening balance	Flat interest	Compound interest	Total interest	Closing balance
1	100.00	9	0.00	9.00	109.00
2	109.00	9	0.81	9.81	118.81
3	118.81	9	1.69	10.69	129.50
4	129.50	9	2.66	11.66	141.16
5	141.16	9	3.70	12.70	153.86
6	153.86	9	4.85	13.85	167.71
7	167.71	9	6.09	15.09	182.80
8	182.80	9	7.45	16.45	199.26
9	199.26	9	8.93	17.93	217.19
10	217.19	9	10.55	19.55	236.74
Total		90	46.74	136.74	

The future value is dependent on the principal amount and the parameters of the compound interest formula. The next few sections investigate how the rate of compounding, m, and the interest rate, i, influence the value.

(i)   Contribution from flat interest and compound interest

The future value of a fixed-interest investment can be viewed as being composed of two components: the component due to flat interest and the component due to compound interest. The contributions of the flat interest and the compound interest to the total interest in each period are shown in Table 5.1 and Figure 5.3. With time, the contribution of compound interest to the total interest becomes dominant. This is illustrated in Figure 5.4. Notice that by year ten the contribution of compound interest to the total interest exceeds that arising from flat interest. The contribution from compound interest becomes more prominent at higher interest rates.

(ii)   The effect of the interest rate

The interest rate is a primary parameter in the compound interest formula. In other words, the future value is highly dependent on the interest rate. This is shown in Figure 5.5. A fixed-interest investment at an interest rate of 25% would be worth 2.63 times the value of a fixed-interest investment at an interest rate of 3% after five years, and 6.93 times after ten years.

(iii)   The effect of the rate of compounding

The rate of compounding, that is, the number of times per period that the interest is calculated, influences the future value, although this effect is not as dramatic as the effect of the interest rate itself. In spite of its less dramatic effect, it is still an important aspect of interest-bearing accounts because the principal amounts can be very large. The effect of the compounding period is shown in Table 5.2.

The last three compounding periods in Table 5.2 are not generally found in practice, although continuous compounding is used in academic finance circles. The formula for continuous compounding is given by the following expression:

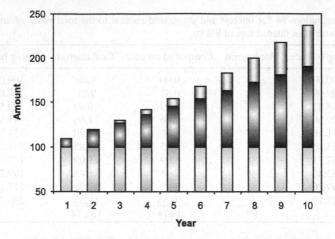

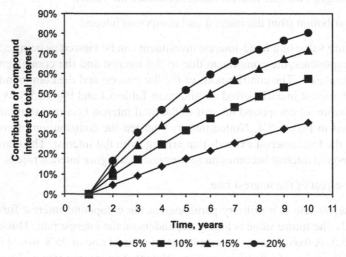

Figure 5.3 The contributions of flat interest and compound interest to the total value of an interest-bearing investment at 9% pa

Figure 5.4 The contribution of the compound interest to the total interest increases with the increasing term of the deposit and increasing interest rate

$$FV = PVe^{rt} \qquad (5.7)$$

where r is the interest rate for continuous compounding and t is the time.

The discussion until now has focused on investment. However, the same concept is applicable to the growth of a population or an economy. The concept of compounding is also applicable to inflation, which is the growth in price due to the erosion of the buying power of money with time. These two topics are discussed next.

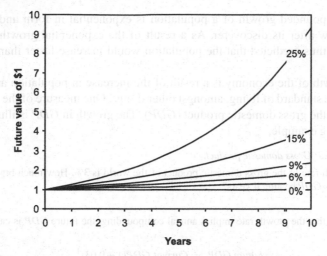

Figure 5.5 Effect of interest rate on the future value of a fixed-interest deposit

Table 5.2 The effect of the compounding period on the effective rate for a fixed interest deposit at a nominal rate of 10% pa

Compounding period	Compounding Times per year, m	Effective annual Rate, e
Year	1	10.0000%
Semi-annual	2	10.2500%
Quarter	4	10.3813%
Month	12	10.4713%
Week	52	10.5065%
Day	365	10.5156%
Hour	8,760	10.5170%
Minute	525,600	10.5171%
Continuously		10.5171%

(iv)   Compounding and growth

Compounding is a general concept, related to many fields. Consider the following two examples for the growth of a population and for the growth in the economy.

Example 5.2: Population growth.

A population of a village of 1000 people is projected to grow at a rate of 7% per annum over the next five years. How many people will live in the village in five years from now?

Solution:

Assuming that the growth rate implies annual compounding, the population in five years is calculated as follows:

$$Population = 1000(1 + 0.07)^5 = 1403$$

The population of the village is expected to be 1403 in five years.

The compounded growth of a population is exponential in form and is called Malthus' law after its discoverer. As a result of the exponential growth in populations, Malthus predicted that the population would increase faster than the food supply.

The growth of the economy is a result of the increase in population and the increase in the standard of living, amongst other things. One measure of the size of an economy is the gross domestic product (*GDP*). The growth in *GDP* is illustrated in the following example.

Example 5.3: Gross domestic product.

The growth rate of the gross domestic product of the world is 3%. How much bigger will the world's *GDP* be after 4 years?

Solution:

Assuming that the growth rate implies annual compounding, the future *GDP* is calculated as follows:

$$Future\ GDP\ =\ Current\ GDP(1+0.03)^4$$

The increase in *GDP* is calculated as follows:

$$\%Increase = (Future\ GDP - Current\ GDP)/(Current\ GDP) = (1+0.03)^4/1 = 12.6\%$$

The world's *GDP* is expected to 12.6% larger in four years.

Sometimes when a growth rate is quoted it might be called CAGR, which stands for compounded annual growth rate. This term is used to emphasise that the growth rate has been calculated with compounding. For example, the growth rate of 3% used in Example 5.3 is the same as the CAGR. The average year on year growth rate for the four years is $12.6\%/4years = 3.15\%$.

(v)   Compounding and inflation

Inflation is the increase in the level of prices. The purchasing power of money is diminished with time by inflation, which means that the purchase of the same item will require more money in the future. Inflation refers to the escalation of the general level of prices. However, not all prices rise in concert – some increase, while others may be constant or decline, even though the majority of prices are increasing.

The most common measure of inflation is the consumer price index (*CPI*), which aggregates the price of a basket of goods and services that represent the purchases of an average consumer. The *CPI* is a simple ratio of the current price of the basket of goods to the price for the same goods in a reference year. For example, in the US, the reference "year" that was chosen was the period 1982–1984, and the index in that period was set to 100. So, the *CPI* in the US is defined as follows:

$$CPI = \frac{price\ of\ basket\ in\ current\ year}{price\ of\ basket\ in\ 1982 - 1984} \tag{5.8}$$

The rate of inflation, which expresses how rapidly inflation is changing, is the growth in the index with respect to its value in the previous year. This can be expressed as follows:

$$Rate\ of\ Inflation_t = \frac{CPI_t - CPI_{t-1}}{CPI_{t-1}} \tag{5.9}$$

where CPI_t refers to the CPI in year t. The CPI is a general measure of price increase for consumers. In Chapter 4, the adjustment of the price of capital equipment for time was discussed. It was mentioned there that there are a number of indices, such as the Chemical Engineering Plant Cost Index, that track the cost of equipment with time. These indices are conceptually the same as the consumer price index, except that they measure the cost or price for a different "basket of goods."

The rate of inflation is commonly quoted, not the CPI itself. The calculation of the price from the inflation rate is illustrated in the following two examples.

Example 5.4: Inflation.

If the price of the basket of goods costs $100 in 2007, how much would the same goods cost in nine years later if the average inflation rate were 4% pa over that period?

Solution:

The price is calculated as follows:

$$Price = 100(1 + 0.04)^9 = \$142.33$$

Example 5.5: Doubling time.

If the price doubles in ten years, what is the inflation rate?

Solution:

The inflation rate is calculated as follows:

$$2 = (1 + Inflation\ rate)^{10}$$

$$Inflation\ rate = 0.07177 = 7.18\%$$

The "rule of 72" is an approximation that can be used to estimate the doubling time from the inflation rate. The doubling time is approximated by the following expression:

$$Doubling\ time = \frac{72}{rate\ 100} \tag{5.10}$$

For example, if the inflation rate were 7.2%, how long would it take the prices to double? Using the "rule of 72," the doubling time is equal to $72/7.2 = 10$ years. For example, if inflation is 15%, the doubling time is equal to $72/15 = 4.8$ years. The compound interest formula can be use to determine the exact doubling time, which in the latter case is 4.96 years.

The use of the "rule of 72" can be applied to other situations in which the value increases due to compound growth. Populations other than those of humans also increase by compounding, leading to exponential growth. The following example illustrates the growth of a population of infectious bacteria.

Example 5.6: Bacterial infection.

A person is infected with a single bacterial cell that causes disease. This bacterium has a doubling time of 10 minutes. How long will it take the population to reach 1 billion bacteria in the patient?

Solution:

The growth rate of the bacterial population is calculated from Equation 5.10 as follows:

$$Rate = \frac{72}{Doubling\ time(100)} = \frac{0.72}{10} = 0.072 \text{ per minute}$$

The growth of the bacteria occurs continuously, and it can be modelled by continuous compounding. The expression for continuous compounding is given by Equation 5.7. As a result, the bacterial population is modelled as follows:

$$N = N_0 e^{rt}$$

This means that the number of bacteria, N, grows exponentially at a rate, r, from the initial number, N_0. The time to reach 1 billion bacteria is calculated as follows:

$$t = \frac{1}{r} \ln\left(\frac{N}{N_0}\right) = \frac{1}{0.072} \ln(1e9) = 299 \min = 4.98 \text{ hr}$$

Just five hours after receiving only a single bacterium, the patient has a significant infection.

(vi)   The nominal interest rate and the real interest rate

The interest that is quoted, that is, the nominal rate, includes the effect of inflation. So the interest charged compensates the lender for the loss of use of the money for the loan period, for the diminished purchase power of the principal amount of the loan when it is returned to the lender and for the risk of loaning the money to another party.

The real interest rate is the interest rate that does not include the effect of inflation. The relationship between the real rate of interest and the nominal rate can be determined from the future value formula. Consider an investor who places an amount in a bank for year at an annual interest rate of i. The future value of the money deposited includes both principal and interest, and is given by the following expression:

$$FV = PV(1+i) \qquad\qquad (5.11)$$

The future value of the deposit can be seen as being determined by two factors: (i) earning at the "real" interest rate; and (ii) being escalated by general inflation. If the *real* interest rate is r, and the rate of inflation is h, then the future value of the investment is given by the following expression:

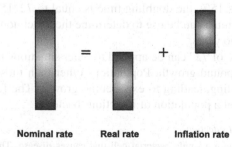

Nominal rate   Real rate   Inflation rate

Figure 5.6 The nominal interest includes a contribution from the real interest rate and the inflation rate

$$FV = PV(1+r)(1+h) \tag{5.12}$$

These two expressions are equivalent to each other. By rearranging, we obtain the nominal interest rate, which is given by the following expression:

$$i = r + h + rh \tag{5.13}$$

The term rh, called the Fisher effect after the economist Irving Fisher, is a multiple of two smaller numbers, and is therefore often neglected. This means that the nominal interest rate is a sum of the real interest rate and the inflation rate, or put differently, the real interest rate is the nominal interest rate less the inflation rate (Figure 5.6).

The application of the real interest rate is illustrated in the following two examples.

Example 5.7: Real rates.

The interest rate in Brazil is 19% and the inflation rate is 14%. What is the real interest rate?

Solution:

$$i = r + h + rh$$
$$0.19 = r + 0.14 + r(0.14)$$
$$r = 4.3\%$$

The real interest rate is 4.3%.

Example 5.8: House prices.

House prices have grown at a real rate of 2.8% for the last thirty years. The inflation rate has averaged 6%. What is a house that was purchase thirty years ago worth today? The average prime interest rate for deposits over the period was 7%. What would the same amount have earned in the bank account?

Solution:

The nominal interest rate is calculated as follows:

$$i = r + h + rh = 0.028 + 0.06 + 0.028(0.06) = 8.968\%$$

The price increase is the nominal rate compounded annually. The increase in price is calculated as follows:

$$Price = (1 + 0.08968)^{30} = 13.15 .$$

The price of the house would be worth 13.15 times the original purchase price. The value of a deposit in the bank account earning 7% would be worth $(1 + 0.07)^{30} = 7.61$ times the original amount. An investment in a house would have been a significantly better investment than in a bank account.

In the next section, the compounding of multiple cash flows is considered.

5.7.2 *Multiple Equal Payments: Annuities*

The discussion up to this point has focused on determining the future value of a single investment. A common investment opportunity is an annuity, in which equal instalments are made for a number of periods (or years). Two types of annuity are discussed: an ordinary annuity and an annuity due.

(i)   Ordinary annuity

An *ordinary annuity* is one in which the first payment is at the end of the first year and payments are made in equal instalments thereafter. The payment schedule is illustrated in Figure 5.7.

It is general practice to assume that the payments are made at the end of the period. This is known as the *end-of-year convention*.

The future value of the annuity is the sum of the future values of each of the instalments. For example, say the investor makes instalments of $100 in a three-year annuity investment. What is the investment worth at the end of the three years if the annuity earns 8% pa compounded annually?

The cash flow diagram for this investment is shown in Figure 5.8. The end-of-year convention is used, that is, the instalments are treated as if they were received at the end of the year. The future value at the end of year 3 of the instalment made in the third year earns no interest, so its value is $100. The future value of the second instalment must be compounded for one period, that is, the future value is calculated as follows:

$$FV = 100(1 + 0.08) = 108$$

The future value of the first instalment is calculated as follows:

$$FV = 100(1 + 0.08)^2 = 117$$

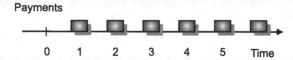

Figure 5.7 Payment schedule for an ordinary annuity

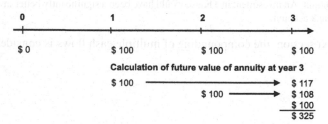

Figure 5.8 Cash flow profile for a three-year *ordinary annuity* with $100 instalments

Therefore the future value of the investment is the sum of individual terms, that is, equal to $100 + $108 + $117 = $325.

Generalizing this into a formula, the future value of an annuity, *FVA*, is expressed in the following form:

$$FVA = A\left[(1+i)^{n-1} + (1+i)^{n-2} + \ldots + 1\right] \qquad (5.14)$$

where A is the instalment, i is the interest rate earned per period, and n is the total number of periods. This formula can be expressed in a more compact form as a summation, as shown in the following expressions:

$$FVA = A\sum_{t=1}^{n}(1+i)^{n-t} \qquad (5.15)$$

$$FVA = A\sum_{t=0}^{n-1}(1+i)^{t} \qquad (5.16)$$

An examination of the *FVA* formula reveals that it is a geometric progression, and the formula for the summation of a geometric progression can be used to derive an analytic expression for calculating the value of the annuity. The sum of a geometric series is given by the following series summation:

$$B + Bx + Bx^2 + Bx^3 + \ldots + Bx^{p-1} = B\left(\frac{1-x^p}{1-x}\right) \qquad (5.17)$$

If the substitutions $x = 1+i$, $p = n-1$ and $B = A$ are made into Equation 5.17, it can be shown that the future value of the annuity can be calculated by using the following equation:

$$FVA = A\left(\frac{(1+i)^n - 1}{i}\right) \qquad (5.18)$$

The term in parentheses on the right-hand side of this equation is called the future value interest factor of an annuity (FVIFA). It determines the value at the end of the period of an annuity that earns an interest i.

The future value of an annuity can be calculated by calculating the sum of the values of the individual terms, or by using the FVIFA formula. Manual calculation using a spreadsheet is as easy as using the FVIFA formula, and is probably more preferred when others who are not as familiar with the annuity formulae need to check or audit the work.

The application of the annuity equation is illustrated in the following example.

Example 5.9: Future value of an ordinary annuity.

A three-year annuity earns 8%. What is it worth at the end of the period if the annuity amount is $100?

Solution:

The cash flow profile for the annuity is the same as that shown in Figure 5.8. The substitution of the values into the annuity formula gives the following result:

$$FVA = A\left(\frac{(1+i)^n - 1}{i}\right) = 100\left(\frac{(1+0.08)^3 - 1}{0.08}\right) = 324.6$$

The value at the end of the period is $324.6, which is the same as the value shown in Figure 5.8 that was calculated term by term.

(ii)   Annuity due

The previous discussion concerned an ordinary annuity. An *annuity due* is specifically structured so that the first instalment is due at the beginning of the period, rather than at the end of the period. The cash flow profile for this investment is shown in Figure 5.9.

We wish to calculate the value at the end of the third year of a three-year annuity of this type that earns 8% pa compounded annually. The procedure is the same as before: the value is the sum of the future values of each of the cash flows, which are calculated using the future value formula. These calculations are given by:

First Year:

$$FV = 100(1+0.08)^3 = 125.97$$

Second Year:

$$FV = 100(1+0.08)^2 = 116.64$$

Third Year:

$$FV = 100(1+0.08) = 108$$

The total value of the annuity due is therefore the sum of these values, which is equal to $350.61.

This calculation can be generalised into a series formula, given as follows:

$$FVA\ due = A\sum_{t=0}^{n-1}(1+i)^{n-t} = A\sum_{t=1}^{n}(1+i)^t \tag{5.19}$$

A comparison of this formula and that for the geometric series yields that the future value for an annuity due is given by the following expression:

$$FVA\ due = A\left(\frac{(1+i)^{n+1} - (1+i)}{i}\right) \tag{5.20}$$

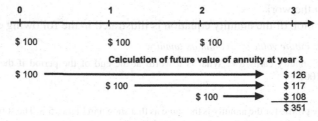

Figure 5.9 Cash flow profile for a three-year *annuity due* with $100 instalments

The future value of an annuity due is more valuable than an ordinary annuity because of the timing of the cash flows. This emphasises again that the value for an investor depends not only on the amount of the cash flows, but also on the timing of the cash flows.

5.7.3 Multiple Periods: Growth Annuities

Growth annuities are annuities in which the amounts placed in the annuity increase each year. These amounts can increase by a constant amount or by a fixed percentage. The annuity that increases by a constant amount is called an arithmetic growth annuity and the annuity that increases by a fixed percentage is called a geometric growth annuity. The formulae that are derived can be used in determining the present value of an engineering project so it is worthwile examining their derivation in detail. These two annuities are discussed next.

(i)   Arithmetic growth annuity

The ordinary annuity and the annuity due both had constant amounts deposited in the account. The arithmetic growth annuity increases the deposit by a constant amount each year. The cash flow profile of the arithmetic growth annuity is shown in Figure 5.10.

The cash flow for each year is increased by a constant amount. This means that the cash flow in year t is given by the following expression:

$$CF_t = A + (t-1)G \tag{5.21}$$

where CF_t represents the cash flow in year t, A represents a base amount that is constant and equivalent to the amount deposited in year 1 and G represents the amount by which the deposit is increased each year. The year t must be greater than one.

The cash flows for the arithmetic growth annuity are separated into those that form an ordinary annuity and those that result in growth. This separation of the cash flows is the basis for the derivation of the formula for the annuity. The formula for the ordinary annuity has already been discussed, so the growth portion needs to be derived.

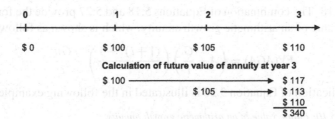

Figure 5.10 Cash flow profile for a three-year arithmetic growth annuity

Notice the first contribution of the growth amount is in year 2, and the last in the final year. The future value of the cash flow from the growth portion is given by the following expression:

$$FV(Growth) = G(1+i)^{n-2} + 2G(1+i)^{n-3} + ...G(n-2)(1+i) + G(n-1)$$
(5.22)

where n is the length of the annuity.

The first term in Equation 5.22 represents the contribution from the growth term to the cash flow in year 2. The final term represents the contribution from the growth term to the cash flow in the final year. The constant value G is a common factor, so that Equation 5.22 is written as follows:

$$FV(Growth) = G\left[(1+i)^{n-2} + 2(1+i)^{n-3} + ... + (n-3)(1+i)^2 + (n-2)(1+i) \right.$$
$$\left. + (n-1)\right]$$
(5.23)

The multiplication of Equation 5.23 by $(i+1)$ yields the following expression:

$$FV(Growth)(1+i) = G\left[(1+i)^{n-1} + 2(1+i)^{n-2} + ... + (n-3)(1+i)^3 \right. \quad (5.24)$$
$$\left. + (n-2)(1+i)^2 + (n-1)(1+i)\right]$$

The subtraction of Equation 5.23 from 5.24 yields the following expression:

$$FV(1+i) - FV = G\left[(1+i)^{n-1} + (1+i)^{n-2} + ... + (1+i)^3 + (1+i)^2 \right. \quad (5.25)$$
$$\left. + (1+i) + 1\right] - nG$$

The term within the brackets is a geometric series. If the substitutions $x = 1+i$, $p = n-1$ and $B = 1$ are made in Equation 5.17, the term in brackets is simplified. It is given by the following expression:

$$\left[(1+i)^{n-1} + (1+i)^{n-2} + ... + (1+i)^3 + (1+i)^2 + (1+i) + 1\right] = \frac{(1+i)^n - 1}{i}$$
(5.26)

This means that the future value of the growth term is given by the following equation:

$$FV(Growth) = \frac{G}{i}\left(\frac{(1+i)^n - 1}{i} - n\right)$$
(5.27)

The future value of the constant amount that forms the basic amount is given by the future value of an annuity. The formula for the ordinary annuity was given as Equation 5.18. The combination of Equations 5.18 and 5.27 provide the formula for the future value of an arithmetic growth annuity, which is shown as follows:

$$FV(AGA) = \left(A + \frac{G}{i}\right)\left(\frac{(1+i)^n - 1}{i}\right) - \frac{Gn}{i}$$
(5.28)

The application of Equation 5.28 is illustrated in the following example.

Example 5.10: Future value of an arithmetic growth annuity.

A three-year arithmetic annuity earns 8%. What is it worth at the end of the period if the annuity amount is $100, and it is increased by $5 each year?

Solution:

The cash flow profile for the annuity is the same as that shown in Figure 5.8. The substitution of the values into the annuity formula gives the following result:

$$FV(AGA) = \left(A + \frac{G}{i}\right)\left(\frac{(1+i)^n - 1}{i}\right) - \frac{Gn}{i} = \left(100 + \frac{5}{0.08}\right)\left(\frac{(1+0.08)^3 - 1}{0.08}\right) - \frac{5(3)}{0.08}$$

$$= 340$$

The value at the end of the period is $340.04. This is the same as the value shown in Figure 5.10, which was calculated term by term.

(ii)   Geometric growth annuity

The geometric growth annuity is similar to the arithmetic growth annuity, except that the growth portion increases at a constant rate rather than by a constant amount. The cash flow profile for the geometric growth annuity is shown in Figure 5.11.

The cash flow for each year is increased at a constant rate. This means that the cash flow in year t is given by the following expression:

$$CF_t = A(1+g)^{t-1} \tag{5.29}$$

where CF_t represents the cash flow in year t, A represents an initial amount and g represents the rate at which the deposit is increased each year. Equation 5.3 gives the future value of the cash flow. The combination of Equations 5.3 and 5.29 yields the following expression for the future value of the cash flow for year t:

$$FV(CF_t) = A(1+g)^{t-1}(1+i)^{n-t}$$

The future value of the geometric growth annuity is calculated from the following formula:

$$FV(GGA) = A\left(\frac{1 - (1+g)^n(1+i)^{-n}}{i - g}\right)(1+i)^n \quad i \neq g \tag{5.30}$$

$$FV(GGA) = An(1+i)^{n-1} \quad i = g$$

The application of the formula for the future value of a geometric growth annuity is illustrated in the following example.

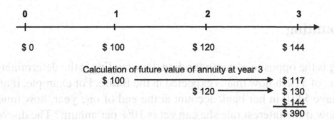

Figure 5.11 Cash flow profile for a three-year geometric growth annuity

Example 5.11: Future value of a geometric growth annuity.

A product that a life insurance company is selling is a retirement annuity. The purchaser is to deposit an amount each year. The amount each year escalates at a constant rate, which the purchaser can choose. If the annuity earns 8%, what is it worth at the end of a period of ten years if the annuity amount is $100 and it is increased at a rate of 15% each year? How much difference would there be in the fund if the purchaser escalated payments by 7%?

Solution:

The future value of the annuity is calculated as follows:

$$FV(GGA) = A\left(\frac{1-(1+0.15)^{10}(1+0.08)^{-10}}{0.08-0.15}\right)(1+0.08)^{10} = 2{,}695.19$$

$$FV(GGA) = A\left(\frac{1-(1+0.07)^{10}(1+0.08)^{-10}}{0.08-0.07}\right)(1+0.08)^{10} = 1{,}917.74$$

If the purchaser chooses to escalate the payments by 7%, the annuity would be worth $1,917 compared with $2,695 if it was escalated at 15%. The value for an escalation of 15% is 40% higher than that for 7%.

5.7.4 Multiple Payments of Unequal Amounts

The cash flows from an investment or an engineering project are not uniform for most investments or projects. The revenues and costs will change during the life cycle of the project. In this case there is no formula that is equivalent to the series summations that were derived in the previous sections. The calculations must be performed term-by-term for each cash flow. However, the principle of the calculation method is the same as before, that is, the future value of the investment is the sum of the future values of each cash flow. This can be expressed in the following form:

$$FV = \sum_{t=0}^{n} CF_t(1+i)^t \tag{5.31}$$

where FV is the future value, CF_t is the cash flows at year t, i is the interest rate earned, t is the year, and n is the total number of years.

5.8 Discounting

Discounting is the opposite of compounding. Discounting is the determination of the present value of a cash flow that is expected in the future. For example, if an investor wished to have $110 in her bank account at the end of one year, how much should she invest now if the interest rate she can get is 10% per annum? The discounting of cash flows is discussed for single and multiple cash flows.

5.8.1 Single Amounts of Future Cash Flow

The present value of an amount that is expected in the future is determined by the compound interest formula given by Equation 5.3. The equation is simply re-arranged to express the present value in terms of the future value. This is given by the following expression:

$$PV = FV\frac{1}{(1+i)^n} \tag{5.32}$$

To answer the question posed earlier, the investor who wishes to have \$110 at the end of the year needs to deposit \$100 (= 110/(1+0.10)) in an interest-bearing account at a fixed rate of 10% pa.

The effects of the interest rate and the period of compounding are the same as those discussed in Section 5.7 for compounding, except that they are inverted. This is illustrated by examining the effect of the interest rate on the present value.

The interest rate that an investor can receive is a prime determinant of value in the compound interest formula, and the present value is sensitive to this parameter. For example, at an interest rate of 3%, an investor needs to deposit 74c to receive a dollar in ten years from now; however, if the interest rate was 25%, the investor need only deposit 11c to receive the same dollar in ten years from now.

The effect of the interest rate on the present value of a fixed interest investment is shown in Figure 5.12.

The calculation of the present value is illustrated in the following example.

Example 5.12: Present value.

A retirement investment is guaranteed to pay out \$100,000 in ten years. If the average interest rate is expected to be 10%, what is the value of the investment now?

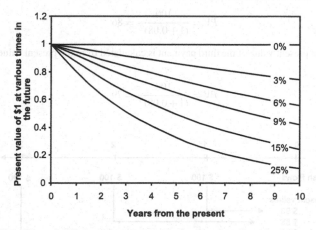

Figure 5.12 The effect of different interest rates on the present value of a dollar as a function of different times into the future when the dollar will be paid

Solution:

The present value of the investment is calculated as follows:

$$PV = FV\frac{1}{(1+i)^n} = 100,000\frac{1}{(1+0.10)^{10}} = \$38,554.32$$

The present value of the retirement investment is \$38,554.32.

5.8.2 Multiple Equal Amounts

In Section 5.7, two types of annuities were discussed, the ordinary annuity and the annuity due, and formulae for the future value of these annuities were derived. The present value of these annuities can be obtained by summing the present values of each of the present values of the expected cash flows. The calculation of the present value of the annuity is illustrated in the following example.

Example 5.13: Present value of an annuity.

An investor expects to receive instalments of \$100 from a three-year ordinary annuity investment. What is the investment worth today if the annuity earns 8% pa compounded annually?

Solution:

The cash flow for this annuity is shown in Figure 5.13. The end-of-year convention is used, which means that the investor expects to receive these payments at the end of the year.

The present value of the first payment is discounted by a year. If the investor can receive 8% pa on other investments, the present value of the first payment is calculated as follows:

$$PV = \frac{100}{(1+0.08)} = 93$$

The present value of the second payment is calculated as follows:

$$PV = \frac{100}{(1+0.08)^2} = 86$$

Similarly the present value of the third payment is calculated from the present value formula as follows:

$$PV = \frac{100}{(1+0.08)^3} = 79$$

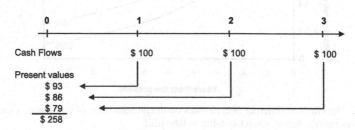

Figure 5.13 Present value of a three-year ordinary annuity

Therefore the total present value to the investor of this annuity is the sum of the present values of these three individual cash flows, that is, equal to $93 + 86 + 79 = \$258$.

A general formula for the present value of an ordinary annuity, PVA, is given by:

$$PVA = A \sum_{t=1}^{n} \frac{1}{(1+i)^t} \tag{5.33}$$

where A is the value of the cash flow payment. The geometric series formula can be used to simplify the summation term in Equation 5.33. If this is done, the present value of an ordinary annuity is given by:

$$PVA = A \left(\frac{1 - 1/(1+i)^n}{i} \right) \tag{5.34}$$

The term in parentheses on the right-hand side of this equation is called the present value interest factor of an annuity (PVIFA).

Another type of annuity is a deferred annuity. In a deferred annuity, the equal instalments or payments are deferred for some time before they commence. For example, an investor has invested in a three-year deferred annuity in which three equal instalments of $100 will be received starting two years from today. The cash flow profile for this investment is shown in Figure 5.14.

The procedure for manually determining the present value of the deferred annuity is the same as that presented before, and is illustrated in Figure 5.14 using a discount rate of 8% pa. Alternatively, the present value interest factors for an annuity can be used to calculate the present value of this annuity. The deferred annuity is an ordinary annuity for the full period less an annuity for the deferred period. As a result, the present value is calculated as follows:

$Present\ value\ for\ deferred\ annuity\ =\ PVIFA(4year, i) - PVIFA(1year, i)$

$Present\ value\ for\ deferred\ annuity\ =\ 100(3.312 - 0.926) = 239$

The interest-bearing investments that have been discussed up to this point have been of a finite life. On the other hand, a *perpetuity* is an annuity that pays for an infinite period. For example, a preference share of a company that pays a fixed

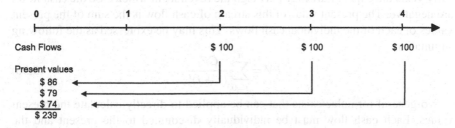

Figure 5.14 Cash flows profile for a three-year annuity deferred for two years

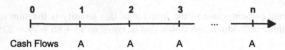

Figure 5.15 Cash flow profile for a perpetuity

dividend may be regarded as a perpetuity. The cash flow profile for a perpetuity is shown in Figure 5.15.

The present value of a perpetuity is given by the following equation:

$$PVP = \frac{A}{i} \tag{5.35}$$

where A is the cash flow or instalment amount and i is the interest rate that is used to value the perpetuity.

The application of the perpetuity formula is illustrated in the following example.

Example 5.14: Present value of a perpetuity.
An investor wishes to purchase 100 preference shares with a face value of $10 each that pay 7% pa. What is the present value of this investment?
Solution:
The amount that these shares pay is the face value times the interest rate, that is, equal to $70 = 100$ shares $\times$ $10 per share $\times$ 7% each year forever, or as long as the company exists. If the investor can deposit money in the bank for 8%, the investor should use that rate to value the preference shares. The present value of the preference shares to the investor is calculated as follows:

$$Present\ Value = \frac{70}{0.08} = \$875$$

The present value of the preference shares is $875.

5.8.3 Uneven Cash Flows

Many investments do not produce a steady stream of cash. For example, the investment in a company or a new venture will produce an uneven flow. In fact, in the early years the expenditure may outweigh the revenue, in which case the cash flows are negative. The present value of this stream of cash flow is the sum of the present values of each of the individual cash flows. This may be expressed as the following summation

$$PV = \sum_{t=0}^{n} \frac{CF_t}{(1+i)^t} \tag{5.36}$$

No general formulae exists that can be applied to directly calculate the present values. Each cash flow must be individually discounted to the present and the summed. This is very easily done using a spreadsheet.

5.9 Compound Interest Formula

In the preceding sections, main compound interest formulae were discussed. These arise from the consideration of single payments and a uniform series of payments. In addition, there are formulae for arithmetic and geometric series of payments. This section summarises the formulae and provides a standard notation for them.

5.9.1 Single Payment Interest Formula

The two interest rate formulae that are relevant to a situation where only a single payment is considered have been encountered earlier. Equation 5.3 gives the future value of a single payment and Equation 5.32 gives the present value of a single payment. These two equations relate the future value and present value as a function of interest rate and time periods:

$$FV = PV(1+i)^n \qquad (5.37)$$

$$PV = FV\frac{1}{(1+i)^n} \qquad (5.38)$$

The term in brackets of the right-hand side of Equation 5.37, $(1+i)^n$, is known as the compound factor, or the "single payment compound factor." The term $1/(1+i)^n$ on the right-hand side of Equation 5.38 is known as the discount factor, or the single payment present value factor.

5.9.2 Multiple Equal Payments

The schedule of cash flows for multiple payments of equal amounts is shown in Figure 5.16. The next four formulae are derived from the annuity formula, encountered earlier.

Equation 5.39 expresses the future value of an annuity and Equation 5.40 expresses the present values of an annuity:

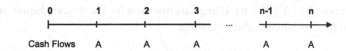

Figure 5.16 Time line with equal cash flows at each year

$$FV = A\left(\frac{(1+i)^n - 1}{i}\right) \qquad (5.39)$$

$$PV = A\left(\frac{1-(1+i)^{-n}}{i}\right) \qquad (5.40)$$

The term on the right-hand side on Equation 5.39 is known as the future value interest factor of an annuity (FVIFA) or the uniform series compound factor. The term on the right-hand side on Equation 5.40 is known as the present value interest factor of an annuity (PVIFA), or the uniform series present value factor.

Two further formulae can be derived from these two by rearranging to make A the subject of each of the formula. The rearrangement of Equation 5.39 yields the following equation:

$$A = FV/\left(\frac{(1+i)^n - 1}{i}\right) = FV\frac{i}{(1+i)^n - 1} \qquad (5.41)$$

The rearrangement of Equation 5.40 yields the following expression:

$$A = PV/\left(\frac{1-(1+i)^{-n}}{i}\right) = PV\frac{i(1+i)^n}{(1+i)^n - 1} \qquad (5.42)$$

The factor $\frac{i}{(1+i)^n - 1}$ on the right-hand side of Equation 5.41 is known as the sinking fund deposit factor, while the factor $\frac{i(1+i)^n}{(1+i)^n - 1}$ on the right-hand side of Equation 5.42 is known as the capital-recovery factor. Of course, they are merely the inverse of the FVIFA and PVIFA, respectively.

5.9.3 A Notation for Interest Rate Factors

The system of notation that is introduced now is slightly more complicated that what has been used up previously, but has the advantage of succinctly clarifying the relationships between the interest rate formulae. There are five variables in the compound interest rate equations: PV, FV, A, i, and n. The notation names the factor that relates the unknown variable to the known variables. For example, consider Equation 5.42. The unknown is A, which must be calculated from PV, i, and n. The notation for the capital recovery factor is $(A|PV, i, n)$, which is read as the factor which calculates A given PV, i, and n. Similarly, the sinking funds factor in Equation 5.41 is given by $(A|FV, i, n)$. Using this notation for the factors, Equations 5.41 and 5.42 can be re-written as the following:

$$A = FV(A|FV, i, n) \qquad (5.43)$$

and

$$A = PV(A|PV, i, n) \qquad (5.44)$$

Similarly, the single payment factors are given by:

$$(FV|PV, i, n) = (1+i)^n \qquad (5.45)$$
$$(PV|FV, i, n) = 1/(1+i)^n \qquad (5.46)$$

The interest rate factors are summarised in Table 5.3.

Like most topics in engineering studies, it is important to understand the logic behind the derivation of the formula. There are two considerations that should be borne in mind: (i) the calculations represented in the formulae can be easily set up on a spreadsheet, and set up in a more general manner than the summation formulae allow; and (ii) there is no consistent or standard notation, so that communication with other professionals may not be clear. For example, those with an insurance or investment background may refer to a formula as an annuity factor, those with a bond finance background may refer to them as discount factors (for zero coupon and coupon bonds), while an engineer may refer to it as a uniform series interest rate factor. In practice, the most transparent method is to construct a spreadsheet that can be easily audited by colleagues who have probably long since forgotten the formula.

The application of the interest rate formula is illustrated in the following four examples.

Example 5.15: Saving for an anticipated expense.

The Chief Executive Officer (CEO) of a company anticipates that a major item of equipment will require replacement in three years time. The CEO also expects the company to make sufficient profit to deposit $1,100,000 each year for the next three years in order to pay for this item. The company is offered an interest rate of 6% pa for this deposit. If the equipment currently costs $2,900,000, and an analysis of the equipment price index suggests that the price will increase by 7% pa, will the company have enough on deposit to purchase the item? If not, how much should the CEO deposit in order to afford the replacement?

Solution:

The deposit profile is that of an annuity, so the uniform series formula is applicable. The future value of the deposit is calculated using formula 3 in Table 5.3:

$$FV = 1,100,000(FV|1,100,000; 0.06; 3) = 1,100,000[(1+0.06)^3 - 1]/0.06 = \$3,501,960$$

The future value of the equipment is a single payment and is calculated using formula 1 in Table 5.3. The replacement price of the equipment can be calculated as follows:

$$FV = 2,900,000(1+0.07)^3 = \$3,552,624$$

Table 5.3 Compound interest formulae

Formula	Find	Given	Factor notation	Factor formula	Payment profile
1	FV	PV	$(FV|PV,i,n)$	$(1+i)^n$	Single
2	PV	FV	$(PV|FV,i,n)$	$1/(1+i)^n$	Single
3	FV	A	$(FV|A,i,n)$	$[(1+i)^n - 1]/i$	Uniform series, *i.e.*, annuity
4	PV	A	$(PV|A,i,n)$	$[(1+i)^n - 1]/[i(1+i)^n]$	Uniform series, *i.e.*, annuity
5	A	FV	$(A|FV,i,n)$	$i/[(1+i)^n - 1]$	Uniform series, *i.e.*, annuity
6	A	PV	$(A|PV,i,n)$	$i(1+i)^n/[(1+i)^n - 1]$	Uniform series, *i.e.*, annuity

The savings are insufficient. The deposit amount required each year could be calculated using formula 5 in Table 5.3. This calculation is given as follows:

$$A = F(A|FV, i, n) = 3,555,624 \frac{0.06}{[(1+0.06)^3 - 1]} = \$1,115,914$$

The company should deposit $1,115,914 each year to meet this requirement.

Example 5.16: Annuity formula.

If an investor deposits $250 pa for 5 years at 12% pa, what is the present value and future value of the investment?

Solution:

Because these are equal instalments, this is an annuity. The present and future values are calculated using formulae 3 and 4 from Table 5.3:

$$FV = A(FV|A, i, n) = 250[(1+0.12)^5 - 1]/0.12 = \$1,588$$
$$PV = A(PV|A, i, n) = 250[(1+0.12)^5 - 1]/[0.12(1+0.12)^5] = 250(3.60477) = \$901$$

Of course, the present value could have been calculated directly from the *FV* using the five-year discount factor (formula 2 in Table 5.3).

Example 5.17: Sales contract.

A salesman has secured a three-year $10 million contract for the supply of broadband connectivity. The details provide for an immediate purchase of $1 million payable in advance, $2 million at the beginning of the second year, $3 million at the beginning of the third year, and $4 million at the beginning of the last year. Assuming a 10% interest rate, is this package worth more than $10 million? What is it worth today?

Solution:

This is not an annuity, since the cash flows are not equal and they start from the present. However, the cash flow profile increases linearly. Formulae equivalent to those in Table 5.3 for a series with linear or arithmetic growth can be derived, or the contributions from the individual cash flows can be determined. It is the latter approach that will be adopted here.

$$FV_0 = 1(1+0.1)^3 = 1.331 \, million$$
$$FV_1 = 2(1+0.1)^2 = 2.420 \, million$$
$$FV_2 = 3(1+0.1)^1 = 3.300 \, million$$
$$FV_3 = 4(1+0.1)^0 = 4.000 \, million$$

$$Total = \$11.051 \, million$$

$$PV = (11.051)\left(1/(1+0.12)^3\right) = \$7.865 \, million$$

The present value of the salesman's contract is $7.86 million.

Example 5.18: Bond.

A bond is a loan that works like this: an investor buys the bond that has a face amount *F* for an amount *P*. The bond pays coupons at a fixed percentage of *F* at regular intervals for a fixed number of years and together with the last coupon pays the initial sum back.

Determine the future value of the bond if the face value is $100, the coupon rate is 6%, and the term is 10 years. Suppose the investor can invest the coupon payments in a bank account at 5%.

Solution:

The coupons are fixed cash flows occurring at the end of the year, so an annuity formula can be used to calculate their future value. The coupon amount is the face value multiplied by the coupon rate, that is, 100(0.06). The future value of the coupons is therefore given by:

$$FV = A\,(FV|A, i, n) = 6[(1 + 0.05)^{10} - 1]/0.05 = \$75.47$$

The future value of the repayment of the principal is simply $100. This means that the total future value of the bond is $175.47.

Notice the structure of this problem: the investor receives payments (the coupon payments), which, unlike a bank account, cannot be reinvested in the same security. The investor receives the coupon, and must invest it in other investments, receiving, possibly, a different return than that which the bond provided (6%). If the bond worked like a bank account, where the interest could be reinvested at the same rate as the principal, the future value of the bond would be $179.08. This example emphasises that, since the process of compounding means earning interest on interest, if the interest portion earns at a different rate of return than the principal, the value of the principal and the interest should be calculated separately at their separate rates.

5.10 Case Study: Zero Coupon and Coupon Bonds

An electrical utility offered securities to the public on July 1, 2002. Under the terms of the deal, the electrical utility promised to pay the owner of the security $1,000,000 on July 1, 2012. The utility was asking investors to pay $233,757 for each of these securities. Such a security, in which the investor pays now to receive a sum in the future, is one of the simplest possible investments.

(i)   What is the nature of the relationship between the utility and the investor? Why would the investor consider buying this security?

(ii)   At what rate was the utility selling the security?

(iii)   The security is actively traded on the Bond Exchange. Was the price in 2007 higher or lower than the original $233,757?

(iv)   What price would you expect the security to be trading at in 2007?

(v)   If the price of the security on the bond exchange on July 1, 2005 was $337,282.00, and an investor purchased it and held it until maturity, what rate of return would she earn? Why is this rate different from the rate calculated in part (i)?

(vi)   If the security paid an annual amount, called a coupon, to investors that was worth 5% of the face value of the security, at what price would the issuer offer the security if the issuer wanted the interest rate (or yield) to remain constant? What is the advantage to the seller of the coupon bond of paying coupons? And to the buyer?

Solution:
(i) The investor is loaning money to the utility for ten years. The investor would be willing to do this if she reached a favourable assessment of the security based on the risks and returns that she would receive from other investments in which she was able to invest.
(ii) The cash flow profile for this bond is given in Figure 5.17.

The future value and the present value are both known with certainty. The compound interest formula is given as follows:

$$FV = PV(1+i)^n$$

All the values are known, except for the interest rate. The substitution of the values into the formula yields the following expression for the interest rate:

$$1,000,000 = 233,757(1+i)^{10}$$

The value of i is obtained by taking the logarithm of both sides:

$$\ln(1,000,000) = \ln(233,757) + 10\ln(1+i)$$
$$1+i = \exp((\ln(1000000) - \ln(233757))/10) = \exp(0.1453) = 1.1564$$
$$i = 15.64\%$$

The investment in this security yields a return of 15.64% (nominal annual compounded annually). The interest rate for a bond is commonly known as the bond's yield. If the investor was considering buying this security or putting her money in a fixed deposit account with a bank for ten years, the interest on the bank account would have to earn more than 15.64% for her to prefer the bank account.
(iii) Because no interest payments or coupons are paid during the term of the investment, the price of the security must reflect the interest that this security earns for the investor. As time passes between the offer date and the maturity date, the bond

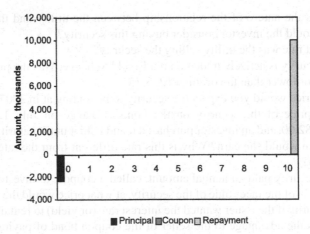

Figure 5.17 Cash flow profile for the ten-year, zero-coupon bond with a face value of $1,000,000 and a purchase price of $233,757

collects interest, which it does not pay out. It is effectively like a bank account, where interest is earned and paid into the account of the account-holder. Therefore, its price should be higher, reflecting the earned but unpaid interest.

(iv) The bond is five years old in 2007. Therefore, the value can be calculated for the issue price of $233,757.00, the age of 5 years, and the expected yield of 15.64%.

$$FV = PV(1+i)^n = 233,757(1+0.1564)^5 = \$483,398$$

The bond is expected to have a value of $483,398 after five years.

(v) This is calculated in a manner similar to that in part (ii).

$$1,000,000 = 337,282(1+i)^7 \rightarrow i = 16.79\%$$

The interest rate, i, is referred to as the yield of the bond. The investor would receive a yield of 16.79%. The yield is different from the issue rate calculated in part (ii) because the economic conditions have changed since the bond was issued. If the economic conditions were exactly the same, the purchase price would be $361,612. However, interest rates have increased in the economy as a whole, and the seller of the security must lower his price in order to attract a buyer. In fact, this is the mechanism by which the market determines interest rate.

From an examination of the prices of bonds, one can determine the market expectations for future interest rates by performing similar calculations with different maturities. It should be noted that in some markets bonds are sold on price, while in others they are sold on yield.

(vi) The face value of the bond is $1,000,000. If the bond pays a 5% coupon, this means every year the bond pays an amount equal to $0.05(1,000,000) = \$50,000$ to the owner of the bond. The cash flow profile for this bond is shown in Figure 5.18.

The present value of the bond is given by Equation 5.36, which is given as follows:

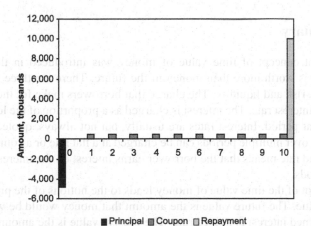

Figure 5.18 Cash-flow profile for a ten-year bond with a 5% coupon

$$PV = \sum_{t=0}^{n} \frac{CF_t}{(1+i)^t}$$

The cash flows are all known except the first one. The interest rate is 15.64% as in part (ii). The present value of the bond is zero. The present value of the cash flows can be divided into their component parts: the purchase price at time zero, the coupons and the face value at year 10. The coupons are all equal and they have the distribution of a ten-year annuity. The present value of an annuity can be used to determine their value. Therefore, the present values for each of the three components are as follows:

1) $PV(Purchase) = CF_0$

2) $PV(Coupons) = Coupon[(1+i)^n - 1]/[i(1+i)^n]$
$$= 50,000 \frac{(1+0.1564)^{10} - 1}{0.1564(1+0.1564)^{10}} = \$244,936.00$$

3) $PV(Bullet) = \dfrac{1,000,000}{(1+0.1564)^{10}} = \$233,840.19$

The substitution of these values into Equation 5.36 yield the following expression for the purchase price:

$$0 = CF_0 + 244,936 + 233,840.19$$
$$CF_0 = -\$478,776.20$$

This means that the utility should sell the coupon bond for $478,776.20. The advantage to the seller is that they will raise more money immediately from the sale. The advantage to the buyer is that they receive some of the money, in the form of coupons, earlier than ten years.

5.11 Summary

The important concept of time value of money was introduced in this chapter. Money today is worth more than money in the future. There are three reasons for this: inflation, risk and liquidity. The charge that borrowers make for the use of the money is the interest rate. The interest is charged as a proportion of the loan amount for a particular period. Interest rates are usually, but not always, quoted as annual rates. Interest over multiple periods can be charged at a flat rate or a compound rate. The compound rate means that the borrower earns interest on the interest earned in previous periods.

The concept of the time value of money leads to the notions of the present value and future value. The future value is the amount that money would be worth in the future if it earned interest at a fixed rate. The present value is the amount of money that must earn interest at a fixed rate in order to have a particular value at a date in the future. The present value and the future value formula allow the conversion of the value of money at different times.

Different forms of interest rate investments were discussed. These are the annuity, the annuity due, the arithmetic growth annuity and the geometric growth annuity. Interest rate formulae for each of these different investments were derived. A case study involving a bond, which is a method borrowing large amounts from many different investors, was discussed.

5.12 Looking Ahead

In this chapter, the concept of the time value of money was explored. This examination led to the notions of compounding, discounting, present value and future value. In the next chapter, these concepts are used to derive criteria for assessing the economic suitability of an engineering project. These criteria are central to the business decisions that face engineers, scientists and managers in an organization. The criteria are of two types: those that include the time value of money and those that do not. Examples of the criteria that do not include the time value of money are the payback period and the return on investment. There are six main criteria that include the time value of money. The advantages and disadvantages of each of these criteria are discussed.

5.13 Review Questions

1.   Why does money loose value with time?
2.   What are the determinants of the time value of money?
3.   What is the charge for the use of money?
4.   Why would anyone pay interest?
5.   What is the difference between flat interest and compound interest?
6.   Why is the term simple interest confusing?
7.   What is the nominal interest rate?
8.   How does the nominal interest rate differ from the period rate and the effective rate?
9.   What is the difference between an annuity and a perpetuity?
10.   How is inflation measured?
11.   How is the inflation rate related to the interest rate?
12.   What does the term "present value" mean?
13.   Why does the timing of cash flows affect their value?
14.   What is an annuity due and is it different from a geometric growth annuity?
15.   What is CAGR? Is it the same as nominal or effective interest rate?
16.   What is the formula for the effective rate of an interest rate that is quoted as nominal annual compounded quarterly?

5.14 Exercises

1. What is the present value of a payment of $1,000 in five years time if the interest rate is 5%?

2. What is the value in 10 years of $1,000 if the amount is invested at an interest rate of 10%?

3. If an amount of $250 were invested at a flat or simple interest rate of 10%, what would it be worth after 3.5 years?

4. An investment earns interest at a rate of 6%. If the compounding period is monthly, what is the effective interest rate?

5. If an amount is deposited in an interest-bearing account at a fixed rate of 8%, how long will it take for the amount to double in value? How long will it take to triple in value?

6. If an investor pays $250 into an ordinary annuity for four years, what is the present value of the investment if the interest rate is 5%? What is the value of the investment at the end of the annuity period?

7. If the present value of a five-year ordinary annuity is $1,000, and the annuity payments are $300, what is the interest rate?

8. What will an investor get in five years if she deposits $5 each year at 8% pa compounded annually?

9. A rich uncle has set up a trust fund that will pay you $100,000 in five years. If the interest paid by the trust fund is 11%, how much has your uncle deposited today?

10. You have been offered an investment that doubles your money in ten years. What interest rate have you been offered?

11. You have been offered an investment that will pay you 9% per year. If you invest $15,000, how long until you have $45,000?

12. If an investor deposits $250 pa for 5 years at 12% pa, what is the value of the investment?

13. Which do you prefer: $4,500 cash now, or $1,200 each year for four years? Assume a discount rate of 10% pa.

14. You plan to make a series of deposits into an interest-bearing account. You plan to deposit $1,000 today, $2,000 in two years and $8,000 in five years. If you withdraw $3,000 in three years, and $5,000 in seven years, how much will you have after eight years if the interest rate is 9%? What is the value today?

15. The going rate on a credit card is 9% APR. What is the effective annual rate if the balance must be paid monthly?

16. Show that the compound interest formula for continuous compounding is given by the following formula:

$$FV = PVe^{iT}$$

where i is the interest rate, and T is the time.

17. Show that the present value of an ordinary annuity is given by:

$$PVA = A\left(\frac{1-(1+i)^{-n}}{i}\right)$$

where i is the interest rate, and n is the number of years.

18. Derive the formula for the present value of an *annuity due*.

19. A company has taken out a loan for an amount of $500,000. The loan is re-payable in four years. If the amount that must be repaid is $750,000, what is the interest rate?

Chapter 6
Evaluation Criteria for Investment Decisions

6.1 Creating Value for the Investor

Value is created for an investor if the investor earns more than the investment costs. More specifically, value is created for an investor if the investor earns more than he or she could on other investments that he or she could make. This is because there is an *opportunity cost* to investing in the project. The opportunity cost is the "cost" that is incurred if the investor places cash into a project, so that the cash is not available for investment in other, possibly more lucrative, investments that have the same amount of risk attached to them. Thus, in determining the criteria for the evaluation of the investment, the analytical techniques chosen must select for investments that not only create value, but also create it at a greater rate than the other investments with similar risk.

A large number of criteria have been developed that aim to determine and quantify if value is created for the investor. Of these criteria, the discussion in this chapter will be restricted to the most common criteria, that is, the payback period, return on investment, equivalent annual charge, net present value, profitability index, internal rate of return, the benefit-cost ratio and the modified internal rate of return. The first two methods do not account for the time value of money, which was discussed in the previous chapter, whereas the last six do. Those criteria that include the time value of money are called *discounted cash flow* techniques. Each of these criteria has value in decision-making. They assist in summarising the value of the investment, and often companies do not chose to invest on the basis of one criterion, but will evaluate the project using two or more criteria.

In this chapter, these criteria will be presented, the advantages and disadvantages discussed, and their application under differing conditions will be reviewed. Thus, the criteria for decision-making on the economic attractiveness of a project are established in this chapter. The application of the criteria to a variety of applications is discussed in the next chapter.

6.2 Non-discounted Cash Flow Techniques

6.2.1 Payback Period

The payback period determines the point in the project at which the investor gets her investment back. In other words, the payback period is the period at which the cash flow generated by the investment is equal to the cash invested in the project. All positive cash flows after that are excess earnings for the investor. The payback period is the oldest of the decision-making criteria. It is the easiest to compute, and has intuitive value: the longer the investor has to wait for the project to return the initial investment, the less lucrative the project.

The calculation of the payback period is illustrated in the following example.

Example 6.1: Payback period.

Suppose an investment had the cash flow profile shown in Table 6.1. Determine the payback period. (The negative sign on a cash flow means that it is an out-flow, whereas positive values are in-flows. This means that the investor has invested $15,000 and receives the amounts following this $15,000 in Table 6.1.)

Solution:

The payback period is determined by calculating the time it would take until the sum of the returns from the investment is equal to the cost of the investment. This can be determined from the cumulative cash flow. The payback period is reached when the cumulative cash flow changes sign from negative to positive. The cumulative cash flow, also called net cash flow, is shown in Table 6.2 below.

The cumulative cash flow changes sign from negative to positive during the third year. This means that the payback period is between the end of the second year and the end of the third

Table 6.1 Cash flow profile for an investment

Year	Cash flow
0	−15,000
1	7,000
2	6,000
3	3,000
4	2,000
5	1,000

Table 6.2 Cumulative cash flow for the investment

Year	Cash flow	Cumulative cash flow
0	−15,000	−15,000
1	7,000	−8,000
2	6,000	−2,000
3	3,000	1,000
4	2,000	3,000
5	1,000	4,000

year. To calculate the date more accurately, linear interpolation between the points at 2 and 3 years can be used:

Since $y = mx + c$, the root is given by the following expression:

$$0 = (1000 - (-2000))/(3 - 2)(Payback - 2) - 2000$$
$$Payback = 2.66 \, \text{years}$$

This means that the investor will wait for 2.66 years before the investment returns all of the initial investment.

The method of payback period does not provide a decision point for the investor. In other words, the decision-maker or investor does not know whether a payback period of 2.66 years in the previous example represents a good investment or not. If the project is longer than the payback period, the investment should provide a return to the investor. If the payback period is longer than the life of the project, the investment will not return even the original investment to the investor.

The investor can specify acceptable payback periods for different classes of investments based on their perceived risk. Higher risk projects or activities will have a lower payback period than lower risk ones.

In evaluating alternative investment options, the one with the shortest payback is favoured. This is because the amount invested is returned to the investor earlier for the project with the shortest payback period. Consider the following example.

Example 6.2: Alternative projects.

The cash flow profiles from two alternative projects are given in the Table 6.3 below. Since the projects are mutually exclusive, one of them must be chosen. Which is the better choice?
Solution:

Construct a table containing the cumulative cash flows for each project and determine the point at which the project cash flow changes from negative to positive. This table is given as Table 6.4 below, and the Figure 6.1 illustrates the values.

Table 6.3 Cash flow profiles for two projects

Year	Project A	Project B
0	-100	-100
1	30	40
2	40	40
3	50	20
4	60	20

Table 6.4 Cumulative cash flow for two projects

Year	Project A	Project B
0	-100	-100
1	-70	-60
2	-30	-20
3	20	0
4	80	20

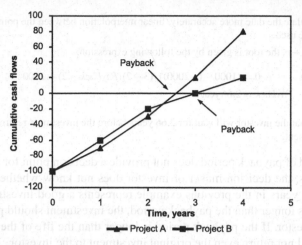

Figure 6.1 The payback period for the two projects, showing that Project A has a shorter payback period than Project B

The payback period for Project A is less than that for Project B. Therefore, project A is preferred.

As discussed in Chapter 2, the payback method is more a measure of liquidity than it is a method for measuring how much value is created for the investor. The faster the project returns its investment to the investor, the earlier the investor can reinvest that amount in another investment. The payback period has a number of advantages:

(i)   It is easy to calculate and it is easy to interpret.

(ii)   It is a measure of project liquidity, so in situations in which liquidity is important, the payback period has an important role. For example, the company may find it difficult or deem it too expensive to raise external funds. It may then require its investments to have a short payback period, so that cash is not tied to one investment for a long period, making it difficult to invest in other opportunities. Other examples are the following: The market is subject to a high degree of technological change so that the product may become obsolete (mobile phones, computers, software), or the product is model-based (cars, fashion items). In both these cases, the company must recover its investment within the model period. If the investment is subject to forces with which the investor is uncomfortable, such as an offshore investment, the investor may wish to limit her risk by getting the investment amount back as soon as possible.

(iii) It may be difficult to forecast the cash flow beyond three to five years into the future with any confidence. However, many managers, particularly those in production environments, believe that if the project can return its investment within a short period, say less than two years, the project will make a positive contribution to the company.

A drawback is that the payback period criterion does not specifically provide a value for the decision point. In other words, if the payback period is 18 months, it is not clear whether this is a good investment or a poor one. The payback period does not indicate whether the investor should make the investment or not. Another disadvantage is that it does not account for the time value of money.

6.2.2 Return on Investment

Return on investment is the name for a group of similar calculations that express a ratio of profit to a measure of value. It is a measure of the profitability of the investment. Usually, the return on investment is calculated from figures provided on the accounting statements. As a result, the return on investment is sometimes called the *accounting return*. The calculation of the return on investment is shown in the following example.

Example 6.3: Return on investment.

Consider an investment with the accounting projections given in Table 6.5. The profit is obtained from the income statement and the assets from the balance sheet. The assets have been depreciated on a straight-line basis over the life of the project, hence their diminishing value. Calculate the return on investment.

Solution:

One formulation of the return on investment is the annual profit divided by the original investment:

$$Return\ on\ investment\ = 30,000/200,000 = 15\%$$

Another formulation of the return on investment is the average profit divided by the average value of the assets on the balance sheet. This definition is sometimes called the average accounting return.

$$Return\ on\ investment\ = 27,500/100,000 = 27.5\%$$

Another definition of the return on investment could be as follows:

$$Return\ on\ investment = (Total\ income - Original\ investment)/(Average\ book\ value)$$
$$= (110,000 - 200,000)/(100,000)$$
$$= -90\%$$

Table 6.5 Profit and asset at different times for an investment

Year	Assets	Profit
0	200,000	
1	150,000	20,000
2	100,000	30,000
3	50,000	30,000
4	0	30,000

The return on investment is calculated using the accounting figures. As was discussed in Chapter 3, the assets are represented at their book values, which are unrelated to their market values or their earning potentials. The income or profits include depreciation and as such are unrelated to the cash earned by the employment of the assets. In addition, the definition of return on investment is not consistent or agreed upon, as shown in Example 6.3, making it difficult to interpret the results with confidence.

The return on investment is a relative measure, that is, it is not dependent on the size of the investment. This is an advantage, since it allows projects of different sizes to be compared. In addition, investors or management must specify an acceptable value for return on investment since the method does not specify what a good return should be. However, the return on investment can be compared with the interest earned in a bank account, which is a very useful comparison.

The return on investment does not account for the time value of money. The discounted cash flow techniques, which account for the time value of money, are discussed in the next section.

6.3 Discounted Cash Flow Techniques

6.3.1 Net Present Value

If creating value for the investors or shareholders in a company is the goal of management, then the returns for each project undertaken by the company must exceed the cost of the project. The cost of the project is the amount that the company invests in the project. Thus, the value of a project is the difference between the cash flows generated by the project and the cash flows consumed by the project. The cash flows generated by the project will occur at some point in the future. For example, a project proposal to produce a new line of products has the cash flows given in the Table 6.6.

The sum of the cash in-flows from the project is 282. The total value that this project will add to the company is equal to the sum of all the in-flow and out-flows, which is equal to $282 - 200 = 82$. On the face of it, the project would create value for the company because the in-flows exceed the out-flows. However, in the previous chapter, it was demonstrated that the timing of the cash flows affects their value.

Table 6.6 Cash flow projections for a proposal for a new product line

Year	0	1	2	3	4	5	6	7	8
Initial cost	−200								
Net cash in-flow		35	35	35	35	35	35	35	35
Cash flow from salvage									2
Net cash flow	−200	35	35	35	35	35	35	35	37

The value of the future cash flows is worth less today due to the time value of money. In order to evaluate the contribution of the cash flows that are anticipated in the future, they must all be compared at a common date. Since the investment is assumed to occur now, the most logical common date is the present. The cash flows are discounted to the present using Equation 5.32, which is given by the following expression:

$$PV = \frac{CF_t}{(1+k)^t} \qquad (6.1)$$

where CF_t is the cash flow anticipated at year t and k is the discount rate. In Chapter 5, the amounts were discounted at the interest rate. In this and following chapters, the calculation of the present value is at the discount rate. The investor or the company specifies the value of the discount rate. It is not equal to the various commercial interest rates that are quoted.

The present value for each of the expected cash flows is discounted to the present. The calculation of these values is shown in Table 6.7 if each cash flow is discounted using a discount rate of 15%. The discount factor is the value of $1/(1+k)^t$ in Equation 6.1.

The value that is created for the company is the sum of all these present values, which, in this case, is -42.29. Since the value is negative, it costs more to invest in this opportunity than it generates. Because the earnings in the future are worth less than their equivalent value today due to the time value of money, this investment does not generate wealth, as the simple analysis presented earlier suggested; instead, wealth is destroyed.

The net present value (NPV) is the sum of all the cash flows discounted to the present using the time value of money. If the net present value is greater than zero, it is expected that value will be created for the investor. If it is less than zero, it is expected that value will be destroyed for the investor. The net present value is a method of quantifying the value that is to be created for the investor.

The net present value can be formulated as an equation in the following manner:

$$NPV = \sum_{t=0}^{n} \frac{CF_t}{(1+k)^t} \qquad (6.2)$$

where CF_t is the cash flow at year t, n is the life of the investment of the engineering project and k is the discount rate.

Thus, the net present value of the project is the present value of the anticipated cash flows generated by the project less the present value of the cash flows consumed by the project.

Table 6.7 Calculation of the present values of the cash flows for the project proposal

Year	0	1	2	3	4	5	6	7	8
Net cash flow	−200	35	35	35	35	35	35	35	37
Discount factor	1.000	0.870	0.756	0.658	0.572	0.497	0.432	0.376	0.327
Present value	−200.00	30.43	26.47	23.01	20.01	17.40	15.13	13.16	12.10

Example 6.4: Net present value of Santa Clara hydroelectric power scheme (HEPS).

In Chapter 1, the cash flows for the Santa Clara HEPS were presented. Determine the net present value for this opportunity if the discount rate is 15%. The cash flows for the first eight years are given in Table 6.8 below. After year eight, the cash flows are identical to those in the eighth year.

Solution:

The free cash flow is the amount of cash that is consumed or generated by the project. *NPV* is calculated from the free cash flow. The calculation of the present values of the free cash flows for the first eight years is shown in the last line in Table 6.9 below.

Since the net present value is the sum of the present values of all the future cash flows, the summation of the last line in the table gives the *NPV*. Note that the life of this project is 35 years, so the table needs to be extended for the entire period. If this is done, and the present values summed, the result is that the *NPV* is equal to $311 million.

Since the *NPV* is positive, the project will create value and should be acceptable.

The *NPV* is an estimation of the value added to the company if the project is accepted and implemented. The *NPV* is an absolute value, that is, it is dependent on the size of the contribution of the project to the wealth of the company. If there are choices between projects, the set of projects that maximize the total *NPV* will maximize the total value of the company.

The net present value is additive. If the company invests in two projects, their total *NPV* should be the sum of the two individual net present values.

In addition, the net present value is a property of the project, not the owner of the project. This means that the discount rate must be a property of the project, rather than the company.

To summarise, the advantages of the net present value are the following:

(i)   It is an absolute measure of value. The *NPV* of a project is the value that the project is expected to add to the value of the company.

Table 6.8 Cash flow for the Santa Clara HEPS

Year	1	2	3	4	5	6	7	8
Revenue	0.0	0.0	0.0	0.0	109.6	156.3	161.2	161.2
Operating expenses	0.0	0.0	0.0	0.0	4.4	4.4	4.4	4.4
Royalty	0.0	0.0	0.0	0.0	2.2	3.1	3.2	3.2
Cash flow	0.0	0.0	0.0	0.0	103.0	148.8	153.6	153.6
Capital	10.0	74.9	77.8	114.2	140.6	60.3	3.8	0.0
Free cash flow	−10.0	−74.9	−77.8	−114.2	−37.6	88.6	149.8	153.6

Table 6.9 Calculation of the present values of the cash flows

Year	0	1	2	3	4	5	6	7	8
Free cash flow	−10.0	−74.9	−77.8	−114.2	−37.6	88.6	149.8	153.6	153.6
Discount factor	1.000	0.869	0.756	0.657	0.571	0.497	0.432	0.375	0.326
Present value	−10.0	−65.1	−58.8	−75.1	-21.5	44.0	64.8	57.7	50.2

(ii)   It accounts for the time value of money.
(iii)   It is straightforward to interpret: it is the net value of the project in today's
        money and represents the amount of value that the project will add to the
        value of the company.
(iv)   *NPV* is additive, so that the value of two projects is the sum of the *NPV*s for
        each of the projects.
(v)   *NPV* is a property of the project. It should not be a function of the owner. This
        means that the *NPV* of different projects, owned by different companies, can
        be compared.
(vi)   The use of *NPV* as an assessment criterion has the advantage of a built in
        decision-point. If the *NPV* of a project is less than zero, the value of the
        resources that the project uses is greater than the value of the benefits the
        project creates.

The main difficulty with net present value is the choice of discount rate. The
values of the cash flow, CF_t, and the value of n in Equation 6.1 are determined
by the project. However, the value of the parameter, k, the discount factor, needs
to be set by management. The method for choosing the value of the discount rate
that is the most popular is by comparison with the company's cost of raising capital
on the debt and equity markets. This is called the weighted average cost of capital
(WACC). The determination of WACC will be discussed in more detail in Chapters 8 and 12. Unlike the payback period, for which management also has to set an
acceptable value for the acceptance or rejection of proposals, there is a method of
benchmarking the discount rate with more objective values traded in the market.

The sensitivity of the *NPV* to the discount rate can be determined by changing
the discount rate. The *NPV* as a function of the discount rate for a particular set of
cash flows is shown in Figure 6.2.

Just as the present values and the future values are sensitive to the discount rate,
so is the *NPV*. The results illustrated in Figure 6.2 show that the *NPV* is exponen-

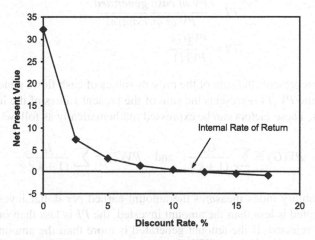

Figure 6.2 The effect of different discount rates on the *NPV*

tially dependent on the discount rate. The shape of the *NPV* curve with respect to the discount rate is dependent on the cash flow profile. In normal investments, the profile will be similar to that shown in Figure 6.2.

The point on the curve where the discount rate is zero represents maximum project value, which is the value if the time value of money is zero. The discount rate on the curve where the *NPV* changes from positive to negative represents the internal rate of return (*IRR*), another decision criterion that will be discussed later in this chapter.

The discount rate that is used in the calculation of the *NPV* need not be constant over the lifetime of the project. For example, the expected inflation rate may change, justifying an adjustment in the discount rate. Similarly, the risk of the project may change over time, which may also justify a change in the discount rate.

If the discount rate is different each year, the *NPV* is modified by substituting the required discount rate for each year. This is expressed as follows:

$$NPV = \sum_{t=0}^{n} \frac{CF_t}{(1+k_t)^t} \qquad (6.3)$$

where k_t is the expected discount rate for each year t.

6.3.2 Profitability Index or Benefit-cost Ratio

The profitability index (*PI*), which is the same as the benefit-cost ratio, is the ratio of the present value to the cash flows generated to the present value of the cash flows consumed. It is a measure of the profitability per dollar invested. The profitability index can be determined from the following equation:

$$PI = \frac{PV \ of \ cash \ generated}{PV \ of \ investment} \qquad (6.4)$$

$$PI = \frac{PV(G)}{PV(I)} \qquad (6.5)$$

where $PV(G)$ represents the sum of the present values of cash flows generated each year (G_t), while $PV(I)$ represents the sum of the present values of the investments each year (I_t). These factors can be expressed mathematically as follows:

$$PV(G) = \sum_{t=0}^{n} \frac{G_t}{(1+k)^t} \quad and \quad PV(I) = \sum_{t=0}^{n} \frac{I_t}{(1+k)^t} \qquad (6.6)$$

The profitability index measures the amount earned per dollar invested. If the amount generated is less than the amount invested, the *PI* is less than one, and the investment is rejected. If the amount generated is more than the amount invested, the *PI* is greater than one, and the investment is recommended.

Table 6.10 Investment costs and present values of cash flows for six projects

Project	Initial investment	Present value of the cash flows generated
A	100,000	120,000
B	150,000	180,000
C	200,000	220,000
D	400,000	320,000
E	10,000	30,000
F	280,000	305,000

Table 6.11 Calculation of the *NPV* and the *PI*

Project	NPV	PI	Ranking based on NPV	Ranking based on PI
A	20,000	1.20	3	2
B	30,000	1.20	1	2
C	20,000	1.10	3	3
D	-80,000	0.80	4	5
E	20,000	3.00	3	1
F	25,000	1.09	2	4

The following example illustrates the calculation of the profitability index.

Example 6.5: Profitability index.

Rank the projects A through F based on *NPV* and *PI*. Table 6.10 provides details of the cash flow and investments for each project.

Solution:

The *NPV* is calculated from Table 6.10 as $PV(G)$ less I, that is, $PV(G) - I$, and the profitability index from $PV(G)/I$, where I represents the initial investment. The result of these calculations is given in Table 6.11.

The worst project is Project D. It has a negative *NPV* and a *PI* less than one, which means that it does not generate sufficient revenues to justify the investment. The project that ranks highest on the *PI* ranking does not have the highest *NPV*. It is the smallest project. The conflict with the *NPV* will be discussed later.

The profitability index can also be expressed in terms of the *NPV* and the present value of the investment:

$$PI = \frac{NPV}{PV(I)} + 1 \qquad (6.7)$$

The profitability index is a relative measure, which means that it cannot distinguish between the sizes of projects. This is in contrast to the *NPV*, which is an absolute measure of the value of a project.

6.3.3 Internal Rate of Return

The internal rate of return (*IRR*), which is also called the rate of return (ROR) and the discounted cash flow rate of return (DCFROR), is the value of the discount rate at which the net present value is zero. The definition of the internal rate of return was illustrated in Figure 6.2. It is the discount rate at which the net present value is zero. This can be expressed mathematically in the following equation:

$$0 = \sum_{t=0}^{n} \frac{CF_t}{(1+IRR)^t} \tag{6.8}$$

Equation 6.8 is non-linear and it cannot be solved directly. One method of obtaining the *IRR* from Equation 6.8 is by trial and error: guess a value for the discount rate, calculate the *NPV*, and repeat the procedure until a guess is made which satisfies Equation 6.8. Another way is to choose a range of discount rates, calculate the *NPV*, and then interpolate between the points on either side of the line for the point at which the *NPV* is equal to zero. A third way is to use a computer search algorithm, such as the secant method. A fourth method is to use built-in functions on calculators and spreadsheets, such as the "goalseek" function in MS Excel.

Two of the methods for determining the *IRR* are illustrated in the example that follows.

Example 6.6: Internal rate of return.

An investment costs $100, and returns two payments of $60 in the following two years. Calculate the *IRR*.

Solution:

The cash flow profile for this investment is shown in Figure 6.3. The formula for the *IRR* can be expanded to give the following expression for this investment:

$$0 = -100 + \frac{60}{(1+IRR)^1} + \frac{60}{(1+IRR)^2}$$

This is a special case of the *IRR* expression since it only contains the first two discounted terms. It can be solved using the quadratic formula. However, this is not done here. Rather a more general method, that of trial and error with linear interpolation, is used here. The first step is to choose a range of discount rates, and calculate the *NPV* at those discount rates. These results are shown in Table 6.12.

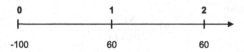

Figure 6.3 Cash flow profile for the investment considered in Example 6.6

Table 6.12 *NPV* as a function of discount rate

Discount rate	*NPV*
0%	20.0
5%	11.6
10%	4.1
15%	−2.5
20%	−8.3

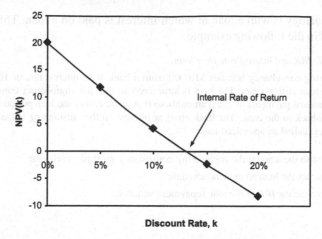

Discount Rate, k

Figure 6.4 The net present value as a function of the discount rate for Example 6.6

These results are illustrated in Figure 6.4.

The results plotted in the figure indicate that the *IRR* lies between 10 and 15%. By linear interpolation, this is equal to 13.15% as shown in the following calculation:

$$0 = (4.1 - (-2.5))/(10 - 15)(10 - IRR) + 4.1$$

$$\Rightarrow IRR = 13.15\%$$

By trial and error, or goalseek on MS Excel, *IRR* = 13.066%.

It is not possible to determine whether this is a good project or not simply from the value of the *IRR*. The *IRR* must be compared with that of other investments that can be made, or with a specified rate that the project must exceed. The specified *IRR* is sometimes called the minimum attractive rate of return (MARR) or hurdle rate.

The *IRR* is a relative measure, that is, it does not depend upon the size of the project or the investment. It is usually expressed as an annual percentage, in much the same way as an interest rate. Managers can compare the *IRR* for a project with current interest rates and with the rates of return of other projects. As a result, *IRR* has some attraction in the ranking of opportunities. Unfortunately, like the profitability index it can rank them in a fashion that is inconsistent with the *NPV*. The ranking of projects will be considered in Chapter 7.

The *IRR* can be interpreted as being analogous to the yield on a bond or the rate of interest on a loan. Bonds were discussed briefly in Example 5.18 and in Section 5.10. The purchaser pays for the bond, which is an investment. In return, the issuer of the bond pays amounts at regular intervals throughout the life of the bond, and at maturity, pays back the face value of the bond. This is similar to interest payments and the repayment of the principal on a loan. Thus, the yield of the bond is the interest rate at which the investor earns returns on her investment. A review of the Section 5.10 shows that the yield of a bond is calculated in the same manner as the internal rate of return.

Another analogy is with a loan in which interest is paid on a loan. This analogy is considered in the following example.

Example 6.7: IRR and interest rate on a loan.

An engineering consultancy borrows $10,000 from a bank at an interest rate of 10%. The term of the loan is five years. The loan is structured so that the engineering consultancy must make annual payments of equal amounts so that at the end of the loan period the loan is fully paid back to the bank. The bank charges interest on the outstanding balance. This type of loan is called an amortized loan.

(i)   Establish the amount the engineering consultancy must pay each year.

(ii)   Construct the loan repayment schedule.

(iii)   Determine the *IRR* of the loan repayment schedule.

Solution:

(i)   The present value of the loan at the start is $10,000. The repayments will be in equal payments. This means that they have the cash flow profile of an annuity. Reference to Table 5.3 confirms that the annuity formula that is required is Equation 6 in Table 5.3, and given as follows:

$$A = P(A|P, i, n) = P\left(i(1+i)^n / [(1+i)^n - 1]\right)$$

The substitution of the values for P, n and i yield the following calculation:

$$A = 10,000\left(0.1(1+0.1)^5 / [(1+0.1)^5 - 1]\right) = \$2,637.97$$

The annual payment required for the loan is $2,637.97.

(ii)   The loan repayment schedule is shown in Table 6.13 below. The opening balance is at the beginning of the period, whereas the closing balance is at the end of the period. The interest is charged on the opening balance at 10%, and these two figures added together constitute the end of period amount shown in the fourth column. The closing balance is the end of period amount less the annual payment.

The most confusing part of the loan schedule is clarifying the dates. The schedule of payments is shown in Figure 6.5 on page 177.

Interest accrues during the year on the outstanding balance. Each year the payment exceeds the interest, so that the balance is reduced each year, until finally the loan is paid back in full.

(iii)   The cash flow profile from the bank's point of view is given in Table 6.14. The negative amount is the loan, which is an out-flow for the bank, while the positive

Table 6.13 Loan repayment schedule

Year	Opening (begin)	Interest charged	Amount (end)	Payment (end)	Closing (end)
1	10,000.00	1,000.00	11,000.00	2,637.97	8,362.03
2	8,362.03	836.20	9,198.23	2,637.97	6,560.25
3	6,560.25	656.03	7,216.28	2,637.97	4,578.30
4	4,578.30	457.83	5,036.13	2,637.97	2,398.16
5	2,398.16	239.82	2,637.97	2,637.97	0.00

Table 6.14 Discount cash flows for the loan repayments

Year	Amount	Discount factor	DCF
0	−10,000	1.000	−10,000
1	2,637	0.909	2,398
2	2,637	0.826	2,180
3	2,637	0.751	1,982
4	2,637	0.683	1,802
5	2,637	0.621	1,638
		Total	0

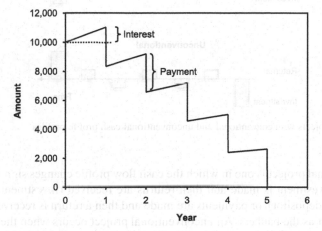

Figure 6.5 Loan repayments on a loan of $10,000

amounts are the in-flow from the engineering consultancy making their payments. The years are numbered as at the end of each year. The cash flows for the engineering consultancy are same, but with the opposite sign.

The third column is the discount factor, which is given by $1/(1+k)^n$, where k is the discount rate and n is the year. The fourth column is the discount factor multiplied by the cash flow in that year to give the discounted cash flow for the year. The figures in the table were calculated using a discount rate of 10%. The sum of the figures in column four is the net present value. Since the net present value is zero, the *IRR* is 10%.

Example 6.7 indicates that one way of viewing the *IRR* is as the interest on an amortized loan. It is a major advantage for the *IRR* that its interpretation is similar to the effective interest rate, which managers intuitively understand. Importantly, management or investors must specify a value for the *IRR* that is acceptable. That is, they must set a MARR or a hurdle rate that the *IRR* must exceed. For example, if management specified that the project *IRR* must be greater than 12%, then the project in the Example 6.6 is acceptable since the *IRR* is greater than 12%.

The cash flow profile has a possible influence on the calculation of the *IRR*. The cash flow profiles for projects are classified as conventional or unconventional.

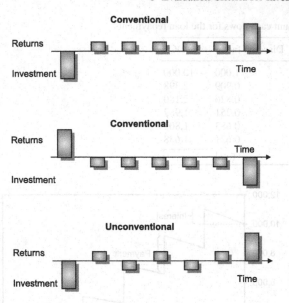

Figure 6.6 Projects with conventional and unconventional cash profile profiles

A conventional project is one in which the cash flow profile changes sign only once. Either an investment is made and then returns are received (investment, bank account as the depositor), or payments are made and then a return is received (lender, bank account as the banker). An unconventional project occurs when the cash flow profile changes sign more than once. The difference between these two types of cash flow profiles is illustrated in Figure 6.6.

From the theory of equations, it is known that the number of real roots of an equation is the number of times the equation changes sign. This means that if the cash flow profile changes sign more than once, it is possible that multiple values for the *IRR* exist. In this case, it is best to plot the net present value as a function of the discount rate. On the other hand, if the cash flow profile does not change sign at all, the *IRR* has no meaning.

A project with an unconventional cash flow profile is considered in Example 6.8.

Example 6.8: Multiple values of IRR.

Mining projects typically involve mine closures at the end of operations that might be costly. If the investment proposal for such an opportunity had a cash flow profile similar to that given in Table 6.15 below, determine the *IRR*.

Solution:

The *NPV* as a function of the discount rate is shown in the Figure 6.7. There are clearly two intersections with the x-axis, so that the definition of *IRR* does not provide a unique answer. Many calculator and spreadsheet methods will not identify multiple solutions for *IRR*. Under these conditions, the *IRR* method breaks down completely, and under these circumstances it is not possible to specify a single meaningful value for the *IRR*.

Table 6.15 Cash flow profile for a mine with closure costs

Year	Cash flow
0	-50
1	30
2	30
3	50
4	50
5	-120

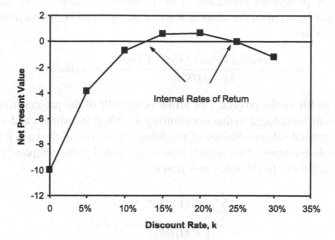

Figure 6.7 The *NPV* as a function of the discount rate, indicating that there are two possible values for the *IRR* for this cash flow profile

To summarize the discussion, the *IRR* has intuitive appeal because it can be interpreted in the same way as the interest rate on a loan or the yield on a bond. However, the *IRR* is more difficult to use than this suggests. There are four main problems with the *IRR*:

(i) One of the attractions of the *IRR* is in comparing projects, or ranking them. Unfortunately, the *IRR* may rank projects in an order that is in conflict with the other DCF techniques, particularly with the *NPV*. The goal of management is to maximize value for the shareholders, which the *NPV* does. The ranking of projects by any other technique should be consistent with the goal of management and hence the *NPV* criterion. The ranking of projects is discussed in more detail in Chapter 7.

(ii) The *IRR* can produce multiple values if the cash flow profile is unconventional. These multiple values occur rarely in practice but when they do, they mostly go unrecognised.

(iii) The effort to calculate the *IRR* is significantly more than the other DCF techniques.

(iv)   If the cash flows do not change sign, for example in cost saving projects, the *IRR* is undefined.

6.3.4 Modified Internal Rate of Return

The modified internal rate of return is the return earned by the project as if the cash flows from the project are reinvested at the company's discount rate. The *MIRR*, which has also been called the Growth Rate of Return, can be found from the following expression:

$$0 = \frac{Terminal\ Value\ of\ Cash\ Flows}{(1+MIRR)^n} - Investment \qquad (6.9)$$

where n is the life of the project. The *MIRR* is the *IRR* of the project in which the cash flow profile is reduced to that constituting an initial investment and a terminal value. The terminal value is the sum of the future values for all the cash flows other than the initial investment. This future value is calculated at the company's discount rate. Equation 6.9 can be rewritten as follows:

$$0 = \frac{\sum_{t=1}^{n} CF_t(1+k)^{n-t}}{(1+MIRR)^n} - CF_0 \qquad (6.10)$$

The calculation of the *MIRR* is explained in the following example.

Example 6.9: The modified internal rate of return.

A businessperson invests $120,000 and earns a return from this investment. The cash flows from the investment are given in Table 6.16. The earnings, for reasons such as the limitations on the expansion opportunities and scale of the investment, cannot be reinvested in the project. They are reinvested in other businesses that earn a return of 10%. Determine the *MIRR* for this project.

Solution:

Each of the cash flows from the project is reinvested until the end of the project in other investments that yield a return of 10%. The future values of these cash flows are shown in Table 6.17. For example, the future value of the cash flow at year 2 is calculated using

Table 6.16 Cash flow profile for an investment

Year	Cash flow
0	−400,000
1	150,000
2	120,000
3	100,000
4	50,000
5	60,000

Table 6.17 Calculation of future values of the cash flows

Year	Cash flow	Compound factor	Future value
1	150,000	1.4641	219,615
2	120,000	1.331	159,720
3	100,000	1.21	121,000
4	50,000	1.1	55,000
5	60,000	1	60,000
Terminal value			615,335

Equation 5.3 as follows:

$$FV = PV(1+i)^n = 30,000(1+0.1)^3 = 39,930$$

The terminal value is the sum of the earnings reinvested in other businesses at 10% until the end of the project.

The cash flows for the project are then reduced to two: that at the beginning of the project and the terminal value at the end of the project. The modified *IRR* is the *IRR* calculated from the terminal value and the original investment:

$$0 = \frac{615,335}{(1+MIRR)^5} - 400,000$$

The rearrangement of this equation gives the following calculation for the value of the *MIRR*:

$$MIRR = \sqrt[5]{\frac{615,335}{400,000}} - 1 = 8.996\%$$

The *MIRR* has a value of 8.996%.

The project yields a return less than of the company's other investments (10%), and, therefore, the proposal to invest in this project is not acceptable.

The *MIRR* is often advocated because of the *reinvestment assumption* that is erroneously presumed to be the cause of the conflict between the recommendations signalled by the *NPV* and the *IRR*. The conflict between the *IRR* and the *NPV* is discussed later, as is the reinvestment assumption.

6.3.5 Discounted Payback Period

The discounted payback period is the same as the payback period, except that the time value of money has been included. The payback period is the time taken for the cash in-flows from the project to be equal to the cash out-flows. At that point the investor has "got her money back." The discount payback period is the same concept, except that is incorporates the time value of money. This rectifies one of the main criticisms of the payback period as a decision criterion, that is, that it does not account for the time value of money.

Table 6.18 Cash flow profile for an investment

Year	Cash flow
0	−120,000
1	39,000
2	30,000
3	21,000
4	37,000
5	46,000

Table 6.19 Calculation of the discounted payback period

Year	Cash flow	Discount factor	Discounted cash flow	Cumulative cash flow	Cumulative discounted cash flow
0	−120,000	1.000	−120,000	−120,000	−120,000
1	51,000	0.909	46,364	−69,000	−73,636
2	47,000	0.826	38,843	−22,000	−34,793
3	25,000	0.751	18,783	3,000	−16,011
4	43,900	0.683	29,984	46,900	13,974
5	46,000	0.621	28,562	92,900	42,536

Example 6.10: Payback period and the discounted payback period.

A businessperson invests $120,000 and earns a return from this investment. The cash flows from the investment are given in Table 6.18. Determine the payback period and the discounted payback period if the discount rate is 10%.

Solution:

The calculation of the discounted cash flows and the cumulative discounted cash flows are shown in Table 6.19. The results from this table are shown in Figure 6.8. The discounted

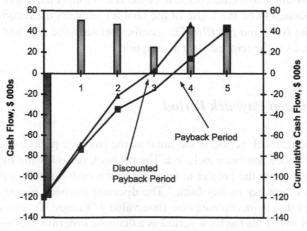

Figure 6.8 Discounted payback period

payback period is the time at which the cumulative discounted cash flow changes sign from negative to positive. This occurs between years 3 and 4. Figure 6.8 indicates that the discounted payback period is close to 3.5 years. This is in contrast to the payback period, which is about 4 years.

6.3.6 Benefit-Cost Ratio

The benefit-cost ratio is the ratio of the present value of all the benefits to the present value of all the costs. This can be expressed as the following formula:

$$B/C = \frac{PV(Benefits)}{PV(Costs)} \qquad (6.11)$$

The benefits are the cash in-flows to the project, and the costs are the cash out-flows to the project. The costs are presented as positive, which is contrary to the treatment of costs in the other methods.

The *benefit-cost ratio is the same as the profitability index* discussed earlier. The use of the name benefit-cost ratio is more acceptable in the public sector where the aim is to provide a benefit to the citizens rather than to make a profit.

6.3.7 Equivalent Annual Charge

The equivalent annual charge (*EAC*) distributes the present value of the project equally over the life of the project as if it were an annuity. The equivalent annual charge is used mainly to compare alternatives, such as in studies concerning the replacement of ageing equipment.

The *EAC* is given by the following expression:

$$EAC = PV(A|P, k, n) = PV\left(\frac{k(1+k)^n}{(1+k)^n - 1}\right) \qquad (6.12)$$

where *PV* is the present value or net present value of the project, *k* is the discount rate, and *n* is the life of the project. The calculation of the *EAC* is the same as that used in Example 6.7 part (i) to determine the payment that must be made in order to repay a loan by the end of the loan period.

The interpretation of the *EAC* is illustrated in Figure 6.9. It spreads the net present value over the life of the project as if it were an annuity.

The equivalent annual charge can also be used to calculate the annual payment that must be made each year in order to repay the investment. If the *EAC* of the cost of the fixed capital is determined, it indicates the amount each year that must be recovered in order to pay for the capital and the interest accrued at the discount rate. In this form, it is known as the *capital recovery factor*.

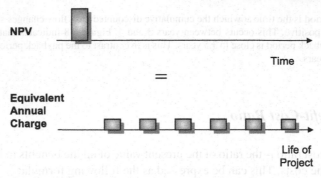

Figure 6.9 The relationship between *NPV* and equivalent annual charge

The calculation of the equivalent annual charge is illustrated in the following examples.

Example 6.11: Calculation of equivalent annual charge.

A 30-year project has an *NPV* of $15 million determined at 9%. Determine the equivalent annual charge for this project.

Solution:

The equivalent annual charge is an annuity can be determined from the annuity formula (Table 5.3):

$$A = PV(A|P, k, n) \tag{6.13}$$

where k is the discount rate. The annuity factor $A|P,k,n$ is given by the following expression:

$$A|P, k, n = k(1+k)^n / [(1+k)^n - 1]$$

Since the discount rate is 9%, $k = 0.09$ and project has a life of 30 years, $n = 30$. Thus the annuity factor is calculated as follows:

$$0.09(1+0.09)^{30}/[(1+0.09)^{30} - 1] = 0.09733$$

The calculation of the equivalent annual charge is as follows:

$$EAC = 0.09733(15) = \$1.46 \text{million}$$

The equivalent annual charge is $1.46 million.

The equivalent annual charge is also known as the *annual worth, equivalent uniform annual worth, equivalent uniform annual charge* and *annual equivalent*. These names, such as annual worth are misleading, in the sense that they do not convey the meaning of this measure. The term annual worth seems to be a misnomer. A better name is to replace the word "annual" in any of the previous names with "annuity," so that "annual worth" becomes "annuity worth" because the method treats the amounts spent or earned as if they were an annuity. The term capital recovery can be changed in a similar fashion. However, this text has chosen to use the names "equivalent annual charge" and "capital recovery," as they seem to be the least misleading of the terms in use. This annuity nature of the measure is clearly illustrated in Figure 6.9.

The equivalent annual charge can be used to determine how much the revenue for a project must exceed its operating expenses in order for it to pay back the investment capital. This is considered in the following example.

Example 6.12: Charges for capital recovery.

A municipality wishes to install a new electricity distribution network. It is estimated that this will cost $5 million. The project will last 20 years. The municipality estimates that its discount rate is 7%. How much must the municipality charge customers for the capital cost of the system? In other words, what is the charge, in addition to the costs of the electricity and other operating expenses incurred by the municipality in distributing the electricity, for the capital cost of the system?

Solution:

The *EAC* for a $5 million project at 7% for 20 years is given by:

$$Equivalent\ Annual\ Charge = PV(A|P,k,n)$$
$$= (-5)0.07(1+0.07)^{20}/[(1+0.07)^{20} - 1] = (-5)0.09439$$
$$= -\$471,964$$

Note that costs have been represented as negative numbers. This is done to be consistent throughout the chapter. In replacement studies, the costs are sometimes represented as positive numbers.

The municipality must pass on an additional capital charge of $471,964 to its customers each year for the next twenty years in order to pay for the capital investment for the distribution network. This exceeds the simple allocation of these costs in equal allotments across the life of the equipment by an amount equal to $471,964 − $5,000,000/20 = $221,964. This represents the effect of the interest charged on the capital investment. Effectively, customers are buying the distribution network and paying for the cost over twenty years at an interest rate of 7%.

As a matter of interest, a mortgage loan on a home or a building is analogous to the equivalent annual charge discussed in the previous example. Instead of purchasing a distribution network and paying it off over 20 years, the owner of a mortgage purchases a property and pays it off over 20 or 30 years, depending on the term. The interest on a mortgage is paid on the outstanding balance, and payments are made in equal instalments so that both the interest and the capital are repaid at the end of the term of the loan. The repayments for a mortgage loan are considered in the next example.

Example 6.13: Mortgage loan.

An engineering consultancy wishes to purchase an office for $300,000. The current interest rate for mortgages is 5% over a 25-year term. What are the monthly repayments?

Solution:

Since the payments are made monthly, the rate that must be used in the calculation is the nominal monthly interest rate. This is equal to $5\%/12 = 0.004166$.

The equation for the equivalent monthly charge is then given by:

$$A = PV(A|P,i,n) = (300,000)(0.05/12(1+0.05/12)^{25(12)}/[(1+0.05/12)^{25(12)} - 1]$$
$$= (300,000)(0.005845)$$
$$= \$1,753.77 \text{ per month.}$$

The monthly mortgage repayment is $1,753.77 per month (assuming the interest rate remains constant).

Mutually exclusive projects are projects where only one of the alternatives can be chosen. When an investment doesn't necessarily earn revenue by itself, such as in the case of the replacement of a piece of equipment, the project is often referred to as a *service-producing project*. Such projects are compared on the basis of their costs and the project that is the least cost alternative is chosen. The *EAC* can be used to compare different service-producing projects, such as the replacement of equipment.

If the *EAC* is used to compare projects of different durations, or useful lives, an additional assumption is introduced into the analysis. It implicitly introduces the assumption that the project can be replicated with the same cost structure indefinitely. This assumption may not be valid for a number of reasons, such as the escalation of costs, technology obsolescence and changes in the performance of alternatives, amongst others. In spite of this assumption, the equivalent annual charge is often used to assess these types of options.

The use of the equivalent annual charge in assessing mutually exclusive options is illustrated in the following example.

Example 6.14: The use of EAC to compare mutually exclusive projects.

The company is considering the replacement of a piece of equipment. The capital and operating costs for the two alternatives are given in Table 6.20. The capital cost for project B is much less than Project A, while the life of Project B is much less than that of Project A. The operating costs are similar, but as Project A ages, it is anticipated that the operating costs will almost double due increased maintenance requirements. Determine the preferred option if the discount rate is 5%. In addition, determine if the choice of the project is dependent on the discount rate.

Table 6.20 Cost profile for two projects with unequal lives

Year	Project A	Project B
0	−19,000	−12,000
1	−2,000	−1,700
2	−2,000	−1,700
3	−2,000	−1,700
4	−2,000	−1,700
5	−2,000	
6	−2,000	
7	−3,500	
8	−3,500	
9	−3,500	
10	−3,500	

Table 6.21 Calculation of present values of project costs

Year	Project A	$PV(A)$	Project B	$PV(B)$
0	−19,000	−19,000	−12,000	−12,000
1	−2,000	−1,905	−1,700	−1,619
2	−2,000	−1,814	−1,700	−1,542
3	−2,000	−1,728	−1,700	−1,469
4	−2,000	−1,645	−1,700	−1,399
5	−2,000	−1,567		
6	−2,000	−1,492		
7	−3,500	−2,487		
8	−3,500	−2,369		
9	−3,500	−2,256		
10	−3,500	−2,149		
Total PV		−38,413		−18,028

Solution:

The methodology used is to determine the present value of the costs and the equivalent annual charge can be calculated for both options. The option with the lowest *EAC* is the preferred option. Since the options have unequal lives, the annual cost of replacing equipment B at the end of its life with another model of the same equipment will be the same as that for the first model. Thus, in comparing projects with unequal lives using the *EAC*, the assumption is made that the projects can be repeated at the same costs as the first implementation. Thus, the difference in useful life is neatly taken into account by the annual equivalent charge. However, this can only be a true assessment of the situation if the equipment can be replaced for the same cost as that for which it was originally installed.

The calculation of the present value of the costs is shown in Table 6.21.

The equivalent annual charge can be calculated for the two projects:

$$EAC(A) = (-38,413)0.05(1+0.05)^{10}/[(1+0.05)^{10} - 1] = (-38,413)0.1295 = -4974.65$$

$$EAC(B) = (-18,028)0.05(1+0.05)^{4}/[(1+0.05)^{4} - 1] = (-18,028)0.2820 = -5084.11$$

Clearly, Project A costs slightly less than Project B and it is the recommended alternative. Note that since the total *PV* is that of costs, the preferred choice is the one with the lower costs. Project B has the higher *NPV*, but this is biased because it has a much shorter life. One way to reduce this bias is to choose a period, called a study life, that is the lowest common multiple of the two unequal lives.

(It should also be noted that since the cash flow profile does not have a sign change, that is, is change from positive or negative or negative to positive over the course of the project, the *IRR* method couldn't be used. Under these conditions, the *IRR* is undefined. It was the failure of *IRR* in this situation that gave rise to the classification of projects as either service-producing or revenue-producing projects, because different techniques were needed for the assessment of these different categories if *IRR* was the preferred method of analysis.)

The recommended method for the analysis of projects is to use the *NPV* criteria. If the projects have unequal lives, they should be evaluated using *NPV* over a common period or study life. Replacement studies are considered in more detail in Chapter 8.

The difference between the equivalent annual charges for each of the options as a function of discount rate is shown in Figure 6.10 on page 188, indicating that the preferred choice changes from Project A to Project B if the discount rate increases above 8%.

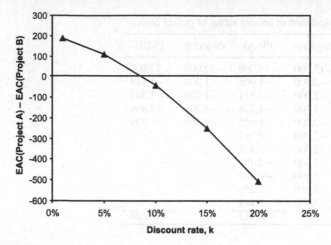

Figure 6.10 The difference in the *EAC* for the two alternatives

The equivalent annual charge has the advantage of being relevantly easy to interpret: it is the cost incurred or value added for each year of the life of the project. It can be used to determine the annual charges that need to be levied in order to recover the cost of an investment, and it can be used to compare alternatives of equal lives. It has the disadvantage that if it is used to compare alternatives of different lives, it introduces the assumption that the projects can be reproduced for the same costs. This assumption is difficult to justify in practice. Replacement decisions are considered in more detail in Chapter 7.

The decision criteria discussed in Sections 6.2 and 6.3 form the core of the methods that are used to assess the economic viability of projects. In the next section the discounted cash flow techniques are compared.

6.4 Comparison Between Discounted Cash Flow Techniques

6.4.1 Relative and Absolute Measures

The net present value, the internal rate of return and the profitability index are the most commonly used techniques for the assessment of project proposals in companies. They are all discounted cash flow techniques, so they account for the time value of money. They all require management to set a parameter: for *IRR* it is the hurdle rate, while for *NPV* and *PI*, it is the discount rate. However, *PI* and *IRR* are relative measures, that is, they compare the earnings and the growth in earnings with the cost of the investment, while the *NPV* is an absolute measure, that is, a measure of the value that the project is expected to add to its owner's portfolio of assets.

6.4.2 Agreement and Conflict Between Measures

When evaluating a single project with conventional cash flows, the discounted cash flow techniques will always make the same recommendation concerning the economic attractiveness of the project. The profile of *NPV* against discount for conventional cash flows for an investment is downward sloping, that is, it has a negative slope. This means that if the *NPV* is positive at the discount rate, the *IRR* will always be greater than the discount rate. In addition, if the *NPV* is positive, the present value of the cash flows generated is greater than the present value of the cash flows invested. If the *NPV* is positive, the profitability index, which is a ratio of the present values of the cash flows generated to those invested, must be greater than one. Thus, the discounted cashflow techniques will make the same recommendation.

However, in comparing multiple projects, or cash flows that are not conventional, the different discounted cash flow criteria may conflict with one another in their recommendations. The slope of the profile of *NPV* against the discount rate, even though downward sloping for conventional cash flows, need not have the same value. Projects with different cash flow profiles will have different slopes for their profiles of their plots of the *NPV* versus the discount rate. Different slopes give rise to the possibility of the profiles intersecting with one another. Such an intersection, called Fisher's intersection after the economist Irving Fisher, is shown in Figure 6.11. The existence of Fisher's intersection at positive values for the *NPV* indicates that conflicting recommendations from *NPV* and *IRR* analysis might exist.

For the projects shown in Figure 6.11, Project B is the preferred option if the discount rate is less than that at Fisher's intersection, while Project A is preferred if the discount rate is greater than the rate at the intersection. The *IRR* for Project A is greater than that for Project B. This means that if the discount rate were less

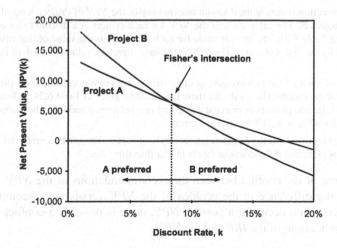

Figure 6.11 The effect of discount rate, k, on the *NPV* for two different projects

Table 6.22 Cash flow profile for two projects

Year	Project A	Project B
0	−37,000	−31,000
1	20,000	1,000
2	15,000	5,000
3	10,000	10,000
4	5,000	15,000
5	1,000	20,000

Table 6.23 Calculation of *NPV* for two projects

Year	Project A	Project B	Discount factor	PV (A)	PV (B)
0	−37,000	−31,000	1.000	−37,000	−31,000
1	20,000	1,000	0.909	18,182	909
2	15,000	5,000	0.826	12,397	4,132
3	10,000	10,000	0.751	7,513	7,513
4	5,000	15,000	0.683	3,415	10,245
5	1,000	20,000	0.621	621	12,418
NPV				5,128	4,218

than the rate at Fisher's intersection, the *NPV* and *IRR* methods would recommend different options.

The calculation of Fisher's intersection is demonstrated in the following example.

Example 6.15: Fisher's intersection.

Two projects have the following cash flow profiles. Determine the discount rate at Fisher's intersection.

Solution:

Fisher intersection occurs at the discount rate that makes the *NPV* of Project A equal to the *NPV* of Project B. The calculation of the *NPV* for each project at a discount rate of 10% is shown in Table 6.23. The present value for each cash flow is the value of that cash flow multiplied by the discount factor. The *NPV* is the sum of present values for each of the cash flows.

The *NPV* of each of the two projects at other discount rates was calculated in the same manner as these results. The results for these calculations, given in Table 6.24, indicate that the *NPV* of the two projects intersect at a discount rate between 8 and 9%. The intersection point was found to be 8.21% by trial and error.

For these two projects, if the discount rate is less than 8.21%, Project B is preferred, while Project A is preferred if the discount rate is higher than this value.

The source of the conflict between the recommendations of the *NPV* and *IRR* methods is the difference in the profiles of the *NPV* versus the discount rate. If Fisher's intersection occurs at a positive *NPV*, there is possibly a conflict between the recommendations of the *IRR* and the *NPV*.

Table 6.24 *NPV* for the two projects as a function of discount rate

Discount rate	NPV (A)	NPV (B)
5%	9,188	11,137
6%	8,322	9,616
7%	7,484	8,168
8%	6,673	6,788
9%	5,888	5,473
10%	5,128	4,218
11%	4,391	3,021
12%	3,678	1,878

The source of the conflict is not, as is far too commonly stated in finance and accounting texts, a result of the "reinvestment assumption." It is argued later that the reinvestment assumption is a fallacy.

For the intersection of the *NPV*-discount rate to be at a positive *NPV* there must be a difference in the slope of these profiles. The factors that affect the slope of the *NPV*-discount rate profile are the factors that affect the value of the *NPV*: the size, timing and duration of the cash flow for the projects. Altering these factors alters the *NPV*-discount rate profile, and hence alters the point of intersection between different projects.

The obvious question is which technique is correct if there is a conflict between the two techniques? The basis for an answer lies in the goal of financial management and engineering economy: to maximise shareholder value. If shareholder value is the aim of the company's management and directors, the project that creates greater value, not greater rates of return, is the preferred choice.

There are two methods that can be used to determine which is the preferred option when the *NPV* and *IRR* conflict with each other. The first method is incremental analysis, which is the analysis of the difference between the two projects. In other words, the cash flows for the one project are subtracted from those for the second, and the *NPV* and *IRR* of this "incremental project" are determined. The second method is that in which the cash flows for the two projects are adjusted so that they are both equivalent by borrowing money.

Both of these methods of determining the preferred option are demonstrated in the following example.

Example 6.16: Shareholder wealth.

Consider two projects, each with the cash flows given in Table 6.25.

Table 6.25 Cash flow profile for two projects

Year	Project A	Project B
0	−17,500	−17,500
1	20,000	0
2	5,000	28,000

(i)   If the company's discount rate is 10%, determine the *NPV* and *IRR* for each project.

(ii)   Determine the *NPV* and *IRR* of the difference between the two projects.

(iii)   If it was argued that Project A was the preferred option because it generated cash early, which could be used for other investments, show that this is not a valid argument if maximizing shareholder value is the goal.

(iv)   What is the wealth created for the shareholders?

Solution:

(i)   The calculation of the *NPV* at a discount rate of 10% is shown in Table 6.26 below. The *IRR* for each option is found by trial and error, and is also shown in Table 6.26.

On the basis of *NPV*, project B is preferred. On the basis of *IRR*, project A is preferred.

(ii)   Incremental analysis involves the subtraction of the cash flows of one project from those of the other, and determining the *NPV* and *IRR* for the incremental project. For the analysis of this problem, the incremental analysis is performed on the difference between project B and project A.

The incremental cash flows for the two projects and the *NPV* and *IRR* for the incremental cash flows are given in Table 6.27.

The result for the *NPV* means that value created by project B exceeds that created by project A by 826.4, so choose B. The *IRR* of the incremental cash flows exceeds the specified rate (in this case, 10%), so choose project B.

(iii)   Project A delivers cash flow earlier, and the argument may be made that this is a significant factor in the choice between projects. This earlier cash can be used for other profitable opportunities. However, the project cash flows can be made equivalent by borrowing the cash at the company's discount rate and investing the borrowed money in the external opportunities. In year 2, the borrowed amount and the interest owed need to be repaid. If the cash flows were made equivalent on this basis, the cash flows for the two projects would be those shown in Table 6.28.

Table 6.26 Calculation of the *NPV* and *IRR* for the projects

Year	Project A	Project B	Discount factor	PV (A)	PV (B)
0	−17,500	−17,500	1.000	−17,500	−17,500
1	20,000	0	0.909	18,182	0
2	5,000	28,000	0.826	4,132	23,140
NPV				4,814	5,640
IRR				35.39%	26.49%

Table 6.27 Incremental cash flow analysis

Year	B-A	Discount factor	PV(B-A)
0	0	1	0
1	−20,000	0.909	−18181.8
2	23,000	0.826	19008.3
NPV			826.4
IRR			15%

Table 6.28 Adjusted cash flows

Year	Adjusted cash flow	Explanation
0	−17,500	Investment
1	20,000	Loan
2	6,000	Earnings less loan repayment = 28,000−20,000−0.1(20,000)
NPV	5,640	
IRR	39.96%	

The *NPV* of the adjusted Project B is still 5,640 as it originally was, but now the company has the additional capital earlier for investment in other opportunities. Project B still dominates. The *IRR* for the adjusted cash flow for Project B now exceeds that for Project A, emphasising that Project B is the better option.

(iv) The shareholder wealth created, measured at the present, is the *NPV* for each project. Measured at the end of the project's life, this is the terminal value. The terminal value for each of the projects is calculated from the following expression:

$$TV = \sum_{t=0}^{n} CF_t (1+k)^t$$

where *TV* represents the terminal value.

The terminal value, calculated in this fashion, is equal to 6,407 for Project A and 7,507 for project B. These results clearly indicate that *NPV* is the better decision criterion, since it is the criterion that results in maximum shareholder value either now or at the end of the project.

The *NPV* is the method that is preferred in all cases. It is the method that measures the contribution of the project to shareholder value. It is easy to interpret, has the least disadvantages and incorporates the least number of assumptions. The strongest point in its favour is that it is not easy to apply it incorrectly. This is not true of the other measures, in which there are situations that are commonly found where the method can be misapplied.

6.4.3 The Reinvestment Assumption in the NPV and IRR

It is commonly stated that the reason for the conflict between the *NPV* and *IRR* methods under some conditions is a result of the "reinvestment assumption." In this section, it is argued that the reinvestment assumption does not exist. The equation for the *NPV* was presented as Equation 6.3. The *IRR* is essentially a special case of the *NPV*, that is, the value of the discount rate, *k*, when the *NPV* is zero.

There is no assumption in the discounting formula concerning the reinvestment of proceeds. Neither is there an assumption concerning the reinvestment of the interest earned in the period between the time at which the cash flow occurs and the end of the project. This is because the cash flows are discounted to the present in both the *NPV* and *IRR* formula.

The *NPV* formula is not concerned with the way in which the cash flow that is generated by the project is used. The mathematical procedure of discounting simply determines the value of that cash flow at the present. If that cash flow received at a particular time was invested in some other venture, or even wasted, it does not make any difference to the value of the *NPV*, or by extension, to the *IRR*.

The reinvestment assumption is investigated further by considering a project with a cash flow profile shown in Figure 6.12. The discount rate for each period is different from that in the other periods.

The future value of the project can be calculated from the cash flow profile. This is given by the following expression:

$$FV = CF_0(1+k_1)(1+k_2)(1+k_3) + CF_1(1+k_2)(1+k_3) + CF_2(1+k_3) + CF_3$$

The future value of the project is the sum of the values of the cash flows earned during each period invested at the discount rate that the project earns for each period. The future value clearly contains a reinvestment assumption. The future value of the projects is also the present value of the project invested at the project rate of return until the end of the project. This is given by the expression:

$$FV = NPV(1+k_1)(1+k_2)(1+k_3)$$

Since the future values should be equivalent, these two expressions can be equated, given as follows:

$$NPV(1+k_1)(1+k_2)(1+k_3) = CF_0(1+k_1)(1+k_2)(1+k_3)$$
$$+ CF_1(1+k_2)(1+k_3) + CF_2(1+k_3) + CF_3$$

The simplification of this expression leads to the following expression for the *NPV*:

$$NPV = CF_0 + \frac{CF_1}{(1+k_1)} + \frac{CF_2}{(1+k_1)(1+k_2)} + \frac{CF_3}{(1+k_1)(1+k_2)(1+k_3)}$$

This is simply the usual formula for the *NPV*, given by Equation 6.3. None of the cash flows are reinvested beyond the period at which they occur. Each cash flow is reinvested at the period rate, and then discounted back at the same rate, which eliminates the reinvestment of the cash flow. This analysis clearly shows that the

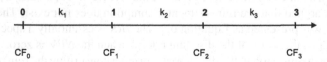

Figure 6.12 The cash flow profile for a project with a life of three periods

reason for the conflicting recommendations of the *IRR* and *NPV* methods is not the "reinvestment assumption," as stated in many texts, but a result of the difference in the cash flow profiles that leads to a Fisher intersection.

6.5 Case Study: Decision Criteria

Consider a project with the cash flows given in Table 6.29.

Compare the various decision criteria with respect to their recommendation concerning the project if the discount rate is 12%.

Solution:

Table 6.30 below shows the calculation of the different criteria for the assessment of the project.

Table 6.29 Cash flows for a project

Year	Project A
0	−15,000
1	3,000
2	3,000
3	3,000
4	3,000
5	3,000
6	3,000
7	3,000
8	3,000
9	3,000
10	3,000
11	3,000
12	3,000

Table 6.30 Calculation of the various assessment criteria

Year	Project A	Discount factor	PV(A)	Compound factor	FV(A)	CF for MIRR	Cumulative CF	Cumulative DCF
0	−15,000	1.000	−15,000	0.00	0	−15,000	−15,000	−15,000
1	3,000	0.893	2,679	3.478	10,436	0.0	−12,000	−12,321
2	3,000	0.797	2,392	3.105	9,318	0.0	−9,000	−9,930
3	3,000	0.712	2,135	2.773	8,319	0.0	−6,000	−7,795
4	3,000	0.636	1,907	2.475	7,428	0.0	−3,000	−5,888
5	3,000	0.567	1,702	2.210	6,632	0.0	0	−4,186
6	3,000	0.507	1,520	1.973	5,921	0.0	3,000	−2,666
7	3,000	0.452	1,357	1.762	5,287	0.0	6,000	−1,309
8	3,000	0.404	1,212	1.573	4,721	0.0	9,000	−97
9	3,000	0.361	1,082	1.404	4,215	0.0	12,000	985
10	3,000	0.322	966	1.254	3,763	0.0	15,000	1,951
11	3,000	0.287	862	1.120	3,360	0.0	18,000	2,813
12	3,000	0.257	770	1.000	3,000	72,399	21,000	3,583

The explanation of the different columns in the table in the order of the columns is as follows:

(i)   Problem definition.

(ii)   Problem definition.

(iii)   The discount factor, $1/(1+k)^t$, where t is the year and k is the discount rate, which is 12%.

(iv)   The present value of the cash flow occurring at year t. This is given by the value in column 2 multiplied by that in column 3.

(v)   This is the compound factor between the current year in the row and the end of the project. The compound factor is used to determine the future value of a cash flow, and in defined as $(1+k)^{n-t}$, where n is the life of the project, t is the year and k is the discount rate.

(vi)   The future value at the end of the project of the cash flow at year t. This is obtained by multiplying the values in column 2 with those in column 5.

(vii)   The cash flow profile to determine the *MIRR*. It consists of the initial investment, which remains unchanged from column 2, and the terminal value, which in this case is the sum of the future values at the end of the project of all cash flows except the initial capital investment. The terminal value is thus the sum of the values in column 6. All values in between this initial capital investment and the terminal value are zero.

(viii) Column 8 is the cumulative cash flows, that is, a running total of the values in column 2.

(ix)   Column 9 is the cumulative discounted cash flows, that is, a running total of the values in column 4.

The net present value is the sum of the values in column 4. The internal rate of return is obtained by changing the discount rate until the net present value is zero. The profitability index is obtained from the *NPV* and the investment in fixed capital in year 0, and the *MIRR* is the *IRR* using column 7 as the cash flow profile. The results of these calculations are given in Table 6.31.

The payback period and the discounted payback period can be obtained from columns 8 and 9, respectively. The values in these two columns are plotted in the Figure 6.13.

The recommendation for the project is that it should be accepted because the *NPV* is positive, the *IRR* is greater than the discount rate, the profitability index is greater than one and the *MIRR* is greater than the discount rate. The payback

Table 6.31 Assessment measure for the project

Measure	Value
NPV	3,583
IRR	16.9%
PI	1.239
MIRR	14.0%
Payback period	5.0 years
Discounted payback period	8.0 years

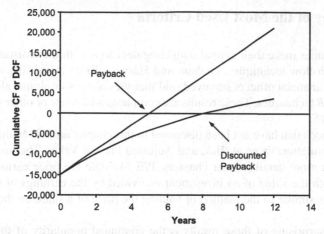

Figure 6.13 The cumulative cash flow and discounted cash flow for the project, indicating the payback period and the discounted payback period

period of 5 years and the discounted payback period of 8 years are more difficult to interpret. One way is to say that these payback periods are shorter than the life of the project and hence the project should be recommended for acceptance.

It is interesting to compare the *NPV*, the *IRR* and the *MIRR*. These measures for the assessment of a project are plotted as a function of the initial capital cost or investment in Figure 6.14. These results indicate that the *NPV* is a linear function of the initial capital cost, while the *IRR* and the *MIRR* are dependent in a non-linear manner on the initial capital cost. The *IRR* and the *MIRR* are equal to one another when the *NPV* is zero, that is, when the discount rate is equal to the *IRR*. As the initial capital cost gets smaller, the *IRR* gets very large. It is in highly profitable ventures that the *IRR* becomes substantially different from the *MIRR*.

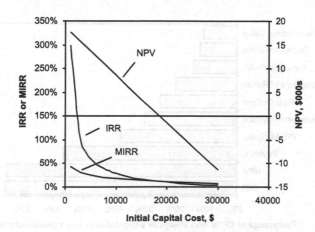

Figure 6.14 The effect of the initial capital cost on the *NPV*, the *IRR* and the *MIRR*

6.6 Survey of the Most Used Criteria

Most companies make their capital budgeting decisions with the assistance of discounted cash flow techniques. Graham and Harvey (2001) found that about 75% of the chief financial officers surveyed said that they always or almost always used *NPV* and *IRR* techniques. These results are compared with those of other techniques in Figure 6.15.

The methods that have not been discussed in this chapter are P/E Multiples, Real Options, Simulation/Value at Risk, and Adjusted Present Value. The last three are discussed in more detail in later chapters. P/E Multiple are price earnings multiples, in which the value of an investment is divided by the earnings of the investment. This is similar to the method of valuing the price of a share on the stock exchange.

Perhaps surprising of these results is the continued popularity of the payback method. This rule of thumb provides a measure that is easy to understand and to implement. The other surprising result was the use of real options, which is a technique that employs much more complex mathematics than any of the techniques discussed in this chapter. The application of real options to capital projects is examined in Chapter 15.

Similar surveys to that of Graham and Harvey have been published in the past. In the 1950s, the payback period and the return on investment dominated the other methods. With the passage of time, both these methods decreased in usage as the primary technique used for decision-making, to be replaced mainly be the *IRR*. By the 1980s the internal rate of return and the net present value were popular choices as the primary technique for decisions on capital expenditure. These surveys suggest that managers preferred *IRR* to *NPV* during this period. In contrast, the academic literature, dating back to Fisher's work over a century ago, is overwhelmingly in favour of *NPV*. The results shown in Figure 6.15 indicate that the *NPV* has increased in usage to the point now where both *IRR* and the *NPV* are used equally.

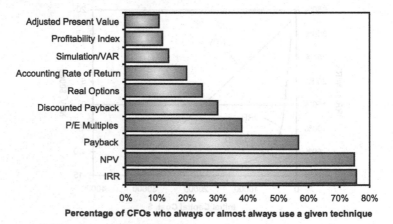

Figure 6.15 The results of the survey by Graham and Harvey (2001) of the popularity of various decision-making criteria

6.7 Summary

The main assessment or decision criteria for the evaluation of projects have been discussed in this chapter. Two of these criteria, the payback period and the return on investment, do not account for the time value of money, while all the others do. The discounted cash flow techniques are either absolute or relative. The absolute techniques, such as the *NPV*, provide an absolute measure of the value created by the project. Relative measures, such as the *IRR*, determine the returns to the project compared with the investment size.

The *payback period* is the time it takes for the cash in-flows to accumulate to the amount of the original investment. At this point the investor breaks even in the sense that the investment has returned all the money originally invested in the project. This measure has the advantage that it is easy to compute and intuitively easy to interpret in terms of the risk that the project entails.

The *return on investment* is the measure devised at Du Pont for the purposes of making capital-budgeting decisions. There are a number of ways of defining the return on investment, such as the annual profit divided by the original investment. The method is easy to compute and historically it has been of great benefit to its users.

The discounted cash flow techniques that are discussed are the net present value, internal rate of return, the profitability index, equivalent annual charge, the modified internal rate of return and the discounted payback period. The benefit-cost ratio is the same as the profitability index. These methods all provide insight into the performance of a project. However, they do not all recommend the same choices.

The *net present value* is the only method that is clearly aligned with the goals of financial management: to create value for the investors in or owners of the company. The net present value is straightforward to interpret: it measures the absolute value in today's terms of the value added to the company by the project. Another advantage is that is has a built-in decision point: if the net present value is greater than zero, it creates value for the project's owner and if it is less than zero it destroys value for its owner. A disadvantage of the net present value is related to the choice of the discount rate, which is the subject of much discussion in later chapters.

The *profitability index* is a relative measure of the anticipated performance of a project. It is defined as the ratio of the present value of the cash generated by the project to the present value of the cash invested in the project. This measure is the same as the benefit-cost ratio. If the cash generated by the project exceeds the cash consumed by the project, the profitability index will be greater than one, and the project is acceptable. If the profitability index is less than one, it should not be recommended. The disadvantage of the method is the choice of discount rate, and the comments made on this topic in the previous paragraph are applicable here.

The *internal rate of return* is a relative measure of the anticipated performance of the project. The internal rate of return is the value of the discount rate when the net present value is zero. It has the advantage that the discount rate need not be provided. However, there is no in-built decision point that specifies whether the investment is acceptable or not. The project's owners or the financial managers need to specify an acceptable rate, which is often referred to as the hurdle rate. If the project has an internal rate of return greater than the hurdle rate, it is recommended, otherwise not. The internal rate of return can be interpreted by analogy with the yield on a bond or the interest rate on an amortized loan. This "interest rate interpretation" has enormous intuitive appeal. There are several disadvantages to the internal rate of return. It does not rank projects in a manner consistent with the net present value. In addition, the internal rate of return is undefined for projects only involving costs.

The *modified internal rate of return* is the same as the internal rate of return, except that cash in-flows from the project are reinvested at the project owner's discount rate to the end of the project.

The *discounted payback period* is the same as the payback period, except that the cash flows are discounted to their present values prior to the determination of the payback period.

The *equivalent annual charge* distributes the value of the project as a uniform series over the life of the project. It is mathematically equivalent to the payment required on an amortized loan, or the payment on an ordinary annuity. It can be used to determine the charges required, in addition to the normal operating expenses, in order to pay for the capital cost of the project. In this form, it may be called the *capital recovery factor*. The equivalent annual charge is often used to choose between mutually exclusive projects, for example, equipment replacement decisions. If the equivalent annual charge is used to compare projects of unequal lives, as is often recommended, then an additional assumption is introduced that the project can be replicated with the same cost structure.

6.8 Looking Ahead

The application of the decision criteria to a variety of situations is discussed in the next chapter.

6.9 Review Questions

1.   What does net present value mean? What does it measure?
2.   Are the net present values of projects owned by different companies equal to the sum of their net present values?
3.   Must the project lives of alternatives be the same in order to use the equivalent annual charge?
4.   What does the internal rate of return mean?
5.   Which of the discounted cash flow techniques is consistent with the goal of financial management?
6.   What is the reinvestment assumption?
7.   Under what conditions will the net present value and the internal rate of return conflict with one another?

8.   Why could the internal rate of return have multiple values?

9.   What is the basic concept behind the modified internal rate of return?

10.   When will the modified internal rate of return be equal to the internal rate of return?

11.   What are the assumptions concerning the profitability index?

12.   Discuss the difference between annual worth and the capital recovery factor.

13.   What is the preferred method for evaluating projects? Explain your answer.

14.   Provide arguments for and against on whether projects should be ranked on the basis of *IRR*.

6.10 Exercises

1.   Which do you prefer: $5,500 cash now, or $1,200 each year for five years? Assume a discount rate of 10% pa. Use as many of the assessment criteria as possible to determine your answer.

2.   A proposed expansion has the following cash flow profile:

Year	Cash flow
0	−100
1	50
2	60
3	70
4	200

Calculate the payback period, the *NPV*, the *PI*, the *EAC*, the discounted payback period and the *IRR* for this project if the discount rate is 7%.

3.   Consider the following two projects:

Year	Project A	Project B
0	−20,000	−40,000
1	10,000	20,000
2	10,000	20,000
3	10,000	20,000

(i)   Compare the two projects using *NPV* and *IRR* if the required rate of return or discount rate is 11%.

(ii) Explain why the *IRR* cannot distinguish between these two projects?

4.   Consider the cash flow profile of the following two cost saving projects:

(i)   Determine the *NPV* and *EAC* for the projects and recommend a choice of projects based on these criteria. The discount rate is 13%.

Year	Project A	Project B
0	−20,000	−40,000
1	−10,000	−20,000
2	−10,000	−20,000
3	−10,000	−20,000

(ii) Explain why the *IRR* cannot be used to choose between these projects.

5. A company is considering a project with the following contributions to its cash flow profile:

	Amount
Capital investment	$50 million
Revenue (per year)	$35 million
Expenses	
Raw materials (per year)	$13 million
Salaries and wages (per year)	$3 million
Maintenance (per year)	$1 million
Utilities (per year)	$2 million

If the company's discount rate is 10%, determine the following:

(i) Determine the tax paid if the tax rate is 35%, and the capital is depreciated for tax purposes on a straight-line basis over ten years.

(ii) Determine the working capital and the change in working capital if the days receivable is 45 days, the days payable is 30 days and the raw materials inventory days is 25 days. Assume that all the working capital is released in the tenth year.

(iii) Determine the free cash flow for the 10-year duration of this project.

(iv) Determine the *NPV* and the *IRR* for this project.

(v) Determine whether this project meets the company's investment criteria.

6. A company owns office buildings that it lets to tenants. A particular building that cost $40 million has 7,500 m^2 of office space. The building is expected to last for 30 years. The maintenance costs are $2 million each year. If the company's discount rate is 15%, determine the annual revenue that the company must make from rentals in this building in order to meet its investment criterion that the *NPV* must be greater than $1 million. Determine the minimum rental per square metre that the company must charge to meet this investment criterion.

7. A municipality has rehabilitated an old quarry into a rather challenging golf course. The cost of building the golf course was $15 million and annual maintenance is expected to cost $2 million. If the golf course attracts 5,000 golfers a month, what fees should the course charge in order to meet the cost of finance, which is estimated to be 8% pa, over a period of 25 years.

8. A project has a capital cost of $100,000 and annual maintenance costs of $20,000. The salvage value of the project at the end of 5 years is $30,000. Determine the total costs as an equivalent annual charge for the project if the discount rate is 20%.

9. A project has revenues of $40,000 and expenses of $20,000. The capital investment for the project is $75,000. Determine the *EAC*, *NPV* and the *IRR* for this project if the project has a life of 15 years and the company's discount rate is 15%. Recommend whether the company should invest in this project.

10. You wish to make an endowment to your university that would pay for the annual costs of tuition for a less privileged student. The costs are $10,000 per year and the endowment earns 5% pa after inflation. How much money must you pay as a lump sum if the endowment is to last 50 years?

11. The interest rate on a mortgage loan for a house that cost $500,000 is 7% pa. The repayments are paid monthly and the term of the loan is 30 years. Determine the monthly repayments. Determine the total interest paid during the term of the loan.

12. Two alternative routes for a road are being considered. The costs for these alternatives are given in the following table:

	Project A	Project B
Road construction	$100 million	$200 million
Road maintenance (per year)	$7.5 million	$4 million

If the discount rate is 10%, determine the following:

(i) The *EAC* and *NPV* of the two alternatives if the projects have a life of 25 years.

(ii) If only one of the two can be chosen, recommend the most economical choice.

13. A project has the following costs and returns:

	Amount
Capital investment	$100,000
Operating costs (per year)	$20,000
Revenue (per year)	$40,000
Salvage value after 15 years	$10,000

Determine the *IRR* for this investment.

14. Explain the difference between capital recovery and depreciation.

15. Derive the expression for profitability index given by Equation 6.6 from Equation 6.4.

16.   A project with a life of 5 years has an *EAC* of 1,000. If the discount rate is 10%, determine the *NPV* of the project.

17.   A company wishes to invest $1,000 in a project that is indefinite. If the company's discount rate is 10%, determine the annual return that the company must make to justify the investment.

18.   Parents need to save for their children's education. Current costs of a year's tuition and living expenses are $15,000. The average degree is four years long. The average interest rate that the couple can get on their investments is 7%, and college fees are expected to rise by 7% per annum over the period. If the child has just been born, and college education will begin when the child is eighteen, how much money must the couple save each year to meet this expense.

Chapter 7
Mutually Exclusive, Replacement and Independent Projects

7.1 Classification of Asset Allocation Decisions

Although the pattern of usage is changing to favour *NPV* as the primary decision criterion, *IRR* was favoured over other discounted cash flow techniques in the past. One of the drawbacks of *IRR* is that if the investment is driven by cost considerations, like those of equipment replacement choices, the *IRR* is undefined. As a result, in classifying projects, there was a primary split of investment types into "revenue-producing investments" and "service-producing investments." *Revenue-producing* investments are conventional projects that generate cash flows, and all the discounted cash flow techniques, including *IRR*, are applicable. *Service-producing* investments involve the saving of cost, and because *IRR* is inapplicable, different techniques are needed. Of course, this distinction between revenue-producing and service-producing alternatives is not required if *NPV* is the preferred criterion.

Projects are also classified on the basis of the decision that is required. Revenue and service-producing alternatives might be mutually exclusive, or mutually non-exclusive (independent). Only one option can be chosen from alternatives that are mutually exclusive, while several options can be adopted from alternatives that are mutually non-exclusive. As discussed in Chapter 2, independent projects require yes/no decisions, while mutually exclusive projects require either/or decisions. Service-producing projects are about *cost savings*, which by definition implies a comparison between alternatives. Because service-producing alternatives are always regarded as requiring a choice between alternatives (either/or decisions), they are always mutually exclusive.

The classification of projects adopted in this chapter is shown in Figure 7.1. Projects are separated into mutually exclusive and independent projects. The mutually exclusive options are usually of two types: replacements, which are projects in which there is a current incumbent, or projects in which there is no current incumbent. There is no need to further classify independent projects for the application of the assessment criteria.

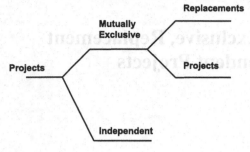

Figure 7.1 Classification of projects

In a company setting, projects can be distinguished on a completely different basis, such as the distinction between mandatory or discretion projects.

In the sections that follow, the application of the evaluation criteria to independent (or non-mutually exclusive) and mutually exclusive decisions will be discussed. Because different criteria are used in different companies, it is important to be familiar with all of the decision criteria.

7.2 Mutually Exclusive Alternatives

Mutually exclusive options are those requiring a decision between options, where one of the options is required. These are very common decisions in practice. Replacement decisions, where the decision is to replace an item now or later, fall into this category. Trade-off studies, where a decision is needed on the sizing or the materials of construction, are also of this type.

An example of a mutually exclusive replacement decision is the following. A filter is nearing the end of its life and the costs of annual maintenance and unplanned maintenance are increasing. Does management choose to replace the filter now or later? The options are: (i) stay with the current equipment; and (ii) purchase new equipment, for which there may be one or a number of different competing options. Replacement decisions typically involve service-producing alternatives and are analysed on the basis of cost savings.

Other common examples of mutually exclusive decisions are production expansion, operational improvement and project implementation decisions. For these alternatives to be mutually exclusive means that, of those alternatives available, only one can be chosen. An example of a mutually exclusive decision that is typical of an income-producing alternative is the expansion of production, where there are two possible production alternatives, only one of which can be implemented.

It must also be recognised that, unless the project is mandatory, the decision makers always have the option of doing nothing, and the do-nothing alternative should always be included in the analysis.

The objective of the economic analysis of mutually exclusive options is to recommend the most preferred option. The service or the output provided by each of the alternatives must, however, be the same. If they are not the same, it does not make good sense to compare the two alternatives with each other. The two alternatives must also meet a minimum design capacity, and capacity in excess of this design requirement might be disregarded. This is not to say that extra capacity over the current market requirements should be neglected in the assessment. The engineering design may call for spare production capacity because management anticipates the market demand to increase.

The service life of the different options will affect the assessment. For example, if one of the projects has a life of three years, and the other six, the assessment will be biased. As a result, the discussion of the assessment of mutually exclusive options is divided into the assessment of projects with equal lives, and those with unequal lives. The choice of the most preferred alternative amongst projects with equal lives is discussed next.

7.2.1 Ranking Mutually Exclusive Options with Equal Lives

Mutually exclusive projects with equal lives can be ranked using any of the discounted cash flow techniques. A possible conflict exists between the different discounted cash flow techniques when comparing mutually exclusive options. It was argued in Chapter 6 that the reason for the conflict is the difference in the cash flow profiles for different projects. It was also argued that the NPV dominates, that is, it always gives the correct result, whereas IRR may under some situations provide results that do not maximise shareholder value. In order to get the correct result using IRR, incremental analysis must be used.

The ranking of projects with equal lives using NPV and incremental techniques is discussed in the following sections.

(i)   Ranking mutually exclusive projects with equal lives using NPV

If the projects are of equal duration, the NPV of the alternatives should be used to rank their preference. The case of project alternatives with unequal lives requires further discussion, which is provided in Section 7.2.2. The ranking of projects with equal lives is the subject of the next example.

Example 7.1: Ranking of mutually exclusive projects with equal lives.

Rank the projects shown in Table 7.1 in the order of preference based on NPV, IRR and PI if the discount rate is 10%.

Solution:

The calculation of the NPV, IRR and PI for the five projects is shown in Table 7.2.

The values in Table 7.2 are calculated as follows:

(i)   Column 1 is the same as column A in the previous table.

(ii)   Column 2 is the set of discount factors given by $1/(1+k)^t$ where t is the year and k is the discount rate of 10%.

(iii)   Columns 3 to 7 are the present values of the cash flows for each year, that is the cash flow multiplied by the discount factor.

Table 7.1 Cash flow profile for Projects A to E

Year	Project A	Project B	Project C	Project D	Project E
0	−113	−113	−113	−113	−113
1	30	0	45	30	100
2	40	0	45	30	30
3	50	70	45	50	25
4	60	110	45	70	25

Table 7.2 Calculation of the *NPV*, *IRR* and *PI* for the Projects A to E

Year	Discount factor	PV(A)	PV(B)	PV(C)	PV(D)	PV(E)
0	1.000	−113.0	−113.0	−113.0	−113.0	−113.0
1	0.909	27.2	0.0	40.9	27.2	90.9
2	0.826	33.0	0.0	37.1	24.7	24.7
3	0.751	37.5	52.5	33.8	37.5	18.7
4	0.683	40.9	102.4	30.7	47.8	17.0
NPV		25.8	42.0	29.6	24.4	38.5
IRR		19.0%	20.0%	21.6%	18.2%	31.3%
PI		1.229	1.372	1.262	1.216	1.341
Rank NPV		4	1	3	5	2
Rank IRR		4	3	2	5	1
Rank PI		4	1	3	5	2

(iv)   The *NPV* is the sum of the present values of the cash flows.

(v)   The *IRR* is the discount rate at which the *NPV* is zero. This is obtained by trial and error, or by using a root search algorithm (for example, goalseek) or by using an in-built routine in the spreadsheet.

(vi)   The *PI* is the profitability index and is obtained from Equation 6.7.

(vii)   The last three rows rank the projects on the basis of *NPV*, *IRR* and *PI* in ascending order.

The three discounted cash flow techniques are not in agreement. Interestingly, the *NPV* and the *PI* agree; it is the *IRR* that is out of line. Project B is the most preferable on the basis of *NPV* and *PI*. Further investigation is required. The inconsistent result can be resolved using incremental analysis.

While *NPV* always yields the result that is consistent with the goal of maximizing shareholder value, *IRR* is a widely used decision criterion that may give results that are inconsistent with the *NPV*. It was shown in Chapter 6 that in order to get the correct result using internal rate of return, incremental analysis should be used. The application of incremental analysis to mutually exclusive proposals will be discussed next.

(ii)   Selecting mutually exclusive options with equal lives using incremental analysis

Incremental analysis is the analysis of two projects based on the difference between the cash flow for each year of each project. This difference in the cash flow for the two projects is the incremental cash flow. The *NPV* and *IRR* of the incremental cash flows are calculated. In addition, the difference in the cash flow profiles is calculated so that the incremental cash flow profile is such that costs precede in-flows. If the incremental *NPV* is positive, the first project is chosen. If the *IRR* is greater than the discount rate, then the first project is chosen. The discounted cash flows for the incremental analysis will choose the correct option.

In Example 7.1, Project B and Project E were ranked differently by *IRR* and *NPV*. In the next example incremental analysis is applied to resolve the conflict between the techniques for these two projects.

Example 7.2: Incremental analysis of two projects with equal lives.

Perform an incremental analysis on Projects B and E from the previous example.

Solution:

The calculation of the incremental cash flows is shown in Table 7.3

The first three columns represent the project cash flows, while the fourth column is the difference in the cash flows for Project B and Project E. The fifth column is the discount factor determined at a discount rate of 10%. The sixth column is the present value of the difference in cash flows for each project, obtained by multiplying the values in column 4 with those in column 5.

The *NPV* is the sum of the figures in column 6 and the *IRR* is the value of the discount rate that makes the *NPV* equal to zero.

These results indicate that the *NPV* of the incremental cash flows is positive, so Project B is preferred. For conventional cash flows in which out-flows precede in-flows, the *IRR* criterion is that the following relationship holds for the proposal to be acceptable:

$$IRR > k \qquad (7.1)$$

where k is the discount rate.

The *IRR* of the incremental cash flows of Project B less that Project E is greater than the discount rate of 10%, which indicates that Project B is preferred.

Note that the incremental cash flows must be calculated so that the costs precede income. If this is not the case, the *IRR* value of the incremental cash flows is not a rate of return, but a rate of investment. This problem is illustrated by calculating the incremental cash flows of Project E with Project B, and then recalculating the *IRR*. The details of this calculation are given in Table 7.4.

In this case, the *NPV* of the incremental cash flows is negative, indicating that Project B is preferred. However, the *IRR* remains 11.3%, which is confusing.

Table 7.3 Selection between Projects B and E based on incremental analysis

Year	Project B	Project E	Project B − Project E	Discount factor	PV(B − E)
0	−113	−113	0	1.000	0.000
1	0	100	−100	0.909	−90.909
2	0	30	−30	0.826	−24.793
3	70	25	45	0.751	33.809
4	150	25	125	0.683	85.377
NPV					3.483
IRR					11.3%

Table 7.4 Incremental analysis with revenue preceding costs

Year	Project B	Project E	Project E − Project B	Discount factor	$PV(E-B)$
0	−113	−113	0	1.000	0.000
1	0	100	100	0.909	90.909
2	0	30	30	0.826	24.793
3	70	25	−45	0.751	−33.809
4	150	25	−125	0.683	−85.377
NPV					−3.483
IRR					11.3%

Because the cash flows are no longer conventional cash flows for an investment-type project, the sign of the inequality given by Equation 7.1 must be changed from the greater than sign to a less than sign. This is given as the following expression:

$$IRR < k$$

This means that if the income precedes costs, the *IRR* must be less than the discount rate for the proposal to be acceptable. Again, it is worth pointing out that *NPV* again provides the correct choice without any such difficulties.

If the *IRR* of the incremental project is used as the assessment criterion, then the relative size of each project is important. For different sized projects, the incremental cash flows must be calculated as those of the larger project minus the smaller project, so that the incremental cash flow profile is *conventional for an investment project*. In other words, the costs of the investment must precede the returns. Then if the incremental *NPV* is positive, or the incremental *IRR* is greater than discount rate, the bigger project is the preferred option. If in doubt, choose the project the makes the incremental *NPV* positive.

In the discussion until this point, the projects that were to be selected were of equal duration, that is, they had the same project life. In the following section, the ranking of mutually exclusive projects of different durations are discussed.

7.2.2 Ranking Mutually Exclusive Projects with Unequal Lives

The life of the project can have a significant effect on its value. This makes it difficult to compare projects with unequal lives. The situation is shown schematically in Figure 7.2. A recommendation is required on which of the two projects should be chosen. To be fair to both alternatives, the lives of the projects must be modified so that they are equivalent.

There are three modifications that can be made to make the lives equal: (i) replicating the shorter option until the lives are equivalent, using the lowest common multiple of the two lives; (ii) ignoring the extra life of the longer option; and (iii) obtaining the actual costs to extend the shorter life to that required. Some judgement must be executed concerning the nature of the projects and the modifications

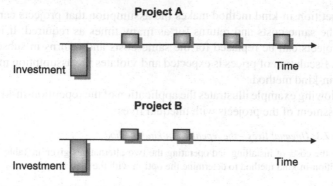

Figure 7.2 Evaluating projects with unequal lives

that are adopted. As a result, this introduces additional assumptions into the analysis that may or may not be true. This is not about the mathematics of the techniques, but about the projects themselves. Economic and business analysis, like engineering analysis, often involves the integration of different considerations, and all the assumptions need to be thoroughly justified. This is no less true in this case.

Each of the three different options for making the lives of the projects equal, that is, for obtaining a common study period, is discussed in the following sections.

(i)   Repetition-in-kind

A common study period can be obtained by repeating the projects until the lives of both options are equal. This means that the common study period is the lowest common multiple of the life of the original project. This method is known as the repetition-in-kind method, which is illustrated in Figure 7.3.

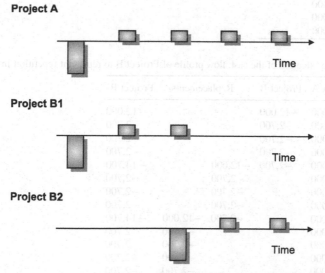

Figure 7.3 The repetition of Project B so that the modified Project B has the same life as Project A

The repetition-in-kind method makes the assumption that projects can be replicated for the same costs and returns for as many times as required. It is unlikely that the projects can be replaced for the same costs and returns in subsequent installations. Escalation of prices is expected and violates the assumption made in the repetition-in-kind method.

The following example illustrates the application of the repetition-in-kind method to the assessment of the projects with unequal lives.

Example 7.3: Unequal lives – the repetition-in-kind method.

Consider the costs of installing and operating the two alternatives given in Table 7.5. Use the repetition-in-kind method to determine the option with the least costs.

Solution:

The net present value of the costs for Projects A and B is $-25,814$ and $-20,559$ respectively. This suggests that Project B is preferred. However, the life of Project B is significantly less than that of Project A, a fact that has not been accounted for in this assessment.

Table 7.5 Cash flow profile for Projects A and B

Year	Project A	Project B
0	−19,000	−12,000
1	−1,000	−2,700
2	−1,000	−2,700
3	−1,000	−2,700
4	−1,000	−2,700
5	−1,000	
6	−1,000	
7	−1,000	
8	−1,000	
9	−1,000	
10	−1,000	
11	−1,000	
12	−1,000	

Table 7.6 Modification of the cash flow profile of Project B to represent repetition-in-kind

Year	Project A	Project B	Replacements		Project B'
0	−19,000	−12,000			−12,000
1	−1,000	−2,700			−2,700
2	−1,000	−2,700			−2,700
3	−1,000	−2,700			−2,700
4	−1,000	−2,700	−12,000		−14,700
5	−1,000		−2,700		−2,700
6	−1,000		−2,700		−2,700
7	−1,000		−2,700		−2,700
8	−1,000		−2,700	−12,000	−14,700
9	−1,000			−2,700	−2,700
10	−1,000			−2,700	−2,700
11	−1,000			−2,700	−2,700
12	−1,000			−2,700	−2,700

Table 7.7 Calculation of the *NPV* of Project A and the modified Project B'

Year	Project A	Project B'	Discount factor	PV(A)	PV(B')
0	−19,000	−12,000	1.000	−19,000	−12,000
1	−1,000	−2,700	0.909	−909	−2,455
2	−1,000	−2,700	0.826	−826	−2,231
3	−1,000	−2,700	0.751	−751	−2,029
4	−1,000	−14,700	0.683	−683	−10,040
5	−1,000	−2,700	0.621	−621	−1,676
6	−1,000	−2,700	0.564	−564	−1,524
7	−1,000	−2,700	0.513	−513	−1,386
8	−1,000	−14,700	0.467	−467	−6,858
9	−1,000	−2,700	0.424	−424	−1,145
10	−1,000	−2,700	0.386	−386	−1,041
11	−1,000	−2,700	0.350	−350	−946
12	−1,000	−2,700	0.319	−319	−860
NPV (costs)				−25,814	−44,191

The repetition-in-kind method can be used to modify the shorter project using the least common multiple of the lives of the two projects, and to calculate the option with the lowest costs. The modification of the project is shown in Table 7.6.

Project B' represents the cash flows for Project B repeated three times. The net present value of the costs of the two projects can now be determined. These calculations are shown in Table 7.7.

The costs of Project A are clearly lower than those of Project B' (*NPV* of Project A is greater than Project B'). This can also be analysed using incremental analysis. The incremental cash flows of the costs of the two options are given in Table 7.8.

The *NPV* of the incremental project is positive, which indicates that Project B' costs more than A, so Project A is the recommended option. An investment of the difference in costs between A and B' yields an incremental *IRR* of 45%. Since the incremental *IRR* is greater

Table 7.8 Incremental analysis of Projects A and B'

Year	Project A	Project B'	CF(A)-CF(B')	PV(A-B')
0	−19,000	−12,000	−7,000	−7,000
1	−1,000	−2,700	1,700	1,545
2	−1,000	−2,700	1,700	1,405
3	−1,000	−2,700	1,700	1,277
4	−1,000	−14,700	13,700	9,357
5	−1,000	−2,700	1,700	1,056
6	−1,000	−2,700	1,700	960
7	−1,000	−2,700	1,700	872
8	−1,000	−14,700	13,700	6,391
9	−1,000	−2,700	1,700	721
10	−1,000	−2,700	1,700	655
11	−1,000	−2,700	1,700	596
12	−1,000	−2,700	1,700	542
NPV				18,378
IRR				45%

than the discount rate of 10%, choose A over B'. Therefore both the *NPV* and the incremental *IRR* recommend the choice of Project A.

Note that, in order to be consistent with other methods, costs have been represented as negative numbers. Costs may sometimes be represented as positive numbers in replacement studies.

The equivalent annual charge is sometimes used to compare alternatives, particularly in replacement studies. The comparison of two projects with different lives using the *EAC* assumes that the projects can be repeated until the total life is the lowest common multiple. Therefore, it is conceptually a repetition-in-kind method, and will suffer from the same deficiencies for the repetition-in-kind mentioned above, primarily the assumption that the project can be repeated for the same costs and returns. This assumption that the project can be repeated for the same costs and returns is referred to as the "assumption of repeatability."

The calculation of the equivalent annual charge for mutually exclusive projects is demonstrated in the following example.

Example 7.4: Application of the EAC for projects with unequal lives.

Compare Projects A and B given in Table 7.5 using the equivalent annual charge.

Solution:

The equivalent annual charge is defined as follows:

$$EAC = PV(A|P, k, n) = PV\left(\frac{k(1+k)^n}{(1+k)^n - 1}\right) \tag{7.2}$$

The *EAC* can be calculated from the present value of the project, which is then substituted into Equation 7.2. An alternative method of calculation is to separate the initial capital outlay from the operating expenses, and to calculate the *EAC* of each component. This can be expressed in the following form:

$$EAC(Total) = EAC(Capital) + EAC(Operating\ Expenses) = CR + AOC \tag{7.3}$$

where *AOC* represents the annual operating charge and is the equivalent annual charge of the operating expenses. *CR* is the *capital recovery*, and is the equivalent annual charge of the initial capital outlay.

Since the operating expenses are constant for each year in this example, their cash flow profile represents an annuity. In this special case, the *EAC* of the operating expenses, that is, *AOC*, is simply the annual cost.

The *EAC* is therefore calculated as follows:

$$EAC(Project\ A) = -19,000(A|P, k, n) - 1,000$$

$$= -19,000\left(\frac{0.1(1+0.1)^{12}}{(1+0.1)^{12} - 1}\right) - 1,000 = -3,788$$

$$EAC(Project\ B) = -12,000(B|P, k, n) - 2,700$$

$$= -12,000\left(\frac{0.1(1+0.1)^4}{(1+0.1)^4 - 1}\right) - 2,700 = -6,485$$

The equivalent annual charge for these two options is $-3,788$ for Project A and $-6,485$ for Project B. The recommended choice is the project with the lowest costs, that is, Project A. This agrees with the repetition-in-kind method discussed in Example 7.3.

(ii)   Neglect extra life

The additional life of the alternative may not be required. For example, consider the replacement of a boiler with an estimated life of twenty years in a plant that has a planned life of eight years. In this case, the additional life is of little value (other than that it can be sold at the end of eight years). The nature of the problem and the project may justify setting a fixed period for the evaluation of alternative projects. The period of evaluation is called the study life.

This method can be executed in two ways: (i) the study period is that of the shorter lived project so that the longer project is truncated, or shortened, to that of the shorter project; and (ii) the study period is shorter than both projects so that the lives of both projects are truncated. This method is schematically illustrated in Figure 7.4. This is a valid method of analysis if the service is only required for the study period.

Example 7.5: Neglect the additional life.

Choose between Project A and B using the truncation method for the example shown above.

Solution:

In this case, the costs beyond year 4 are irrelevant, so that the cash flows considered are shown in Table 7.9.

The incremental *NPV* is negative. The cause of the negative *NPV* is that the costs of Project A are higher than those of Project B in the early years. This means that the recommendation is to choose Project B. The incremental *IRR* is less than the discount rate of 10%, which means that investment in the incremental cash flows is not recommended, that is, the recommendation is to choose Project B.

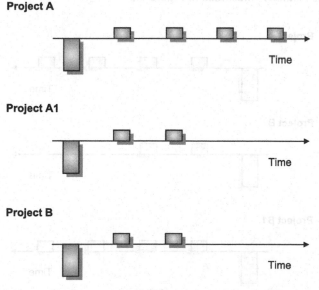

Figure 7.4 Neglecting the additional life of Project A to obtain the modified Project A1 that has the same life as Project B

Table 7.9 Calculation of the incremental *NPV* and *IRR* for the projects

Year	Project A	Project B	Discount factor	PV(A)	PV(B)	CF(A)-CF(B)	PV(A-B)
0	−19,000	−12,000	1.000	−19,000	−12,000	−7,000	−7,000
1	−1,000	−2,700	0.909	−909	−2,455	1,700	1,545
2	−1,000	−2,700	0.826	−826	−2,231	1,700	1,405
3	−1,000	−2,700	0.751	−751	−2,029	1,700	1,277
4	−1,000	−2,700	0.683	−683	−1,844	1,700	1,161
NPV				−22,170	−20,559		
NPV (incremental)							−1,611
IRR (incremental)							−1.15%

The truncation of the life of the project has resulted in a completely different recommendation to that in the previous section. For this method to be valid, the extra life must be of zero value to the owner. If this is not true, the incorrect recommendation may be provided.

(iii)   Actual costs to extend life

This method of getting a common study life is based on obtaining or estimating actual costs for maintenance, repairs and operating costs to extend the life of the shorter-lived option to that of the longer-lived option. This may require significant engineering experience and judgement, or for vendor participation if the equipment is supplied. This is the recommended method if it can be implemented. Obviously, if the service or equipment is not required for the longer period, this method shouldn't be used. This method is illustrated in Figure 7.5.

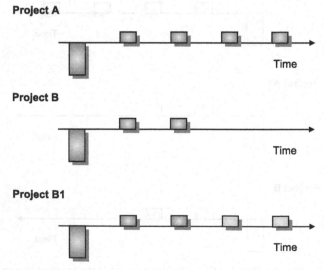

Figure 7.5 The method of getting actual costs to extend life of the shorter Project B to the same life as Project A, that is to Project B1. (Both projects could be modified to a common study period.)

7.2.3 Selection of Mutually Exclusive Alternatives

The ranking and selection of mutually exclusive projects has been discussed in the previous sections. Once the difference in the lives of the alternatives has been addressed, the net present value or another assessment criterion can be used to determine the preferred option. In most cases, the different discounted cash flow methods will concur with one another. If there is a size or a timing disparity, these methods may conflict with one another. The conflict can be examined by determining the Fisher intersection and can be resolved by incremental analysis. For projects involving only costs, sometimes called service-producing projects, the *IRR* method cannot be used because it is mathematically undefined. In this case, the *NPV* or another technique, such as the equivalent annual charge can be used. There can be conflict between the *EAC* and the *NPV* for projects of equivalent lives.

Replacement decisions are an application of mutually exclusive decisions and are examined next.

7.3 Replacement Studies: Mutually Exclusive Decisions with an Incumbent

Equipment replacement is required when the service obtained from the equipment deteriorates to the extent where maintenance costs are too high, where the level of reliability required from the equipment is not being met, or the productivity required from the equipment falls below expectations. Other reasons may be the obsolescence of the equipment due to technological changes and improvements in efficiencies of newer models. For example, a compressor that has been performing for several years may need to be replaced because it has broken down more frequently than expected and maintenance costs on the compressor appear to be unjustified. Is it economically sound to replace it now, or later?

Replacement decisions are amongst the most common capital investment decisions. They are often classified as mandatory decisions, since the service must be provided or else the operations are compromised. They are mutually exclusive decisions where one of the alternatives is to retain the currently installed equipment. As a result, these decisions are similar to those discussed in Section 7.2.2, in which alternatives of different service lives were assessed. An extension of the problem is to assess the optimal live of currently installed equipment and to determine whether it is due for replacement.

The discussion of the previous section indicated that the life of the projects being compared might affect the assessment. There are two options: (i) a study period is unspecified so that the planning horizon is unlimited; and (ii) the planning horizon is limited to a particular study period. If the study period is unspecified, the economic service life or the expected life of the equipment can be used to define the planning horizon. The economic service life is calculated from the equivalent annual charge. The option with the lowest costs at each option's respective economic service life is the recommended option.

The incumbent is often referred to as the *defender* and the best of the new alternatives is referred to as the *challenger* in equipment replacement studies. If the defender and the challenger have different lives, which they almost certainly will, the use of the *EAC* as the decision criteria introduces the assumption that the subsequent installation cycles can be repeated with the same costs as the first implementation. This assumption, called the repeatability assumption, must be tested and found to be true for this analysis to hold. If it does not hold, the actual costs to make the lives and service of the alternatives the same are required. It is best to remove this assumption from the analysis by determining the costs to extend the life of the current incumbent to be the same as that for the challenger. The use of the *EAC* in projects of equal life contains no assumption about their repetition.

The challenger is the best option from the alternatives that are available. The costs of the defender may include modifications, repairs and improvements in order to optimise its performance. Such possible improvements to the defender should always be included in the analysis.

The economic service life is discussed next.

7.3.1 Economic Service Life

The economic service life (ESL) is the minimum value of the total *EAC* as a function of the remaining life, n. The total cost of either the defender or the challenger can be separated into the cost of the initial capital outlay and the operating expenses. This is given in terms of the equivalent annual charge of each of these components as follows:

$$EAC(Total) = CR + AOC \qquad (7.4)$$

where CR is the capital recovery and AOC is the equivalent annual charge of the operating costs. The capital recovery is given by the following expression:

$$CR = P(A|P,k,n) = P\left(\frac{k(1+k)^n}{(1+k)^n - 1}\right) \qquad (7.5)$$

where P is the cost of the equipment.

The *EAC* for the operating costs for a project of life n is given by the following expression:

$$AOC = PV_n(Opex)(A|P,k,n) \qquad (7.6)$$

where AOC is the annual operating charge, and $PV_n(Opex)$ represents the present value of the operating expenses for a project of life n. The substitution of the present value annuity factor yields the following expression for the annual operating charge:

$$AOC = PV_n(Opex)\left(\frac{k(1+k)^n}{(1+k)^n - 1}\right) \qquad (7.7)$$

The annual operating charge should not be confused with operating costs or operating expenses that are actually incurred each year. The *AOC* is the equivalent annual charge of the present value of the operating expenses. In other words, in order to calculate the *AOC*, first determine the present value of the anticipated operating costs, and then spread this present value out evenly across the life of the project as an annuity. If the operating costs are constant and the discount rate is constant, then the operating costs per year are the same as the annual operating charge; otherwise they are not the same.

The economic service life is the minimum value of the *EAC(Total)*, given by Equation 7.4, as a function of the remaining life, n. Both the *CR* and the *AOC* are dependent on the value of n. The *CR* decreases with increasing n because the capital used to purchase the equipment is in active use for longer. The *AOC* may increase or remain constant with increasing n, depending on whether it becomes more expensive to operate the equipment as it ages or not. The economic service life represents the optimal use of economic resources for the equipment.

The calculation of the economic service life is illustrated in the next example.

Example 7.6: Calculation of economic service life.

Consider a machine that costs $10,000 to purchase and install. The operating costs for each year after installation are expected to rise, as shown in Table 7.10. Determine the economic service life for this machine.

Solution:

The economic service life is the minimum value of the *EAC* as the remaining life is changed. The first step in determining the *AOC* is calculating the PV_n(Opex). The results of this calculation are shown in Table 7.11. The calculation of the values in each of the columns is as follows:

(i)   Column 2 is the operating costs in each year.

(ii)   Column 3 is the discount factor, given by $1/(1+k)^t$, where t is the year and k is the discount rate, which is 10%.

(iii) Column 4 is the present value of the operating expenses for the year, which is obtained by multiplying the values for column 2 and column 3 for each row.

(iv)   Column 5 is a cumulative total of the values in column 4. This means that column 5 is the sum of the present values up to the year n.

The values in column 5, that is, PV_n(Opex), are used to determine the *AOC*. The results of this calculation, and that of the capital recovery and the total *EAC*, are shown in Table 7.12.

Table 7.10 Annual operating costs for a machine

Year	Operating costs
0	0
1	−2,500
2	−3,000
3	−4,000
4	−5,000
5	−6,000
6	−8,000
7	−10,000
8	−12,000

Table 7.11 Calculation of the present value of the operating costs

Year	Operating costs	Discount factor	PV(Opex)	Cumulative PV_n(Opex)
0	0	1.0000	0	0
1	−2,500	0.9090	−2,273	−2,273
2	−3,000	0.8264	−2,479	−4,752
3	−4,000	0.7513	−3,005	−7,757
4	−5,000	0.6830	−3,415	−11,172
5	−6,000	0.6209	−3,726	−14,898
6	−8,000	0.5644	−4,516	−19,414
7	−10,000	0.5131	−5,132	−24,545
8	−12,000	0.4665	−5,598	−30,143

Table 7.12 Calculation of the EAC

Remaining life, n	A|P,k,n	PV_n(Opex)	AOC	CR	EAC(total)
1	1.1000	−2,273	−2,500	−11,000	−13,500
2	0.5761	−4,752	−2,738	−5,762	−8,500
3	0.4021	−7,757	−3,119	−4,021	−7,140
4	0.3154	−11,172	−3,525	−3,155	−6,679
5	0.2637	−14,898	−3,930	−2,638	−6,568
6	0.2296	−19,414	−4,458	−2,296	−6,754
7	0.2054	−24,545	−5,042	−2,054	−7,096
8	0.1874	−30,143	−5,650	−1,874	−7,525

The values in each of the columns are calculated as follows:

(i)   Column 1 is the remaining life of the machine. This means that if n is equal to five then the machine has a life of five more years. It is important not to confuse this with the actual year of operation, which is what is represented in column 1 of the previous table.

(ii)   Column 2 is the capital recovery factor given by $k(1+k)^n/((1+k)^n-1)$, where k is the discount factor.

(iii)   Column 3 is the same as column 5 of the previous table.

(iv)   Column 4 is the annual operating charge, which is obtained by multiplying the values in column 2 by those in column 3. The AOC is the equivalent annual charge of the present value of the operating expenses for a remaining life of n years. The AOC is increases with increasing remaining life because the operating costs increase with each year.

(v)   Column 5 is the capital recovery of the equipment, which is obtained by multiplying the fixed capital cost of $10,000 with the values in column 2. It diminishes as the remaining life increases because there is more operational life over which to spread the capital cost.

(vi)   The total EAC, given in column 6, is the sum of the values in columns 4 and 5.

The values of the AOC, CR and total EAC as a function of the remaining life are shown in Figure 7.6. The economic service life is the value of the remaining life at the minimum value of the EAC. This is shown in Table 7.12 and in Figure 7.6 to be five years.

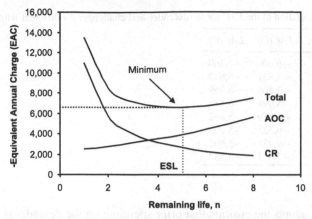

Figure 7.6 The annual operating charge, *AOC*, the capital recovery, *CR*, and the total *EAC* as a function of the remaining life. The economic service life, ESL, is the minimum of the total ESL

The market value of the equipment impacts on the analysis because at the end of its life it can be sold, thus reducing the costs. The selling of the incumbent equipment alters the capital recovery, which can be expressed in the following form:

$$CR = P(A|P,k,n) - S(A|F,k,n) \qquad (7.8)$$

where P is the initial cost or current market value of the equipment and S is the end of life value, that is, either the salvage value or market value at year n. The factors $A|P,k,n$ and $A|F,k,n$ are given in Table 5.3.

The economic service life is the value of the remaining life that minimizes the equivalent annual charge. In other words, it is the period that spreads both the capital and operating costs in the form of an annuity in the most efficient manner. The *EAC* of an option at the ESL is the best possible performance for the equipment.

The application of the *EAC* as the assessment criterion is discussed next.

7.3.2 Selection of Defender or Challenger Based on Equivalent Annual Charge

The replacement study determines whether the challenger should replace the defender now. The viewpoint that is adopted for establishing the costs for each of the defender and the challenger is that of an external advisor or consultant. From this perspective, the fact that the defender is already paid for and in the possession of the company is irrelevant. The external assessor takes the view that both the defender and the challenger must be purchased now, and their relative performance must be assessed as mutually exclusive options. The defender is acquired for its current market value, while the challenger is acquired for the purchase price. Both the defender and the challenger will be sold at the end of the study period.

Table 7.13 Calculation of the *EAC* for the defender and challenger for different lengths of service

Remaining life	EAC(D)	EAC(C)
1	−6,000	−7,300
2	−5,821	−6,843
3	−5,866	−6,459
4	−5,998	−6,135
5	−6,179	−5,862
6	−6,523	−5,695
7	−6,955	−5,607
8	−7,437	−5,582

This view adopts the position that prior spending on the defender is regarded as sunk costs that must be ignored. The values that should be used in the assessment are current values for the purchase of a defender in that state. In addition, it is the market value of the defender, not the accounting or book value. Book value is the original purchase price depreciated according to a formula. Book value does not account for the state of the equipment, and is unrelated to its market value.

If the study period or planning horizon is unspecified, the equivalent annual charge is obtained at the economic service life of both the defended and challenger. If the study period is specified, the *EAC* over that period is used to assess the costs. The option with the lowest *EAC* at either the ESL or the study period is the recommended option. The use of the *EAC* to assess the alternatives whose lives are different implies that the assumption of repeatability holds, which means that the project can be repeated for the same costs and the same returns. The selection of the most attractive option is illustrated in the following example.

Example 7.7: Selection of challenger or defender.

The *EAC* for the defender, given by *EAC*(D), and the challenger, *EAD*(C), are given in Table 7.13. The *EAC* for each of these options was calculated using the method outlined in Example 7.6. A study period has not been specified. Recommend which of the options should be implemented.

Solution:

The economic service life of either the defender or the challenger is the remaining life at which the *EAC* is a minimum. For the defender, the ESL is 2 years, while for the challenger it is 8 years. The *EAC* at a remaining life of two years is –$5,821 for the defender, while the *EAC* for the challenger at a remaining life of eight years is –$5,582. This means that the costs of the challenger are lower than the costs of the defender and that the challenger should replace the defender. The optimal use of the defender is worse than the optimal use of the challenger.

If the study period is defined, then the *EAC* for that length of remaining life must be used. This is the subject of the following example.

Example 7.8: Specified study period.

The *EAC* for the defender and the challenger are given in Table 7.14. Select the best option if the study period is 3 years.

Table 7.14 *EAC* for the defender and the challenger

Remaining life	EAC(D)	EAC(C)
1	−6,000	−7,300
2	−5,821	−6,843
3	−5,866	−6,459
4	−5,998	−6,135
5	−6,179	−5,862
6	−6,523	−5,695
7	−6,955	−5,607
8	−7,437	−5,582

Solution:

The *EAC* for the defender is less negative than that for the challenger at a study period of 3 years. This means that the costs of the defender are lower than the costs of the challenger. Therefore, the defender is the recommended option.

7.3.3 Replacement for Service Required for Defined Period

If the service is required of the equipment for a defined period, it is not required forever. The consequence of this is that the assumption underpinning Section 7.2.3 is invalid and the method of *EAC* cannot be used. In this case, the methods of Section 7.2.2 parts (i) to (iii) must be used. The method is based on the *NPV*, and the option with the least costs is the recommended choice. This is illustrated in the following example.

Example 7.9: Replacement with a defined study period.

A company providing analytical services must ensure that it provides the best service at a competitive price. Because technological progress in the automation of analytical instrumentation is rapid, the company expects to replace instruments regularly, depending on the usage and the rate of technological progress. For the particular instrument considered in this problem, the replacement cycle is about four years.

The current market value of the defender is $10,000 and the cost of purchasing the challenger is $16,000. The value of the defender is expected to decline at 25% per annum, whereas the value of the challenger is expected to decline at 20% per annum. The operating costs of the defender and the challenger are given in Table 7.15.

Table 7.15 Operating costs for two projects

Year	Project D operating costs	Project C operating costs
0	0	0
1	2,500	2,500
2	3,000	2,500
3	4,000	2,500
4	5,000	2,500

Table 7.16 Calculation of the cash flows and *NPV* for the defender

Year	Discount factor	Capital	Salvage value	Operating costs	Total cash flow	PV(CF)
0	1.000	−10,000			−10,000	−10,000
1	0.909			−2,500	−2,500	−2,273
2	0.826			−3,000	−3,000	−2,479
3	0.751			−4,000	−4,000	−3,005
4	0.683		3,164	−5,000	−1,836	−1,254
NPV						−19,011

Determine whether the defender should be replaced now or remain in service for another four years. The discount rate is 10%.

Solution:

The viewpoint that is adopted is that of an external advisor. Assume that both instruments need to be purchased now at the market price and that they will both be sold in the market at the end of the four-year study period.

The market value of the defender in four years from now is expected to be equal to $10,000(1.0 - 0.25)^4 = \$3,164$ because it declines at a rate of 25% per year. Similarly, the market value of the challenger declines at a rate of 20% per year, so it is expected to be $16,000(1.0 - 0.2)^4 = \$6,554$.

The cash flows for the defender are given in Table 7.16.

(i)   Column 2 is the discount factor, $1/(1+k)^t$, where k is the discount rate and t is the year.

(ii)   Column 3 is the capital to purchase the defender at the current market value.

(iii)   Column 4 is the expected value of the instrument in four years when it will be sold.

(iv)   Column 5 is the operating costs of the instrument.

(v)   Column 6 is the sum of columns 3, 4 and 5 to provide the cash flow.

(vi)   Column 7 is the discounted cash flow, obtained by multiplying the values in column 2 with those in column 6. The net present value is the sum of the values in column 7.

The *EAC* for the defender is calculated as follows:

$$EAC = P(A|P, k, n) = P\left(\frac{k(1+k)^n}{(1+k)^n - 1}\right)$$

$$= -19,011\left(\frac{0.1(1+0.1)^4}{(1+0.1)^4 - 1}\right) = -19,011(0.3154) = -5,997$$

Table 7.17 Calculation of the cash flows and *NPV* for the challenger

Year	Discount factor	Capital	Salvage	Operating costs	Total cash flow	PV(CF)
0	1.000	−16,000			−16,000	−16,000
1	0.909			−2,500	−2,500	−2,273
2	0.826			−2,500	−2,500	−2,066
3	0.751			−2,500	−2,500	−1,878
4	0.683		6,554	−2,500	4,054	2,769
NPV						−19,448

The *EAC* for the defender is −$5,997.

The cash flows for the challenger are obtained in a similar manner to those of the defender and are provided in Table 7.17.

The *EAC* of the challenger is equal to −$6,135. The costs of the defender are slightly lower than the cost of the challenger, so the defender is retained.

In the previous example, the defender was better than the challenger, but only just. The salesperson, sensing that this may be the case, may offer the company a trade-in value that is slightly higher than the market value of the defender, and this would tip the decision in favour of the challenger. The effect of the salesperson either offering the company a higher trade-in value or a discount on the price of the challenger is the same. The *NPV* varies linearly with both the capital for the defender and the challenger. If the salesperson offers either a discount for the purchase of the challenger or a trade-in at a price higher than the market value, these are the values that should be used. It is not correct for the analyst to second-guess the motives and hence the "true price" of the alternatives. The analysis is a *commercial* assessment using values for the cash flows that are realistic and achievable, not an assessment that attempts to get to the heart of " true costs" and "true prices."

It is useful to know the market value for the defender at which the challenger becomes attractive. The effect of the market value of the defender on the difference between the net present value of the defender and the challenger considered in Example 7.9 is shown in Figure 7.7. These results indicate that the breakeven value is at $10,558. If the market value for the defender is below this value, the defender is preferable, and if it is above this value, the challenger is preferable.

Notice in Example 7.9 that there is no mixing of the cash flows between those associated with the defender and those associated with the challenger. The sale of or the trade-in offer for the defender does not diminish the purchase price of the challenger. This is correctly accounted for by adopting the perspective of an exter-

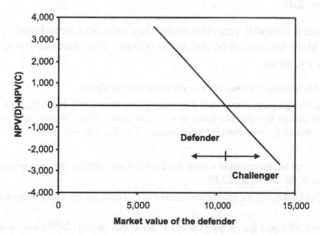

Figure 7.7 Effect of the market value of the defender on the difference between the *NPV* of the defender, D, and the challenger, C

nal assessor, one in which the service of both the defender and the challenger are required and must be obtained from the market now. This perspective reduces replacement decisions to the mutually exclusive decisions discussed in Section 7.2.2.

7.4 Non-mutually Exclusive or Independent Projects

Independent projects are those that do not affect the other investments, except that they may consume the resources available to the others. All independent options may be chosen, if there are sufficient funds for all of them and if they are all assessed as profitable. Examples of independent projects are investing in a new operation, embarking on a research and development initiative, and investing in a marketing campaign.

The value of the project depends on the amount, timing and risk of the cash flow. The topics concerning risk are discussed in Chapters 11–15.

In this section, the effect of size and timing of cash flow on the value of the project is discussed in order to establish some of the important influences on the value of a project. Although the net present value is the preferred method for the selection of independent projects, the other discounted cash flow criteria will be used in this section in order to demonstrate their application. Following this, the ranking of projects is examined, and finally, the selection of the optimal set of projects under conditions of limited capital, called capital rationing, is discussed.

7.4.1 Effect of Starting Times Delays and Project Life

(i)   Starting date

If two projects have different start times, they must both be indexed to a common date. The most obvious common date is the present. The calculation is illustrated in the following example.

Example 7.10: Selection of projects with different starting dates.

Consider the two projects whose cash flow profiles are shown in Table 7.18, both of which are identical, except that the one starts now, and the other starts in two years time. How much additional value does starting earlier create? The discount rate is 10%.

Solution:

The absolute and the incremental values for the *NPV* and *IRR* for the two options are calculated. This is shown in Table 7.19.

The methods of obtaining the results represented in this table are the same as those discussed before.

Both the total *NPV* and the incremental *NPV* agree that there is 20% increase in present value in starting earlier. This is highly significant, emphasizing that the timing of the cash flow is critical to the calculation of the discounted cash flow criteria. The starting date of the analysis does impact in comparing two projects at a common date. The difference between

Table 7.18 Cash flow profile for two projects with different start dates

	Project A	Project B
0	−133	0
1	30	0
2	30	−133
3	30	30
4	30	30
5	30	30
6	30	30
7	30	30
8	30	30
9	30	30
10	30	30
11	0	30
12	0	30

Table 7.19 Calculation of the absolute and incremental cash flow for Project A and B

Year	Project A	Project B	Discount factor	PV(A)	PV(B)	A-B	PV(A-B)
0	−133	0	1.000	−133.00	0.00	−133.00	−133.00
1	30	0	0.909	27.27	0.00	30.00	27.27
2	30	−133	0.826	24.79	−109.91	163.00	134.71
3	30	30	0.751	22.53	22.53	0.00	0.00
4	30	30	0.683	20.49	20.49	0.00	0.00
5	30	30	0.621	18.62	18.62	0.00	0.00
6	30	30	0.564	16.93	16.93	0.00	0.00
7	30	30	0.513	15.39	15.39	0.00	0.00
8	30	30	0.467	13.99	13.99	0.00	0.00
9	30	30	0.424	12.72	12.72	0.00	0.00
10	30	30	0.386	11.56	11.56	0.00	0.00
11	0	30	0.350	0.00	10.51	−30.00	−10.51
12	0	30	0.319	0.00	9.55	−30.00	−9.55
NPV				51.33	42.42		8.91
IRR				18.38%	18.38%		18.38%

the two options is the result of discounting the NPV of the project that starts later for two years. This calculation is as follows:

$$NPV_B = NPV_A \frac{1}{(1+k)^2} = \frac{51.33}{1.1^2} = 42.42$$

This means that there is a cost to the company for delaying a decision. If the project is delayed and reconsidered after two years, its NPV would be the same as the NPV for Project A, that is, equal to 51.33. The difference in value in this case is not an inherent part of the project, but simply results from the delayed decision.

Unfortunately, the IRR for the two projects and for the incremental cash flows is the same and does not reflect the effect of the different starting dates. The IRR method does not suggest that starting earlier is better.

The profile of NPV against discount rate for these two options is shown in Figure 7.8.

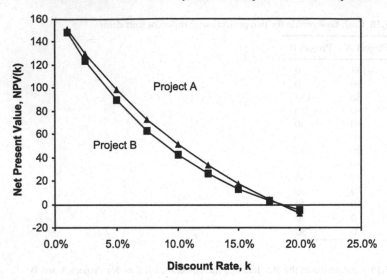

Figure 7.8 Effect of the different starting dates on the project's *NPV* and *IRR*

The value created for the shareholders or owners of the company at the end of the project is the future value of the two projects. The future value is determined at the end of the second project, that is, the terminal value, once both projects are complete. The future value is 161.1 for Project A and 133.2 for Project B, indicting that starting earlier creates 21% more value. This is an agreement with the *NPV* criterion.

(ii)   Project delays

As a matter of both practical value and interest it is worth examining the effect of project delays. A common engineering situation is that the project is initiated and then is delayed due to range of foreseeable and unforeseen reasons. The effect of delays (without budget overruns) is illustrated in the following example:

Example 7.11: Effect of a delay during a project.

Consider two projects, both of which start now, but the revenue for the second project is delayed due to construction or commissioning troubles. Determine the effect on the value to the investors and shareholders. The projects are the same as those in the previous example.

Solution:

The calculation of the net present value and the internal rate of return for these two projects are shown in Table 7.20.

These results clearly illustrate the destruction of value due to the delay. Both *NPV* and *IRR* are significantly less on the delayed project.

It must be emphasised that this illustrates the loss of value due to a delay in the revenue. This does not take into account the budget overruns that plague many projects. A delay in construction could easily result in budget overruns due to price escalation during the delay. This can be aggravated to such an extent that a project with a positive *NPV* never makes money.

Table 7.20 Calculation of the *NPV* and *IRR* for a project that is delayed after construction

Year	Project A	Project B	Discount factor	PV(A)	PV(B)
0	−133	−133	1.000	−133.00	−133.00
1	30	0	0.909	27.27	0.00
2	30	0	0.826	24.79	0.00
3	30	30	0.751	22.53	22.53
4	30	30	0.683	20.49	20.49
5	30	30	0.621	18.62	18.62
6	30	30	0.564	16.93	16.93
7	30	30	0.513	15.39	15.39
8	30	30	0.467	13.99	13.99
9	30	30	0.424	12.72	12.72
10	30	30	0.386	11.56	11.56
11	0	30	0.350	0.00	10.51
12	0	30	0.319	0.00	9.55
NPV				51.33	19.34
IRR				18.4%	12.3%

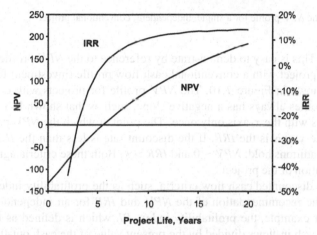

Figure 7.9 The effect of different project lives on the *NPV* and *IRR* of a project

(iii)   Project lives

The sensitivity of the net present value and internal rate of return to the project life is shown in Figure 7.9 using the cash flows for Project A in Example 7.11. The *NPV* is calculated at a discount rate of 10%. It is clear that project life is an important parameter for both *NPV* and *IRR*. Lengthening a project can add significant value *if the annual cash flows remain the same.*

7.4.2 Ranking

The *NPV*, *IRR* and other discounted cash flow techniques, such as *PI*, always agree on the attractiveness of a single, independent project that has a conventional cash

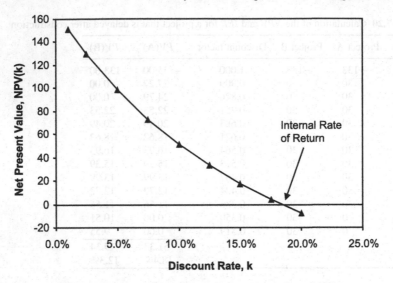

Figure 7.10 The *NPV* profile for a single, independent, conventional project

flow profile. This is easy to demonstrate by reference to the *NPV* profile. The *NPV* profile for a project with a conventional cash flow profile (investment followed by returns) is shown in Figure 7.10. The *NPV* profile for projects with conventional cash flow profiles always has a negative slope, such as that shown in Figure 7.10, and intersects with the *x*-axis only once. The point at which the *NPV* profile intersects with the *x*-axis is the *IRR*. If the discount rate is less than the *IRR*, then the following conditions hold: $NPV > 0$ and $IRR > k$. Both these criteria agree on their recommendation of the project.

The other discounted cash flow criteria, such as the profitability index will also agree with the recommendation of the *NPV* and *IRR* for an independent project. Consider, for example, the profitability index, *PI*, which is defined as the present value of the cash in-flows divided by the present value of the cash out-flow. This is expressed as follows:

$$PI = \frac{PV(Inflows)}{PV(Outflows)}$$

The *NPV* is the difference in the present values of the in-flows and the present value of the out-flows. This is expressed as follows:

$$NPV = PV(Inflows) - PV(Outflows)$$

If the *NPV* is positive, then the value of $PV(Inflows)$ is greater than the value of $PV(Outflows)$, which means that the profitability index must be greater than one.

If there is no shortage of funds, all independent projects with an *NPV* that is greater than zero are recommended for investment. If there is a shortage of capital or a limitation on capital spending so that not all projects can be chosen, it is said that capital is rationed. The selection of independent projects under conditions of capital rationing is discussed in the next section.

Because *IRR* is a popular method of assessing projects, projects are often ranked in an order of preference using *IRR*. In other words, the projects with the highest values of the *IRR* are the most preferred, while those with the lowest values of *IRR* are the least preferred. Any relative measure of the project's benefits to its costs, such as the *IRR*, *PI* or the B/C ratio, can be used as a ranking criterion. Projects are also ranked using *IRR* in order to determine the minimum attractive rate of return (MARR). The determination of MARR is discussed in Chapter 8 and 12. The ranking of projects by various methods is illustrated in the following example.

Example 7.12: Ranking independent projects.

Rank the projects whose cash flow profiles are given in Table 7.21 using *IRR*, *PI*, B/C ratio, and *MIRR*.

Solution:

The calculation of the discounted cash flows for each year is shown in Table 7.22.

The *PI* and B/C ratio are calculated from the present values of the cash flows, shown in Table 7.23.

Notice that the *PI* and the B/C ratio are identical, as would be expected from Chapter 6.

Also shown is the *IRR*, determined by trial and error from the cash flows, and the modified internal rate of return (*MIRR*). To calculate the *MIRR*, the following steps are required: (i) the future values of cash flows representing the returns are determined at the discount rate and summed to get the terminal value (TV); (ii) the *IRR* of the investment in year 1 and the TV in year 5 is obtained; (iii) the value obtained from the *IRR* calculation using the modified cash flows is the *MIRR*.

The projects can be ranked on the basis of their attractiveness using the relative measures. These rankings are given in Table 7.24.

Table 7.21 Cash flow profiles for four independent projects

Year	A	B	C	D
0	−10,000	−25,000	−35,000	−50,000
1	6,000	10,000	5,000	17,000
2	6,000	10,000	10,000	17,000
3	6,000	10,000	17,000	17,000
4	6,000	10,000	20,000	17,000
5	6,000	10,000	25,000	17,000

Table 7.22 Calculation of the present values for the cash flow profiles for the four projects

Year	Discount factor	PV(A)	PV(B)	PV(C)	PV(D)
0	1	−10,000	−25,000	−35,000	−50,000
1	0.9091	5,455	9,091	4,545	15,455
2	0.8264	4,959	8,264	8,264	14,050
3	0.7513	4,508	7,513	12,772	12,772
4	0.6830	4,098	6,830	13,660	11,611
5	0.6209	3,726	6,209	15,523	10,556

Table 7.23 Calculation of the *PI*, B/C ratio *NPV*, *IRR* and *MIRR*

	PV(A)	PV(B)	PV(C)	PV(D)
PV(CF)	22,745	37,908	54,766	64,443
NPV	12,745	12,908	19,766	14,443
IRR	53%	29%	26%	21%
PI	2.274	1.516	1.565	1.289
B/C	2.274	1.516	1.565	1.289
TV	36,631	61,051	88,201	103,787
MIRR	29.6%	19.6%	20.3%	15.7%

Table 7.24 Ranking of the four projects

	PV(A)	PV(B)	PV(C)	PV(D)
IRR	1	2	3	4
PI	1	3	2	4
B/C	1	3	2	4
MIRR	1	3	2	4

The *IRR* conflicts with the other measures in the ranking of Projects B and C. If there is no capital rationing, there is no need to resolve the conflict between the measures for independent projects, because all four of the projects have positive values for their *NPV* and hence all are acceptable. If capital rationing exists, a more detailed analysis is required and is discussed next.

7.4.3 Selection of Projects Under Capital Rationing

Companies and investors generally have more opportunities than resources. The owners of the opportunities would like to maximise the value of their investments. This is achieved by selecting the portfolio of projects that maximise value within the budget constraint. The selection of independent projects within the confines of a budget is known as capital rationing. The budget is finite and companies must select the set of projects that will create the most value for their owners within the confines of the available resources. In this situation, the largest project *NPV* is not necessarily the optimal choice, but the largest combination of project *NPV*s that sum to the capital budget is.

A limited budget is a constraint set by management or external forces. For companies experiencing financial stress, the capital budget is constrained by external forces. For a profitable large company, it is not a constraint set by external forces, since a profitable company can usually raise additional funds by issuing more debt and more equity. Management regards the issuing more stock as expensive, so they may be reluctant to raise funds in this manner. There may be other reasons for constraining the budget. For example, the lack of human resources to execute the projects can result in failure no matter how positive the opportunity. A third reason is that the limitation of the size of the budget creates managerial discipline that as-

Table 7.25 Project investment and values

Project	Investment	PV(CF)	NPV
A	20,000	53,000	33,000
B	5,000	13,000	8,000
C	25,000	65,000	40,000

sists in ensuring that the proposals submitted for approval are not overly optimistic or lacking in rigour.

Independent projects can be selected using either an absolute measure of value, such as *NPV*, or a relative measure of value, such as the *IRR*, as the decision criterion. These two methods are discussed next.

(i)   Selection based on absolute value (*NPV*)

Making the best use of the available funds means finding the combination of projects that yield the highest value, given by *NPV*, for the given funds. For a small number of projects, doing this is close to trivial. Consider the three projects in Table 7.25.

If the capital budget was $25,000, the company can choose project C and get 40,000, or it can choose projects A and B and get 41,000 (= 33,000 + 8,000). Clearly the choice of A and B is superior.

Consider the projects shown in Table 7.26. There are $2^4 = 16$ combinations of possible projects. If the budget was $25,000, what is the best combination? The possible combinations are identified in Table 7.27, and are ordered in terms of ascending investment requirements.

The combinations with a combined investment above 25,000 are not feasible since they exceed the budget. All other combinations are possible. From this list, the project combination that maximizes *NPV* is the optimal choice. The project combination AD creates the maximum wealth for the specified budget.

In order to come to a decision, the combination of all possible projects must be examined. This is unwieldy with any reasonable number of projects. If there are 20 projects, this entails a million possible combinations. The problem can be formulated in mathematically concise terms as follows:

Find the maximum value of *Z*, given by the following expression:

$$max(Z) = \sum_{i=1}^{N} X_i \, NPV_i \tag{7.9}$$

Table 7.26 Project investment and *NPV*

Project	Investment	NPV
A	10,000	12,000
B	5,000	8,000
C	8,000	7,000
D	15,000	19,000

Table 7.27 Possible combination of projects that can be selected from those given in Table 7.26

Project	Investment	NPV
None	0	
B	5,000	8,000
C	8,000	7,000
A	10,000	12,000
BC	13,000	15,000
D	15,000	19,000
AB	15,000	20,000
AC	18,000	19,000
BD	20,000	27,000
CD	23,000	26,000
ABC	23,000	27,000
AD	25,000	31,000
BCD	28,000	34,000
ABD	30,000	39,000
ACD	33,000	27,000
ABCD	38,000	46,000

where X_i is the proportion of the project in the budget, N is the total number of projects under consideration, and Z is the value of the selected projects. This maximization is subject to the constraint that the total investment is less than the budget amount. This is expressed as follows:

$$Budget \geq \sum_{i=1}^{N} X_i I_i \qquad (7.10)$$

In most formulations of the capital budgeting problems, the project is treated as indivisible, that is, it is either accepted in total or rejected. This means that X_i can only take on values of zero or one. It is not really true that projects are indivisible, since projects can, and are often, developed in joint venture with others. The joint venture partners are all part owners of the project. However, project indivisibility creates the constraint given as follows:

$$X_i = \{0, 1\} \, \forall i \qquad (7.11)$$

Equation 7.11 says that X_i belongs to a set that only contains zero and one for all values of i.

This system of equations conforms to the definition of a linear programme (LP) and can be solved using LP techniques. It can also be solved by brute force. A brute force implementation would be a computer program, such as a spreadsheet macro, that loops through all the possible values for X_i and finds the largest Z amongst those values of Z that do not violate the budget constraint.

The use of LP is illustrated in the following example. The LP used is an add-in to the Microsoft Excel spreadsheet package.

Table 7.28 Investment requirements and *NPV* for six projects

Projects	Investment	NPV
A	12,000	17,000
B	5,000	12,000
C	8,000	9,000
D	10,000	12,000
E	15,000	19,000
F	17,000	25,000
Total	67,000	94,000
Budget	30,000	

Example 7.13: Creating a capital budget using linear programming.

Consider the six projects with the investment requirement and *NPV* given in Table 7.28. If the budget is $30,000, determine the set of projects that should be funded.

Solution:

A spreadsheet can be constructed that calculates the summations in Equations 7.9 and 7.10. This is shown in the table below. In order to determine the optimal portfolio of projects, the values of X_i must be varied to maximize Z. This can be achieved using the "solver" function in MS Excel. This is done as follows:

(i) Install "solver" using the Add-In Manager.

(ii) From the menu choose Tools > Solver.

(iii) In the "solver" dialog box enter the cell reference for Z in the "Set Target Cell" edit box.

(iv) Choose the "Max" radio button.

(v) Enter the reference to the cells for the X_i in the "By Changing Cells" edit box.

(vi) Press Add to add constraints. Enter the values of X_i as the first constraint and that they are binary, so that the resulting constraint after accepting is similar to $F$35:$F$40 = binary.

(vii) Add another constraint, representing the budget constraint, so that the cell with the total investment given by the summation in Equation 7.10, is less than the budget. This will result in a constraint written similar $G$41 <= $G$42.

(viii) Press "Solve."

The results are shown in Table 7.29.

Table 7.29 Determination of the set of selected projects

Projects	I_i	NPV_i	X_i	$X_i I_i$	$X_i NPV_i$
A	12,000	17,000	0	0	0
B	5,000	12,000	1	5,000	12,000
C	8,000	9,000	1	8,000	9,000
D	10,000	12,000	0	0	0
E	15,000	19,000	0	0	0
F	17,000	25,000	1	17,000	25,000
Total	67,000	94,000		30,000	46,000

The projects that are selected are those in column 4, denoted X_i, that have a value of one. The projects that are not selected have a value of zero. The results indicate that the company should invest in projects B, C, and F. This will consume the entire budget, which is $30/67 = 44\%$ of the value of the projects on offer, for an overall NPV of \$46,000, which is $46/94 = 48\%$ of the value on offer.

Additional constraints that can be added to the capital budgeting process are related to possible interrelationships between projects. These relationships can be added to the analysis so that there is no need to perform separate calculations for different classes of projects. These interrelationships concern mutually exclusive projects and contingent projects.

(i)   *Mutually exclusive projects* are projects where only one of those under consideration can be chosen. This constraint can be expressed in terms of the following inequality:

$$1 \geq \sum_{j \in J} X_j \quad \text{and} \quad X_j = \{0, 1\} \tag{7.12}$$

where J represents the set of mutually exclusive projects.
If it is necessary to select at least one of them (the replacement has to happen), then the inequality becomes an equality:

$$1 = \sum_{j \in J} X_j \tag{7.13}$$

(ii)   *Contingent projects* are projects where the acceptance of the project is dependent or contingent on the acceptance of another project. For example, project B requires the acceptance of D in order for it to be accepted. In this case, the contingent decision can be expressed in the following form:

$$X_B \leq X_D \tag{7.14}$$

The following example illustrates the application of these relationships in the selection of a capital budget.

Example 7.14: Capital rationing with mutually exclusive and contingent projects.

Consider the projects given in Table 7.30, in which the investment requirement and NPV of each project is provided.

Projects I and J are mutually exclusive projects, for example, a replacement that must be done. Project K is contingent upon the acceptance of project B. Determine the projects that should be selected.

Solution:

Projects I and J represent the replacement of equipment, so that Equation 7.12 describes their relationship. This is programmed on the spreadsheet by creating a cell that is the sum of X_I and X_J and making a constraint in "solver" that the value of this cell must be equal to one.

Project K is contingent upon the acceptance of Project B, so an equation of the form of Equation 7.14 describes their relationship.

Table 7.30 Investment requirements and *NPV* for 11 projects

Projects	Investment	NPV
A	12,000	17,000
B	5,000	12,000
C	8,000	9,000
D	10,000	12,000
E	15,000	19,000
F	17,000	24,000
G	12,000	18,500
H	13,000	19,000
I	5,000	11,000
J	4,000	11,000
K	2,000	4,000
Total	103,000	156,500
Budget	30,000	-

Table 7.31 Selection of projects for acceptance

Projects	I_i	NPV_i	X_i	$X_i I_i$	$X_i NPV_i$
A	12,000	17,000	0	0	0
B	5,000	12,000	1	5,000	12,000
C	8,000	9,000	0	0	0
D	10,000	12,000	0	0	0
E	15,000	19,000	0	0	0
F	17,000	24,000	1	17,000	24,000
G	12,000	18,500	0	0	0
H	13,000	19,000	0	0	0
I	5,000	11,000	0	0	0
J	4,000	11,000	1	4,000	11,000
K	2,000	4,000	1	2,000	4,000
Total	103,000	156,500		28,000	51,000

The constraints are set up in "solver" as described in Example 7.13, and the programme is run in exactly the same manner. The results are shown in Table 7.31.

Projects that have been selected have a value of one in column 4. This means that Projects B, F, J and K have been selected. The mutually exclusive projects were I and J, and J has been selected. This is in accordance with a decision that compared Projects I and J on their own. Project K is contingent on B, which means that only if B is chosen is there a possibility of K being selected. In this case, both B and K have been chosen.

The overall result is to only invest $28,000 of the $30,000 in the budget from a set of projects on offer whose capital requirements totalled $103,000.

The above discussion assumes that all of the capital will be spent in one year. This may not be true, and is probably highly unlikely. The capital budgeting process is strategic in nature, and must anticipate the capital requirements for some years in advance. The extension of the formulation to multiple periods is relatively straight-forward. Past commitments for capital must be excluded from the budget amount, and the constraint given by Equation 7.10 must account for investment out-flow in future years. This is given by the following expression:

$$Budget_t \geq \sum_{i=1}^{N} X_i\, I_{it} \tag{7.15}$$

where $Budget_t$ represents the budget in year t, and I_{it} represents the investment cost for project i in year t. Thus, there will be t constraints, one for each year in which the budget is under consideration.

(ii)　Selection based on relative measure of value (IRR, PI)

The selection of the project based on *IRR* or *PI* consists of two steps: (i) ranking the project in terms of the *IRR* or *PI*; and (ii) determining the set of projects with the highest *IRR* or *PI* whose sum of capital investment is equal to the budgeted amount. These steps are illustrated in the flowing example.

Example 7.15: Capital rationing by IRR.

A company is considering the approval of the projects given in Table 7.32. Determine the projects for approval if the budget constraint is 300.

Solution:

The projects are ranked in terms of the *IRR*, and those with the highest *IRR* at that constraint are selected. The ranked projects are shown in Table 7.33.

Table 7.32 Investment requirements and internal rates of return for eight projects

Project	Investment	IRR (%)
A	10	52.8
B	25	28.6
C	35	25.5
D	50	20.8
E	60	19.9
F	70	23.1
G	80	25.4
H	90	34.3

Table 7.33 Ranking of the eight projects in therms of the internal rate of return

Project	Investment	IRR	Cumulative investment
A	10	52.8	10
H	90	34.3	100
B	25	28.6	125
C	35	25.5	160
G	80	25.4	240
F	70	23.1	310
D	50	20.8	360
E	60	19.9	420

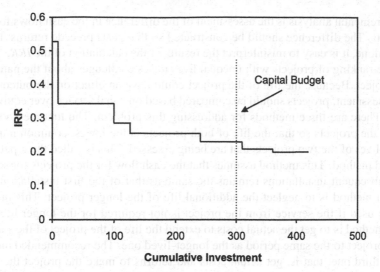

Figure 7.11 Selection of project for approval

The cumulative investment must be less than the budget of 300. This means that projects A, H, B, C and G are selected. These projects have a total investment of 240. Project F, which requires an investment of 70 is too large, so is not selected. But the budget is under-utilized, so either D or E should be chosen.

The selection of the projects based on *IRR* is shown in Figure 7.11.

The *IRR* of the best project that is rejected is regarded as the minimum attractive rate of return, the hurdle rate, or the opportunity cost of capital. This method can be used to determine the hurdle rate.

7.5 Summary

Three topics were examined in this chapter: mutually exclusive projects, replacement studies, and independent projects.

Mutually exclusive projects are those in which one of the alternatives must be chosen. The assessment of the projects must not only determine if the projects meet the economic requirements set by the company, but must also recommend which of the alternatives is the preferred choice. To do this, the projects must be ranked in order of preference.

The ranking of mutually exclusive projects with equal lives can be performed using any of the assessment criteria. The primary criteria used are the *NPV* and the *IRR*. A possible conflict can arise between the recommendations made by the *NPV* and *IRR*, depending on the cash flow profiles of the two projects. If this is the case, the conflict is resolved by using incremental analysis.

Incremental analysis is the assessment of the difference in the cash flows for two projects. The difference should be constructed so that costs precede returns. If this is not done, it is easy to misinterpret the results of the calculation of the *IRR*.

The ranking of projects with unequal lives raises challenges about the nature of the project. Because the life of the project could have an effect on the outcome of the assessment, projects should be compared based on equal service over equivalent lives. There are three methods for addressing this problem. The first method is to repeat the projects so that the life of both projects is the lowest common multiple of the lives of the two projects that are being assessed. This is called the repetition-in-kind method. This method assumes that the cash flow for the projects for second and subsequent installations remains the same as that of the first installation. The second method is to neglect the additional life of the longer project. This method can be used if the service from the project is not required for the longer life. The third method is to get the actual costs to extend the life of the project of the shorter-lived project to the same period as the longer-lived one. The recommended method is the third one, that is, get actual costs and returns to make the project the same duration.

The most common situation requiring a decision on capital expenditure is the replacement of existing equipment. These are mutually exclusive decisions that are mandatory, since the service is required. Any of the methods of ranking mutually exclusive decisions can be used, and if the alternatives have unequal lives, it is recommended that actual costs be obtained in order to assess the alternatives fairly.

Replacement studies are one of the most common capital budgeting decisions. The incumbent equipment, called the defender, is compared against the replacement, called the challenger. Replacement studies are often conducted using the *EAC*. If the time horizon is specified, the *EAC* of the two alternatives is compared. If the time horizon is unspecified, the two alternatives are compared at the minimum *EAC* with respect to their remaining service. This minimum point represents the optimal use of the capital and operating expenses, and is referred to as economic service life. The option with the lowest *EAC* is chosen in both instances.

Independent projects pose no new challenges to the method of analysis than those discussed for mutually exclusive projects, except for the budget constraint. All projects with a positive *NPV* are acceptable if there are sufficient funds available. The projects are chosen so that the maximum total *NPV* is obtained for the budget amount.

7.6 Review Questions

1. What is the difference between a revenue-producing project and a service-producing one?
2. How are mutually exclusive projects different from independent projects?
3. Provide some examples of mutually exclusive projects.

4. What types of decisions are required for mutually exclusive projects and for independent projects?
5. Why is there sometimes a conflict between *NPV* and *IRR*?
6. How can a conflict between *NPV* and *IRR* be resolved?
7. What is incremental analysis?
8. Should income precede costs or costs precede income in the incremental analysis?
9. What three methods can be used to ensure that projects with unequal lives are compared fairly?
10. What assumption is made in applying the *EAC* to projects of different lives?
11. What assumption is made in the repetition-in-kind method of equalizing the lives of projects?
12. When and why is the economic service life determined?
13. What is the capital recovery?
14. What is the annual operating charge?
15. How does the annual operating charge differ from the operating costs each year? Are they the same?
16. What is capital rationing?
17. What is the objective in optimising the capital budget?
18. What solution techniques are required to solve the optimum budget?

7.7 Exercises

1. Two mutually exclusive projects are being considered. The cash flow profiles are given in the table below:

Year	Project A	Project B
0	−24,000	−24,000
1	12,000	0
2	12,000	5,000
3	12,000	10,000
4	12,000	25,000

   (i) Determine the *NPV* of the two projects if the discount rate is 12%.
   (ii) Determine the *IRR* of the two projects.
   (iii) Determine the *EAC* and the *PI* for the projects.
   (iv) Recommend which of these two projects should be adopted.

2. The company is evaluating the replacement of equipment. The cash flow profiles for the defender and the challenger are given in the table below:

Year	Defender	Challenger
0	−20,000	−40,000
1	−5,000	−5,000
2	−7,500	−5,000
3	−10,000	−5,000
4	−12,500	−5,000

(ii)   Determine the *NPV* and *EAC* for each of the options if the discount rate is 6%.

(ii)   Recommend which of the options should be chosen.

3.   An engineer, Dr. Howard, has discovered a method of converting cellulose material, such as sugar-cane bagasse, into ethanol that can be used as a green fuel. Dr. Howard has developed three different flowsheets for the arrangement of the equipment and sequencing of the chemistry. The costs for each of these alternatives is given in the table below:

Alternative	Capital cost	Operating cost	Value Of fertilizer sales
1	50,000	5,300	21,000
2	62,000	6,700	28,000
3	70,000	7,600	31,000

The discount rate is 10% and the life of the operation is 10 years.

(i)   Determine the preferred choice without taxes.

(ii)   If the capital equipment is depreciated on a straight-line basis over ten years, and tax is levied at 40%, determine the preferred choice.

4.   A pharmaceutical company is evaluating the options for the manufacture of a chemical component in one of its drugs. There are three different locations, and the company needs to choose one of these. The costs of the different options are given in the table below:

	East Coast	West Coast	Midlands
Land	$3,000,000	$4,000,000	$3,500,000
Buildings	$6,000,000	$6,500,000	$7,000,000
Equipment and installation	$9,000,000	$10,000,000	$11,000,000
Electricity per year	$4,000,000	$5,000,000	$4,000,000
Transportation per year	$8,000,000	$6,000,000	$9,000,000
Labour per year	$15,000,000	$12,000,000	$16,000,000

The company will be charged property taxes by the local government at a rate of 1% of the value of the land and buildings. The land will have a salvage value of 90%, and the buildings of 50% of their original costs.

If the company uses a discount rate of 12%, and evaluates these types of choice over a life of 15 years, determine the preferred option.

5. Energy costs have risen significantly over the last few years and the company has decided to take steps to reduce the cost of energy by installing an energy recovery system. Two options are available. Their costs are as follows:

	Option A	Option B
Equipment and installation costs	$1,000,000	$750,000
Operating costs	$15,000	$20,000

If the market value of the equipment declines at a rate of 15% each year, the operating costs of both options increase by 50% each year and the company's discount rate is 9%, determine the following:

(i) The economic service life for the two options.
(ii) The *EAC* for the two options at the economic service life.
(iii) The preferred option based on part (ii).

6. The company is considering the purchase of a new machine for the insertion of electronic components into a printed circuit board. The machine was purchased three years ago at a cost of $500,000, and is operating adequately, although maintenance costs and defect rates are rising. A new machine is available for $400,000. The new machine should have lower costs and better defect rates. The estimation of the costs are given in the table below:

Year	Defender		Challenger	
	Operating costs	Salvage value	Operating costs	Salvage value
0		150,000		350,000
1	40,000	120,000	15,000	320,000
2	45,000	100,000	15,000	280,000
3	50,000	80,000	15,000	250,000
4	55,000	60,000	15,000	220,000

The company's discount rate is 11%.

(i) Determine the *EAC* for the two alternatives if the life of the machines is expected to be four years.
(ii) Recommend whether the machine should be replaced now.

7. Two alternatives for the climate maintenance system in an office building are under consideration. The life of the building is 25 years, and the discount rate

for the owners of the building is 10%. The costs for the two systems are given in the table below:

	Heat pump	Air conditioning
Investment	$500,000	$300,000
Energy costs per year	$ 20,000	$ 30,000
Maintenance costs per year	$ 10,000	$ 20,000

(i)   Determine the *NPV* and *EAC* for the two alternatives.

(ii)   Recommend which alternative should be chosen.

8.   An analytical company provides chemical analyses. The competitiveness of its business depends on having the most up-to-date equipment and properly trained staff. A particular instrument costs $35,000 and can be traded-in for a value that declines at a rate of 15% each year. Maintenance costs are fixed by a service contract at $3,000 per year. The maximum life of the instrument is ten years. If the company's discount rate is 12%, determine the following:

(i)   What is the economic service life of the instrument? Can you generalize this result?

(ii)   Determine the *EAC* of the instrument.

(iii)   If ten thousand samples are processed each year, and the personnel cost of operating the instrument is $5,000 per month, determine the minimum amount per sample that the company should charge its customers.

9.   Process plants need to be painted regularly to prevent corrosion. In addition, piping, safety areas and the like that need to be colour-coded must be properly maintained. There are two options for paint. The first costs $1 per litre and the other costs $0.50 per litre. The more expensive option lasts longer. If the company's discount rate is 10%, determine the following:

(i)   The time that the more expensive paint must last if the cheaper paint lasts three years. This can be determined by equating the *EAC* for each of the options.

(ii)   What implicit assumptions are made in this analysis?

(iii)   Determine the answer using a method of analysis other than *EAC*.

10. The following independent projects are under consideration:

Project	Initial investment	Present value of the cash flows
A	100	120
B	150	180
C	200	220
D	400	320
E	10	30
F	280	305

(i)   Rank the projects in terms of *NPV* and *PI*.
(ii) If the capital budget is 800, determine which projects should be approved.

11. The following projects have been submitted for approval:

Project	Initial investment	*IRR*
A	200	13%
B	250	11%
C	275	12%
D	300	14%

If the capital budget is 700, determine which projects should be approved.

12. The following projects have been submitted for approval:

Project	Investment	*NPV*
A	12,000	17,000
B	5,000	12,000
C	8,000	9,000
D	10,000	12,000
F	15,000	25,000
G	17,000	24,000
H	12,000	18,500
I	11,000	19,000
J	8,000	11,500
K	8,000	11,000
M	2,000	4,000

If the budget is $85,000, determine which projects should be approved.

13. Two projects have the following cash flow profiles:

Year	Project A	Project B
0	−100	−100
1	60	20
2	60	60
3	60	100

   (i)   The *NPV* and *IRR* of the projects if the discount rate is 8% pa.
   (ii)   Determine the preferred project.
   (iii) Do these projects have a "Fisher's intersection?"

14. Determine Fisher's intersection for the following two projects:

Year	Project A	Project B
0	−10,000	−60,000
1	5,000	30,000
2	5,000	30,000
3	5,000	30,000
4	5,000	30,000

15. The benefit-cost ratio is often used in evaluating public sector investments. A contractor wishes to win the tender for the building of a road. If the benefits are expected to be $150,000 each year for 25 years, determine the maximum cost that the contractor can charge for the work if the public discount rate is 10% pa.

16. If the costs in Question 14 are in real terms, and the inflation rate is 6% pa, determine the maximum amount that the contractor can charge.

17. A toll road that will last 25 years is estimated to cost $214 million to build. The benefits are estimated from the time saved by road users and the reduction in traffic congestion on other roads. If the public sector discount rate is 12%, and the inflation rate is 7%, determine the annual benefits that are required to justify this investment.

Chapter 8
Practical Issues in the Evaluation of Projects

8.1 Introduction

In the previous chapters, the discussion has been confined to describing the analytical techniques and demonstrating their application. The discount rate, for example, was prescribed. Some practical issues in the application of discounted cash flow techniques are examined in this chapter. The topics chosen in this chapter represent the main points of discussion amongst practicing engineers in assessing projects. These are the issues of whether the analysis should be performed using prices that are escalated for inflation or not, whether the tax position of the company as a whole should affect the tax position of the project, how the depreciation should be calculated, and what discount rate should be used.

8.2 Inflation and Price Escalation

The escalation of prices for the same goods and services is inflation. In other words, it is the loss of purchasing power of money. It is the erosion of the value of money, a frictional loss on productive effort. Inflation is central to the competing theories of macroeconomics, such as the opposing Keynesian and monetarist views. Keynes believed that government intervention in the economy was necessary in order for the economy to create full employment, and that government spending can spur economic recovery. On the other hand, the monetarists, led by Friedman, believe that the government should not intervene in the markets, that government spending is less efficient than private enterprise, and that inflation, not employment, needs to be controlled. This macroeconomic debate is unresolved, although supporters of the monetarist view seem to be in the majority. Many of the governments around the world steer their economies by adjusting the interest rate in an attempt to control inflation.

Inflation is generally measured by the *consumer price index*, *CPI*, which is a weighted average of a standard set of goods and services. These goods and services are surveyed monthly for price changes, which are then aggregated into the *CPI* as a measure of inflation. The *producer price inflation* is a similar measure, but of a general set of raw materials.

In Chapter 5, inflation was discussed in the context of nominal and real interest rates. Nominal rates include inflation and real rates exclude inflation. In the same manner, assessments of engineering projects can be made on the basis of nominal amounts, or real amounts. "Nominal amounts" refers to preparing the free cash flows for a project in prices and values that are inflated each period by the rate of inflation. "Real amounts" refers to preparing these same free cash flows without the effects of inflation. These two methods are also referred to as the "constant dollar" method and the "escalated dollar" method. Nominal amounts are also called "money of the day" or "monetary" amounts.

The problem of using the nominal method for the calculation of the free cash flows is the prediction of the inflation rate. In the absence of taxes, the real method and the nominal method will produce the same result if all figures are inflated by the same amount. However, inflation has an effect on after-tax values, since depreciation (a component of the tax calculation) is based on historical costs, and is not escalated. This is illustrated in the following two examples.

Example 8.1: The influence of inflation in the absence of taxes.

Evaluate the project with the cash flows in real terms given in Table 8.1. The discount rate in real terms is 10% and the expected inflation rate is 15%.

Solution:

The *NPV* in real terms is calculated using the real discount rate. This is shown in Table 8.2.

Table 8.1 Cash flow profile for a project in real terms

Year	Free cash flow
0	−10,000
1	6,000
2	6,000
3	6,000
4	6,000
5	6,000

Table 8.2 Calculation of the net present value using the real discount rate

Year	Free cash flow	Discount factor	PV
0	−10,000	1.000	−10,000
1	6,000	0.909	5,455
2	6,000	0.826	4,959
3	6,000	0.751	4,508
4	6,000	0.683	4,098
5	6,000	0.621	3,726
NPV			12,745

Table 8.3 Calculation of the net present value using the nominal discount rate

Year	Free cash flow	Nominal CF	Discount factor	PV
0	−10,000	−10,000	1.000	−10,000
1	6,000	6,900	0.791	5,455
2	6,000	7,935	0.625	4,959
3	6,000	9,125	0.494	4,508
4	6,000	10,494	0.391	4,098
5	6,000	12,068	0.309	3,726
NPV				12,745

The nominal discount rate, k, is related to the real discount rate, r, by:

$$k = r + h + rh$$

where h is the inflation rate. This means that the nominal rate is equal to 26.5% = 10% + 15% + (10%)(15%).

In order to calculate the NPV using a nominal discount rate, the real cash flows must be escalated by the inflation rate of 15%, to get the nominal cash flows. Then the cash flows must be discounted at the nominal discount rate to get the NPV. These calculations are given in Table 8.3.

The NPV for both the real cash flow analysis and the nominal cash flow analysis are exactly the same.

Example 8.2: The influence of inflation with taxes.

A company wishes to invest in an opportunity to produce ball bearings. The equipment will cost $100,000, and can be depreciated over three years on a straight-line basis. The process will generate revenues of $80,000 for three years, after which the equipment will be scrapped for an estimated $10,000. The operating costs are expected to be $20,000. There will be an increase in working capital for inventory of $3,000, which can be recovered at the end of the project. Assess the project using NPV as a decision criterion if the tax rate is 30%, the real discount rate is 5% and the inflation rate is 7%.

Solution:

The free cash flows in real terms are given in Table 8.4.

The tax is calculated as follows:

$$\text{Tax} = \text{TaxRate}(\text{Operating Profit} - \text{Depreciation}) = 0.3(60 - (100 - 10)/3)$$
$$= 0.3(60 - 30) = \$9,000$$

Table 8.4 The calculation of the annual free cash flow for the project (amounts in thousands)

Year	0	1	2	3
Revenue	0	80	80	80
Operating costs	0	20	20	20
Profit	0	60	60	60
Tax	0	9	9	9
Investment	−100			
Change in working capital	−3	0	0	3
Salvage				10
Free cash flow	−103	51	51	64

Table 8.5 Free cash flow for each year escalated at the inflation rate (amounts in thousands)

Year	0	1	2	3
Revenue		85.60	91.59	98.00
Operating costs		21.40	22.90	24.50
Profit		64.20	68.69	73.50
Tax		10.26	11.61	13.73
Investment	−100.00			
Change in working capital	−3.00			3.68
Salvage				12.25
Free cash flow	−103.00	53.94	57.09	75.70

Table 8.6 Calculation of the tax charged (amounts in thousands)

Year	0	1	2	3
Profit		64.20	68.69	73.50
Add profit on sale of equipment				2.25
Less depreciation		30.00	30.00	30.00
Taxable income		34.20	38.69	45.75
Tax at 30%		10.26	11.61	13.73

The *NPV* for this cash flow stream is $47,116.

The free cash flows in nominal terms are given in Table 8.5.

The *NPV* for this cash flow stream must be determined at the nominal discount rate, which is equal to 5% + 7% + (5%)(7%) = 12.35%. The *NPV* at this discount rate is $43,617, which is lower than that calculated using real values. The reason is that tax is calculated using depreciation that is not escalated. The details of the tax calculation are given in Table 8.6.

The main reason for the difference in the *NPV* of the two scenarios is the difference in the *NPV* of the tax for each of the two (obviously discounted at the real and nominal rates, respectively). It is easy to show that if the depreciation were escalated, there would be no difference between the two (if there were no salvage gain and no change in working capital).

The effect of inflation on the *NPV* of a project is shown in Figure 8.1. (The results were calculated for a project that has the same revenue and operating costs as the previous example. However, the investment required is $400,000, the life of the project is 10 years, the equipment is depreciated to zero over the life of the project, and there are no working capital requirements.) The results of Figure 8.1 clearly show the erosion of value by inflation. As discussed in the previous example, the reason for this is the tax calculation. The total amount of tax paid over the life of the project in nominal amounts is shown in Figure 8.2. This figure demonstrates that the erosion of the value is due to tax collection. In the light of this, it is not surprising that the taxes are sometimes referred to as "fiscal take."

A common error is to escalate the depreciation, along with the figures for all the other variables, for inflation. This will lead to incorrect tax calculations, and, perhaps, the incorrect recommendation. Depreciation is calculated based on the historical cost of the fixed capital. It is not subject to inflation.

Another common mistake is to co-mingle the nominal and real methods. This is illustrated in the following example.

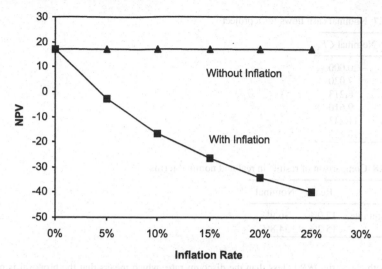

Figure 8.1 The effect of inflation on the *NPV* of a project

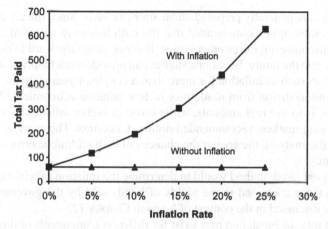

Figure 8.2 The effect of the inflation rate on the total taxes collected from the project over the life of the project in nominal amounts

Example 8.3: Co-mingling the real and nominal methods.

A project has the nominal cash flows shown in Table 8.7. The real discount rate is 17%. Is the project acceptable on the basis of the project's *IRR*? The inflation rate is 17%.

Solution:

The project's *IRR* is 34.8%.

Wrong answer: The project is acceptable because the *IRR* is greater than the discount rate of 17%.

Correct answer: The calculated *IRR* is the nominal *IRR*, while the discount rate is the real discount rate. Either escalate the real discount rate to a nominal equivalent, or discount the nominal *IRR* to a real rate before comparing the two to make a recommendation. Both solutions are shown in Table 8.8.

Table 8.7 Nominal cash flows for a project

Year	Nominal CF
0	−20,000
1	7,020
2	8,213
3	9,610
4	11,243
5	13,155

Table 8.8 Comparison of results in real and nominal terms

	Real	Nominal
Discount rate	17.0%	36.9%
IRR	15.2%	34.8%

In both cases, the *IRR* is less than the discount rate, which means that the proposal is not acceptable.

Cash flows are generally prepared on an after-tax basis. Since the tax calculation affects the results, it is recommended that the cash flows are prepared in nominal amounts. This, however, raises other issues. It is not straightforward to estimate the inflation rate into the future. Economic studies can provide some help, but these rarely forecast factors such as inflation for more than a couple of years into the future.

The recommendation from academics in low inflation economies, like the US in the 1990s, is to use real amounts, while those in higher inflationary economies of the emerging markets recommended nominal amounts. The more assumptions included in the analysis, the greater the chances of both arithmetic errors and errors of judgement.

The most unbiased method would to determine the inflation expectations for the future from those contained in the yields of bonds sold by the government. This topic will be discussed in the context of loans in Chapter 12.

Different rates of escalation may exist for different components of the cash flow analysis. For example, due to local shortage of labour, the wage costs may rise at a greater rate than the average rate of inflation. In another situation, the prices for steel may increase at a different rate to the price of labour. It is good practice to derive these escalation figures as deviations or variances from the general inflation rate. For example, it is better to state in the analysis that the cost of local labour is expected rise at a rate that is 2% above inflation for the first five years, rather than to state that the cost of labour is expected to rise at a rate of 7%.

Inflation often makes the repayment of debt easier, and benefits those in debt. However, projects are generally evaluated on the entity basis. The advantage of debt, because it is tax deductible, is included in the after-tax weighted average cost of capital that is used as the discount rate. There should be no interest or debt in the calculation of the project free cash flows. This is not a theoretical position; rather it is a practical position. It is simply done in order to separate the investment and financing decision.

8.3 Taxation

The principles of taxation and their application to the determination of project cash flow were discussed in Chapters 3 and 4. The purpose of this discussion is to elaborate on the methods of depreciation, the choice of the method for the calculation of depreciation, and the effect of the depreciation method on the value of the *NPV* and *IRR*.

8.3.1 Tax Position of the Company

Since the cash flows that are important are those prepared on an after-tax basis, the tax position of the company can influence the evaluation. Tax losses from other businesses of the company may reduce the tax liability of the project under investigation. There are four tax positions that are important:

(i)   The company or the investor does not have other tax losses against which income can be expensed. In this case, the project must be evaluated on a stand-alone basis. Losses within the project are carried forward (if allowed) and expensed against taxable income when it is realized by the project. This is the usual method for the analysis of projects.

(ii)   The company may be in an assessed loss position and the project's taxable income can be offset against these other losses. Not all the projects that are under investigation may be able to claim the loss against which to offset their income. It is better to assess the different projects on the same basis, which means handling the project's tax internally within the project and disregarding the company's tax position.

(iii)   The company or the investor has sufficient losses in other existing businesses so that the company believes that it will not be liable for tax for a long time into the future. In this case, tax calculations can be left out of the analysis.

(iv)   The project may not be liable for taxes, such as those in a tax-free zone, or be owned by the government or a charitable organisation.

The project should be assessed on its own merits, and when ranked against other projects, the basis for the comparison must be fair.

8.3.2 Methods of Calculation of Depreciation

The tax payable by the project or the company is dependent on the depreciation, or as is it called in the United Kingdom, the capital allowance. The concept of depreciation was discussed in Chapter 3, and it was shown how the depreciation affects the income statement, the balance sheet and the cash flow statement. Also discussed were the book value of the asset, and the treatment of the profit or loss made on the disposal of the asset.

It was mentioned in the discussion of depreciation in Chapter 3 that there are different methods for the calculation of the depreciation. Up to this point, the calculation of the depreciation has been on a straight-line basis. In this section, the different methods for the calculation of depreciation *for tax purposes* are discussed.

The tax codes of different countries usually prescribe asset classes that specify the depreciation period. For example, vehicles are in an asset class that is to be depreciated over a period of five years. The US tax code also separates assets into personal property and real property. Real property is real estate and improvements to real estate, such as buildings and structures. Personal property is all tangible assets that are not real property. For example, equipment, machinery vehicles and computers are all regarded as personal property. An example of the asset classes specified by the US tax code is given in Table 8.9.

The UK system does not rely on a depreciation period, since the calculation of the capital allowance is on a declining balance method, which is discussed later. This method provides a tax incentive for long-lived assets compared with other methods of depreciation.

The main forms for the calculation of depreciation are following:

(i)   Straight-line depreciation (SL)
(ii)   Declining balance
(iii) Sum-of-years-digits (SOYD)
(iv)   Modified Accelerated Cost Recovery System (MACRS)

These methods of calculation of the depreciation charge are presented in the following sections.

(i)   Straight-line depreciation

The straight-line method spreads the cost of the fixed capital expenditure evenly over the life of the equipment. The annual depreciation charge in year t, d_t, can be expressed as the following equation:

$$d_t = \frac{P - S}{N} \tag{8.1}$$

Table 8.9 Asset classes for the US tax code

Property class	Asset description and examples
3-Year property	Special manufacturing tool and devices, motor vehicles and tractors
5-Year property	Computers, automobiles, trucks, aircraft, research equipment, manufacturing equipment
7-Year property	Office furniture, fixtures, railroad cars and engines, agricultural machinery
10-Year property	Petroleum refining, ship building
15-Year property	Roads, land improvements, pipelines, telephone distribution
20-Year property	Municipal sewers, farm buildings, water utilities
27.5-Year property	Residential rental property
39-Year property	Non-residential real property, but not the land itself

where P is the initial cost of the item, S is the estimated salvage value after N years, and N is the depreciation period.

The initial cost of the item is the delivered and installed cost. It is also known as the first cost basis, or unadjusted basis. The salvage value is the estimated trade-in or market value at the end of the depreciation period. It may be negative if there are disposal costs. The depreciation rate is equal to $1/N$ for the straight-line method.

The book value at year t, B_t, is the initial cost, P, less all of the accumulated depreciation. This can be expressed as follows:

$$B_t = P - \sum_{i=0}^{t-1} d_i \qquad (8.2)$$

The book value can also be expressed as the book value of the previous year less the annual depreciation charge, given by the following expression:

$$B_t = B_{t-1} - d_t \qquad (8.3)$$

The calculation of the depreciation is illustrated in the following example.

Example 8.4: Straight-line depreciation and book value.

An asset is purchased for $7,000, it costs $1,000 to deliver it to site and a further $2,000 to install it. It can be depreciated for tax purposes over seven years and will have a salvage value at the end of seven years of 5% of the original value. Create a depreciation schedule for the asset.

Solution:

The initial cost is equal to the purchase price plus the delivery and installation costs, that is, a total $10,000. The salvage value is 5% of this total, that is, $500. The annual depreciation is equal to $(10,000 - 500)/7 = \$1,357.14$. The book value over the life of the asset is calculated using Equation 8.3. The results of this calculation are given in Table 8.10.

The book value for each year is the book value of the previous year less the annual depreciation charge. For example, for the first year, the book value is equal to $10,000 less $1,357.14, which amounts to $8,642.86.

Table 8.10 Calculation of depreciation and book value for the straight-line method

Year	Annual depreciation charge	Book value
0		10,000.00
1	1,357.14	8,642.86
2	1,357.14	7,285.71
3	1,357.14	5,928.57
4	1,357.14	4,571.43
5	1,357.14	3,214.29
6	1,357.14	1,857.14
7	1,357.14	500.00
Total	9,500.00	

If assets are purchased at different points in time, a separate depreciation schedule for each asset must be constructed to determine the total depreciation in any particular year even if the assets are to be depreciated at the same rate.

(ii)   Declining balance depreciation

The declining balance method is a fixed percentage method, that is, the book value is obtained by multiplying the book value of the previous year by a fixed rate of depreciation. The rate of depreciation is commonly given by $2/N$, where N is the depreciation period. In this case, the method is called double-declining balance (DDB). Another depreciation rate that may be found is 150% method, that is, the rate of decline is equal to $1.5/N$.

The annual depreciation charge for the double-declining balance method is given by the following expression:

$$d_t = \frac{2}{N} B_{t-1} = \frac{2}{N}(P - \sum_{i=0}^{t-1} d_i) \tag{8.4}$$

For the 150% declining balance depreciation, the "2" in the equation above would be replaced with "1.5."

The calculation of the declining balance is illustrated in the next example.

Example 8.5: Double-declining balance depreciation.

Calculate the depreciation schedule for the same asset discussed in Example 8.4.

Solution:

The annual depreciation charge is equal to 2/7 times the book value in the previous year, since the depreciation period is seven years. The book value is the book value in the previous year less the depreciation charge. This can be expressed by the following two equations:

$$d_t = \frac{2}{7} B_{t-1} \tag{8.5}$$

$$B_t = B_{t-1} - d_t \tag{8.6}$$

The book value at the beginning, B_0, is the initial cost. With these rules, the depreciation schedule shown in Table 8.11 can be constructed.

Notice that the book value at the end of the period is not equal to the salvage value. This means that the equipment is not fully depreciated. If it were disposed of at the end of the seventh year for the salvage value, there would be a loss since the book value is higher than the sale price. The treatment of this loss is discussed later.

Interestingly, the declining balance method is the easiest to calculate in a tabular form (such as a spreadsheet) when assets are purchased at different points in time. Because the depreciation is at a fixed rate, the newly acquired assets are simply added to a pool consisting of the sum of the book value of all the assets. The depreciation calculation needs to only differentiate between assets with different depreciation rates if this method is used.

The capital allowance in the UK is calculated on a declining balance method. There are three main rates. Plant and machinery is depreciated at 25%, while "long-lived" assets are depreciated at a rate of 6%, both of which are calculated on the

Table 8.11 Depreciation schedule using the double-declining balance method

Year	Annual depreciation charge	Book value
0		10,000.00
1	2,857.14	7,142.86
2	2,040.82	5,102.04
3	1,457.73	3,644.31
4	1,041.23	2,603.08
5	743.74	1,859.34
6	531.24	1,328.10
7	379.46	948.65
Total	9,051.35	

Table 8.12 Depreciation schedule with a quarter declining balance

Year	Annual depreciation charge	Book value
0		10,000.00
1	2,500.00	7,500.00
2	1,875.00	5,625.00
3	1,406.25	4,218.75
4	1,054.69	3,164.06
5	791.02	2,373.05
6	593.26	1,779.79
7	444.95	1,334.84
Total	8,665.16	

basis of a declining balance. Industrial buildings are depreciated at a rate of 4% on a straight-line basis. The calculation of the capital allowance (depreciation charge) for plant and machinery in the UK is illustrated in the following example.

Example 8.6: Declining balance depreciation.

Calculate the depreciation schedule for the same asset discussed in Example 8.4 if the asset is plant and machinery for a UK based company.

Solution:

The annual depreciation charge is equal to 0.25 times the book value in the previous year and the book value is the book value in the previous year less the depreciation charge. The depreciation schedule is shown in Table 8.12.

The capital allowance calculated at fixed rate of 25% depreciates the asset less rapidly than the double-declining balance method for this set of conditions. The double-declining balance method is dependent on the depreciation period, while the fixed percentage method used in the UK is not, so general conclusions cannot be reached.

(iii)   Sum-of-years-digits depreciation

The sum-of-years-digits method is not as commonly used as the other methods, but is worth discussing for the sake of completeness. The depreciation charge is calculated as follows:

$$d_t = \frac{N - t + 1}{\sum\limits_{i=1}^{N} i} (P - S) \tag{8.7}$$

where N is the depreciation period, t is the current year, P is the initial cost, and S is the salvage value. The summation term is an arithmetic series of the years. For example, if the depreciation period is 7 years, the summation is given by the following expression:

$$\sum_{i=1}^{7} i = 1 + 2 + 3 + 4 + 5 + 6 + 7 \tag{8.8}$$

The summation of an arithmetic series is the following:

$$\sum_{i=1}^{N} i = \frac{N(N+1)}{2} \tag{8.9}$$

The substitution of this expression into that for the depreciation charge yields the following expression:

$$d_t = \frac{2(N - t + 1)}{N(N+1)} (P - S) \tag{8.10}$$

The calculation of the sum-of-years-digits method is illustrated in the following example.

Example 8.7: Sum-of-years-digits depreciation.

Calculate the depreciation schedule for the same asset discussed in Example 8.4.

Solution:

The depreciation period is 7 years, which means that the depreciation charge in year t is given by the following expression:

$$d_t = \frac{2(7 - t + 1)}{7(7+1)} (10,000 - 500)$$

The calculation of the depreciation charge and the book value is given in Table 8.13.

The depreciation schedule for sum-of-years-digits depreciation ends on the salvage value, unlike the declining balance method.

(iv)   Modified accelerated cost recovery system (MACRS)

The US tax code specifies a complex system for *personal property* that is based on a declining balance depreciation with conversion to straight-line depreciation when the depreciation rate for straight-line depreciation is greater than the double-declining balance. This system is called the modified accelerated cost recovery system, or MACRS. MACRS depreciation assumes that there is no salvage value, and it uses a half-year convention, which means that it assumes all assets are purchased halfway through the year.

Table 8.13 Depreciation schedule with sum-of-years digits depreciation

Year	Annual depreciation charge	Book value
0		10,000.00
1	2,375.00	7,625.00
2	2,035.71	5,589.29
3	1,696.43	3,892.86
4	1,357.14	2,535.71
5	1,017.86	1,517.86
6	678.57	839.29
7	339.29	500.00
Total	9,500.00	

The US tax code treats real property differently. Real property is depreciated on a straight-line basis over 39 years using the half-year convention. The half-year convention means that a portion of the depreciation occurs in the fortieth year.

The MACRS depreciation rate is commonly provided in tabular form for assets with different depreciation periods. This rate is the fraction of the original cost. This depreciation table is given in Table 8.14.

The application of the MACRS system is illustrated in the following example.

Table 8.14 Depreciation rate (%) for MACRS depreciation schedule for personal property

Year	3-Year class	5-Year class	7-Year class	10-Year class	15-Year class	20-Year class
1	33.33	20	14.29	10	5	3.75
2	44.45	32	24.49	18	9.5	7.22
3	14.81	19.2	17.49	14.4	8.55	6.68
4	7.41	11.52	12.49	11.52	7.7	6.18
5		11.52	8.93	9.22	6.93	5.71
6		5.76	8.92	7.37	6.23	5.29
7			8.93	6.55	5.9	4.89
8			4.46	6.55	5.9	4.52
9				6.56	5.91	4.46
10				6.55	5.9	4.46
11				3.28	5.91	4.46
12					5.9	4.46
13					5.91	4.46
14					5.9	4.46
15					5.91	4.46
16					2.95	4.46
17						4.46
18						4.46
19						4.46
20						4.46
21						2.23

Table 8.15 Depreciation schedule using MACRS depreciation

Year	Depreciation rate (%)	Annual depreciation charge	Book value
0			10,000
1	14.29	1,429	8,571
2	24.49	2,449	6,122
3	17.49	1,749	4,373
4	12.49	1,249	3,124
5	8.93	893	2,231
6	8.92	892	1,339
7	8.93	893	446
8	4.46	446	0

Example 8.8: MACRS depreciation.

Calculate the depreciation schedule for the same asset discussed in Example 8.4.

Solution:

The asset is in the seven-year depreciation class. The depreciation charge is the depreciation rate, given in Table 8.14, multiplied by the initial cost of the item. The calculation of the depreciation charge and the book value of the asset are shown in Table 8.15.

8.3.3 Comparison Between Depreciation Methods

The effects of the different methods for the calculation of the depreciation on the book values of the asset are compared in Figure 8.3. As expected, the book value of the straight-line method varies linearly from the initial cost to the salvage value. The other methods all accelerate the depreciation. As discussed previously, the DDB method does not necessarily depreciate to the salvage value, while the MACRS method depreciates to a salvage value of zero in the half year after the end of the depreciation period.

In calculating the book value of an asset for the purposes of the company's financial accounts, the company may use any depreciation method. Companies may have two sets of calculations of the asset value, those for use in the preparation of the balance sheet, and those for use in the calculation of tax. However, for the purposes the calculation of tax, the tax authorities specify both the method and the depreciation period.

8.3.4 Effect of the Depreciation Method on NPV and IRR

The method of depreciation does affect the cash flows to a company on a year-by-year basis. This is important from an accounting point of view. It affects the liquidity of the company, that is, the amount of cash that the company has available to it.

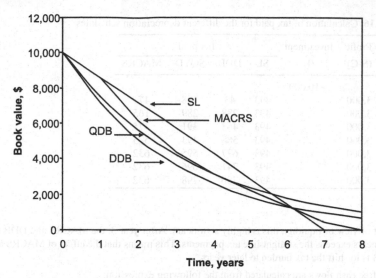

Figure 8.3 Effect of different depreciation methods on the book value. The 25% declining balance method is referred to as QDB

This is a prime variable in the management of a company. As a result, managers and accountants wish to depreciate assets as quickly as possible, claiming the tax relief as soon as possible. In addition, depreciation is a historical cost, so its value is eroded by inflation. If the asset is depreciated quicker, then the loss of value due to inflation is less.

Over the lifetime of the project, however, the importance of the method of depreciation may not be as important. This is considered in the next example.

Example 8.9: Effect of depreciation.

Consider a project in which an investment of $10,000 is to be made in order to earn a profit before tax of $3,000 each year. The project is to last seven years. The asset is to be depreciated over seven years with a salvage value of $500. Determine the tax paid and the after tax cash flows using the four different depreciation methods discussed. The tax rate is 30%, and the discount rate is 10%.

Solution:

The calculation of the annual depreciation charges for an asset costing $10,000 with a depreciation period of seven years and a salvage value of $500 has been calculated in the previous examples. The tax paid by the project is given by the following expression:

$$tax_t = T(s_t - c_t - d_t) \tag{8.11}$$

where T is the tax rate, $s_t - c_t$ is the profit before tax in year t, and d_t is the depreciation charge in year t. The tax paid for each of the different depreciation methods is given in Table 8.16. SL refers to the straight-line method, DDB to the double-declining balance method and SOYD to the sum-of-years-digits method.

Notice that on a year-by-year basis the method of depreciation can have a large effect on the tax paid. Compare the tax paid on a straight-line basis with the tax calculated using MACRS in year 2. The MACRS tax is almost half that of the straight-line tax. From the

Table 8.16 Calculation of tax paid for the different depreciation schedules

Year	Profit	Investment	Tax paid			
	(S-C)	(I)	SL	DDB	SOYD	MACRS
0		−10,000				
1	3,000		493	43	188	471
2	3,000		493	288	289	165
3	3,000		493	463	391	375
4	3,000		493	588	493	525
5	3,000		493	677	595	632
6	3,000		493	741	696	632
7	3,000		493	786	798	632

point of view of liquidity, this is highly significant. After year 3, the MACRS and DDB tax payments exceeds the straight-line tax payments. This means that the effect of MACRS and DDB is to shift the tax burden to later years.

The free cash flows are calculated from the following expression:

$$free\ cash\ flow_t = s_t - c_t - tax_t - I_t \tag{8.12}$$

where $s_t - c_t$ is the annual profit of \$3,000, tax_t is the tax charge calculated in the previous table, and I_t is the capital investment.

The results of the calculation of the free cash flows are shown in Table 8.17.

The net present value and the internal rate of return for the project using the different methods of depreciation vary in a narrow range. The *NPV* using MACRS depreciation is 5.4% higher than the *NPV* using the straight-line method.

This example demonstrates that the assessment of the project is not very sensitive to the depreciation method. A project that needs to use MACRS depreciation rather than straight-line depreciation in order to get it into a positive *NPV* region is probably not a good project. The method of depreciation does not have a great influence on the economic assessment of the project. It is important to realise, though, that this

Table 8.17 Comparison of free cash flow profiles using the different depreciation methods

Year	Free cash flow			
	SL	DDB	SOYD	MACRS
0	−10,000	−10,000	−10,000	−10,000
1	2,507	2,957	2,813	2,529
2	2,507	2,712	2,711	2,835
3	2,507	2,537	2,609	2,625
4	2,507	2,412	2,507	2,475
5	2,507	2,323	2,405	2,368
6	2,507	2,259	2,304	2,368
7	2,507	2,214	2,202	2,368
NPV	2,206	2,338	2,393	2,326
IRR	16.4%	17.2%	17.3%	16.9%

conclusion does not imply which method is preferred for the purposes of running the company. Obviously, liquidity and other accounting issues are important in the year-by-year operation of the company. In the long-term, which is the viewpoint of project assessment, the depreciation method is a not a great influencer of the value of the project.

The economic assessment of projects can be more easily computed by using the straight-line method than using MACRS method, and this method should be used in the initial assessment of projects. The assessment of the projects should not be affected, and those that are, should be reviewed again, since they are marginal projects in any case.

8.3.5 Disposal of a Depreciable Asset

When an asset is sold, one of three cases can occur. These are as follows: (i) the asset is sold for an amount that exceeds its original purchase price; (ii) the asset is sold for an amount that exceeds the book value but is less than the original purchase price; and (iii) the asset is sold for an amount that is less than the book value. In the first case, a capital gain occurs, in the second, depreciation recapture occurs, and in the third a capital loss occurs.

(i)   Capital gain

The capital gain is the amount by which the sale price exceeds the purchase price of the asset (also called the first cost). The capital gain, cg_t, is expressed as follows:

$$cg_t = S - P \qquad (8.13)$$

where S is the price at which the asset is sold and P is the original cost.

It is not possible to predict capital gains of a depreciable asset, and they are not considered in the project financials. Land, which is usually not considered a depreciable asset, usually results in capital gains on sale. The tax rate for capital gains differs from country to country.

(ii)   Depreciation recapture

If the asset is sold for a value that is greater than its book value, then depreciation recapture occurs. The value of the depreciation recapture, dr_t, is given by the following expression:

$$dr_t = S - B_t \qquad (8.14)$$

where S, is the selling price and B_t is the book at the time of the sale. The depreciation recovery can be determined when the salvage value can be estimated. Depreciation recovery will inevitably occur when the method of depreciation used reduces the book value to zero. Any amounts that accrue from depreciation recapture are taxed at the corporate tax rate.

(iii)   Capital loss

A capital loss occurs when the asset is sold for a price for less than the book value. Thus, the capital loss, cl_t, is given by the following expression:

$$cl_t = B_t - S \tag{8.15}$$

Like capital gains, capital losses are difficult to predict, and are seldom included in the project financials.

The tax rates for capital gains may be different to that for ordinary income. As a first approximation, it is sufficient to assume that capital gains and losses are taxed at the company's effective tax rate. This means that the taxable income can be expressed as follows:

$$Taxable\ income = s_t - c_t - d_t + cg_t - cl_t + dr_t \tag{8.16}$$

The tax is simply the taxable income multiplied by the tax rate.

8.4 Choice of the Discount Rate

The calculation of the net present value and the other discounted cash flow techniques require a value for the discount rate. The choice of a value for the discount rate is essentially a strategic function and is done from the viewpoint of the entire organisation. In large organisations, the calculation of the discount rate is done by the corporate finance or financial planning department, and prescribed to the projects departments in different divisions. The results generated by using the discount cash flow techniques are sensitive to the value of the discount rate used, and, as a result, it is important to understand the concepts behind the calculation of the discount rate.

Terms such as the cost of capital and the hurdle rate are often used interchangeably with the discount rate. In addition, there are other terms in use, such as the minimum attractive rate of return (MARR), and the risk-adjusted discount rate (RADR). The value of the discount rate that is used can be the financial cost of capital, the economic cost of capital or the risk-adjusted discount rate. While the justification for the different methods for the calculation of the discount rates is discussed in Chapter 12, it is valuable to review these concepts here.

(i)   Financial cost of capital: Weighted average cost of capital

The funding of a company comes from two main sources: the equity that owners initially put into the company and that which they leave in the company by not taking dividends, and the debt that the company raises. The price of the debt is the payment of interest. The return that debt-holders expect for the loan is the interest rate. The return that shareholders expect is a cost of equity. The cost of equity is dependent on the market conditions and the assessment by investors of the company's risk relative to the market risk. The determination of the cost of equity is discussed in

Chapter 12. The "cost" of these investments in the company is the return they must provide to the investors.

The weighted average cost of capital, WACC, is the combination of these two components, weighted by their relative contribution to the company's total capital. The weighted average cost of capital is given by the following expression:

$$WACC = \left(\frac{E}{E+D}\right) R_E + \left(\frac{D}{E+D}\right) R_D \qquad (8.17)$$

where E is the amount of equity, D is the amount of debt, R_E is the cost of equity, and R_D is the after-tax cost of debt. The after-tax cost of debt is relatively easy to obtain. It is the interest rate paid by the company on its loans adjusted for the fact that interest is tax-deductible. Thus, the cost of debt is usually estimated by the following expression:

$$R_D = R_{DBT} (1-T) \qquad (8.18)$$

where R_{DBT} is the interest rate on the company's debts before tax, and T is the tax rate.

The cost of equity, expressed in the same form as an interest rate, is a little more difficult to obtain. The method for estimating the cost of equity is discussed in Chapter 12. A prerequisite for that chapter is an understanding of financial risk and return, which is presented in Chapter 11.

The weighted cost of capital can be used directly in the discounted cash flow techniques as the discount rate. The weighted cost of capital can be regarded as a financial cost of capital, that is, the cost of raising the monies needed to start the business and to keep it running. An alternative cost of capital is an economical cost of capital, that is, the cost of the opportunities that the company foregoes because it does not have the resources to execute the projects.

(ii)   Economic cost of capital: Minimum attractive rate of return

The company's resources are limited and usually there are more projects than resources will allow. As a result, even good projects can be rejected. As was discussed in Chapter 7, under conditions of capital rationing, the company may wish to select the projects with the highest profitability. The opportunity cost of capital is the return on the most profitable project that is not accepted. This is the *minimum attractive rate of return* (MARR). The opportunity cost of capital and the minimum attractive rate of return are terms that are used interchangeably to mean the same thing. They can be used directly in discounted cash flow analysis to assess the economic viability of the project.

The weighted average cost of capital is the *financial* cost of capital, the cost of raising funds from debt and equity sources. The opportunity cost of capital cannot be lower than the financial cost of capital. The opportunity cost of capital is the *economic* cost of capital.

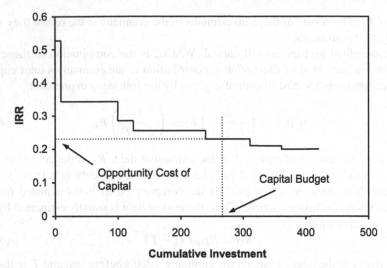

Figure 8.4 Opportunity cost of capital

The MARR or opportunity cost of capital can be obtained by ranking (independent) projects in terms of the *IRR* of each project as a function of their investment requirement. This is illustrated in Figure 8.4. The *IRR* at the limit of the capital budget represents the return of the worst project selected on the one side and the return of the best project rejected on the other. The opportunity cost of capital is the *IRR* at this budget cut-off.

This procedure seems slightly pointless from a practical implementation point of view. The projects need to be ranked prior to determining a discount rate (which will be used to select projects). If the projects can be ranked there is no need to determine a discount rate.

The ranking of projects for appraisal in this fashion is static, as if they are all ready now, either to be accepted or rejected. In a dynamic world, the company's planning of projects occurs over a horizon in which opportunities are in various states of development and it may be several years before they are ready for the investment decision.

Surveys of best practice in large companies have found that the opportunity cost of capital is not the method that is most used. Rather, the weighted average cost of capital is used. The weighted average cost of capital may be adjusted for risk of projects that have a risk profile different from that of the company as a whole. The adjustment of the discount rate to account for risk is discussed in the next section.

(iii)   Risk-adjusted discount rate

Both the weighted cost of capital and the opportunity cost of capital assume that the projects all have a similar risk to each other and to the company as a whole. This assumption may not be valid for some individual projects where the risk of the project is different to that of the company as a whole. The sources of additional risk are numerous. An example is the opening of a project in a new country, where the company does not have local experience. The country risk that this specific project has may be accounted for by increasing the discount rate to obtain the risk-adjusted discount rate. This is given by the following equation:

$$RADR = Company's\ Discount\ Rate\ for\ Project\ of\ Normal\ Risk$$
$$+ Project\ Risk\ Premium \tag{8.19}$$

where the *Project Risk Premium* is a factor to account for the project-specific risk. The company's discount rate for a project of normal risk might be described by the WACC, or by some other mechanism.

Increasing the required rate of return without a proper calculation of the risk premium creates a bias towards more risky projects, an effect that is clearly not intended, and which can have deleterious effects on the selection process. The discussion of the choice of discount factor is extended in Chapters 12 and 13 to account for issues of risk.

8.5 Summary

The techniques for the assessment of a project's cash flows were outlined in the previous chapter. Three topics on the application of the discounted cash flows techniques for the assessment of a project were discussed in this chapter. They concern the following influences on the evaluation of a project: inflation; the different methods for the determination of the depreciation charge or capital allowance; and choice of the discount rate.

There are two schools of thought on the best method for the preparation of the cash flows for a project. One insists that they should be prepared on the basis of real values, that is, without the effects of inflation included. The other school believes the opposite, that inflation should be included in the cash flows. Both have pros and cons. The disadvantage of the "real values" school is that tax is calculated on nominal amounts, so that the tax calculation will be incorrect. The disadvantage of the "nominal values" school is that the inflation rate must be estimated. A common error is to co-mingle the two, by using a real discount rate to evaluate the nominal cash flows, or vice versa.

The company can be in four positions with respect to tax: the company has no losses against which profits or earnings can be expensed; the company is in an assessed loss position; the company has losses such that it does not expect to pay taxes for long time; the company (or the project) may be in a tax-free zone. Generally, projects are evaluated as if they all have to pay their own taxes, that is, they are evaluated on a stand-alone basis.

The depreciation of assets creates a depreciation charge, which is used to reduce the taxable income or profits. The depreciation charge depends on the class of the asset, which is related to the life of the asset, the initial cost and the rate of depreciation. There are four main methods of depreciation: the straight-line method, the declining balance method, the sum-of-years digits method and the MACRS method. The method used in the UK is the declining balance method and that used in the US is the MACRS method. It is shown that the method of calculation of the depreciation charge, while important in the year-by-year operation of the company, is not important to the project's *NPV* or *IRR*.

The choice of the discount rate was the final topic of the chapter. Three different methods were discussed. These are the following: (i) a financial cost of capital, which is based on the cost of debt and the cost of equity; (ii) the economic cost of capital, which is based on the return from the worst project that is accepted; and (iii) a risk-adjusted discount rate, in which a base rate is obtained from (i) or (ii) and a premium is added to it to account for additional risk. The economic cost of capital is also called the minimum attractive rate of return (MARR). The financial cost of capital (WACC) is the most commonly used method. It may be adjusted for risk, that is, by the third method. Typical risks that are accounted for in this manner are country risk or project-specific risk. The risk of a project is discussed in detail in Chapter 13.

8.6 Review Questions

1. What is the inflation rate?
2. How is the rate of inflation measured?
3. How does inflation affect the project cash flows?
4. Should the project cash flows be calculated with inflation or without inflation?
5. What is the relationship between real and nominal interest rates?
6. Should the discount rate used to assess projects be the nominal rate or the real rate?
7. Does inflation make a project more or less attractive?
8. How does inflation affect the tax paid?
9. What does the "tax position of a company" mean?
10. Are all projects eligible for tax?
11. What is a tax-free zone?
12. What are the methods of depreciation?
13. What sort of depreciation is important in the assessment of a project?
14. What are depreciation classes?
15. Do all countries adopt the same depreciation methods?
16. What is the advantage of the declining balance method?
17. Under what conditions does the method of depreciation influence the attractiveness of a project?
18. What is the relationship between MARR and the opportunity cost of capital?
19. Is the weighted average cost of capital the same as the MARR?

20. How is the risk of a project accounted for?
21. Why should the project's cash flows ignore sunk costs?

8.7 Exercises

1. The nominal interest rate is 20% and the inflation rate is 8%. What is the real interest rate?

2. A company wishes to evaluate a 10-year project with the following cash flow profile:

	Amount
Initial capital outlay	$50,000
Annual cash flow (real)	$30,000

If the discount rate for the company is 15% and the inflation rate is 8%, determine the following:

(i)   The NPV using nominal values
(ii) The NPV using real values
(iii) Comment on the results

3. The capital cost for a petroleum refinery was $202 million ten years ago. The inflation rate has been 7% pa over that period. Determine the capital cost for a similar refinery today.

4. A company expects a real discount rate of 12%. If the inflation rate is 7%, what should the nominal discount rate be?

5. A company wishes to invest in a five-year project with the following cash flow profile:

	Amount
Initial capital outlay	$50,000
Annual before tax cash flow (real)	$25,000

If the discount rate for the company is 10% and the inflation rate is 5%, determine the following:

(i)   Determine the cash flow after tax in nominal terms if the tax rate is 35% and the depreciation is calculated on a straight-line basis over five years.
(ii) Determine the NPV for the free cash flow in nominal terms.

(iii)   Determine the cash flow after tax in real terms if the tax rate is 35% and the depreciation is calculated on a straight-line basis over five years.

(iv)   Determine the *NPV* for the free cash flow in real terms.

(v)   Comment on the results.

6.   What is the value of a property that costs $20,000 now in five years time if the real interest rate is 20% and the inflation rate is 5%.

7.   An investor wishes to purchase property that should be worth $50,000 in three years. The property taxes are 1% of the purchase price and capital gains are taxed at 15%. The investor requires a return of 12% on investment and the inflation rate is 6%. What should the investor pay for the property?

8.   A project has the cash flow profile given in the table below in nominal terms:

Year	0	1	2	3	4	5	6
Free cash flow	−10,000	2,000	3,000	4,000	4,500	5,000	4,500

If the inflation rate is 5% and the nominal discount rate is 11%, determine the *NPV* and *IRR* in real and nominal terms.

9.   Determine the depreciation schedule for an asset that costs $100,000 under the following conditions:

(i)   Straight-line depreciation over 5 years
(ii)   Double-declining balance
(iii) Quarter-declining balance
(iv)   MACRS depreciation over five years

10. Determine the book value after 5 years for an asset that is depreciated using quarter declining balance if the asset originally cost $25,000.

11. Determine the depreciation charge in the third year of an asset's life, if the asset originally cost $10,000 and can be depreciated over ten years, using the following depreciation calculation methods:

(i)   Straight-line depreciation
(ii)   Sum-of-years-digits depreciation
(iii) Double-declining balance depreciation
(iv)   150% declining depreciation

12. A company has an ordinary income of $95,000. In the same year it sold an asset that had a book value of $5,000 for $30,000. Determine the company's tax liability.

13. A company sells an apparatus that was originally purchased for $75,000 for $40,000 seven years after purchase. If the asset was depreciated for tax purposes using quarter-declining balance, determine the depreciation recapture.

14. The half-year convention assumes that the asset is purchased in the middle of the year. Determine the depreciation schedule for an asset purchased and

installed for $20,000 if it is depreciated over five years using the half-year convention.

15.   The gross earnings of a project are given in the table below in nominal terms:

Year	0	1	2	3	4	5	6
Gross earnings	−10,000	2,000	3,000	4,000	4,500	5,000	4,500

(i)   Determine the depreciation schedule for the assets if they can be depreciated on a straight-line basis over 6 years.

(ii)   Determine the tax charged if the tax rate is 30%.

(iii) Determine the free cash flow.

(iv)   Determine the *NPV* and *IRR* of the project if the inflation rate is 5% and the nominal discount rate is 11%, determine the *NPV* and *IRR* in real and nominal terms.

Chapter 9
Sensitivity, Scenario and Other Decision Analysis Techniques

9.1 Introduction

The engineering economic analysis of a project provides a single result, such as the *NPV* or IRR of the project. This result is important. However, it is also important to use the project financials and the measure of profitability, such as the *NPV*, to develop an understanding of the performance of the project under a variety of conditions. This chapter discusses the analysis and interrogation of the economics of the project to develop an understanding of the project.

The first topic of the chapter is a visual technique for the construction of financial models using influence diagrams. The influence diagram can be used in a number of ways: (i) to check that all influences of value have been included; (ii) to check that the relationships between variables are correctly represented; and (iii) to communicate the assessment model to team members and the investment committee.

Once a model of the project financials and the *NPV* has been developed, the results must be thoroughly examined to enhance the understanding of the study. Two methods are presented. The first is sensitivity analysis, in which the sensitivity of the model outputs to various inputs is investigated, and the second is scenario analysis, in which a few possible forecasts of the most sensitive inputs are investigated. The main difference between the two methods is that in sensitivity analysis one parameter is changed at a time, while in scenario analysis several parameters are changed simultaneously to represent a new scenario.

9.2 Influence Diagrams

Financial models should be transparent and accurate. An influence diagram is a visual representation of the model. It is a useful technique that allows the model to be built in parts, and for the effects of various parts to be seen without getting immersed in the details of the model. It is particularly useful in the context of spreadsheet implementations of a model, where the equations are hidden making the auditing

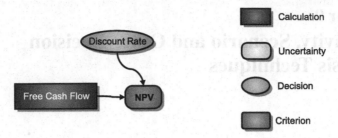

Figure 9.1 Initial influence diagram for *NPV* model

more difficult. Developing the influence diagram prior to constructing the model can save the time and effort that is often wasted in debugging a spreadsheet.

The construction of an influence diagram is best explained by example. The diagram begins with the output. For example, suppose that the purpose of the model is to determine the *NPV* of a project. The next step is to ask, "what influences the *NPV*?" or "what is the *NPV* dependent upon?" The answer to this particular question is that the *NPV* is determined by the free cash flows and the discount rate. The representation of these two influences on the *NPV* is shown in Figure 9.1.

The next question is "what does the free cash flow depend on?" or "what does the discount rate depend on?" Suppose that the discount rate in this example does not depend on anything; it is specified by the company's financial management. It is then marked as an input. The free cash flows are dependent on the cash flows and the capital investment. This is shown in Figure 9.2.

The influence diagram is developed further by following a similar procedure. This process continues until all of the outer variables are the inputs to the model. Some of these outer variables are known with certainty, and others are known within a particular range, that is, they are uncertainties. Some of the uncertainties can be resolved by decision. For example, an uncertainty such as the production rate of oil from a field can be resolved by deciding, based on reasoned logic and engineering principles, what the rate should be.

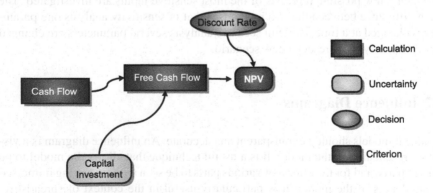

Figure 9.2 Next level of the influence diagram

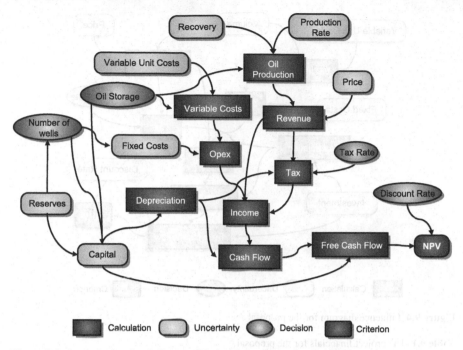

Figure 9.3 An example of an influence diagram for an oil production project

Consider a project for the production of oil. The decision that is required is whether to invest or not. The decision will be made on the basis of *NPV*, which depends on the discount rate and the free cash flow. The free cash flow depends on the capital investment, and the cash flows, as was shown in Figure 9.3. Developing this further involves splitting the cash flows into their component parts: revenue; operating costs; depreciation; taxes; and changes in working capital. The depreciation depends on the depreciation period and the investment. The investment depends on the number of wells, reserves and the storage of oil. A diagram for such a project is shown in Figure 9.3.

The outcome of the influence diagram is a picture of the dependent and independent variables, and the relationship between them. Consider the following example.

Example 9.1: Influence diagram.

A company is considering a proposal to develop a new product line. The company has conducted competitor analysis and market surveys and believes that the product can sell for $100 and that entry into the market at 13,000 units a year would be optimal. The capital investment is expected to cost in the region of $1,000,000 and can be depreciated for tax purposes over five years. The fixed costs in running the operations at this volume are $200,000 per year, while the variable costs are $30 per unit. The discount rate is 10%.

Construct an influence diagram and from that a model of this proposal.

Solution:

The first step is to construct an influence diagram to assist in the building of the model. This is shown in Figure 9.4. The influence diagram illustrates the relationships between

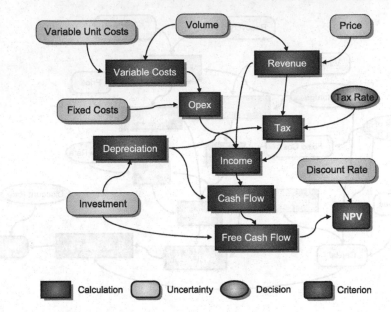

Figure 9.4 Influence diagram for the proposal

Table 9.1 The project financials for the proposal

Year	0	1	2	3	4	5
Revenue		1,300	1,300	1,300	1,300	1,300
Price		100	100	100	100	100
Volume		13	13	13	13	13
Operating costs (Opex)		590	590	590	590	590
Fixed costs		200	200	200	200	200
Variable costs		390	390	390	390	390
Income		710	710	710	710	710
Depreciation		200	200	200	200	200
Tax		153	153	153	153	153
Income after tax		357	357	357	357	357
Investment	−1,000					500
Free cash flow	−1,000	557	557	557	557	1,057
Discount factor	1.00	0.91	0.83	0.75	0.68	0.62
Present value	−1,000	506	460	418	380	656
NPV	1,422					

variables clearly. In particular, independent variables, marked on the diagram as rectangles with rounded corners, do not have influence arcs as inputs. Dependent variables have influence arcs as inputs. This symbolizes the dependency of the variable on the value of another variable. The dependent variables are shown as rectangles. There may be some variables, such as the tax rate in the current example, which are given by some other process or authority. These given variables are shown as ovals in Figure 9.4.

The next step is to build the model from the influence diagram. This is shown in Table 9.1.

A strategy or alternative strategies may be evident from the influence diagram. For example, there may be a group of independent variables over which management can exercise control. Changes to the values of these independent variables can be grouped into different decision strategies, as discussed in Chapter 2. The link between the influence diagram, the decision table and decision strategies is discussed in the case study later.

In the next section, the sensitivity of the financial model to a particular variable is discussed.

9.3 Sensitivity Analysis

Often those tasked with developing mathematical models will posit the view that it is not the single result that is important in modelling, but the insight that is to be gained. If this is so, how can this insight be attained? One of the methods of trying to develop insight is through sensitivity analysis, determining how sensitive the model outputs are to changes in the model inputs. For example, it might be required to determine the *NPV* at two or three different discount rates. This establishes how much the *NPV* changes with a change in discount rate, that is, the sensitivity of the *NPV* with respect to the discount rate.

The sensitivity is the rate of change of a variable with respect to the change of another variable with the values of all other variables held constant. This can be expressed as the partial derivative of the first variable with respect to the second one. This is given as the following equation:

$$S = \frac{\partial c}{\partial b} \tag{9.1}$$

where c and b are dependent and independent variables, and S is the sensitivity of c with respect to b. The interpretation of the sensitivity is as the slope of the dependent variable with respect to the independent variable. The sensitivity can be obtained by plotting the value of the independent variable at a variety of values of the dependent variable.

The following example demonstrates the calculation of the sensitivity.

Example 9.2: Calculation of sensitivity.

A manufacturer can produce in the region of 500 pairs of shoes a day. The shoes wholesale at a value of about \$50 a pair. Determine the sensitivity of the manufacturer's revenue to price and production.

Solution:

The revenue of the manufacture is given by the production multiplied by the price. The sensitivity measures the effect a change in production or price on the revenue.

If the price increases by 10% to \$55 a pair, the revenue will be equal to 500 pairs per day times \$55 per pair = \$27,500. The sensitivity of the revenue with respect to price is calculated as follows:

$$S_{price} = \frac{27,500 - 25,000}{55 - 50} = 500$$

where S_{price} represents the sensitivity of the revenue with respect to the price.

If the production increases by 10% to 550 pairs per day, the revenue is equal to $27,500. The sensitivity of the revenue with respect to the production rate is calculated as follows:

$$S_{production} = \frac{27,500 - 25,000}{550 - 500} = 50$$

where $S_{production}$ represents the sensitivity of the revenue with respect to the rate of production. The units of S_{price} are pairs/day, while the units of $S_{production}$ are $/pair.

The following example demonstrates a graphical representation of the sensitivity.

Example 9.3: Graphical representation of sensitivity analysis.

Determine the sensitivity of the output to the input parameters for Example 9.1, and represented the results graphically.

Solution:

The independent variables are specifically identified in the influence diagram. The values of these variables for the base case are given in Table 9.2.

The sensitivity analysis investigates how each of the independent variables affects the *NPV*. The model inputs are changed for –40% of the base value to +40%, and the effect of the change on the net present value of the projects is calculated. The results of these calculations are given in Table 9.3, and are shown in Figure 9.5.

It is clear from the sensitivity plot that the project's financials are most sensitive to the price and the sales volume. The influence of these parameters is positive, that is, an increase in the

Table 9.2 Values of the parameters and independent variables

Variable	Value
Discount rate	10%
Price	100
Volume	13
Fixed costs	200
Variable costs	30
Investment	1,000

Table 9.3 Results of the sensitivities calculations

Percentage change	−40%	−30%	−20%	−10%	0%	10%	20%	30%	40%
Discount rate	6%	7%	8%	9%	10%	11%	12%	13%	14%
NPV	1,720	1,640	1,564	1,492	1,422	1,355	1,292	1,230	1,172
Price	60	70	80	90	100	110	120	130	140
NPV	42	387	732	1,077	1,422	1,767	2,112	2,457	2,802
Volume	8	9	10	12	13	14	16	17	18
NPV	456	698	939	1,180	1,422	1,663	1,905	2,146	2,388
Fixed costs	120	140	160	180	200	220	240	260	280
NPV	1,634	1,581	1,528	1,475	1,422	1,369	1,316	1,263	1,210
Variable costs	18	21	24	27	30	33	36	39	42
NPV	1,836	1,732	1,629	1,525	1,422	1,318	1,215	1,111	1,008
Capital investment	600	700	800	900	1,000	1,100	1,200	1,300	1,400
NPV	1,731	1,654	1,576	1,499	1,422	1,345	1,267	1,190	1,113

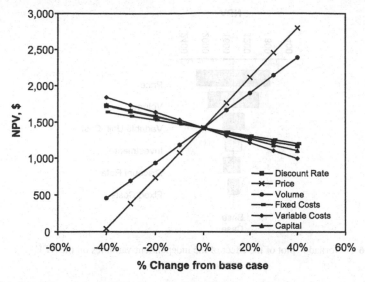

Figure 9.5 Sensitivity of the *NPV* for Example 9.1 with respect to the mode inputs

sales volume indicates an increase in the *NPV*. All of the other model inputs or parameters have a negative sensitivity. An increase, for example, in the fixed costs means a decrease in the *NPV*. A change in the value of these other parameters does not have as great as effect on the *NPV* as a change in, say, the price. This means that the *NPV* is more sensitive to the price, than to the fixed costs, the variable costs, the discount rate and the capital.

The sensitivity diagram for this example may seem trivial and perhaps obvious to some. However, it has lead to the insight that more attention should be paid to obtaining reliable estimates of the price and sales volume than to the capital or operating costs. Such insight, or at least the numerical confirmation of this result, can prove to be highly valuable in a business venture.

An alternative form for the graphical representation of the sensitivity is the tornado and volcano diagram. In these diagrams the sensitivity of the independent variables is ranked so that the most important variables are clear. These diagrams are discussed in the example below.

Example 9.4: Tornado and volcano diagrams.

Illustrate the sensitivity of the outputs to the inputs parameters for Example 9.1 by using tornado and volcano diagrams.

Table 9.4 The effect of an increase and decrease of 20% in each parameter on the *NPV*

	Low	High
Discount rate	1,564	1,292
Volume	939	1,905
Price	732	2,112
Fixed costs	1,528	1,316
Variable unit cost	1,629	1,215
Investment	1,576	1,267

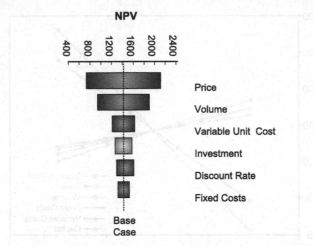

Figure 9.6 A "tornado" plot of the effect of the independent variables on the *NPV*

Solution:

The sensitivity analysis investigates how each of the independent variables affects the *NPV*, which is the only output of the model. This is done by increasing and decreasing the value of the variable by 20%, and recalculating the output variable, the *NPV*. Each variable is changed one at a time in turn. The results of this analysis are shown in Table 9.4.

These results can be visualized in the form of tornado or volcano diagrams. These diagrams are shown in Figures 9.6 and 9.7.

The sensitivity of the *NPV* to an input variable is clear from the tornado or volcano plots. The price and the sales volume have the most influence on the economic outcome. As a result, careful attention to these two variables is essential in the plan-

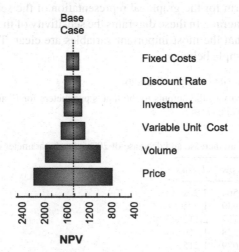

Figure 9.7 A "volcano" plot of the effect of the independent variables on the *NPV*

ning of this business opportunity. However, even with a 20% reduction in price or a 20% reduction in sales volume, the project is still economically attractive.

The example used in this section was particularly straightforward. In a more realistic case, the sales volume would have impacted on the capital required, perhaps in a non-linear manner (such as the rule of two thirds, as discussed in Chapter 4), or it might have been necessary to build in other interdependencies between the inputs and the outputs of the model. In such cases, it is more difficult to guess or intuit the sensitivity of the model results to the changes in parameters, and as a result the insight gained will be proportionally greater.

9.4 Scenario Analysis

Scenario analysis is a variation along the theme of sensitivity analysis. The changes in a set of parameters are grouped into a scenario and the effect of these scenarios is investigated. For example, in a scenario that envisages low market penetration of the company's product, the sales volume and the price may be decreased and the marketing and advertising costs increased. The grouping and the changes in the values of these parameters should generally make sense; they should constitute a defensible scenario.

Different strategies, on the other hand, represent different choices that the management must make, and the analysis of the alternatives may require different models, not just changes in the inputs to a model. For example, there may be two options for the processing technology, one that can be licensed and one that has been developed in-house. These represent two different implementation strategies between which management can choose.

Consider the following example.

Example 9.5: Scenario analysis.

Develop a scenario analysis for the Example 9.1. Consider three additional scenarios in addition to the base case.

Solution:

After extensive discussion with experts in each of the functional departments within the organization, the project team decided that the three possible scenarios, in addition to the base case, are the following: (i) Things go much better than expected. There is evidence that the economy is improving and, since it is anticipated that interest rates will drop, it is also expected that consumer spending will increase. The team believes that there is sufficient demand in the market that both the price and the volume may increase. There is sufficient production and sales capacity to increase production without a change in manufacturing cost (investment); (ii) The production plan at the moment is based on the use of second-hand equipment. While the sales and price expectations are met, the second-hand equipment does not meet expectations, and significant modifications are required. This increases the investment cost and the variable costs of production; (iii) The sales do not materialize, despite marketing's best efforts to estimate demand. The sales volume is less that half of that expected.

The variation in the values used to calculate the *NPV* for the different scenarios is shown in Table 9.5.

Table 9.5 Results for the four different scenarios

	Base	Better than expected	Production issues	Sales issues
Discount rate	10%	10%	10%	10%
Price	100	120	100	100
Volume	13	15	13	6
Fixed costs	200	200	200	200
Variable costs	30	30	65	30
Investment	1,000	1,000	1,500	1,000
NPV	1,422	2,589	−172	122

The results reveal that the team believes that the sales issues are not a threat to the project but the production issues are a threat. As a result, the team recommended that measures are put in place to mitigate the effects of this possible adverse scenario.

9.5 Strategy Space

The strategy space is a method of distinguishing alternatives by the parameters that they are most sensitive to. For example, in the commodities industries there is often a trade-off between capital costs and operating costs (Figure 9.8). A simpler, cheaper operation to build may have higher operating costs, while a more expensive, state-of-the-art operation might have lower operating costs. The advantages of such alternatives can be succinctly pictured in two dimensions, and the value of the trade-off is clearly shown.

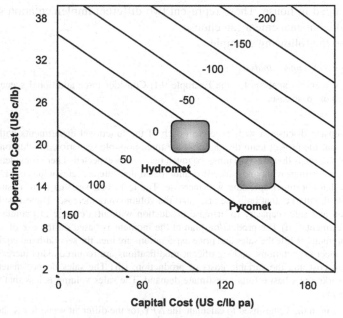

Figure 9.8 Strategy space in which the capital and operating costs represent a trade-off between two options. The *lines* represent contours of constant values of the *NPV*

9.6 Decision Analysis

Decision analysis is a procedure or set of procedures for the formal analysis of decisions. Some of the procedures of decision analysis were presented in Chapter 2 as a framework for decision-making. Decision analysis is a normative process, that is, it represents methods for assessing decision situations and making the decisions logically and formally. It is not a descriptive approach, that is, it is not a representation of how people actually make decisions.

The first step in decision-making is assessing the situation and creating a decision frame that places the decision in the correct context. An important part of this is the creation of alternatives that might be solutions to the problem. This first step is generally regarded as the most important step in the process. The second step is the analytical step that involves evaluation, analysis and interrogation. The third step is the action of deciding.

In this section, a technique for structuring influence diagrams and scenarios, taken from the traditions of decision analysts, is illustrated.

The first part of the framing exercise is to unpack the issues into facts, decisions and uncertainties. In the company context, an "issue-raising" meeting might be held, where concerns are voiced and recorded. The result of such a meeting is to get as broad a view of the issues and the context as possible, so that the context can be envisaged clearly. The issues are classified as fact, decisions and uncertainties, and can be listed in a table.

The facts form the context for the decision. The uncertainties form the variables that must be described in the influence diagram, that is, they form the outer layer of the influence diagram. The uncertainties are the subject of the sensitivity analysis in order to determine which are the most important so that attention can be focused on them.

The decisions are further classified into policy, strategic and tactical decisions. Policy decisions are those that have already been made, strategic decisions are those that are required for this situation now, and tactical decisions are those that can be made at a later stage. The decisions to focus on are the strategic decisions that require attention now. As discussed in Chapter 2, the strategic decisions are used to construct a strategy table, from which the scenarios are chosen.

Thus, the methodology of decision analysis creates a logical method for linking together the decision context, the influence diagram, the financial model and the scenario analysis. This is illustrated in the following example.

Example 9.6: Framing of an oil exploration decision.

Tyler Exploration has been offered participation in an oil exploration opportunity by Jameson Fuels, which owns all of the prospecting rights for the Permit 75 in the Lana Basin in Algeria. Permit 75 covers a large tract of land with good potential holding oil reserves. The sharing of exploration opportunities between competitors is a mechanism for sharing the risk of exploration. In the terminology of oil exploration, Tyler Oil has been offered a "farm-in" opportunity by Jameson, or Jameson has "farmed-out" the opportunity. The terms of the offer are a "two-for-one promote," that is, Tyler pays all of the costs to drill the exploration well to earn a 50% working interest in the lease. The cost to drill the exploration well is $11 million, so the "entry ticket" for this opportunity is $5.5 million, or Jameson's share of the spending on the exploration well that Tyler would be paying.

An initial assessment suggests that the opportunity will meet the *NPV* criterion used by Tyler to assess its projects. In Tyler's assessment, the geology looks promising. There is already a successful exploration well on an adjacent lease area and both lease areas are in a region that is known as being rich in oil and gas. The oil-in-place is a function of the size of the reservoir, which is unknown until appraisal drilling is performed. Even though there may be oil in the reservoir, it is unclear that it is moveable or what the production rate might be. Some of these factors can be resolved by an exploration well, but most of them will require appraisal drilling to resolve them.

The costs of processing the oil and gas products from the well are not known. Jameson does have processing facilities with spare capacity in the vicinity and this might be an option. However, Tyler's engineers are sure that Jameson will extract as much value out of this advantage as possible. They surmise that it would be better to use Alliance Oil's processing facilities because that would probably represent a more fair deal. Alliance Oil is a third party operating in the area. The other costs that are unknown are the appraisal costs.

While Jameson has made an initial offer, there is no reason that these terms cannot be modified. The "two-for-one promote" can be replaced with a different ratio. Alternatively, Tyler might consider joining in the project during the appraisal stage rather than at the exploration phase. However, Tyler may be asked to pay a cash sum to "farm in" at the appraisal stage.

Frame the decision problem from Tyler's point of view; in addition, create the influence diagram and the scenarios for analysis.

Solution:

The problem statement or opportunity statement can be phrased as follows: "Should Tyler 'farm-in' to Jameson's lease, and if so, at what stage and under what terms?"

The main questions for Tyler's consideration are the following:

1. What is the value of the offer, and does it meet Tyler's investment criterion?
2. Are the terms of the offer justified?
3. What would Tyler lose by "farming-in" at the appraisal phase?

The offer from Jameson Fuels has been divided into facts, uncertainties and decisions, as shown in Table 9.6. Facts are known data or information, decisions are choices that can be controlled, and uncertainties are factors that are unknown, but will impact on the opportunity.

The decisions are further categorized as policy, strategic and tactical decisions, as shown in Figure 9.9. This diagram is known as the decision hierarchy. The policy decisions are those that have already been made, the strategic decisions are those that are required now, and the tactical decisions are those that can be made later.

The decisions that are required now are those in the strategic category, which are solely associated with the terms of the deal. They form the basis for building the strategy table, from which the scenarios will be chosen. As described in Chapter 2, the strategic decisions become the headings for the columns of the strategy table, and the options for each decision are listed in the columns. In this case, the headings of the strategy table are the terms of the deal. This is shown in Table 9.7.

Decision strategies are scenarios that are formulated from the strategy table. For example, one of the scenarios is the deal as proposed. Other possible strategies are listed in Table 9.8.

This illustrates the development of a set of scenarios in a structured manner. The different conditions chosen in these scenarios will influence the value of the project and the share of

Table 9.6 The facts, uncertainties and decisions for the "farm-in" opportunity

	Item
Facts	
	1. Jameson Fuels owns a prospecting lease in Algeria.
	2. Jameson Fuels has processing facilities in the vicinity.
Uncertainties	
	1. The chances of finding moveable oil.
	2. The size of the reservoir.
	3. The production rate from the reservoir.
	4. The costs of the processing of the products from the reservoir.
	5. The costs of the appraisal drilling.
	6. Price of oil.
Decisions	
	1. Agree to the offer by Jameson and drill an exploration well.
	2. Agree to the terms of the offer:
	• Share of exploration costs.
	• Share of appraisal costs.
	• Share of development costs.
	• Cash payments.
	• Working interest in production revenue.
	• Processing.
	3. Drill appraisal wells.
	4. Develop oil field.
	5. *NPV* is the decision criterion.

Table 9.7 Strategy table for "farm-in" opportunity

Share of exploration	Share of appraisal costs	Share of development costs	Cash payments costs	Working interest in	Processing production revenue
100%	100%	50% of all	None	0%	Jameson
50%	50%	None	$2 million	50%	Tyler
None	None		$4 million		Third party

the risks taken by each party. The risks in this context are the uncertainties associated with the project, which are listed in the table of facts, decisions and uncertainties. The influence diagram links the decision criterion, in this case the net present value, *NPV*, with the uncertainties. The influence diagram can assist in clarifying the dependencies and relationships between variables, and allows the analysts the opportunity to ensure that all of the important relationships are represented. For instance, it is clear that the unknown operating costs of the production facilities should impact on the *NPV*. However, this factor was overlooked in tabulating the uncertainties. Building the influence diagram corrected this omission (Figure 9.10).

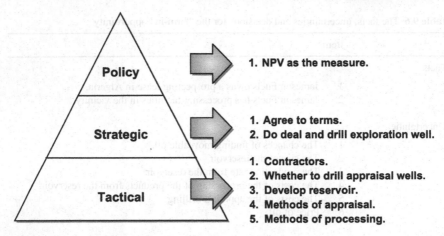

Figure 9.9 Decision hierarchy for framing the Tyler "farm-in" opportunity

Table 9.8 Decision strategies for Tyler's "farm-in" opportunity

Strategic theme	Share of exploration costs	Share of appraisal costs	Share of development costs	Cash payments	Working interest in production revenue	Processing
Proposed deal	100%	50%	50% of all	None	50%	Jameson
Conservative	50%	50%	None	None	50%	Third party

The sensitivity analysis is directly related to the influence diagram. The sensitivity analysis determines how sensitive the decision criteria are to the uncertainties, and the influence diagram is a visual representation of the relationship between the decision criteria and the uncertainties.

9.7 Summary

Three techniques commonly used to gain a deeper understanding of the issues in the investment problem were presented in this chapter. They were influence diagrams, sensitivity analysis and scenario analysis. In addition, the framing of a decision problem in terms of the issues, and uncertainties, and the link between the three techniques through the use of the decision hierarchy and the decision strategy table, were discussed.

The influence diagram is a method of representing the relationships between the outputs of a model with the inputs. The diagram consists of variables joined with vectors that represent that causal link between the variables. The diagram can be

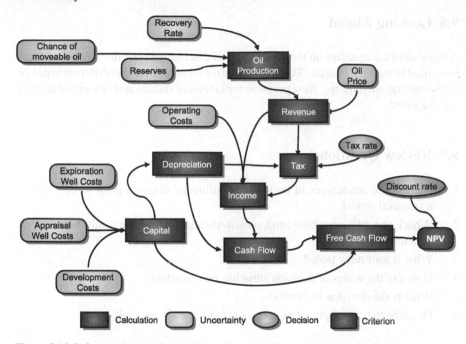

Figure 9.10 Influence diagram for the Tyler oil opportunity

used to verify that the correct relationships are accounted for, that the direction of causality is correct, and that all the input variables have been included.

Sensitivity analysis determines how influential the inputs are and the extent of their effect on the outputs from a financial model. Sensitivity analysis determines the effect that each variable has if the values of all other variables are held constant. A popular form of representing the outcome of the sensitivity analysis is a tornado or volcano diagram. Once the sensitivities of the inputs are known, attention can be focussed on verifying the accuracy of the data or assumptions that inform these variables.

One method that is common in business and particularly business consulting is to determine the two variables that have the most influence on the situation. The opportunity space is then divided into regions and the trade-off and comprises made in changing positions in the opportunity space is examined. In capital decisions, the capital and operating costs might represent such a trade-off in comparing different options. This rather simplified representation can be quite powerful, hence its popularity with management consultants.

Scenario analysis involves selecting a set of values for the inputs to a model, or decisions that the decision-maker can influence, and determining the outcome for that particular set of values. The choice of these scenarios can be structured logically by constructing a strategy table.

9.8 Looking Ahead

A number of case studies on the application of discounted cash flow techniques are presented in the next chapter. These case studies represent various different types of decisions that are required. New projects, replacement studies and new opportunities are discussed.

9.9 Review Questions

1. Discuss the advantages of building an influence diagram prior to developing a financial model.
2. What is the difference between a sensitivity analysis and the results represented on a tornado diagram?
3. What is a strategy table?
4. How can the issues of decision situation be unpacked?
5. What is the decision hierarchy?
6. Describe the development of scenarios from the strategy table.

9.10 Exercises

1. A model is represented by the following set of equations:

$$A = B - C - D$$
$$B = P(V)$$
$$V = f(I)$$
$$C = V(E) + F(I)$$

The symbols A, B, C, D, P, V, I, E, F represent variables, while $f()$ means "a function of."
Draw an influence diagram for the model.

2. The pricing of a product depends on the volume produced, the demand and competitive forces. Draw an influence diagram for this situation.

3. Draw an influence diagram for the calculation of the weighted average cost of capital.

4. The various components of the capital cost of a project, that is, the initial capital outlay, were discussed in the Chapter 4. Construct an influence diagram for the capital costs.

5. The revenue from the sale of an item is dependent on the price and the volume. What is the sensitivity of the revenue with respect to the price and the volume? What are the units of measure for each of these sensitivities?

6. A five-year project is being considered for investment. The product is expected to sell at \$500 and the demand is expected to grow at 10% pa for the next five years. The production volume will initially be 10,000 units. The variable costs are \$244/unit, and the fixed costs are \$1.3 million. The capital investment required is \$3,500,000. The tax rate is 30%. The company's discount rate is 10%. Determine the sensitivity of the *NPV* to the following inputs:

   (i)   Growth in demand
   (ii)   Initial production volume
   (iii)   Price
   (iv)   Capital investment
   (v)   Tax rate
   (vi)   Discount rate
   (vii)   Fixed costs
   (viii) Variable costs

7. A company exports all of its goods. All of its revenues are earned in dollars, while all of its expenses are incurred in pounds. If the company's profit margin is 12%, determine the sensitivity of the company's profit to the exchange rate.

Chapter 10
Case Studies on the Application of the Decision-making Criteria

10.1 Introduction

This chapter contains a number of examples to illustrate the methods discussed in Chapters 6 to 9. The case studies cover decisions concerning capital investment in new projects, equipment replacement and outsourcing. The case studies are intended to represent more realistic situations than the examples in the chapters, and to illustrate the use of a combination of techniques, including the application of decision analysis and discounted cash flow criteria.

10.2 Santa Clara HEPS

10.2.1 Introduction

An aluminium producer in Brazil is experiencing power supply problems. The investment in infrastructure has lagged economic growth in the country and is now one of the main constraints for future prosperity. Infrastructure projects, such as power supply, can take up to ten years to implement, so even though the government is committed to infrastructure development, the problem could persist for some time. The government has in the meantime decided that the best way to deal with the current crisis is to cut electricity supply to industrial users one day a week. The outages are expected to last for the next 5 years. For an aluminium producer, this is not acceptable, because the plant cannot be readily switched on and off. The company has invested in generator sets, which though extremely costly, can keep the aluminium smelters hot for the duration of the one-day a week power outage.

Both the company and the local government have investigated alternative arrangements. The local government is soliciting bids for the following proposal:

(i)   The company (or a consortium of companies) finance the construction of a hydroelectric power plant.

291

(ii)   80% of the power produced would be for the consortium's use, and 20% would be provided to the local government for free.

(iii)   No local taxes would be charged on the project owner's income, but federal tax would be levied at 30% of taxable income. Depreciation can be charged on a straight-line basis over fifteen years on the full capital outlay commencing from the time when production exceeds 80% of the design capacity.

(iv)   The land on which the dam is to be built is in a conservation area and a conservation royalty would be charged at 5% of revenue.

(v)   The owners will own the dam and hydroelectric power scheme for 25 years, after which ownership would be transferred to the local government.

(vi)   The estimated size of the power plant is 460 MW.

(vii)   The owners of the project would not be subjected to the weekly outages, but be guaranteed supply in proportion to their share of ownership in the scheme, until the scheme is in full production.

The aluminium company currently needs about 40% of the power that the dam produces. The future needs of the company could be met by a number of other power projects planned to alleviate the power crisis. However, the company's senior managers believe there is significant value in the final aspect of the proposal and wish to bid on the proposal. They have approached an iron producer and a manufacturing company in the region, both of whom are interested in participating in the bid, to form a consortium. A project team, which has been constituted by the consortium, has estimated the capital and operating costs of the project. The capital costs are given in Table 10.1 and the team estimates the operating costs to be $12.9 M a year.

The divisional business analyst has been tasked with the duty of determining whether the company should bid as part of the consortium, what the range of bid prices that might be offered are, and the value of the bid to the aluminium company. The analyst has also been asked to advise the company's senior management of the strengths and weaknesses of the proposal so that they can make a fair presentation to the company's board of directors.

The case study is discussed in three headings that form the three steps of the decision-making process discussed in Chapter 2: frame, evaluate and decide.

Table 10.1 Projected capital expenditure profile for the Santa Clara Hydroelectric Scheme

Year	Capital expenditure US $ M	Available power MW
1	10	0
2	75	0
3	78	0
4	114	0
5	141	314
6	60	448
7	4	460
Total	482	460

Table 10.2 Facts, uncertainties and decisions for the Santa Clara HEPS project

	Item
Facts	
	1. There are power supply problems to the aluminium works.
	2. The power supply to industrial users to be cut one day a week.
	3. Company has purchased generator sets to restart after weekly power shutdown.
	4. Government is soliciting bids for a hydroelectric power scheme.
	5. The winners of the bid would not be subject to weekly shutdown.
Uncertainties	
	1. The capital investment.
	2. The operating costs.
	3. The price of electricity to be charged by the hydroelectric power scheme.
	4. The amount of electricity produced.
	5. The royalty charged by the local government.
	6. A newly elected government changes previous agreements or commitments.
Decisions	
	1. Whether to bid.
	2. The bid price.
	3. The tax rate.
	4. WACC is used as the discount rate.
	5. NPV is the decision criteria.

10.2.2 Decision Frame

The facts, uncertainties and decisions for the evaluation of the Santa Clara opportunity are given in Table 10.2.

The main questions or issues concerning the Santa Clara HEPS are the following:

1. Is this a good opportunity?
2. If it is a good opportunity, what should the company bid for its stake?
3. What are the strengths and weaknesses of this opportunity?

10.2.3 Evaluation

The first requirement is to determine whether this is a good project, independent of the financing or ownership structure. The robustness of the project must be tested. If the project passes the initial tests, then the bid value can be examined by looking at the context of the decision.

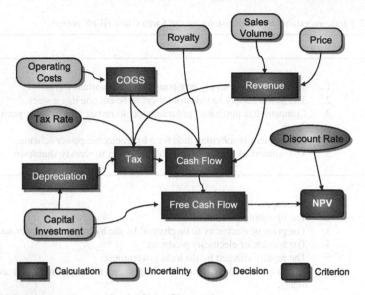

Figure 10.1 Influence diagram for the Santa Clara project

The business analyst constructed an influence diagram for the model, shown in Figure 10.1. The uncertainties and decisions are represented in the diagram. The decision criterion is the *NPV*, which is determined from the free cash flow and the discount rate. The free cash flow is determined in the manner described in Chapter 4, that is, from the revenues, the operating costs, the taxes, the capital investment and the working capital requirements. The components of each of these variables are shown in Figure 10.1.

The construction of the influence diagram gives the analyst her first insight into the issues: surprisingly, there has been no discussion between the consortium members of price for the electricity to be charged. The consortium has not decided on a price to charge its owners for the electricity. After a hasty call to the consortium representatives, a price of 7 USc/kWhr is recommended. The current price in Brazil is 4 USc/kWhr, but due to widespread power shortages, and the need to re-capitalize the industry, it is widely anticipated that the price will rise to about USc10/kWhr. The analyst decides that the price must be treated as a variable, and the sensitivity of the project to the price negotiated between the consortium members must be investigated using the recommended 7 USc/kWhr as the base case.

The project financials are shown in Table 10.3 for the first ten years of operation. No escalation for inflation has been made.

The project financials given in Table 10.3 consist of five main items. These are the revenues, the costs, the taxes and royalties, the capital investment and the change in working capital. The revenue is calculated from the fraction of the energy produced by the power plant per year in kWhr that is paid for, multiplied by the price in US c/kWhr. Only 80% of the power produced is paid for because the plant provides 20% to the local government free of charge. The costs of both operations and

Table 10.3 Project financials for the Santa Clara Hydroelectric Scheme

Year	0	1	2	3	4	5	6	7	8	9	10
Revenue	0.0	0.0	0.0	0.0	154.0	219.8	225.7	225.7	225.7	225.7	225.7
Available power					314.0	448.0	460.0	460.0	460.0	460.0	460.0
Total possible sales					192.5	274.7	282.1	282.1	282.1	282.1	282.1
Costs	0.0	0.0	0.0	0.0	12.9	12.9	12.9	12.9	12.9	12.9	12.9
Royalty	0.0	0.0	0.0	0.0	7.7	11.0	11.3	11.3	11.3	11.3	11.3
Tax	0.0	0.0	0.0	0.0	31.7	49.2	50.8	50.8	50.8	50.8	50.8
Depreciation	0.0	0.0	0.0	0.0	27.9	31.9	32.1	32.1	32.1	32.1	32.1
Profit before tax	0.0	0.0	0.0	0.0	105.6	164.0	169.3	169.3	169.3	169.3	169.3
Cash flow	0.0	0.0	0.0	0.0	101.8	146.7	150.7	150.7	150.7	150.7	150.7
Capital investment	10.0	75.0	78.0	114.0	141.0	60.0	4.0	0.0	0.0	0.0	0.0
Change in working Capital	0.0	0.0	0.0	0.0	0.0	0.0	0.0	0.0	0.0	0.0	0.0
Free cash flow	−10.0	−75.0	−78.0	−114.0	−39.2	86.7	146.7	150.7	150.7	150.7	150.7
Cumulative FCF	−10.0	−85.0	−163.0	−277.0	−316.2	−229.6	−82.9	67.8	218.5	369.2	519.8

administration are estimated to be $12.9 million per year. The royalty charged for conservation is at a rate of 5% of revenue. The tax is calculated from the taxable income, which is given by the following expression:

$$Taxable\ Income\ (Profit) = Revenue - Costs - Conservation\ Royalty$$
$$- Depreciation \tag{10.1}$$

The tax is determined from the taxable income using the following expression:

$$Tax = Tax\ Rate(Taxable\ Income) \tag{10.2}$$

The tax rate used is 30%. The depreciation schedule is given in Table 10.4. It has been assumed that the depreciation commences only once the project produces taxable income, that is, in year 4.

It has been assumed that there are no working capital requirements. The capital investment was given in Table 10.1. The capital costs can be checked against other hydroelectric schemes. A quick search of the capital costs for other hydroelectric power schemes using the Internet reveals that Mozambique is considering the con-

Table 10.4 Depreciation schedule for the Santa Clara HEPS

Year	4	5	6	7	8	9	10	11	12	13	14	15	16	17	18	19
Capital investment	418	478	482													
Depreciation	28	32	32	32	32	32	32	32	32	32	32	32	32	32	32	5
Book value	390	418	390	358	326	294	262	229	197	165	133	101	69	37	5	0

struction of a 2,500 MW scheme and the capital costs are estimated to US $1,300 million. The "rule of two thirds" discussed in Chapter 4 can be used to estimate the capital costs. This calculation is given as follows:

$$Capital\ Expenditure = 1,300 \left(\frac{460}{2,500} \right)^{0.66} = 425$$

The analyst estimates that the capital cost of the project at Santa Clara should be $425 million. This is surprisingly close to the estimate that the project team has provided.

The free cash flows are discounted at the company's discount rate of 10% in order to determine the *NPV*. The value of the *NPV* at 10% is $584 M. The values of other measures are given in Table 10.5.

All of the indicators for this project are extremely positive. The *IRR* and the *MIRR* are higher than the discount rate, which suggests that this project should be recommended. The *NPV* is positive and the *PI* is greater than one, which also suggests that the project should be recommended.

The free cash flow and the cumulative free cash flow are shown in Figures 10.2 and 10.3. The free cash flows shown in Figure 10.2 demonstrate the significant pe-

Table 10.5 Assessment measures for the project

Measure	Value
NPV	$584 M
IRR	27.5%
Profitability index	2.61
Payback period	6.5 years
MIRR	11.59%

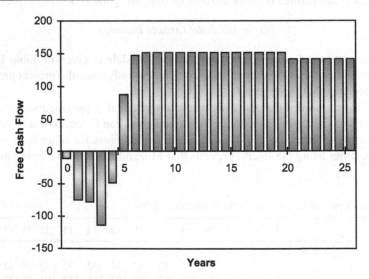

Figure 10.2 Free cash flow over the duration of the project

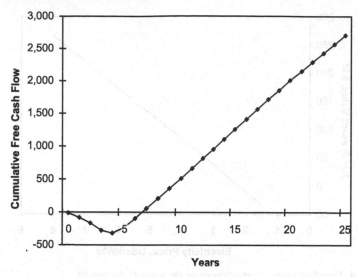

Figure 10.3 Cumulative free cash flow for the project

riod of investment prior to the realization of revenue, while the cumulative free cash flow indicates the payback period, just less than 7 years, for this project. However, after the payback period the project is expected to generate significant value for its owners.

The cash flows in Figure 10.2 decrease after year 20. A check of the calculations reveals that the book value of the hydroelectric scheme is zero at this point, and that the tax saving because of depreciation ceases after this point. Hence, with higher tax charged, the free cash flows decrease after year 20.

The sensitivity of the project with respect to the electricity price is given in Figure 10.4. An examination of these results suggests that the owners could lower the price charged to 2.7 USc/kWhr and the project would still be viable.

The effect of the price on the *NPV* and the *IRR* is shown in Figure 10.5. By trial and error, the analyst finds that the price should be at least 7.35 USc/kWhr for this project to be recommended by the *NPV* method.

Although the project is sensitive to the value of the discount rate, as shown in Figure 10.6, consultation with the management suggests that this is not really a variable, but is set by head office in London, and all projects must be evaluated on the same basis.

The electricity bill at the aluminium smelter is about US $70 million per year and the cash flow in the previous financial year amounted to US $10 million. With all else remaining the same, the value of the additional day per week should be one seventh of this, that is, US $1.4 million. While this is not a trivial amount, the decision to invest in a hydroelectric scheme is not going to hinge around this potential gain. Therefore, the value of the seventh condition in the local government's proposal is deemed to be of marginal interest.

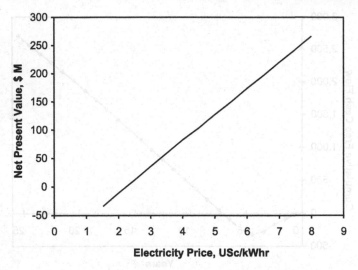

Figure 10.4 The effect of the electricity price on the value of the project

At the investment committee meeting, one of the members pointed out that this is a joint venture, rather than fully funded by the company. In his opinion, the project should pay back the investment before offering a lower price to any of the owners. The committee accepted this view. A decision is taken that a provision to this effect must be included in the joint venture contract. The discussion at the investment committee moved onto the point of what the bid price should be for this opportunity. The current price of electricity is 4 USc/kWhr, which would mean the *NPV*

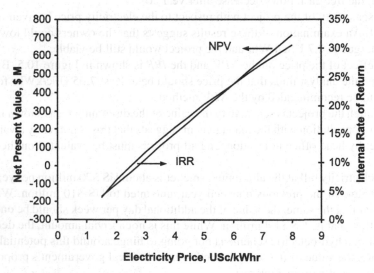

Figure 10.5 The effect of the electricity price on the *NPV* and *IRR*

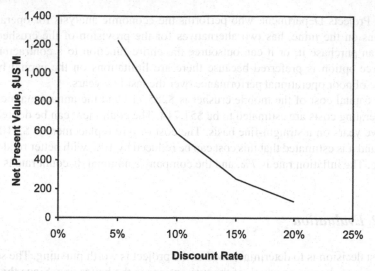

Figure 10.6 The effect of the discount rate on the *NPV*

of the project is US $171 million. After much discussion, a consensus is reached that it should be between a tenth and a third of the *NPV*, that is, within the range of US $17.1 million to US $58.0 million. Since the bid is by auction, rather than by closed tender, the investment committee recommends that the opening bid should be a lower number, and if necessary, to increase the offer up to a maximum of US $58 million.

10.2.4 Decide

This proposal was taken to the company's board of directors for approval. The other consortium members also approved this bid. The bid was accepted. Financing is in the form of project finance, a topic that is discussed in Chapter 17.

10.3 Mobile Crusher

10.3.1 Introduction

A mining company operates an open pit mine. Very large trucks move ore from the pit to the processing facilities. The road that the trucks travel on is gravel, and, because of poor road building, there is excessive wear on the tyres. A proposal has been made to crush the gravel prior to laying the road. The finer gravel should reduce tyre wear significantly. The proposal calls for a mobile crusher that can provide the correctly sized gravel at the point where road repair or road building is happening.

The Projects Department, who performs the economic analysis for operational decisions on the mine, has two alternatives for the provision of the crusher. The mine can purchase it, or it can outsource the entire function to a contractor. The outsource option is preferred because there are limitations on the capital budget because of poor operational performance over the last few years.

The capital cost of the mobile crusher is $285,714 and the annual maintenance and operating costs are estimated to be $51,714. The equipment can be depreciated over five years on a straight-line basis. The cost of tyre replacement is $3,101,071 a year and it is estimated that this cost can be reduced by 10% with better road maintenance. The inflation rate is 7%, and the company's nominal discount rate is 15%.

10.3.2 Evaluation

The first decision is to determine whether the project is worth pursuing. The second decision is to determine which of the two options is the better one. Since this is an after-tax evaluation, all values are computed in nominal terms. It is assumed that all costs and savings will escalate at the rate of inflation.

For example, the savings in tyre costs at the end of the first year is calculated as follows:

$$Tyre\ savings = 3,101,071(0.1)(1+h)^1 = 310,107(1+0.07) = 331,815$$

where h is the inflation rate. The tyre savings in subsequent years is calculated in a similar manner.

The project financials, that is, the cash flows for this project, are shown in Table 10.6 below.

Table 10.6 Free cash flows from the road maintenance project if pursued in-house (amounts in nominal terms)

Item	0	1	2	3	4	5
Savings in tyre costs		331,815	355,042	379,895	406,487	434,941
Crusher maintenance costs		55,334	59,208	63,352	67,787	72,532
Net savings		276,480	295,834	316,542	338,700	362,409
Depreciation		57,143	57,143	57,143	57,143	57,143
Taxable savings		219,338	238,691	259,400	281,557	305,266
Tax		65,801	71,607	77,820	84,467	91,580
Cash flow		210,679	224,227	238,722	254,233	270,829
Capital investment	−285,714					
Free cash flow	−285,714	210,679	224,227	238,722	254,233	270,829
Discount factor	1.000	0.870	0.756	0.658	0.572	0.497
PV(FCF)	−285,714	183,199	169,548	156,964	145,359	134,650
NPV	504,005					
IRR	74%					

This project involves savings, not revenues. However, the savings are treated in much the same way as revenues. The savings arise from the lower cost of replacing tyres on the large trucks moving material from the open pit mine. The costs for the project arise from maintaining the gravel crusher. The tax is calculated from Equations 10.1 and 10.2. The depreciation is calculated on a straight-line basis over five years. The capital investment is the cost of the mobile crusher, and it has been assumed that there are no working capital requirements.

The calculation of the net present value for the five-year project is shown in Table 10.6. The *NPV* is positive and the *IRR* is greater than the discount rate. Both these measures suggest that this project is highly profitable and that it should be recommended for acceptance.

The project can be implemented by outsourcing the function to a contractor. The contractor will purchase the equipment, maintain it, and provide trained personnel for its operation. The contractor has quoted a price of \$125,000 per year for the five-year contract. The calculation of the free cash flows for the outsource option is shown in Table 10.7.

As an outsourced project, the costs are not those of maintaining the mobile crusher, but of paying the contractor for their services. There is no capital investment, and hence no depreciation.

The project as an outsourced contract is more favourable than as an in-house operation because the *NPV* is higher. The recommendation is to outsource the operation. Notice that the *IRR* is undefined. This is because the free cash flow profile does not change sign. If the measure for decision-making is the *IRR*, an incremental analysis of the two options is required.

The incremental cash flow is calculated so that the resulting profile is conventional. This is done by subtracting the free cash flows of outsourcing from the in-house operation. The calculation of the incremental *NPV* and the incremental *IRR* are given in Table 10.8.

The negative *NPV* and the *IRR* less than the discount rate of 15% indicates that outsourcing is favoured by both measures.

Table 10.7 Free cash flows from the road maintenance project if outsourced (amounts in nominal terms)

Item	0	1	2	3	4	5
Saving on tyre costs		331,815	355,042	379,895	406,487	434,941
Contractor costs		133,750	143,113	153,130	163,850	175,319
Net savings		198,065	211,929	226,764	242,638	259,622
Tax		59,419	63,579	68,029	72,791	77,887
Cash flow		138,645	148,350	158,735	169,846	181,736
Investment	0					
Free cash flow	0	138,645	148,350	158,735	169,846	181,736
PV(FCF)	0	120,561	112,174	104,371	97,110	90,355
NPV	524,571					
Nominal IRR	-					

Table 10.8 The incremental free cash flows from the road maintenance project (in-house less outsource). Amounts in nominal terms

Item	0	1	2	3	4	5
Incremental free cash flow	−285,714	72,034	75,876	79,988	84,387	89,094
PV(incremental free cash flow)	−285,714	62,638	57,373	52,593	48,248	44,295
NPV	−20,566					
IRR	12.0%					

10.3.3 Recommendation

The recommendation is to outsource the service. The project itself, either as an in-house operation or as one outsourced to a contractor, is highly desirable.

10.4 Mobile Crusher Contractor

10.4.1 Introduction

This case study is a continuation of the previous one, except from the viewpoint of a contractor. The *NPV* of the outsourced option is greater than that of the in-house operation. This case study examines how this increase in value arises.

10.4.2 Evaluation

The contractor's project financials for the case study above are easily calculated. If the contractor's discount rate was the same as the mine's (15%), the *NPV* for the contractor is calculated in Table 10.9. The revenues for the contractor are the same as the costs for the mine, since the mine pays the contractor for the services. The operator's costs are the same as those for the mine as an operator given in Table 10.6. Other figures are calculated in a similar manner to those for the mine.

It is clear that the contractor will lose money on this deal, since the *NPV* is negative, and the *IRR* is less than the discount rate of 15%. The contractor faces a dilemma: if the contractor prices the contract higher by charging more, he faces the possibility of losing the contract. This is investigated further in Figure 10.7, which shows the loss in value to the mine and the gain in value to the contractor as a function of the contract price if outsourcing is used.

The value lost to the mine is exactly that gained by the contractor. Outsourcing is zero sum gain. Why would the contractor enter into this contract? This can be examined further by considering the contractor's actual cash flow rather than using the cash flow on an entity basis.

It is emphasised, as has been done before, that the choice of the entity basis for the analysis of the project's economic viability is made not on a theoretic foundation,

Table 10.9 Project financials from the road maintenance project for the contractor (amounts in nominal terms)

Item	0	1	2	3	4	5
Revenue		133,750	143,113	153,130	163,850	175,319
Operating costs		55,334	59,208	63,352	67,787	72,532
Operating profit		78,416	83,905	89,778	96,063	102,787
Depreciation		57,143	57,143	57,143	57,143	57,143
Taxable income		21,273	26,762	32,635	38,920	45,644
Tax		6,382	8,029	9,791	11,676	13,693
Cash flow		72,034	75,876	79,988	84,387	89,094
Investment	−285,714					
Free cash flow (FCF)	−285,714	72,034	75,876	79,988	84,387	89,094
Discount factor	1.000	0.870	0.756	0.658	0.572	0.497
PV(FCF)	−285,714	62,638	57,373	52,593	48,248	44,295
NPV	−20,566					
IRR	12%					

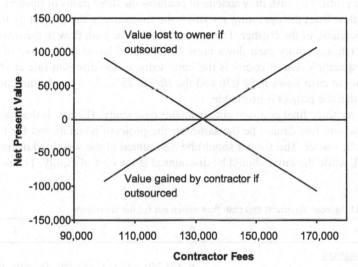

Figure 10.7 The value gained by the contractor and lost by the owner due to outsourcing as a function of price

but on a practical one, that is, the need to separate the evaluation and the financing decisions. The entity basis ignores the interest paid in the cash flow in establishing the project financials.

The contractor can fund the purchase of the mobile crusher by a full loan at an interest rate of 15%. If this is an amortized loan over five years, the repayments are calculated as follows:

$$A = P(A|P, i, n) = Pi(1+i)^n/[(1+i)^n - 1]$$
$$= 285,714(0.15)(1.15)^5/[(1.15)^5 - 1] = 85,233.$$

Table 10.10 Loan schedule for the contractor

Year	0	1	2	3	4	5
Opening	285,714	285,714	243,338	194,606	138,564	74,116
Interest		42,857	36,501	29,191	20,785	11,117
Payment		85,233	85,233	85,233	85,233	85,233
Repayment of principal		42,376	48,732	56,042	64,448	74,116
Closing		243,338	194,606	138,564	74,116	0

Part of the payment of $85,233 is interest and part is capital. Because only the interest is tax-deductible, a loan schedule for this loan is given in Table 10.10.

The interest is paid on the outstanding balance, and the principal repayment is the difference between the repayment and the interest portion.

The cash flow for the contractor's business is shown in Table 10.11.

The cash flow statement is similar to the project financials presented in Table 10.9, except that interest has been accounted for in the calculation of the tax and hence the profit. The cash flow statement contains the three items of interest, that is, the cash flow from the operating activities, the financing of the project by the loan, and the purchase of the crusher. The net cash flow is the cash flow to the owner. The discount rate for equity cash flows must be evaluated based on the cost of equity. If the contractor's cost of equity is the same value as the discount rate at 15%, the *NPV* of these cash flows is $2,036 and the *IRR* is 25%. As a result, the contractor assesses that the project is profitable.

There are three final points to note from this case study. The first is that the choice of the discount rate cannot be the same for the project financials and for the cash flows to the owner. The former should be discounted at the weighted average cost of capital, while the latter should be discounted at the cost of equity. The second is

Table 10.11 Income statement and cash flow statement for the contractor

Year	0	1	2	3	4	5
Income statement						
Revenue	0	133,750	143,113	153,130	163,850	175,319
Operating costs	0	55,334	59,208	63,352	67,787	72,532
Operating profit	0	78,416	83,905	89,778	96,063	102,787
Depreciation	0	57,143	57,143	57,143	57,143	57,143
Interest	0	42,857	36,501	29,191	20,785	11,117
Profit before tax	0	−21,584	−9,739	3,444	18,135	34,527
Tax	0	0	0	1,033	5,441	10,358
Profit after tax	0	−21,584	−9,739	2,411	12,695	24,169
Cash flow statement						
Cash flow from operating activities	0	78,416	83,905	88,745	90,622	92,429
Cash flow from financing activities	285,714	−85,233	−85,233	−85,233	−85,233	−85,233
Cash flow from investment activities	−285,714					
Net cash flow	0	−6,817	−1,328	3,512	5,389	7,196

that the assessment of the equity cash flow is very sensitive to the type of loan. If the loan for the purchase of the crusher were an instalment loan, the conclusion would be the reverse. The third is that there is a possible conflict between the entity basis and the equity basis. This conflict is discussed in more detail in Chapter 12.

10.5 Filtration Joint Venture

10.5.1 Introduction

An underground mine needs to pump water from the workings. This water is dirtied with the fine material produced during the underground blasting operations. Up until now, the water has been pumped to the surface to a settling pond, where the fines have settled out and the water evaporates. Recent research work has lead to the discovery that the fine material contains a significant amount of precious metals, such as platinum, palladium and gold. The mine has issued a request for proposals for the separation of the fines on a continuous basis prior to the disposal of the water in the settling ponds. These solids are to be sold to a toll treatment facility.

The mine's request for proposals indicates that there are to be two separation facilities, one for each of the mine's shafts, and these facilities are to be owned and operated by the winning bidder. The mine will pay a treatment charge to be specified in the proposal. Terms for the payment of the treatment charge will be 90 days after invoicing.

A company selling filtration equipment, Alex Filters, is interested in bidding for the project. However, Alex has no desire to move into operations, so the managing director has contacted a waste treatment company, WasteTek, which has operations in the vicinity of the mine, with the suggestion that they participate in the proposal to own and operate the solid-liquid separation plants at the mine. The two plants would use Alex Filter's propriety filtration systems, which are renowned as expensive but highly reliable.

The consortium of the two companies Alex and WasteTek attend a compulsory site visit, and it is clear that there are at least twenty other interested parties. For the proposal to be successful, it has to be the best possible price. After the site visit, the sales manager of Alex filters was asked to quote for two filters with the same duty as those required for this project by an engineering company. It is apparent that this engineering company wishes to bid on the same project, and either wishes to use Alex Filter's equipment in the bid or they are attempting to get information on the pricing of their rivals.

The managing director of Alex Filters has consulted Engineering Management Solutions, an independent consultancy with corporate finance experience, to determine whether Alex and WasteTek should bid for this opportunity and to establish the charge structure. In addition, he has requested advice on the basis of the commercial arrangement between Alex and WasteTek.

10.5.2 Project Financials

The project financials for the first ten years of operation of the filtration plants are shown in Table 10.12. The project financials are prepared from the five main items that determine free cash flow: revenue, costs, taxes, capital investment and working capital requirements. Each of these five items is discussed in further detail below.

(i)   Revenue

The request for proposals distributed by the mine suggested two mechanisms for the pricing of the contract. The first is a flat fee paid on a monthly basis. The second is a fee based on the amount of material that is treated. The advantage of the former is that the revenues are fixed, making financing of the operation much easier. The disadvantage is that if the amount of fine material rises, the costs incurred will rise without a rise in the income. The advantage of the second method is that if the amount of material treated rises, the revenue will increase, covering costs. However, if the amount of material decreases, there would be a loss of revenue, which may create financial stress.

The preferred pricing mechanism from the viewpoint of Alex Filters is the monthly charge. The initial calculations suggest a value in the range of $120,000 per month. The actual price tendered in response to the request for proposals is discussed later. These values are escalated for inflation at a rate of 5% per annum.

(ii)   Costs

The project manager at Alex Filters has estimated the annual costs for the operation of the two filter plants at the mine site. These costs are given in Tables 10.13 to 10.17.

The costs consist of spares, maintenance, staff, transport, and consumables. Nine operators and a foreman will operate the two plants in three shifts of eight hours each day. The rental of a light delivery vehicle is included in the budget for transportation between the two filtration plants, which are 4 kilometres apart. Planned maintenance of the filters and the auxiliary plant equipment is required.

In addition to these operating costs, there are administrative functions and management that is required. These costs are detailed in Table 10.18. These costs are estimated to total $72,000 annually.

Table 10.12 Project financials for the filtration plants. All amounts in thousands.

Year	0	1	2	3	4	5	6	7	8	9	10
Revenue	0	1,440	1,512	1,588	1,667	1,750	1,838	1,930	2,026	2,128	2,234
Costs	0	709	758	811	870	933	1,001	1,076	1,157	1,246	1,342
Tax	0	140	147	153	159	165	171	176	180	184	187
Capital investment	−2,483	0	0	0	0	0	0	0	0	0	0
Change in working capital	0	303	14	14	15	15	16	17	17	18	18
Free cash flow	−2,483	289	594	609	623	637	650	662	672	680	687

Table 10.13 Estimation of annual costs of spare parts for the filter plants

Spares	Amount
Filter spares	95,000
Auxiliary plant spares	60,000
Pump spares	5,000
Hydraulic unit spares	7,500
Electrical spares	7,500
Programmable logic controller spares	5,000
Total	180,000

Table 10.14 Estimation of annual costs of maintenance of the filter plants

Service and maintenance	Amount
Service visits	31,200
Electrical and instrumentation service	3,200
Auxiliary plant maintenance	28,800
Total	34,400

Table 10.15 Estimation of annual costs of staff for the filter plants

Operating staff	Number	Unit cost	Amount
Foreman	1	56,000	56,000
Operators	9	32,674	294,066
Total			350,066

Table 10.16 Estimation of annual costs of transportation for the filter plants

Vehicles	Amount
Light delivery vehicle rental	3,675
Fuel	3,251
Insurance	410
Total	7,336

Table 10.17 Estimation of annual costs of consumables for the filter plants

Chemicals and consumables	Amount
Chemicals supply	44,015
Hydraulic oil	700
Total	44,715

Table 10.18 Administration and management overheads for the filtration plants

Item	Amount
Management	48,000
Administration	24,000
Total	72,000

The general inflation rate is 5%, but labour costs are increasing at about 10%. The costs are escalated by these factors each year.

(iii)   Taxes

The taxes are calculated using an average tax rate of 29% of the taxable income, which is estimated as follows:

$$Taxable\ Income\ =\ Revenue\ -\ Costs\ -\ Depreciation$$

The depreciation allowed for tax purposes (also called the capital allowance) is calculated on a straight-line basis over ten years. The depreciation schedule is given in Table 10.19.

The annual depreciation amount is equal to $248,388, which is the capital costs divided by ten.

(iv)   Capital costs

The project manager at Alex Filters has estimated the capital costs for the construction of the two filter plants. These estimates are given in Table 10.20.

Quotations for the civil and structural engineering work have been obtained from a contractor, and the electrical and instrumentation costs have been estimated by Alex Filters based on their experience with other installations. The engineering design costs are estimated to be 4% of the capital expenditure, and the project management and the contingency costs are estimated to be 5% of the capital expenditure each.

The filters have been priced with a mark-up of 35%. The normal mark-up is 55%. The cost of the filters can be offered to this project at a discount, depending on the

Table 10.19 Depreciation schedule. All amounts in thousands.

	0	1	2	3	4	5	6	7	8	9	10
Investments	2,483										
Depreciation		248	248	248	248	248	248	248	248	248	248
Book value		2,235	1,987	1,738	1,490	1,241	993	745	496	248	0

Table 10.20 Estimated capital costs for the construction of the filter plants

Item	Amount
Civil	178,695
Structural	845,910
Electrical and instrumentation	352,824
Filters	801,417
Engineering design	87,154
Project management	108,942
Contingency	108,942
Total	2,483,884

Table 10.21 Calculation of the change in working capital

Year	0	1	2	3	4	5	6	7	8	9	10
Accounts receivable	0	355,068	372,822	391,463	411,036	431,588	453,167	475,826	499,617	524,598	550,828
Accounts payable	0	52,316	56,371	60,772	65,551	70,744	76,387	82,523	89,198	96,462	104,368
Working capital	0	302,752	316,451	330,691	345,485	360,844	376,780	393,302	410,419	428,136	446,459
Change in working capital	0	302,752	13,699	14,240	14,794	15,359	15,936	16,522	17,117	17,717	18,323

agreement with WasteTek. Alex Filters will need to make up for the discount from the proceeds of the operating company. This is discussed later.

(v) Change in working capital

The working capital is the net amount of money invested in raw materials stock, finished product inventory, amounts owed to suppliers and amounts owed by customers. The working capital is mainly affected by the accounts payable (creditors) and accounts receivable (debtors). These are estimated using Equation 3.36 and 3.37, and are given as follows:

$$Accounts\,Payable = (Annual\,COGS)(Creditor\,Days)/365 \qquad (10.3)$$
$$Account\,Receivable = (Annual\,Revenue)(Debtors\,Days)/365 \qquad (10.4)$$

The debtor days are estimated to be 90 days and the creditor days to be 30 days. The working capital required is the difference between the accounts receivable and account payable. The working capital is expressed as follows:

$$Working\,capital = Accounts\,Receivable - Accounts\,Payable \qquad (10.5)$$

The working capital for the first year is calculated as follows:

$$Accounts\,Payable = (708,517 - 72,000)(30)/365 = 52,316$$
$$Account\,Receivable = (1,440,000)(90)/365 = 355,068$$
$$Working\,capital = 355,068 - 52,316 = 302,752$$

The working capital for subsequent years is calculated in a similar manner. The change in working capital is the difference between the current year and the previous year. This difference in working capital is the additional capital that is required because monies have not been paid by customers. The change in working capital, shown in Table 10.21 is required in the calculation of the free cash flow. The effect of the working capital is most prominent in the first year.

10.5.3 Assessment

The project financials are shown in Table 10.22.

The project is assessed over a period of ten years. The equipment is known to last for twenty-five years and the life of the mine is estimated to be at least that. The choice of ten years is conservative. The free cash flow and the cumulative free cash flow are shown in Figure 10.8.

The net present value, profitability index and internal rate of return for the project are given in Table 10.23. The discount rate used by Alex Filters is 15%.

Table 10.22 Project financials for the filtration plant. All amounts in thousands.

Year	0	1	2	3	4	5	6	7	8	9	10
Revenue	0	1,440	1,512	1,588	1,667	1,750	1,838	1,930	2,026	2,128	2,234
Costs	0	709	758	811	870	933	1,001	1,076	1,157	1,246	1,342
Labour	0	350	385	424	466	513	564	620	682	750	825
Maintenance and service	0	34	36	38	40	42	44	46	48	51	53
Spares	0	180	189	198	208	219	230	241	253	266	279
Operator training	0	20	21	22	23	24	26	27	28	30	31
Chemicals	0	44	46	49	51	54	56	59	62	65	68
Consumables	0	1	1	1	1	1	1	1	1	1	1
Vehicles	0	7	8	8	8	8	9	9	10	10	11
Administration	0	72	72	72	72	72	72	72	72	72	72
Tax	0	140	147	153	159	165	171	176	180	184	187
Depreciation	0	248	248	248	248	248	248	248	248	248	248
Taxable income	0	483	506	528	549	569	588	605	621	634	644
Investment	−2,484	0	0	0	0	0	0	0	0	0	0
Civil	−179	0	0	0	0	0	0	0	0	0	0
Structural	−846	0	0	0	0	0	0	0	0	0	0
Electrical	−353	0	0	0	0	0	0	0	0	0	0
Filter	−801	0	0	0	0	0	0	0	0	0	0
Engineering design	−87	0	0	0	0	0	0	0	0	0	0
Project management	−109	0	0	0	0	0	0	0	0	0	0
Capital contingency	−109	0	0	0	0	0	0	0	0	0	0
Change in working capital	0	−303	−14	−14	−15	−15	−16	−17	−17	−18	−18
Free cash flow	−2,484	289	594	609	623	637	650	662	672	680	687
Cumulative free cash flow	−2,484	−2,195	−1,601	−993	−369	268	918	1,580	2,252	2,932	3,619

Table 10.23 Decision measures for the project

Measure	Value
NPV	$411,403
Profitability index	1.17
IRR	19%
Payback period	4.5 years

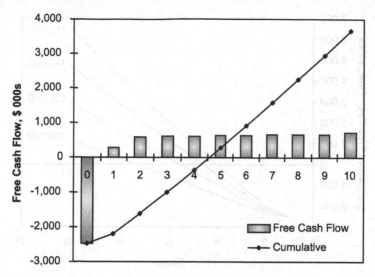

Figure 10.8 Free cash flow and cumulative free cash flow for the project

All of the decision measures shown in Table 10.23 are positive, suggesting that this project is acceptable. The *NPV* is positive, the profitability index is greater than one, and the *IRR* is greater than the discount rate. The payback period seems long since it is about half of the project's life. All of these measures suggest that the project should be acceptable to Alex Filters.

10.5.4 Sensitivity Analysis

The major variable required for the bid is the charge to the mine for the treatment of the water. The effect of the monthly revenues on the cumulative free cash flow is shown in Figure 10.9. Although increasing the revenue dramatically affects the financials, the consortium is likely to loose the bid if the charges are priced incorrectly.

The effect of changing the capital costs by changing the mark-up for the filters on the project financials is shown in Figure 10.10. The effect of the mark-up in the price of the filters on the discounted cash flow measures is shown in Table 10.24.

Table 10.24 Effect of the mark-up on the filter price on the *NPV*, *PI* and *IRR* for the project

Mark-up	NPV	PI	IRR
0%	613,792	1.27	21%
35%	411,403	1.17	19%
60%	266,840	1.10	17%
100%	35,538	1.01	15%

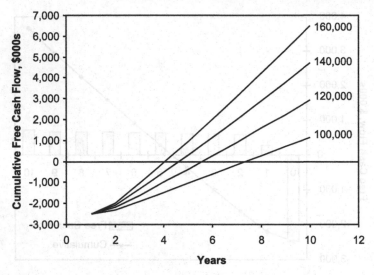

Figure 10.9 Effect of various monthly processing charges on the cumulative free cash flow

The *NPV* of the project is almost zero if the mark-up is 100%, that is, if the price that Alex Filters charges is twice the cost it incurred to manufacture them.

The effect of the debtor days on the working capital requirements is shown in Figure 10.11. The working capital requirements are 12% of the total capital. The results shown in Figure 10.11 indicate that significant savings can be made if the debtor days are reduced from the current proposal of 90 days to 30 days.

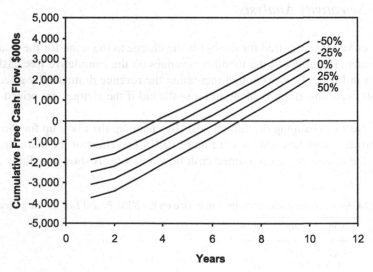

Figure 10.10 Effect of capital expenditure calculated as a percentage increase or decrease of the base case on the cumulative free cash flow

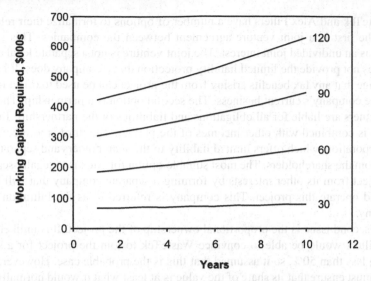

Figure 10.11 Effect of the debtor days (days receivable) on the working capital required

10.5.5 Commercial Arrangement Between Alex and WasteTek

Three issues face Alex in assessing the commercial arrangements with WasteTek. The first is the legal form of the relationship, the second is the proportional owner-ship of the project and the third is the relationship with a commercial bank as the provider of a loan. These relationships are illustrated in Figure 10.12.

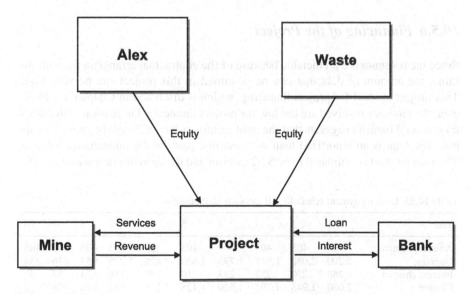

Figure 10.12 Commercial relationships between the various parties

WasteTek and Alex Filters have a number of options to formalize their relationship. The first is a joint venture agreement between the companies. This is also known as an undivided joint interest. The joint venture is not a separate legal entity, and does not provide the limited liability protection that a company does. It has the advantage that any tax benefits arising from the project can be used to offset the tax from the company's current business. The second option is a partnership. The general partners are liable for all obligations and liabilities of the partnership. Taxable income is combined with other incomes of the partner. The third is as a company or a corporation, which offers limited liability to the shareholders and separate taxation from the shareholders. The most suitable option for Alex is to legally separate the project from its other interests by forming a separate company that will build, own and operate this project. This company is referred to as the Filtration Plant Company.

The second issue is the proportional ownership of the project. It is unlikely that Alex Filters would be able to convince WasteTek to join the project for a shareholding less than 50%, so it assumed that this is the probable case. However, Alex Filters must ensure that its share of the value is at least what it would normally earn for the sale of the filters. If it simply sold the filters at their normal price, the profit would be the mark-up, which is equal to $326,503 = ($296,821 per filter) (2 filters) (55% mark-up). The *NPV* of the cash flows from the project, that is, the initial mark-up plus the dividends from the operation, must at least be equal to this figure. In order to calculate the dividends accurately, the financing of the project needs to be determined.

The third issue is the amount of debt that this project can sustain. This is discussed in the next section.

10.5.6 Financing of the Project

Since the revenues are predictable because of the contractual arrangements with the mine, the amount of debt that can be sustained in this project can be very high. This project is ideal for project financing, which is discussed in Chapter 17. However, the amounts involved are too low for project financing. Discussion with Alex's commercial bankers suggests that the debt cannot exceed 80% of the project's capital. The loan is an amortized loan with interest paid on the outstanding balance. The loan amount is estimated to be $2.2 million and the current interest rate is 13%.

Table 10.25 Loan repayment schedule. All amounts in thousands

Year	1	2	3	4	5	6	7	8	9	10
Annual payment	405	405	405	405	405	405	405	405	405	405
Opening	2,200	2,080	1,945	1,793	1,620	1,426	1,205	957	676	358
Interest charged	286	270	252	233	210	185	156	124	87	46
Closing	2,080	1,945	1,793	1,620	1,426	1,205	957	676	358	0
Principal repayment	119	134	152	172	194	220	248	280	317	358

Table 10.26 Forecast Income Statement for the Filtration Plant Company. All amounts in thousands

Year	1	2	3	4	5	6	7	8	9	10
Revenue	1,440	1,512	1,588	1,667	1,750	1,838	1,930	2,026	2,128	2,234
Cost of goods sold	637	686	739	798	861	929	1,004	1,085	1,174	1,270
Operating income	803	826	848	869	890	908	926	941	954	964
Administration	72	72	72	72	72	72	72	72	72	72
Depreciation	248	248	248	248	248	248	248	248	248	248
PBIT (EBIT)	483	506	528	549	569	588	605	621	634	644
Interest paid	286	270	253	233	211	185	157	124	88	47
Tax	57	68	80	92	104	117	130	144	158	173
Profit	140	167	195	224	255	286	318	352	387	424
Dividends paid	126	150	176	202	229	257	287	317	349	382
Retained earnings	14	17	20	22	25	29	32	35	39	42

Table 10.27 Forecast balance sheet for the Filtration Plant Company. All amounts in thousands

Year	1	2	3	4	5	6	7	8	9	10
Assets										
Cash	106	223	324	408	471	512	527	513	465	378
Net accounts receivable	355	373	391	411	432	453	476	500	525	551
Total current assets	461	596	715	819	903	965	1,003	1,013	989	929
Gross fixed assets	2,484	2,484	2,484	2,484	2,484	2,484	2,484	2,484	2,484	2,484
Less accumulated depreciation	248	497	745	994	1,242	1,490	1,739	1,987	2,235	2,484
Net fixed assets	2,235	1,987	1,739	1,490	1,242	994	745	497	248	0
Total assets	2,697	2,583	2,454	2,309	2,145	1,959	1,748	1,509	1,238	929
Liabilities										
Accounts payable	52	56	61	66	71	76	83	89	96	104
Total current liabilities	52	56	61	66	71	76	83	89	96	104
Loans	2,081	1,946	1,793	1,621	1,426	1,206	957	676	359	0
Total long-term liabilities	2,081	1,946	1,793	1,621	1,426	1,206	957	676	359	0
Total liabilities	2,133	2,002	1,854	1,686	1,497	1,282	1,040	766	455	104
Equity										
Investors equity	550	550	550	550	550	550	550	550	550	550
Retained earnings	14	31	50	73	98	127	159	194	233	275
Total equity	564	581	600	623	648	677	709	744	783	825
Total liabilities and equity	2,697	2,583	2,454	2,309	2,145	1,959	1,748	1,509	1,238	929

Table 10.28 Forecast cash flow statement for the Filtration Plant Company. All amounts in thousands

Year	1	2	3	4	5	6	7	8	9	10
Operating activities										
Income from operating activities	731	754	776	797	818	836	854	869	882	892
Adjusted for increase in working capital	−303	−14	−14	−15	−15	−16	−17	−17	−18	−18
Taxation paid	−57	−68	−80	−92	−104	−117	−130	−144	−158	−173
Net cash provided by operating activities	372	672	682	691	698	704	707	708	706	701
Investment activities										
Cash used to acquire fixed tangible assets	2,484	0	0	0	0	0	0	0	0	0
Net cash used for investment activities	2,484	0	0	0	0	0	0	0	0	0
Financing activities										
Equity capital raised	550	0	0	0	0	0	0	0	0	0
Increase in interest-bearing long-term liabilities	2,081	−135	−153	−172	−195	−220	−249	−281	−318	−359
Interest paid	−286	−270	−253	−233	−211	−185	−157	−124	−88	−47
Dividends paid	−126	−150	−176	−202	−229	−257	−287	−317	−349	−382
Net cash provided by financing activities	2,219	−556	−581	−607	−635	−663	−692	−722	−754	−787
Net cash and cash equivalents	106	116	101	84	64	41	15	−14	−48	−86
Cash at beginning of year	0	106	223	324	408	471	512	527	513	465
Cash at end of year	106	223	324	408	471	512	527	513	465	378
Net change in cash position	106	116	101	84	64	41	15	−14	−48	−86

The annual repayment of principal and interest can be calculated using Equation 6 in Table 5.3. This annual payment is calculated as follows:

$$A = P(A|P, i, n) = Pi(1+i)^n/[(1+i)^n - 1] = 2.2(0.13)(1.13)^{10}/[1.13^{10} - 1]$$
$$= \$405,437.02$$

The repayment schedule is given in Table 10.25.

The interest charged is the opening balance for the year multiplied by the interest rate. The principal repayment is the annual payment less the interest charged. The closing balance is the opening balance less the principal repayment. It is necessary to separate the interest charged from the principal repayment, because the interest charged is tax deductible, whereas the principal repayment is not.

The project company's financial statements are forecast on this basis using the methods outlined in Chapter 3. The forecast financial statements for the project company are given in Table 10.26 to 10.28.

The forecast income statements indicate that the Filtration Plant Company is profitable in each year of its life, and that it is able pay 90% of its profit (earnings) as a dividend to its shareholders. There is no point in retaining earnings in the com-

Table 10.29 Calculation of the present value of Alex Filters' share of the project (all amounts in thousands)

	0	1	2	3	4	5	6	7	8	9	10
Profit on sale of filters	208										
Dividends	0	63	75	88	101	115	129	143	159	174	191
Equity capital required	−275										
Terminal value											412
Cash flow	−67	63	75	88	101	115	129	143	159	174	603
Discount factor	1.000	0.870	0.756	0.658	0.572	0.497	0.432	0.376	0.327	0.284	0.247
PV	−67	55	57	58	58	57	56	54	52	50	149
NPV	577										

pany, since there are no other business opportunities for the company to pursue. The shareholders of the company, Alex Filters and WasteTek, can invest their dividends in other business opportunities.

The forecast cash flow statements indicate that the company is solvent for the first ten years of its life.

Alex Filters will receive returns from this project in three ways: (i) from their profit on the sale of the filters; (ii) from dividends received from the Filtration Plant Company; and (iii) from the increased value of Filtration Plant Company (capital gain) The total gain for Alex Filters must be at least what it can make on just selling the filters. The calculation of the present value of Alex Filters is given in Table 10.29.

The profit on the sale of the filters is equal to \$207,774 (= \$296,821 × 2 × 0.35). Alex Filters receives half of the dividend payments from the Filtration Plant Company, and has to invest half of the equity in the company, that is, \$275,000. The value at the end of the ten years (terminal value) to taken as the equity of the Filtration Plant Company from the balance sheet, that is, \$824,899, and one half of this is Alex Filter's share. This means that the business has been valued at book value using the company's net asset value. The addition of these values gives the cash flows to Alex Filters. The net present value of these values is equal to \$576,770, which is greater than the profit of \$326,503 if Alex Filters simply sold the filters. Note that all the figures in Table 10.29 may be subject to tax at different rates, which may or may not alter this conclusion.

10.5.7 Recommendation

The filtration project is profitable both as a project and as a separate company. It is recommended that Alex Filters invest in the opportunity as equal partners with WasteTek in order to submit a bid to the mine. It is recommended that the bid propose a fixed charge to the mine, as this contractual agreement will allow the project company to attract a much high level of debt.

10.6 Bakersfield Water Pumps

10.6.1 Introduction

A water pumping station in Bakersfield, California, delivers water over the Tehachapi Mountains into Southern California. (See Hamilton, 2004. All values in this case study are fictitious.) The plant consists of fourteen pumps. Each pump is a four-stage, 60 MW, 600 rpm unit. The capacity of each pump is 8.92 m^3/s with a static head of 587m. Discharge from the pumping station is through two pipelines with a diameter of 4.26m that feed a tunnel with a diameter of 7.16m. The energy consumption is 840 MW with all fourteen pumps running. There are four Allis-Chalmers pumps, seven Baldwin Lima Hamilton pumps, and three Voith pumps. The Allis-Chalmers pumps were installed in 1971, the Baldwin Lima Hamilton pumps were installed in 1973, and the Voith pumps were installed in 1982. The Voith pumps have a pumping efficiency of 92%, whereas the Allis-Chalmers pumps have an efficiency of 89.7%.

The water flow through the pumping station is projected to double over the next thirty years. The pumps currently operate for 10 hours a day during the off-peak electricity period of 10pm to 8am providing 117 million m^3/yr of water each. The off-peak time is completely used. To meet the increasing demand for water, it will be necessary to use on-peak electricity.

Maintenance of the pumps is a major function. The pumping station only has an availability of 70%, which may be too low as the water demand increases. Over the years, there have been numerous maintenance problems with the pumps, mainly related to electrical issues. All of the motor stators have been rewound. Other problems are related to the impellers, caused by flow-induced erosion, corrosion, and cracking. Officials of the California Department of Water Resources are concerned about the older Allis-Chalmers units. Two major proposals have emerged. The first proposal is to replace the Allis Chalmers units with new Voith units, which are expected to have an efficiency of 92%. The Voith pumps will only require major maintenance every ten years, whereas the Allis-Chalmers pumps undergo a major refurbishment every five years. The second proposal is retain the Allis-Chalmers pumps, redesign the suction bend, and replace the stage one impellers.

Project Management Solutions has been retained by the California Department of Water Resources to assess which of the two solutions is more economically desirable.

10.6.2 Project Financials

The project is an equipment replacement study with cost saving as the measure of economic performance. The alternative with the lowest present value of costs is the preferred alternative. The project financials for the alternative to replace the existing pumps with new pumps are presented in Table 10.30.

The delivery of water increases linearly from 117 million m^3/year for each pump to 234 million m^3/year. The current load of 117 million m^3/year is at the off-peak

Table 10.30 Project financials for the alternative to replacement the existing the pumps with new pumps (all amounts in thousands)

Year	Off-peak energy costs	On-peak energy costs	Regular maintenance	New and refurbish capital	Total costs	Discount factor	PV(costs)
0		0		-32,600	-32,600	1.000	-32,600
1	-3,335	-162	-1,014		-4,511	0.943	-4,256
2	-3,502	-341	-1,055		-4,897	0.890	-4,358
3	-3,677	-537	-1,097		-5,310	0.840	-4,459
4	-3,861	-752	-1,141		-5,753	0.792	-4,557
5	-4,054	-987	-1,186		-6,227	0.747	-4,653
6	-4,256	-1,243	-1,234		-6,733	0.705	-4,747
7	-4,469	-1,523	-1,283		-7,275	0.665	-4,838
8	-4,693	-1,828	-1,334		-7,855	0.627	-4,928
9	-4,927	-2,159	-1,388		-8,474	0.592	-9,861
10	-5,174	-2,519	-1,443	-9,651	-18,787	0.558	-29,881
11	-5,432	-2,909	-1,501		-9,842	0.527	-5,185
12	-5,704	-3,332	-1,561		-10,597	0.497	-5,266
13	-5,989	-3,790	-1,623		-11,403	0.469	-5,346
14	-6,288	-4,286	-1,688		-12,263	0.442	-9,693
15	-6,603	-4,822	-1,756		-13,181	0.417	-5,500
16	-6,933	-5,400	-1,826		-14,160	0.394	-5,574
17	-7,280	-6,025	-1,899		-15,204	0.371	-5,646
18	-7,644	-6,698	-1,975		-16,317	0.350	-5,717
19	-8,026	-7,424	-2,054		-17,504	0.331	-5,785
20	-8,477	-8,205	2,136	14,286	-33,055	0.312	-10,507
21	-8,848	-9,046	-2,222		-20,117	0.294	-5,917
22	-9,291	-9,951	-2,311		-21,553	0.278	-5,981
23	-9,755	-10,924	-2,403		-23,082	0.262	-6,043
24	-10,243	-11,968	-2,499		-24,711	0.247	-6,103
25	-10,755	-13,091	-2,599		-26,445	0.233	-6,162
26	-11,293	-14,295	-2,703		-28,291	0.220	-6,219
27	-11,858	-15,587	-2,811		-30,256	0.207	-6,274
28	-12,451	-16,972	-2,924		-32,347	0.196	-6,328
29	-13,073	-18,457	-3,041		-34,571	0.185	-6,380
30	-13,727	-20,049	-3,162		-36,938	0.174	-6,431
NPV							-234,995

rate for electricity, while all of the increase is at peak electricity rates. The off-peak rate is currently \$13.27/MWhr and the peak rate is \$19/MWhr. The energy inflation is expected to rise at 5% pa. The off-peak cost in the first year is calculated as follows:

$$Offpeak\ energy\ cost = (60MW)(10hr/day)(365days/year)(\$13.27/MWhr)$$
$$(1+h_e)/(91.5\%\ efficiency)$$
$$Offpeak\ energy\ cost = \$3,334,903/year$$

where h_e is the inflation rate for energy prices. The on-peak cost is calculated similarly.

In addition to energy costs, the pumps require regular maintenance. The cost of this maintenance amounts to $975,012 each year. The rate of increase of maintenance is expected to be 4% pa.

The capital cost of the new pumps is expected to be $32.6 million. The pumps will require major refurbishment every ten years. The cost of this refurbishment is expected to be 20% of the capital cost of the pump. The cost of this refurbishment is expected to increase at the same inflation rate as the maintenance costs.

The total costs are sum of the energy costs, the maintenance costs, and the new and refurbishment capital costs. The present value of these costs is determined using a discount rate of 6% pa.

There are no tax considerations since the California Department of Water Resources is not liable for tax.

Table 10.31 Project financials for the alternative to refurbish the existing the pumps (all amounts in thousands)

Year	Off-peak energy costs	On-peak energy costs	Regular maintenance	Refurbish	Total cost	Discount factor	PV(costs)
0		0		−22,600	−22,600	1.000	−22,600
1	−3,402	−162	−1,014		−4,578	0.943	−4,319
2	−3,572	−341	−1,055		−4,967	0.890	−4,421
3	−3,751	−537	−1,097		−5,384	0.840	−4,521
4	−3,938	−752	−1,141		−5,830	0.792	−4,618
5	−4,135	−987	−1,186	−15,123	−21,431	0.747	−16,014
6	−4,342	−1,243	−1,234		−6,819	0.705	−4,807
7	−4,559	−1,523	−1,283		−7,365	0.665	−4,898
8	−4,787	−1,828	−1,334		−7,949	0.627	−4,987
9	−5,026	−2,159	−1,388		−8,573	0.592	−5,074
10	−5,277	−2,519	−1,443	−18,399	−27,639	0.558	−15,433
11	−5,541	−2,909	−1,501		−9,951	0.527	−5,242
12	−5,818	−3,332	−1,561		−10,712	0.497	−5,323
13	−6,109	−3,790	−1,623		−11,523	0.469	−5,402
14	−6,415	−4,286	−1,688		−12,389	0.442	−5,480
15	−6,735	−4,822	−1,756	−22,386	−35,699	0.417	−14,896
16	−7,072	−5,400	−1,826		−14,299	0.394	−5,629
17	−7,426	−6,025	−1,899		−15,350	0.371	−5,700
18	−7,797	−6,698	−1,975		−16,471	0.350	−5,770
19	−8,187	−7,424	−2,054		−17,665	0.331	−5,839
20	−8,596	−8,205	−2,136	−27,236	−46,174	0.312	−14,397
21	−9,026	−9,046	−2,222		−20,294	0.294	−5,970
22	−9,477	−9,951	−2,311		−21,739	0.278	−6,033
23	−9,951	−10,924	−2,403		−23,278	0.262	−6,094
24	−10,449	−11,968	−2,499		−24,917	0.247	−6,154
25	−10,971	−13,091	−2,599	−33,136	−59,797	0.233	−13,933
26	−11,520	−14,295	−2,703		−28,518	0.220	−6,269
27	−12,096	−15,587	−2,811		−30,494	0.207	−6,323
28	−12,701	−16,972	−2,924		−32,597	0.196	−6,377
29	−13,336	−18,457	−3,041		−34,834	0.185	−6,429
30	−14,002	−20,049	−3,162		−37,213	0.174	−6,479
NPV							−235,432

The project financials for the alternative to refurbish the existing pumps are presented in Table 10.31. It is estimated that it will cost $22.6 million to refurbish each of the pumps, and that the pumps will require refurbishment every five years at a cost of 55% of the current refurbishment. The costs for the other items are calculated in a manner that is similar to those represented in Table 10.30.

The present values for these two alternatives suggest that the Allis-Chalmers pumps should be replaced, although there is not much difference between the two alternatives.

The two alternatives have been compared based on the net present value of their costs. They could have been compared based on the equivalent annual charge (*EAC*). The equivalent annual charge is also called annual worth, amongst others. The *EAC* for the refurbishment of the existing pumps is −$17,104,000 per pump and that for replacing the pumps is −$17,072,000. The difference between these two figures is slight, indicating that either of the alternatives could be justified.

10.6.3 Sensitivity Analysis

The effect of the discount rate, the maintenance inflation and the energy inflation on the preferred option was investigated. The results of this investigation are shown in Figures 10.13 to 10.15. These results are shown as the difference in their present values. Since the present values are those of costs, positive values indicate that refurbishing the pumps is the preferred option. For low discount rates, replacement favoured. If the discount rate chosen is greater than 6.1%, the refurbishment of the units is favoured.

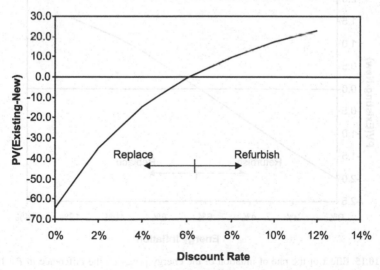

Figure 10.13 Effect of the discount rate on the difference in *PV* for the refurbishment of the existing pumps less that of replacing the pumps with new ones

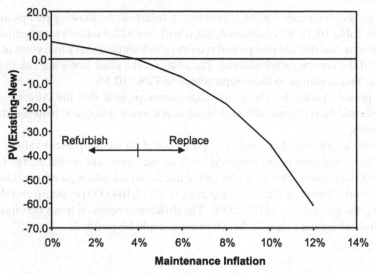

Figure 10.14 Effect of the rate of inflation of the maintenance items on the difference in *PV* for the refurbishment of the existing pumps less that of replacing the pumps with new ones

If the rate of inflation of the maintenance costs is less than 3.9%, the refurbishment of the units is recommended, and if the rate of inflation of the energy costs is higher then 6.3%, replacement of the units is recommended.

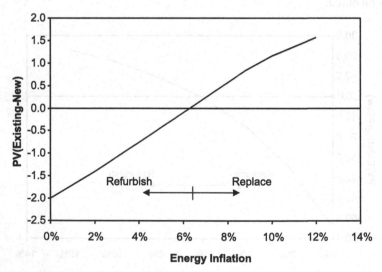

Figure 10.15 Effect of the rate of inflation of the energy prices on the difference in *PV* for the refurbishment of the existing pumps less that of replacing the pumps with new ones

10.6.4 Recommendation

The decision is sensitive to the values for the discount rate, the maintenance inflation and the energy inflation. For the current set of values, the replacement of the units is recommended. However, this is a marginal decision, since the difference between the alternatives is slight.

10.7 Combine Harvester

10.7.1 Introduction

An agricultural services company, Escourt Enterprises, operates a division that harvests the crops for maize and wheat farmers using combine harvesters. Custom rates for harvesting are about $15 per acre. The combining division needs to purchase more combine harvesters in order to meet the demand that it is experiencing. They have proposed purchasing an additional six harvesters at a cost of $217,000 each. The harvesters will operate in teams of three to quickly harvest the crop.

In order to do this, the divisional management needs to make a capital request to the corporate investment committee, who make their decisions based on net present value using the weighted average cost of capital as the discount rate. The value of WACC for the company is 9%.

10.7.2 Project Financials

The project financials are prepared on an entity basis from the five components of the cash flow: revenue, expenses, taxes, capital investment, and change in working capital. Each of these components is discussed below.

(i)   Revenue

The revenue is determined by the price per acre and the acres harvested. The current price charged by Escourt Enterprises for custom harvesting is about $15 per acre. The harvester can harvest a maximum of 50,000 acres in a year.

(ii)   Expenses

The cost of operating a combine harvester consists of staff and labour, maintenance and service, spares, staff training, fuel, insurance, and sales and marketing. The staff and labour, insurance and sales and marketing costs are regarded as fixed costs, since they do not vary with the amount of harvesting that is done with the machine. The fixed and variables costs are given in Tables 10.32 and 10.33, respectively.

Table 10.32 Fixed costs for custom harvesting business

Item	Cost, $
Labour	240,000
Insurance	4,340
Sales and marketing	20,000

Table 10.33 Variable costs for custom harvesting business

Item	Cost, $/acre
Maintenance and service	1.0
Spares	1.0
Operator training	1.0
Fuel	4.0

Table 10.34 Depreciation schedule for the combine harvester

Year	0	1	2	3	4	5	6	7	8	9	10
Investments	217										
Depreciation	0	22	22	22	22	22	22	22	22	22	22
Cumulative depreciation	0	22	43	65	87	109	130	152	174	195	217
Book value	217	195	174	152	130	109	87	65	43	22	0

Table 10.35 Project financials for the harvester project. All amounts in thousands.

Year	0	1	2	3	4	5	6	7	8	9	10
Revenue	0	630	662	695	729	766	804	844	886	931	977
Acres harvested	0	40,000	40,000	40,000	40,000	40,000	40,000	40,000	40,000	40,000	40,000
Price in $/acres	0	16	17	17	18	19	20	21	22	23	24
Costs	0	567	595	625	656	689	724	760	798	838	880
Labour	0	252	265	278	292	306	322	338	355	372	391
Maintenance and service	0	42	44	46	49	51	54	56	59	62	65
Spares	0	42	44	46	49	51	54	56	59	62	65
Operator training	0	42	44	46	49	51	54	56	59	62	65
Fuel	0	168	176	185	194	204	214	225	236	248	261
Insurance	0	0	0	0	0	0	0	0	0	0	0
Sales and marketing	0	21	22	23	24	26	27	28	30	31	33
Tax	0	17	18	19	20	22	23	25	27	29	30
Depreciation	0	22	22	22	22	22	22	22	22	22	22
Investment	−217	0	0	0	0	0	0	0	0	0	0
Harvester	−217	0	0	0	0	0	0	0	0	0	0
Δ(**Working capital**)	0	−78	−4	−4	−4	−4	−5	−5	−5	−5	109
Working Capital	0	78	82	86	90	94	99	104	109	115	120
Free cash flow	−217	−31	44	46	48	50	52	54	57	59	176

(iii)   Taxes

Taxes are calculated using Equations 10.1 and 10.2. The depreciation schedule, which is required to determine the depreciation allowance, is given in Table 10.34. The depreciation is calculated on a straight-line basis with a life of ten years.

(iv)   Capital investment

The capital investment consists of the purchase of the harvester for $217,000. The harvester has an estimated life of ten years.

(v)   Working capital

The customers for the harvesting business on average pay for the services within 45 days. The working capital requirements are calculated from Equations 10.4 and 10.5 with the debtors days of 45 days.

The project financials are given in Table 10.35 for a single harvester. The project financials are used to evaluate the profitability of the project.

10.7.3 Evaluation

The assessment measure calculated from the project financials are given in Table 10.36.

All of the assessment criteria are positive. The *NPV* is positive, the *IRR* and *MIRR* are greater than the WACC, and the profitability index is greater than one. This suggests that the project should be accepted.

The sensitivity of the *NPV* to changes in the parameters is examined next.

10.7.4 Sensitivity Analysis

The sensitivity of the net present value to a percentage change in the various input variables in the project financials is shown in Figure 10.16. The slope of the lines represents the sensitivity. The price and the acreage harvested are the most sensitive variables; these are closely followed by the labour and fuel costs. Changes in the other variables have a much slighter effect on the *NPV*.

Table 10.36 Assessment measure for the combine project

Measure	Value
NPV	85,200
IRR	14.5%
MIRR	17.3%
PI	1.393
Payback	7

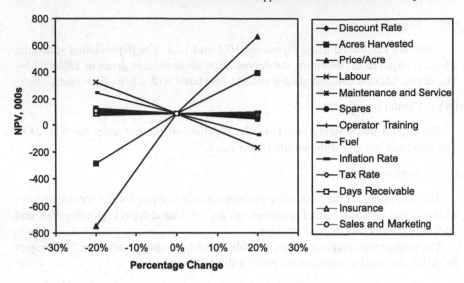

Figure 10.16 Sensitivity of the net present value to the various input parameters

The price charged greatly affects the value of the project. The customers are also sensitive to this price, because they can easily purchase a combine harvester and harvest their crops themselves. The pricing of the harvest is discussed later.

The *NPV* and other assessment measures are sensitive to the acreage harvested each year by the combine. The break-even acreage is shown in Figure 10.17. This means that in order for the project to be profitable, the harvester must process at least 37,763 acres each year. This sets the sales targets and scheduling requirements for asset utilization.

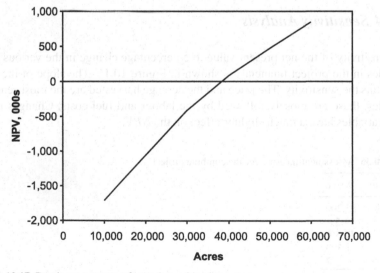

Figure 10.17 Break-even acreage for each combine harvester

Table 10.37 Estimated operating costs for owned harvester

Item	Cost
Labour	2,000
Maintenance and service	1,500
Spares	1,500
Fuel	6,000
Insurance	6,510
Total	17,510

10.7.5 Pricing

Typical farms in the corn region in which Escourt Enterprises operates have 1500 acres of crops in a growing season. The farmer's estimated operating costs are given in Table 10.37.

The annual cost of ownership arising from the capital investment in the harvester can be estimated from the equivalent annual charge (*EAC*). When used for a capital item, the equivalent annual charge is often called the capital recovery. The *EAC* is calculated as follows:

$$A = PV(A|P, k, n)$$
$$A = 217,000k(1+k)^n/[(1+k)^n - 1] = 33,812.96$$

where k is the farmer's discount rate which is equal to 9%.

Thus, the total cost of ownership and operation each year is equal to $49,153 = 17,510 + 33,812$. The cost of harvesting the 1,500 acres is $32.77 per year.

This analysis suggests that the price of harvesting offered by Escourt Enterprises is highly competitive compared with self-ownership and harvesting. In addition, a recent survey of custom harvesting in the region found that the average price charged was $22 per acre. This indicates that Escourt Enterprises may be under-charging for its services.

10.7.6 Recommendation

It is recommended that Escourt Enterprises purchase the fleet of six harvesters.

10.8 PetroGen Oil Field and Petroleum Refinery

A $2.3 billion oil-field project, called PetroGen, is planned in Angola. The project consists of wells in the Atlantic Ocean, a pipeline to the coast, and a coastal refinery. The project will require capital investment of $400 million in the first year, $800 million in the second year, and $600 million in the third year. The project plans to produce 137,000 barrels of oil per day. The total operating costs for the oil production and refining are estimated to be $10 per barrel.

The project is funded by a consortium of oil producers through a bond issue of $1,450 million with a coupon rate of 10% pa. The bonds are amortized (that is, repaid) over fourteen years. The equity contribution of the oil producers is 37% of the total requirement. The total requirement exceeds the fixed capital due to working capital requirements and the debt service reserve account. The debt service reserve is required by the bondholders to ensure that interest and principal payments can be met. In this case, it is equal to six months of the principal and interest that is required for the servicing the bonds.

10.8.1 Project Financials

The project financials, expenses, taxes, capital investment are prepared on the entity basis from its five components: revenue, expenses, tax, capital investment and change in working capital. The project financials for the expected life of 20 years is shown in Table 10.38.

The revenue is calculated based on a price of $21 per barrel, and the costs are assumed to $9 per barrel. The fixed capital is depreciated over 15 years on a straight-line basis. Tax is charged as 35% of the operating income less depreciation. The

Table 10.38 Project financials for the PetroGen opportunity

Year	Revenue	Operating costs	Depreciation	Tax	Investment	Working capital	Change in working capital	Free cash flow
0					−400			−400
1			27		−800			−800
2			80		−600	270	−270	−870
3	1,050	450	120	168		278	−8	424
4	1,082	464	120	174		286	−8	435
5	1,114	477	120	181		295	−9	447
6	1,147	492	120	187		304	−9	459
7	1,182	506	120	194		313	−9	472
8	1,217	522	120	201		322	−9	485
9	1,254	537	120	209		332	−10	498
10	1,291	553	120	216		342	−10	512
11	1,330	570	120	224		352	−10	526
12	1,370	587	120	232		363	−11	540
13	1,411	605	120	240		374	−11	555
14	1,453	623	120	249		385	−11	571
15	1,497	642	120	257		397	−12	586
16	1,542	661	93	276		408	−12	593
17	1,588	681	40	304		421	−12	592
18	1,636	701	0	327		433	−13	595
19	1,685	722	0	337		446	−13	613
20	1,735	744	0	347		460	460	1,104

working capital is assumed to be 15% of the fixed capital. The free cash flow is the given by the expression:

$$Free\ Cash\ Flow = Revenue - Expenses - Tax - Fixed\ Capital$$
$$- \Delta(Working\ Capital) \tag{10.6}$$

where $\Delta(Working\ Capital)$ is the change in working capital. In the last year of the project all the working capital is recovered. All prices and costs are assumed to rise with inflation at a rate of 3% pa.

10.8.2 Assessment

The *NPV* of this opportunity is $1.609 billion at a discount rate of 10%, and the *IRR* is 19.3%. The project is a good investment.

10.8.3 Financing of the Project

The project is financed through non-recourse finance, a topic that is discussed in further detail in Chapter 17. Non-recourse finance, also known as project finance, is a method of isolating the project risks so that the usage of debt is maximized. The debt is raised through the issue of 10% bonds with a maturity of 14 years. The principal repayment and the interest charged are given in Table 10.39.

The schedule of repayment of the principal is defined by the conditions for the issue of the bonds. Interest is charged at 10% on the outstanding amount. In addition, the lenders require that six months of debt service be deposited in a debt service account. The calculation of the amounts in the debt service account and the amount required to fulfil this obligation are also shown in Table 10.39.

The returns to the equity parties are given in Table 10.40. It is worthwile examining the obligations of and the returns to the equity parties on their own.

The operating income is the difference between the revenues and costs, and the change in working capital is as calculated before. The taxation paid is calculated from the following expression:

$$Tax = Tax\ Rate(Revenue - Costs - Depreciation - Interest) \tag{10.7}$$

This means that the returns to the equity parties are not calculated on the entity basis, which is the basis for the project financials. The equity contribution to the investment activities is the balance of the $2.3 billion required for the project that is not contributed by debt. The interest and principal repayments and the debt service account requirements are the same as those in Table 10.39. Thus the net cash flow to the equity parties is given by the expression:

Table 10.39 Debt service

Year	Outstanding loan	Interest	Principal repayment	Debt service	Debt service account	Funding to debt service account
0						
1	800	80	0	-720		
2	1,450	145	0	-505		
3	1,450	145	25	170	-85	-85.00
4	1,425	143	25	168	-84	1.25
5	1,400	140	50	190	-95	-11.25
6	1,350	135	50	185	-93	2.50
7	1,300	130	75	205	-103	-10.00
8	1,225	123	75	198	-99	3.75
9	1,150	115	100	215	-108	-8.75
10	1,050	105	100	205	-103	5.00
11	950	95	125	220	-110	-7.50
12	825	83	125	208	-104	6.25
13	700	70	150	220	-110	-6.25
14	550	55	150	205	-103	7.50
15	400	40	200	240	-120	-17.50
16	200	20	200	220	-110	10.00
17	0	0	0	0	0	110.00
18	0	0	0	0	0	0.00
19	0	0	0	0	0	0.00
20	0	0	0	0	0	0.00

Table 10.40 Returns to equity parties

Year	Operating income	Increase in working capital	Taxation paid	Equity investment	Interest paid	Principal repayments	Increase in debt service account	Net cash flow to equity
0	0	0	0	400	0	0	0	-400
1	0	0	0	230	80	0	0	-310
2	0	270	0	220	145	0	0	-635
3	600	8	117	0	145	25	85	220
4	618	8	124	0	143	25	-1	319
5	637	9	132	0	140	50	11	295
6	656	9	140	0	135	50	-3	324
7	675	9	149	0	130	75	10	302
8	696	9	159	0	123	75	-4	334
9	716	10	169	0	115	100	9	315
10	738	10	180	0	105	100	-5	348
11	760	10	191	0	95	125	8	332
12	783	11	203	0	83	125	-6	368
13	806	11	216	0	70	150	6	353
14	831	11	229	0	55	150	-8	392
15	855	12	243	0	40	200	18	343
16	881	12	269	0	20	200	-10	390
17	908	12	304	0	0	0	-110	702
18	935	13	327	0	0	0	0	595
19	963	13	337	0	0	0	0	613
20	992	-460	347	0	0	0	0	1,104

Net Cash Flow to Equity Parties

$$= Operating\ Income - Tax - \Delta(Working\ Capital) - Interest \qquad (10.8)$$
$$- Principal\ Payments - Debt\ Service\ Account - Capital\ Investment$$

The *NPV* of the equity cash flows should be discounted at the cost of equity. The weighted average cost of capital, WACC, is given by Equations 8.17 and 8.18. The rearrangement of these two equations results in the following expression:

$$WACC = \left(\frac{E}{E+D}\right) R_E + \left(\frac{D}{E+D}\right) R_{DBT}(1-T) \qquad (10.9)$$

WACC is the discount rate and is equal to 10%. The equity contribution is 37% and the debt contribution is 63%. The debt carries interest of 10%, so that R_{DBT}, the cost of debt before tax, is equal to 10%. The tax rate is 35%. The substitution of these values in Equation 10.9 yields the following expression for the cost of equity, R_E:

$$0.1 = (0.37)\,R_E + (0.63)\,0.1\,(1 - 0.35) \quad \Rightarrow \quad R_E = 15.95\%$$

The *NPV* of the cash flows to the equity parties is $696 million and the *IRR* is 27%. This suggests that this is a good project for the equity investors.

10.9 Looking Ahead

The next chapter introduces the concept of risk, which is the next theme of this book.

10.10 Review Questions and Exercises

10.10.1 Santa Clara HEPS

1. Describe how the revenues, costs, taxes, working capital and capital investment have been determined.
2. Determine the free cash flow for the first ten years of the Santa Clara Project if the inflation rate were 9% pa.
3. The depreciation schedule should start in the year that the expense was incurred. Construct separate depreciation schedules for each of the years in which an investment was made, and an overall depreciation schedule. Determine the tax calculation from the revised depreciation schedule.
4. Determine the sensitivity of the net present value to the capital investment, the tax rate and the royalty charge.
5. How would the returns to the investors in the project differ from the project's free cash flows?

10.10.2 Mobile Crusher

1. Construct a decision frame for the mobile crusher investment by doing the following:

    (i)   Create a table of the facts, uncertainties and decisions for the project.
    (ii) Create the decision hierarchy by separating the decisions into policy, strategic and tactical decisions.
    (iii) Create a strategy table from the decision hierarchy.
    (iv) Determine the main questions that need to be answered.

2. Construct an influence diagram for the mobile crusher investment.
3. Justify the treatment of savings as the same as revenue from the owner's point of view.
4. Why is the NPV of the outsourcing option higher than the NPV of the insourced option?
5. Determine the sensitivity of the outsourcing decision the estimation of the tyre savings, the discount rate, the inflation rate and the cost of the mobile crusher.

10.10.3 Mobile Crusher Contractor

1. If the contractor already had a mobile crusher available with a market value of $200,000 and a book value of $120,000, determine the contractor's NPV.
2. Determine the income statement and the cash flow statement for the contractor under the following situations:

    (i) The contractor only needed to borrow half of the money required to finance the crusher. The contractor funds the other half from equity.
    (ii) The loan for the crusher is based on an instalment plan.

3. Explain the differences between the free cash flow in Table 10.9 and the net cash flows in Table 10.11.
4. It was implicitly assumed that the mine paid the contractor immediately. Investigate the effect of payment delays by including working capital into the analysis of the contractor's cash flows.

10.10.4 Filtration Operation

1. Why does Alex Filters prefer a fixed monthly charge? Does it make any difference?
2. Is the inflation rate constant for all items?
3. Why is a change in working capital used in the project financials and not the working capital itself?

4. What is the contingency for in Table 10.20?
5. A number of different methods were discussed in Chapter 4 for the estimation of the capital costs. Discuss the method used to arrive at the values given in Table 10.20 in the light of the presentation in Chapter 4. How accurate do you estimate the value for the capital cost to be?
6. The working capital is about 12% of the fixed capital. Is this reasonable?
7. Is the discount rate reasonable in the light of the inflation rate?
8. Alex Filters stands to benefit from this venture in two ways. What are they? Discuss the trade-offs between them. Which benefit is preferred?
9. What is meant by terminal value in Table 10.29?

10.10.5 Water Pumps

1. Why are the two alternatives compared based on *NPV* rather than *EAC*?
2. If the decision criterion is the *EAC*, is the analyst making the "assumption of repeatability?"
3. Is the decision sensitive to the discount rate?
4. Create a spreadsheet of the water pumps case and determine the sensitivity of the decision to the following items:

   (i) Refurbishment costs
   (ii) Regular maintenance costs
   (iii) Off-peak energy costs
   (iv) On-peak energy costs

10.10.6 Combine Harvester

1. Why is the change in working capital equal to 109 in the tenth year?
2. Would it be economical for a farmer to own his own harvester? Under what conditions would it make economic sense to own the harvester?
3. Determine the cash flow from the farmer's perspective.
4. If the cash flow for the harvesting from the farmer's perspective were determined, would it be appropriate to use a WACC to determine the farmer's *NPV*. Justify your answer.
5. If the company cannot meet the target of 37,763 acres a year for each harvester, what should it do to maintain an economically profitable project?

10.10.7 Petroleum Field and Refinery

1. Create a spreadsheet application that replicates the project financials in Table 10.38. Determine the sensitivity of the *NPV* and *IRR* of the project to the following inputs:

    (i)   The oil price
    (ii)   The costs of production
    (iii)   The inflation price
    (iv)   The tax rate
    (v)   The discount rate
    (vi)   The initial capital investment

    Draw a tornado diagram of the sensitivities.
2. The capital cost estimate is within ±30%. How does this affect the sensitivity assessment?
3. Why would lenders require the debt service account?
4. Explain the change in working capital in year 20.
5. Explain the difference between the free cash flow and the net cash flow to equity partners.

Part III
Risk Assessment

The assessment of the free cash flows for a project focuses on the returns to the investor or owner of the project. These returns must be sufficiently attractive to the investors in the company for them to continue to fund the company's capital requirements. These are anticipated returns. The risk of a project is the chance that the anticipated cash flows do not materialize. In the next five chapters, various aspects of risk are discussed.

The relationship between financial risk and return is examined in Chapter 11. The assessment of risk depends on whose viewpoint is adopted. As a result, a distinction is made between stand-alone risk and portfolio risk. The choice of the discount rate in the assessment of the discounted cash flow analysis depends on the risk status of the company and the project relative to the market as a whole. The effect of risk on the value of the discount rate is examined in Chapter 12. The topic of Chapter 13 is the assessment of risk from the viewpoint of the project, and techniques to quantify risk, such as Monte Carlo simulation, are presented.

The decision to invest in a capital project is not static. Managers have flexibility and options. Two methods have been developed to account for this flexibility in decision-making. They are decision-tree analysis and real option analysis. These topics are discussed in Chapters 14 and 15, respectively.

Chapter 11
Risk and Return

11.1 Introduction

In previous chapters it was argued that to be successful the company must invest in projects that are economically attractive. The minimum condition is that if projects cover their costs they are attractive. The costs that are important in this context are the costs of funds to pursue these projects. The attractiveness of projects is evaluated in terms of the returns that the project is expected to bring to its owner. The company funds the investments in its projects by raising debt and equity. The company must invest in projects so that the company's returns are sufficiently attractive to induce investors to invest in the company's debt and shares.

The analysis assumed that the cash flows in the future are known, or that they represent the most probable outcome without considering the probability of other outcomes. The possibility of other outcomes raises the notion that the anticipated event will not occur or the anticipated cash flow will not materialize. It was assumed without any justification in examples in earlier chapters that each year or time period would be like the previous, that sales, costs, taxes and the like would be constant (except for inflation) year in and year out. However, these factors are unlikely to be constant. This is the risk factor for a project, that the sales volumes are not constant, that labour costs change, that equipment maintenance changes. Change represents risk.

Two of the cornerstones of finance are risk and return. The acknowledgement of risk implies that there is not only one anticipated outcome, but also multiple possible outcomes, and the decision must be made with a range of values in mind. In this chapter, financial risk is examined, and some of the ways to minimize financial risk are discussed. In the chapters following this one, these ideas of risk are incorporated into methods to enhance the discounted cash flow techniques for uncertainty.

11.2 Returns

The return on an investment is the ratio of the net gain to the cost of the investment. The return on investment was defined in the same manner as this in Chapter 6. To an investor investing in a company's stock or shares, the gain comes from an increase in the company's share price and from the payment of dividends. If the company's stock is traded on an exchange, the share price can be monitored continuously. For example, suppose that the prices are monitored daily, the price may move up or down on a daily basis, as shown in Figure 11.1. The return to investors over any one of the periods is the change in price in a period divided by the price at the beginning of the period.

Since the returns to the investor can be in the form of both capital gains and dividend payments, the return can be expressed by the following equation:

$$Return = \frac{P_t - P_{t-1} + D_t}{P_t} \tag{11.1}$$

where P_t is the price and t is the time. The term $P_t - P_{t-1}$ represents the capital gains over the period $[t-1, t]$ and D_t represents the dividend payments that may have been made during the period. If there is no dividend payment during the time period $[t-1, t]$, D_t is simply omitted in the numerator of Equation 11.1. The choice of the period is up to the investor – it can be minutes, hours, days or years. It is the horizon that the investor chooses.

The daily returns for the movement in prices shown in Figure 11.1 are shown in Table 11.1 for a twelve-day period. The price was 0.1 at the beginning of the twelve-day period, and 0.089 at the end, with an average price of 0.097 over the twelve days. The return after the first day was 0.093, because the price increased.

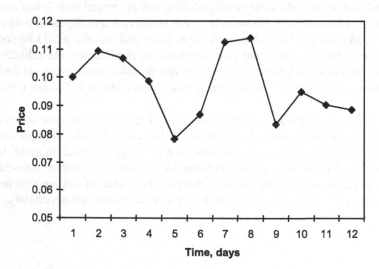

Figure 11.1 Daily share price for a traded stock

Table 11.1 Price and return

Day	Price	Return
1	0.100	
2	0.109	0.093
3	0.107	−0.024
4	0.099	−0.074
5	0.078	−0.207
6	0.087	0.110
7	0.113	0.295
8	0.114	0.014
9	0.083	−0.270
10	0.095	0.141
11	0.090	−0.049
12	0.089	−0.018
Average	0.097	0.001

On average, the daily return was 0.001 per day. If each outcome is equally likely, the expected return is the average of the returns. Thus, the expected return for the price data in Table 11.1 is 0.001 or 0.1% per day.

It is clear, even from this short illustration, that there are significant movements in the price. The change in the price is called the volatility of the price. Prices that move more are more volatile. Movements such as these are the source of risk for the investor. The investor must be convinced that the returns are sufficient to justify this risk of price changes. If the company is considering an investment in a project, the company's managers must be convinced that the project's returns justify its risks.

The movement of the prices on a daily basis can be examined further by determining the distribution of the returns. For example, consider a larger set of prices,

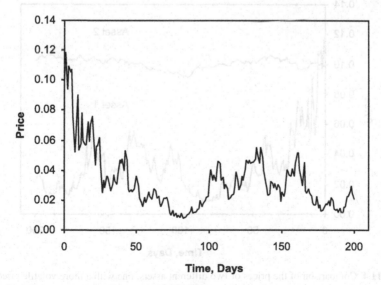

Figure 11.2 Price movement for a period of 200 days

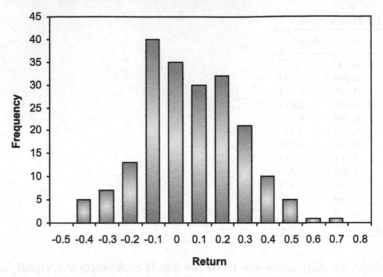

Figure 11.3 Distribution of the price returns

such as that shown in Figure 11.2. The distribution of returns can be determined by calculating the frequency with which a return falls into a range or category. In other words, a number of classes are set up, for example, a class with a range from −0.5 to −0.4, and the analyst counts the number of times the return for each time period (calculated using Equation 11.1, and shown as an illustration in Table 11.1) falls into this class. This is then repeated for the next class, for example, between −0.4

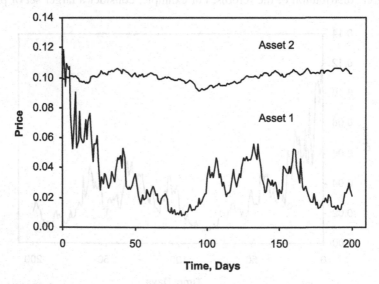

Figure 11.4 Comparison of the prices of two different assets, one with a more volatile price than the other

and -0.3. The results of an exercise such as this using the data shown in Figure 11.2 are shown in Figure 11.3.

The results of Figure 11.3 indicate that there is a wide distribution in the daily returns for the price data shown in Figure 11.2. Some days there is large gain while other days there is a large loss. This variation in possibilities is the volatility of the asset.

The price of the asset shown in Figure 11.2 is compared with that of another asset shown in Figure 11.4. They have different price movements; one is much more volatile than the other. If the distribution of price returns is plotted using the same values for the classes as shown in Figure 11.3, it is clear that the spread, or standard deviation is much less than that of the first asset. The distribution of the returns for the second asset is shown in Figure 11.5.

There would be no variation in the returns if the price were completely certain. If the price were highly volatile, the range of the distribution of returns would be large. The past behaviour of the prices for these two assets indicates that a prediction of the future price of Asset 2 in Figure 11.4 can be made with more certainty than that of Asset 1. Also, the increased width of the first asset's distribution indicates that there is an increased chance of the outcome being lower than the average value for the second asset.

If risk is understood to be the chance of a poor outcome, then the range of the possible returns is a measure of that risk. This range of possible outcomes can be measured by the techniques of probability theory: it is the standard deviation or the variance. The standard deviation is a measure of the risk of an asset. To recapitulate some statistics, the mean, the variance and the standard deviation for a set of N values of the return, R_j, are given by the following expressions:

$$Mean = \bar{R} = \frac{1}{N} \sum_{j=1}^{N} R_j \tag{11.2}$$

$$Variance = \frac{1}{N} \sum_{j=1}^{N} (R_j - \bar{R})^2 \tag{11.3}$$

$$Standard\ Deviation = \sqrt{Variance} \tag{11.4}$$

The mean is also called the expected value.

The returns can be quantified by the mean or expected value and the risk by the standard deviation of the returns. This means that an asset can be compared with other assets in the risk-return space. This is shown in Figure 11.6. Asset 1 has a higher average daily return than Asset 2. However, the risk of Asset 1 is significantly higher than the risk of Asset 2. Which of these two assets is more preferable to an investor depends on the risk appetite of the investor. However, it is fair to say that if they had the same risk, the one with higher return would always be preferred, while if they had the same return, the one with the lower risk would be preferred.

This method of quantifying risk and return is enormously powerful. At a single glance, the impact of the two important variables affecting the various assets can

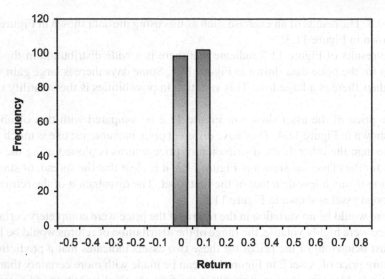

Figure 11.5 Distribution of the price returns for the second asset

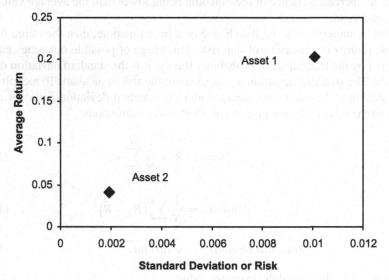

Figure 11.6 Risk versus return

be seen. For example, the risk and return of five different stocks are compared in Figure 11.7. (The values are fictitious.) Clearly, Google is preferred over Vodafone, and Daimler over Google. However, the choice between Rio Tinto and BP is not as clear. By calculating the risk and return position of the individual assets, the analyst can make recommendations to investors with different appetites for risk.

With this fresh understanding of risk, the sources of risks for a company, and particularly the investment risk, is discussed next. After that, some unexpected results from combining assets will be discussed.

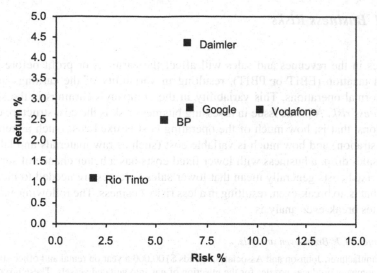

Figure 11.7 The risk and return of five different assets (data fictional)

11.3 Certainty and Uncertainty

Under conditions of certainty, the outcome of an investment decision is known. The choice is whether to invest or not. The investor is required to make the decision based on the net present value of free cash flows. If this were the case, it should be easy to make decisions that work out. However, decision-making is much harder than this. The reason is that uncertainty exists. The free cash flows have been estimated to the best of everyone's ability. However, these are the expected values of the cash flows and in reality, the cash flows might fall in a range about the expected value. The range of the possible outcomes represents the uncertainty in the values.

Uncertainties arise from different quarters. The company has to deal with uncertainty in the demand, the price and the production cost for its product, uncertainty in the investment requirements and uncertainties in the interest rate and other financing activities, amongst others, all of which impact on the range of possible outcomes. As a result, the decision maker needs to take these and other risks into account when making the decision.

There is a formal difference between risk and uncertainty: risk is envisaged as the situation in which the probability of the outcomes can be assessed, whereas uncertainty implies that alternative occurrences cannot be identified or a probability of their occurrence cannot be assessed. In the context of this discussion, the distinction is not of any consequence, and so the terms are used interchangeably.

In the preceding section, risk was related to the change or movement in the price of an asset. The changing conditions within which a business operates are sources of a number of different risks for a company and its management. The principal risks that the company faces are the business risks, the financing risks and the investment risks. These three different sources of risk are discussed next.

11.3.1 Business Risks

Changes in the revenues and sales will affect the earnings or profit before interest and taxation (EBIT or PBIT), resulting in variability of the earnings arising from normal operations. This variability in the company's earnings is the source of *business risk*. A prime issue in assessing business risk is the cost structure of the operations, that is, how much of the operating cost is fixed cost (such as rent and administration) and how much is variable cost (such as raw materials and labour). When sales drop, a business with lower fixed costs has a better chance of survival. Lower fixed costs generally mean that lower sales volumes are needed to meet the costs, that is, to break-even, resulting in a less risky business. The following example illustrates break-even analysis.

Example 11.1: Break-even analysis.

A manufacturer, Johnson and Associates, spends $100,000 a year on rental and office staff. The company produces nozzles for the injection of gas into agitated vessels. These nozzles typically sell for $10 per unit and the cost of manufacture is estimated to be $5 per unit. How many units must the company sell in order to meet its costs?

Solution:

The costs for rentals and offices are fixed costs since they do not depend on the output of the operations. The costs per unit of product are variable costs, since these costs change with the number of units of production. The costs are calculated using these values, and the results given in Table 11.2. These results are shown in Figure 11.8.

The company needs to produce at least 20,000 units in order to meet its costs.

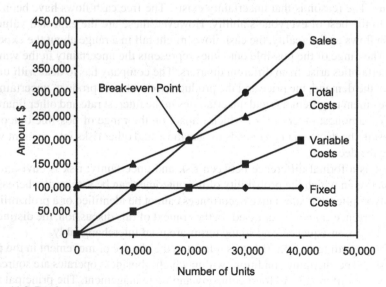

Figure 11.8 Break-even analysis

Table 11.2 Costs and sales as a function of number of units

Units	Fixed	Variable	Total costs	Sales
0	100,000	0	100,000	0
10,000	100,000	50,000	150,000	100,000
20,000	100,000	100,000	200,000	200,000
30,000	100,000	150,000	250,000	300,000
40,000	100,000	200,000	300,000	400,000

The higher the fixed costs in a business, the higher the break-even point. This means that if the level of sales or the costs of goods sold is variable, the chances of the business's profitability being at stake are higher if the proportion of fixed costs to variable costs is higher. If the fixed costs are a smaller proportion of the costs, then the business should be able to more easily meet the costs even if the sales decline. Businesses that experience highly variable sales need to control their fixed costs in order to maintain profitability.

11.3.2 Financing Risks

Businesses are partly funded with debt. The lenders expect the interest payments on the debt irrespective of the conditions under which the business is operating. The interest payments therefore behave in a manner similar to the fixed costs of production. They are fixed financing charges. The higher the debt, the higher the interest charges, and the more susceptible the profit is to changes in business conditions. In addition, changes in the interest rate will affect the quanta of the interest payments, resulting variability in earnings or profit. This is the source of *financing risk*. The company must meet its interest payments in order to avoid business failure and bankruptcy. Higher levels of debt generally make the company more risky.

11.3.3 Investment Risk

Investment risk, which could be seen as part of business risk, is the result of the variability in the returns from an investment that the company has made. For example, the company invests in a project to make a new product, and the returns from this investment are more variable than expected. This type of risk forms the basic context for the discussion of risk in engineering decisions. The source of the investment risk may arise, for example, from market conditions, or from technical implementation reasons or from the actions of competitors. Companies make investments in uncertain environments and the discounted cash flow techniques developed in earlier chapters must be adapted to conditions of uncertainty.

Companies generally own more than one opportunity. The investment in different opportunities is bound to affect the company as a whole unless that company is already very large. In the next section, the effect of owning different opportunities, assets or investments on the portfolio of projects as a whole is examined.

11.4 Portfolio Risk

A *portfolio* is the ownership of a group of assets by an owner. An investor can either personally buy different assets, such as shares in different companies, or purchase an asset that is already a collection of other assets, such as a mutual fund or unit trust. A large corporation is itself a collection of divisions and operations that may be regarded as different assets. The investment in different projects by a company may be seen in a similar light. In Section 11.2, the return and risk for a single project was discussed in terms of the statistics of price movements. When combining assets, each with their own returns and risks, what is the effect on the whole?

Returns are combined in a straightforward manner. The return of the portfolio is the average return weighted by the composition of the portfolio. For example, consider two assets with returns of 2% and 5% that are owned in proportions of 40% and 60%, respectively. The portfolio return is equal to $2(0.4) + 5(0.6) = 3.8\%$. The portfolio return is expressed mathematically as follows:

$$\bar{R}_P = \sum_{i=1}^{N} X_i R_i \qquad (11.5)$$

where R_P is the return on the portfolio as a whole, R_i is the return for each individual asset, and X_i is the proportion of asset i in the portfolio. These proportions sum to the whole, that is, $\sum X_i = 1$.

The combination of risks is not as straightforward as that for the returns. The risk is measured by standard deviation. The combination of standard deviations depends on the standard deviations of the sources and on the covariance between them. The standard deviation of the portfolio of N assets, σ_P, is given by:

$$\sigma_P = \sqrt{\sum_{i}^{N} X_i^2 \sigma_i^2 + \sum_{i=1}^{N} \sum_{\substack{k=1 \\ k \neq i}}^{N} X_i X_k \sigma_{ik}} \qquad (11.6)$$

This means that the risk is dependent on the risks of the individual assets, σ_i and σ_k, and the covariance between each of the assets, σ_{ik}. The covariance is a measure of the correlation of the risks of the individual assets.

The implications of this are demonstrated by an example. Consider the five assets introduced earlier and shown in Figure 11.9. The monthly returns for two of these shares, Google and Vodafone, are shown in Table 11.3. The average return and the standard deviation of the returns is also calculated, and shown in Figure 11.9. Now, construct a portfolio of equal proportions of Google and Vodafone. The return for this portfolio after the first month is $0.5(10.10) + 0.5(17.00) = 13.55$. The other months are determined in a similar manner. The average return for the portfolio is the average of these monthly returns, and the risk is the standard deviation of these monthly returns. A calculation of these two values is shown in Table 11.3.

The return on this portfolio, consisting of 50% Google and 50% Vodafone, is the average of the returns of the two individual components. The standard deviation of the portfolio is calculated from the returns of the portfolio using Equations 11.3 and 11.4. The average return for this portfolio is 2.71%, while the average standard deviation is 4.82. The value of the return is similar to that of the individual assets, whereas the standard deviation is lower than either Google or Vodafone.

A remarkable event has happened when the portfolio is formed: the risk of the portfolio is significantly less than either of the two individual assets! This reduction in the risk is unexpected. This reduction of risk due to the holding of a portfolio is known as the "portfolio effect," the mathematical justification for not "keeping all your eggs in the same basket."

The effect of the construction of two other portfolios, also composed of equal proportions of the two stocks, on the combined risk and return of the portfolio is shown in Figure 11.9. In all three cases shown in Figure 11.9, the risk is less than the average of the risks of the individual assets in the portfolio. This is due to the "portfolio effect."

The reason for this reduction in risk of the combination becomes clear when the returns are compared with one another at the times at which they occur. This is shown in Figure 11.10 for the Google and Vodafone shares. The returns on these two assets are negatively correlated: when the one is increasing the other is decreasing. The combination of the two assets is less volatile, or more "smooth," than either

Table 11.3 Portfolio risk and return

| Month | Returns, % | | |
	Google	Vodafone	Google + Vodafone
1	10.10	17.00	13.55
2	12.00	2.00	7.00
3	−3.00	3.00	0.00
4	4.00	−9.00	−2.50
5	4.00	1.00	2.50
6	−2.80	7.00	2.10
7	−10.00	21.00	5.50
8	−2.00	3.00	0.50
9	3.00	−14.00	−5.50
10	11.00	−8.00	1.50
11	7.50	−0.70	3.40
12	−1.00	10.00	4.50
Average return	2.73	2.69	2.71
Standard deviation	6.72	10.25	4.82

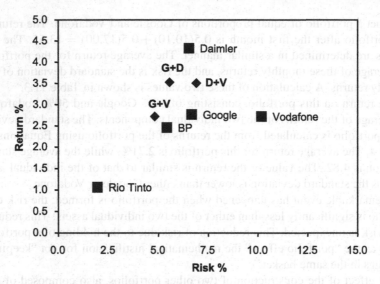

Figure 11.9 Combinations of assets can reduce risk

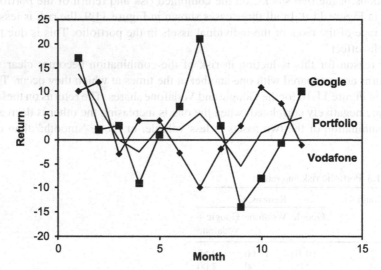

Figure 11.10 The times-series of the returns

of the two shares on its own. This smoothing effect is described by the covariance term σ_{ik} in Equation 11.6 on the overall risk. In other words, it is this term that is responsible for the reduction in the risk of the portfolio.

The covariance term, σ_{ik}, which is responsible for the portfolio effect, is related to the correlation coefficient, ρ_{ik}. The covariance is given by the following expression:

$$\sigma_{ik} = \rho_{ik}\sigma_i\sigma_k \tag{11.7}$$

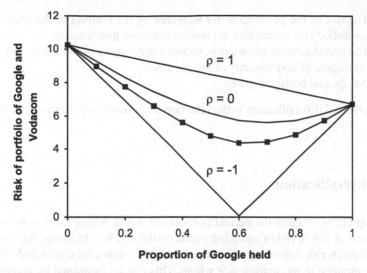

Figure 11.11 Optimal portfolio of two assets, Google and Vodafone

If changes in two variables occur at the same time and same direction, the changes are correlated. If they are perfectly correlated, then correlation coefficient has a value of one. On the other hand, if there is no correspondence in the movement of one variable and that of the other, then the variables are uncorrelated, and the value of the correlation coefficient is zero. If the changes in the variables occur at the same time, but in the opposite direction, then the variables are negatively correlated. If they are perfectly negatively correlated, then the value of the correlation coefficient is minus one.

If the risk can be reduced, can all the risk be eliminated? Is there an optimal portfolio of the two assets that minimizes the risk? The risk of the portfolio as a function of the proportion of the shares held in the portfolio is shown in Figure 11.11 for different values of the correlation coefficient, ρ. When the correlation coefficient is one, the movements are correlated, so that combining the assets will average the risk. The effect of perfect correlation between the assets is illustrated in the curve labelled $\rho = 1$ in Figure 11.11. That this is a straight line also confirms that the source of the portfolio effect is the correlation coefficient, or the second term within the square root sign in Equation 11.6. If the movements in the assets are perfectly negatively correlated, $\rho = -1$, then it is possible to find a combination of the two assets that completely reduces risk.

It is useful at this point to recapitulate a few points about asset risk that have been shown here. These are as follows:

(i)   Risk is related to the movement in price or value.
(ii)   The greater the movements, the higher the risk.
(iii)   The risk can be measured by the standard deviation of price or value returns.
(iv)   The return of the portfolio is weighted by the proportional holding of each of the assets in the portfolio. This means that the returns combine linearly.

(v)   The risk of the portfolio is *not* weighted by the holding of the assets in the portfolio. This means that the risks combine in non-linearly.
(vi)   The correlation, or covariance, between the returns of the assets determines the degree of non-linearity of the portfolio.
(vii)   An optimal portfolio exists.

The effect of diversification in the selection of a portfolio is discussed in the next section.

11.5 Diversification

It is tempting to imagine the perfect portfolio of assets, where there is no risk and only return. A life of riches and leisure time on the beach at Montego Bay beckons. Alas, although risk can be significantly reduced, it cannot be eliminated. There is some covariance in the market as a whole. This can be examined by constructing a portfolio in which there are N shares in the portfolio each held in the same proportion (in other words, the portfolio holds $1/N$ of each stock). Substitution of this proportion of holding into Equation 11.6 yields that the portfolio variance, σ_P^2, is given by the following expression:

$$\sigma_P^2 = \frac{1}{N} \sum_{i=1}^{N} \frac{\sigma_i^2}{N} + \frac{N-1}{N} \sum_{i=1}^{N} \sum_{\substack{k=1 \\ k \neq j}}^{N} \frac{\sigma_{ik}}{N(N-1)} \tag{11.8}$$

The first term on the right contains the average variance of the individual shares in the portfolio, while the second term on the right contains the average covariance. Thus, Equation 11.8 can be rewritten in the following form:

$$\sigma_P^2 = \frac{1}{N} (Average\ variance) + \frac{N-1}{N} (Average\ covariance) \tag{11.9}$$

As the number of shares added to the portfolio increases, that is, as N increases, the contribution of the first term diminishes rapidly. However, as N increases, the variance of the portfolio approaches the average covariance. This is illustrated in Figure 11.12.

The portfolio risk approaches the average covariance in the market, which cannot be further reduced by diversification. This is known as the market risk, the systematic risk, or the undiversifiable risk. An attempt to add assets to the portfolio from a different asset class, such as bonds or shares from an overseas market, will change this limit, because the market risk is the average covariance of all the asset classes that are included. However, such diversification cannot reduce all the risk.

Studies of different stock markets have indicated that the amount of risk that can be reduced by diversification is about 60–80% of the risk on an individual security, depending on the stock exchange. This means that the selection of a portfolio rep-

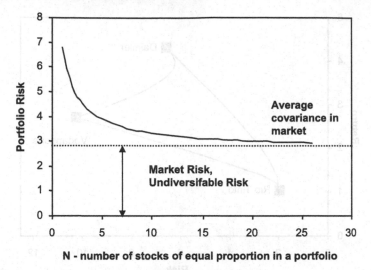

Figure 11.12 The reduction of the portfolio risk by diversification

resents a significant mechanism for the reduction of risk. Interestingly, the results shown in Figure 11.12 indicate that most of the risk is removed with as few as 10 assets in the portfolio. The construction of the most efficient portfolio is the subject of the next section.

11.6 The Attainable Region and the Efficient Frontier

Not all of the risk-return space can be reached, even with all the assets in the market and all of their possible combinations. For example, if three of the shares given in Table 11.2 are combined in pair-wise combinations, as shown in Figure 11.13, the space inside the boundary can be reached by creating new combinations from the pair-wise combinations represented by the lines. In addition, other new combinations may extend the external boundary by a small amount. This internal space that is reachable by combinations of the assets is called the feasible set or attainable region.

If the same method of combining assets is done for the entire market, the attainable region for the market can be found. This is illustrated in Figure 11.14. The optimum portfolio is that which provides the highest return for a given amount of risk, or the lowest risk for a given return. This portfolio must lie on the upper boundary of the attainable region, and is called the efficient frontier.

The assets that have been considered up until now are risky assets, such as shares on the stock exchange. Short-term treasury bills, issued for three months for example, can be considered essentially riskless. The investing public makes these loans to the government to fund the activities of government. These loans meet

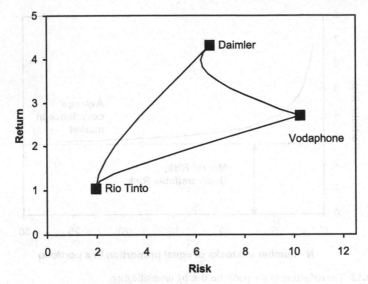

Figure 11.13 Pair-wise combinations of three assets

the shortfall between government expenditure and income, and serve to "balance the budget." The government is unlikely to default on these loans because default will result in a major disruption to the economy. Because there is no default risk for these short-term bills, they are regarded as a riskless asset. Now, if a riskless asset, with a return given by R_F on Figure 11.15, were combined with a risky asset, with

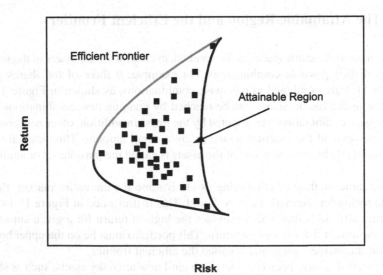

Figure 11.14 The attainable region and the efficient frontier for the market and portfolios of market assets

a risk and return given by point M on Figure 11.15, what would the shape of the portfolio look like in risk-return space?

If the proportion of the risky asset that is in the portfolio is X, the return on this portfolio is given by the following expression:

$$R_P = X R_M + (1 - X) R_F \qquad (11.10)$$

The risk of the portfolio, σ_P, is described by Equation 11.6. The substitution of Equation 11.6 into Equation 11.10 yields the following expression:

$$\sigma_P = \left[X^2 \sigma_M^2 + (1 - X)^2 \sigma_F^2 + 2X(1 - X)\sigma_M \sigma_F \rho_{MF} \right]^{1/2} \qquad (11.11)$$

But since the risk on the riskless asset is zero, that is, $\sigma_F = 0$, Equation 11.11 can be rewritten in the following form:

$$\sigma_P = X \sigma_M \qquad (11.12)$$

An expression that relates the return of the portfolio to the risk of the portfolio can be derived by eliminating X from Equations 11.12 and 11.10. This gives the following expression:

$$R_P = R_F + \left(\frac{R_M - R_F}{\sigma_M} \right) \sigma_P \qquad (11.13)$$

This is a straight line in risk-return space and is referred to as the *capital market line*. In other words, a plot of the portfolio return, R_P, against the portfolio risk, σ_P, is a straight line with y-intercept R_F and slope $(R_M - R_F)/\sigma_M$. This can be interpreted in the following manner:

$$Expected\ return\ =\ Price\ of\ time\ +(\ Price\ of\ risk)(\ Amount\ of\ risk) \qquad (11.14)$$

This interpretation indicates that the expected return is at least the risk-free rate, which accounts for the loss of utility of the amount invested, and an adjustment that depends on the amount of risk.

An investor who can acquire a riskless asset and a risky asset can choose the amount of risk and determine the composition of the portfolio that gives this required amount of risk. Thus, the risk appetite of any investor can be accommodated by a portfolio of a risky and a riskless asset.

If the investor with this portfolio wished to maximize the return on this portfolio, what risky asset should be chosen? After a short period of consideration, it is obvious that the point M should lie on the efficient frontier, and that it should be the point of the tangent to the straight line representing the portfolio. The straight line of the portfolio in risk-return space and the position of the optimal risky asset are shown in Figure 11.15.

All investors in the market who hold risky assets and the riskless asset would wish to hold the risky asset M. This is because it maximizes the portfolio return for any chosen level of risk. This means that all investors will wish to own M and the risk-free asset *irrespective of their risk preferences*. If an investor was more risk averse and wanted to own less risk, the investor would compensate by holding more

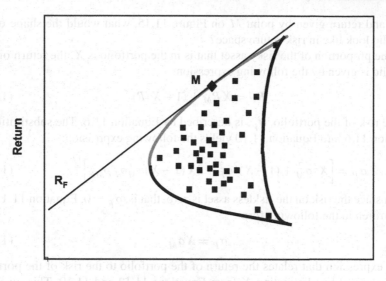

Figure 11.15 A portfolio of a riskless asset and a risky asset

of the riskless asset. The investor would still benefit from the highest return possible for the chosen amount of risk as a result of being on the line connecting R_F and M.

If all investors in the market have the same expectations, and all have the same lending and borrowing rate, R_F, they are all confronted with the same efficient frontier and the same optimal portfolio of risky and riskless assets. Therefore the risky asset, M, held by all investors will be the same. If all investors in the market hold the same risky asset, it must be the market as a whole, that is, a portfolio of all the shares in the market in proportion to the share's contribution to the value of the total market value. This is an important result, because it clearly identifies that the risky asset, M, that all investors wish to own in their portfolio is the market as a whole. Point M is therefore the market portfolio. The market as a whole, and hence, point M, can be determined by a stock market index, such as the S&P 500. Thus an investor who wished to own equities at point M need only purchase an index in the market of her choice.

The straight line representing the portfolio of the riskless asset and the market portfolio M is called the capital market line. In *equilibrium*, all investors will hold combinations of the two portfolios, the market portfolio M, and the riskless asset. This is referred to as the "two mutual fund theorem."

The risk at M is the market risk, or the systematic risk, that was discussed in Section 11.5 and illustrated in Figure 11.11. The risk of the portfolio of all risky assets represented by the point M cannot be reduced any further by diversification.

This analysis has led to the identification of the market portfolio, M, as the optimal portfolio under conditions of equilibrium. In the next section, the focus of the discussion turns from the risk and return on a portfolio of assets to the risk and return of an individual asset.

11.7 Capital Asset Pricing Model

The capital asset pricing model (CAPM) proposes that *equilibrium return* of an individual asset, R_i, is given by the following expression:

$$R_i = a + \beta_i b \tag{11.15}$$

where β_i is the risk of asset i compared with the risk of the market, M.

The form of Equation 11.15 follows that of Equations 11.13 and 11.14. The CAPM proposes that this equation must be applicable to all the assets, both risky and riskless.

The "two mutual fund theorem" can assist in identifying expressions for the values of the variables "a" and "b" in Equation 11.15: (i) the return on the riskless asset is R_F, and it has zero risk, so that the value of β is zero at R_F; and (ii) the return for the market portfolio is R_M, and the β for the market is one (since β is the ratio of the risk of the asset to the risk of the market portfolio). From these two points it is easy to show that the following expression holds:

$$R_i = R_F + \beta_i(R_M - R_F) \tag{11.16}$$

This is the *security market line*, and the body of theory supporting this is called the capital asset pricing model, or CAPM. It represents an equilibrium position for the firm. The quantity $(R_M - R_F)$ is known as the market-risk premium, and the parameter β_i is known as "beta." The market-risk premium is the additional return that an investor expects to make above the risk-free rate as a reward for taking on the risk of investing in the market.

Equation 11.16 states that the return on any asset can be obtained from one factor, the value of β_i for the asset, which is the ratio of the asset's risk to the market's risk. The beta (β) of a stock is an index of its risk to the market risk. Stocks with a beta less than one are less risky than the market, while stocks that have a higher risk than the market have a beta greater than one.

If the not unreasonable assumption was made that the distribution of returns will be same in the future as they are now, then it is possible to draw some conclusions about future risk of investing in the risky asset. The value of the risk of the asset in the past can be determined and indexed to the market as β, and it can used to predict the equilibrium return for the asset in the future within a well-diversified portfolio.

The CAPM is important in the context of discounted cash flow analysis because it is used in Chapter 12 as one of the methods to calculate the contribution of the cost of equity to the weighted average cost of capital from which the discount rate is derived.

Historical values of beta for a stock can be determined from a rearrangement of Equation 11.16 into a form that is known as the market model:

$$R_i = \alpha + \beta_i R_M \tag{11.17}$$

The value of the coefficients α and β_i in the Equation 11.17 can be obtained from a linear regression of the data for R_i and R_M. This is shown schematically in Figure 11.16.

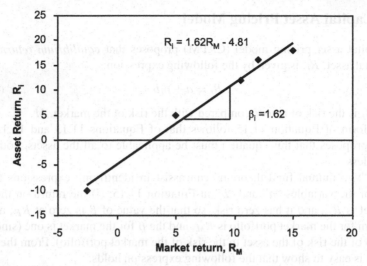

Figure 11.16 Regression analysis using the market model to determine β_i for a stock

The linear regression of the data not only yields the value of beta, but also provides a definition for beta. From the theory of linear regression, the slope of a linear regression of R_i against R_M is given by the following expression:

$$slope = \frac{\sigma_{iM}}{\sigma_M^2} = \beta_i \qquad (11.18)$$

where σ_{iM} is the covariance of the return on the asset i with the return on the market M.

It is interesting to compare the betas for a various companies. A selection of the betas is given in Table 11.4 (these data are not current and are provided for illustration purposes only).

Beta accounts for the risk of investing in the asset i compared with the risk in the market as a whole. An examination of Equation 11.18 yields that the standard deviation of the asset (σ_i) is not the measure of the risk in the context of a portfolio of assets. Rather, it is the covariance with the market, given by σ_{iM}, that is important. The covariance was the source of undiversifiable or systematic or market risk, as was discussed when Equations 11.8 and 11.9 were presented. The risk that beta measures is the risk to investors who are well-diversified; they are interested in market exposure because they have removed the diversifiable risk or unsystematic risk by diversifying their own portfolios. In determining the return for an asset, it is only the systematic risk of the market that is important. Deviations in the return obtained from Equation 11.16 at any particular time are a result of risk that is diversifiable.

Table 11.4 Selected betas for listed corporations in US

Company	Beta
Allergan	0.94
Black and Decker	1.06
Chevron	0.70
Colgate-Palmolive	1.11
Compaq	1.26
Eastman Kodak	0.54
Gillette	0.93
Hewlett-Packard	1.34
McDonald's	0.93
Merck	0.73
Monsanto	0.89
Quaker Oats	1.38
Schering-Plough	0.51
Union Carbide	1.51
Walt Disney	1.42
Whirlpool	0.90

11.8 Portfolio Selection

A single criterion was used in ranking projects in Chapter 7. The internal rate of return, *IRR*, is commonly used, but other relative measures, such as the profitability index, can also be used. None of these criteria accounted for risk. It is necessary to use a measure that included both risk and return in selecting a portfolio from a set of risky assets.

The returns for the shares in the market can be plotted against their value of beta on a particular date and the security market line determined from them. This is shown in Figure 11.17. The shares that are above the market security line are performing better than their level of risk suggests, while those that are below the security market line are performing worse. This suggests that the portfolio should be composed of those stocks that are above the security market line should be incorporated into the portfolio and those that are below should not be. If transaction costs were zero, the portfolio could be regularly reconstructed to optimise the portfolio returns by choosing those shares that are above the security market line.

An examination of Equation 11.16 suggests that $(R_i - R_F)/\beta_i$ should be constant for all assets if they were at equilibrium. This quantity is known as the "excess return to beta." The excess return to beta represents the desirability of including the stock in the portfolio. As such, it represents a measure by which stocks can be ranked that includes both risk and return. If the stock were at the market equilibrium, the excess return would be equal to $R_M - R_F$. If $(R_i - R_F)/\beta_i$ was higher than $R_M - R_F$, then stock i should be included in the portfolio, and if it were less than $R_M - R_F$, then stock i should be excluded from the portfolio.

What is required in order to construct a portfolio is to determine the proportional holding of each stock and a cut-off point below which the proportional holding is

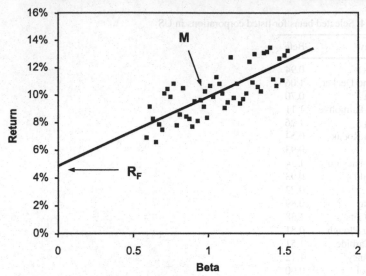

Figure 11.17 The return plotted against the beta for the assets in the market. The straight line represents the equilibrium, which is described by the security-market line

zero. The cut-off point is such that all stocks included in the portfolio have a higher excess return than the cut-off, and all excluded stocks have a lower excess return than the cut-off.

A method for constructing this portfolio is given without any derivation. The method requires that the stocks be ranked from highest to lowest excess return. A candidate for the cut-off point is given by the following expression:

$$C_i = \frac{\sigma_m^2 \sum_{j=1}^{i} \frac{(R_i - R_F)\beta_i}{\sigma_{ej}^2}}{1 + \sum_{j=1}^{i} \left(\frac{\beta_j^2}{\sigma_{ej}^2}\right)} \tag{11.19}$$

where σ_m^2 is the variance in the market, and σ_{ej}^2 is variance of the stock movement not associated with the movement of the market as a whole, that is, the stock's unsystematic or diversifiable risk.

Given R_i, β_i, R_F, σ_m^2 and σ_{ej}^2, it is relatively easy to calculate the C_i in a table of stocks that is ranked in terms of the excess return. The calculation of the value of C_i is shown in Table 11.5. For this example, the values of R_F and σ_m^2 are 5% and 10, respectively.

The calculation of the data in Table 11.5 is as follows:

(i)   Columns 2, 3 and 4 of the table are data.
(ii)   Column 5 is the excess return. It is necessary to have the stocks listed in descending order of excess return for this method to work.

Table 11.5 Calculation of the cut-off rate for inclusion into the portfolio

Stock i	R_i	β_i	σ_{ej}^2	$\dfrac{(R_i - R_F)}{\beta_i}$	$\dfrac{(R_i - R_F)\beta_i}{\sigma_{ej}^2}$	$\dfrac{\beta_j^2}{\sigma_{ej}^2}$	Sum 1	Sum 2	C_i
1	12%	1.00	50	7.00	0.14	0.02	0.14	0.02	1.17
2	15%	1.50	40	6.67	0.38	0.06	0.52	0.08	2.92
3	11%	1.05	20	5.71	0.32	0.06	0.83	0.13	3.59
4	16%	2.00	10	5.50	2.20	0.40	3.03	0.53	4.80
5	10%	0.95	40	5.26	0.12	0.02	3.15	0.55	4.82
6	11%	1.50	30	4.00	0.30	0.08	3.45	0.63	4.73
7	11%	2.00	40	3.00	0.30	0.10	3.75	0.73	4.52
8	8%	0.80	16	3.75	0.15	0.04	3.90	0.77	4.49
9	7%	0.90	20	2.22	0.09	0.04	3.99	0.81	4.39
10	6%	0.60	6	1.00	0.06	0.06	4.05	0.87	4.18

(iii)   Columns 6 and 7 are manipulations of the data for each stock.

(iv)   Column 8 represents the term $\sum_{j=1}^{i} \frac{(R_i - R_F)\beta_i}{\sigma_{ej}^2}$. This summation is cumulative

sum of the data in column 6.

(v)   Column 9 represents the term $\sum_{j=1}^{i} \left(\frac{\beta_j^2}{\sigma_{ej}^2} \right)$. This summation is cumulative sum

of the data in column 6.

(vi)   The final column, column 10, is a calculation of C_i by Equation 11.18. The values for C_i are candidates for the cut-off value C^*.

It is clear from the Table 11.4 that the excess return of stock 5 is higher than the C_i, that is, 5.26 is higher than 4.82, but the excess return of stock 6 is lower than the C, that is, 4.00 is lower than 4.73. This means that stocks 1 to 5 are included in the portfolio, while stocks 6 to 10 are excluded. Thus the cut-off, C^*, is 4.82.

The proportion of each stock in the portfolio is given by the following expression:

$$X_i = \frac{Z_i}{\sum Z_i} \tag{11.20}$$

where the summation only includes the stocks that are to be included in the portfolio and Z_i is given by:

$$Z_i = \frac{\beta_i}{\sigma_{ei}^2} \left(\frac{R_i - R_F}{\beta_i} - C^* \right) \tag{11.21}$$

The Z_i and X_i factors are calculated in Table 11.6 below.

Thus, the portfolio should be composed of 51.6% of stock 1, 24.5% of stock 2, 9.9% of stock 3, 2.1% of stock 4 and 11.9% of stock 5.

The portfolio theory has demonstrated not only which stocks must form part of the portfolio, but also the weight of each stock in the portfolio has been calculated.

Table 11.6 Calculation of the portfolio composition

Stock i	Z_i	X_i
1	15.61	51.6%
2	7.41	24.5%
3	3.00	9.9%
4	0.62	2.1%
5	3.58	11.9%

11.9 Critique of Finance Theory

Some of the most profound developments in the theory and practice of finance over the last fifty years have been covered in a few short pages. A definition of risk was established, the foundations of modern portfolio theory were discussed, and the market equilibrium, described by the CAPM and some of its extensions, were presented. The presentation was brief, and as a result, many of the assumptions and limitations were omitted.

Most of the criticism of the theory that was presented in this chapter revolves around the acceptance and use of the CAPM as a description of the equilibrium. The CAPM has come in for much criticism. Eugene Fama, the renowned financial theorist who was instrumental in providing some of the market analysis that led to the widespread acceptance of the CAPM, has argued since then that there was no empirical justification for the CAPM. The other main criticism of the financial theories stems from the nature of forecasting: the theory requires that certain inputs of future distributions or risks are known, whereas only past distributions can be measured. While the criticisms may be true, there is no accepted substitute for the CAPM.

11.10 Summary

The major concepts concerning risk and return were discussed in this chapter. The periodic return is the increase in value to the owner of the asset over that compared with its value at the beginning of the period. It was argued that risk, in the financial sense, is related to the variance or standard deviation of the returns. Thus, risk is defined as variability or volatility. The risk of loss is also related to variability: the greater the variability, the greater the chances of a poor outcome. Business and financial risk for a company were shown to be a function of the variability of the company's sales or earnings. Traditional views of risk, such as the degree of leverage, or break-even analysis measure the distance of the company's financials from the point of distress that might be occasioned by variable or uncertain results.

The combination of risky assets in a portfolio provides a means of reducing risk. This reduction of risk arises from the correlation of the movements in price. If the movements are perfectly correlated, there is no reduction in risk in combining assets in a portfolio. On the other hand, if the movements in price are less than perfectly correlated, risk might be reduced. Risk is reduced by diversification, that is, by adding more assets to the portfolio. However, there is a limit to the reduction of risk by diversification. This limit is called the market risk, or systematic risk, that is present in the market as a whole.

Not all of the risk-return space can be reached through combinations of assets in the market. The space that can be reached is the feasible set or attainable region. The upper boundary of the attainable region represents the optimum: the greatest return for a given amount of risk, or the lowest risk for a given return. This upper boundary is called the efficient frontier. An investor can create a portfolio that meets his or her risk preferences by forming a portfolio of a riskless asset (government bills or bonds) and a risky asset. This combination is a straight-line in risk-return space, and it must be tangent to the efficient frontier for it to represent an optimum portfolio.

This line represents the optimum combination for any risk preference, and is called the capital market line. The implication of this notion is that all investors, irrespective of their risk preference, will want to own a combination of a riskless asset and the risky asset that forms the tangent point to the capital market line on the efficient frontier. If all investors in the market wish to own this same risky asset, the risky asset at the tangent point on the efficient frontier must be the market as whole (in the proportion of the size of each individual asset).

Following the form of the capital market line, a similar straight-line relationship between the risk of an individual asset and its risk compared with the market risk was proposed. This line, called the security market line, represents the equilibrium position for a company. Regression analysis showed that the risk that is important for the return of an individual asset is not the absolute or stand-alone risk of the asset, but the covariance between the asset and the market. This measure of risk, called beta, represents the systematic risk that cannot be reduced by diversification, that part of the risk associated exposure to the market as a whole rather than exposure to the asset itself. This means that the security market line represents the views of a rational investor who can and will diversify his or her assets. This model is called the capital asset pricing model (CAPM).

Portfolio theory provides a number of methods for selecting a portfolio. A method that uses the excess return to beta was presented. This method identifies assets whose returns are temporarily above the equilibrium and selects them for inclusion in the portfolio. The excess return to beta includes a consideration of both risk and return in the selection of the portfolio. This is in contrast to the selection of projects discussed in previous chapters that only gave consideration to the returns (*NPV*, *IRR*) of the project.

11.11 Looking Ahead

The CAPM is the basis for determining the cost of equity capital used by most large companies in their calculation of the discount rate. The determination of the cost of capital is discussed in the next chapter.

11.12 Review Questions

1. What is risk?
2. A geologist is concerned that the oil well that is currently been drilled will be dry. Is this a diversifiable risk?
3. It is often stated that the price is volatile. How is volatility defined?
4. It is easier to forecast a price with more or less volatility? Explain your view.
5. What is operating leverage?
6. What is the basis of business risk?
7. Why would an investor wish to purchase more than one asset?
8. Define the efficient frontier.
9. What are the components of the expected returns?
10. Describe the "two-mutual fund theorem."
11. What risk does beta measure?
12. What is the beta of the market?
13. Describe the basis for selecting an asset for a portfolio.
14. Why is risk reduced in a portfolio?
15. Can all risk in the market be eliminated?

11.13 Exercises

1. The price of an investment with time is given in the following table:

Month	Price
1	10
2	12
3	9
4	13
5	11
6	10
7	13
8	12
9	14
10	11
11	10
12	9

   Determine the average price, the monthly returns, the average return and the standard deviation of returns.
2. The price of an investment with time is given in the table on top of page 363: Determine the average price, the monthly returns, the average return, the expected return and the standard deviation of returns.
3. Which is more risky: a project with a standard deviation of 0.2 or one with a standard deviation of 0.25?

Month	Price	Dividends
1	110	
2	121	
3	132	
4	130	
5	110	
6	100	5
7	103	
8	120	
9	140	
10	131	
11	130	
12	119	9

4. The fixed costs of a business are $10,000 per year, while the variable costs are $100 per unit. What price must the units be sold for if the company makes 150 units per units and the manager wishes to be 15% above the break-even point?

5. The following information is available for two projects within a company:

	Company A	Company B
Revenue	$6/unit	$6/unit
Variable costs	$3/unit	$4.50/unit
Fixed costs	$60,000	$30,000

(i)   Determine the break-even point for each of these projects.

(ii) If the number of units sold is 25,000, determine the degree of operating leverage, which is defined as:

$$Degree\ of\ Operating\ Leverage = \frac{Sales - Variable\ Costs}{Sales - Variable\ Costs - Fixed\ Costs}$$

(iii) If there is a 10% increase in the degree of operating leverage, determine the increase in EBIT using the result in part (ii).

6. A company reports the following results:

	Amount
Revenue	120,000
Operating expenses	50,000
Overheads	20,000
Interest charged	10,000

If the company produced 20,000 units, find the break-even number of units.

7. An investor purchased 500 shares in a company for $16.50. A 1.5% commission was charged. The shares pay a 5% dividend 5 months after purchase, and a 3% dividend 6 months later. The investor sells the share for $22.50 a year after purchase. What is the investor's return?

8.   The investor has the following shares in her portfolio:

Share	Return	Holding
A	10%	10%
B	21%	30%
C	15%	40%
D	9%	30%

Determine the return on the portfolio.

9.   Show that the variance for a portfolio consisting of two assets is given by the following equation:

$$\sigma_p^2 = X_1^2\sigma_1^2 + X_2^2\sigma_2^2 + 2X_1X_2\sigma_{12}$$

where σ_1 is the standard deviation of returns for asset 1, and σ_{12} is the correlation between the returns for the two assets.

10.   The returns for two assets are given in the table below:

Month	Asset 1	Asset 2
1	17.00	22.00
2	2.00	3.00
3	3.00	5.00
4	−9.00	6.00
5	1.00	−1.00
6	7.00	0.10
7	21.00	6.00
8	3.00	−4.00
9	−14.00	1.50
10	−8.00	9.00
11	−0.70	3.00
12	10.00	1.00
Return	2.69	4.30
Std dev	10.25	6.59

(i)   Determine the standard deviation for a portfolio consisting of equal proportions of the two assets.

(ii)   Determine the standard deviation of a portfolio of the two assets in equal proportions if the correlation coefficient is (a) 1, (b) 0, and (c) −1.

(iii) Determine the risk of the portfolio if the holding of asset 1 is varied from 0 to 1 and if the correlation coefficient is (a) 1, (b) 0, and (c) −1.

(iv)   Determine the risk of the portfolio as the holding of asset 1 is varied from 0 to 1.

11.   If the average variance in the market is 50 and the average covariance is 10, determine the number of assets that must be held in a portfolio in which all assets are held in the same proportion to reduce the risk by 90%.

12. Suppose the risk-free rate is 8%. The expected return on the market is 14%. If a particular share has a beta of 0.6, what is its expected return based on the CAPM? If another share has an expected return of 20%, what must its beta be?

13. Suppose Hahn's shares have a beta of 0.80. The market risk premium is 7% and the risk-free rate is 6 percent. What is the return on Hahn's share expected to be?

14. Suppose two shares have the following returns and betas:

Share	Beta	Expected return
A	1.6	19%
B	0.9	16%

(i) What is the market risk premium obtained from the two shares?
(ii) What would the risk-free rate have to be if they are correctly priced?

15. There are four shares under consideration. They have the following expected return and beta values:

Share	Return, %	Beta
A	17	1.7
B	15	0.8
C	14	1.1
D	18	1.7

(i) Plot the return of the shares against their betas.
(ii) Determine the average return and the average beta for a portfolio of equal weighting.
(iii) Determine the straight-line regression of the data.
(iv) Identify those shares that are undervalued and those that are over-valued.

16. The cash flow for two projects under consideration for different economic states are given in the table below:

State	Project A	Project B
1	900	900
2	800	700
3	400	500
4	300	350
5	200	190

(i) Determine the standard deviation for each project.
(ii) Determine the coefficient of variation for each project.
(iii) Identify the more risky project.

Chapter 12
Cost of Capital

12.1 Introduction

All investments require capital. The upfront expenditure for a research and development project, the purchase price of the new boiler that needs replacing, and the development of a new manufacturing facility, all require capital. The flow of cash for investment projects is a net out-flow before there is a net in-flow. The sources of funds to provide for this out-flow are in essence the company's owners (shareholders) and the company's lenders (debt-holders). The company needs to attract both investors to purchase shares in the company and lenders to loan money to the company. In order to do this, the company offers a return to owners and lenders. This return is a "cost" to the company, called the *cost of capital*. For a project to be an economic success and to add value to the company, the returns from the project must at least cover the costs of raising the capital. As a result, the cost of capital represents the expected return on the portfolio of all the company's activities, investments and projects.

Owners and lenders receive their rewards from investing in the company in different ways. The rewards for lenders are that they will be paid an interest amount on the loan for the duration of the loan and that they will be repaid the principal amount in full. The rewards for the owners of the company are in two forms: (i) dividends that are paid to the owners at regular intervals; and (ii) the increase in the value of the company with time.

The cost of capital for the company reflects the cost of rewarding the owners (cost of equity) and the lenders (cost of debt) for their investment in the company.

The cost of equity and the cost of debt are discussed in the following two sections. The combination of the cost of equity and the cost of debt is the weighted average cost of capital, which is discussed after these two sections.

12.2 Cost of Equity

The owner's equity in a company comes from two different sources: the sale of shares in the company to the owners; and the earnings that the company retains. The retained earnings are an *opportunity cost* to the owners, in the sense that if the company had paid the retained earnings to the owners they could choose to invest it in other, presumably, more profitable opportunities. As a result, the company should earn *at least* the rate on the retained earnings that shareholders could earn on other opportunities of the same risk. Otherwise, shareholders would sell the company's share to invest in these other opportunities themselves. This, of course, would lower the company's share price and hence increase the return to the point at which a new equilibrium is reached where the return is acceptable to investors. This can be easily seen from the definition of return that was used in Chapter 11, given by the following expression:

$$Return = \frac{Capital\ Gain + Dividend}{Price} \tag{12.1}$$

If the retained earnings are not performing, the capital gain will be lower. Investors will sell their shares, resulting in a lower price, so that the returns settle into a new equilibrium that is commensurate with the company's risk.

If this is the rate of return that investors or owners require, how can it be determined so that new projects can be assessed on the basis of their attractiveness to investors? There are three different answers to this question that will be discussed in this section. The first answer is to estimate the equilibrium return for the company, and to use that as the return required by investors. The equilibrium return can be obtained from the capital asset pricing model (CAPM), which was discussed in Chapter 11. A second method that can be used in answer to the question is the use of the growth model, and the third method uses a historical average for the company's stock or the stock for a comparable company. By far the most popular, from both theoretical and analytical points of view, is the CAPM. Each of these will be discussed in turn.

12.2.1 Calculating the Cost of Equity from the CAPM

The capital asset pricing model was discussed in Chapter 11. The CAPM describes the return for a particular stock, R_i, as a function of the risk-free rate, the return on the market, and the stock's beta, β_i. The security market line is given by the following expression:

$$R_i = R_F + \beta_i(R_M - R_F) \tag{12.2}$$

where R_F is the return on the risk-free asset, R_M is the return on the market as a whole, and $R_M - R_F$ is known as the market-risk premium.

The security market line represents the equilibrium return that investors in risky assets expect to earn at different levels of risk. In order to attract investors to buy stock in the company, the company must provide returns that at least meet this requirement.

Equation 12.2 expresses the returns as a function of risk, which is captured in the term β_i. However, beta is not a measure of total risk or the stand-alone risk of the stock, but a measure of the undiversifiable risk of the stock relative to that of the market as a whole. This measure is appropriate from an investor's point of view because the investor can eliminate some of the risk by diversifying her portfolio of assets. This was discussed in Chapter 11, Sections 11.4 and 11.5.

Equation 12.2 can be interpreted in the following fashion. The first term on the right hand side of Equation 12.2 represents the price of time. Investors are rewarded for the loss of opportunity even if there is no risk in the investment. The loss of opportunity, called and opportunity cost, arises because the money they invest in the company is not available to them to make other investments. The second term represents the amount of risk relative to the market, given by β_i, multiplied by the market-risk premium. The market-risk premium is the return that an investor in the risky market expects above the return expected from a risk-free investment. Therefore, Equation 12.2 can be interpreted in the following form:

$$Return = Price\ of\ time$$
$$+ (Amount\ of\ risk\ relative\ to\ market)(Market\ risk\ premium) \quad (12.3)$$

The company as a whole is a collection of projects. In order for the company to attract investors, the projects that it invests in must provide returns that are in accordance with Equation 12.2. The investors represent the equity in the company, so that Equation 12.2 represents the cost of equity capital for the company. Equation 12.2 can be rewritten in terms of the cost of equity, R_E, as follows:

$$R_E = R_F + \beta(R_M - R_F) \quad (12.4)$$

In order to calculate the value of cost of equity capital, it is necessary to estimate the risk-free return, the company's beta and the market-risk premium.

The risk-free rate can be determined from government bills and bonds. Although there is some volatility to treasury bills and bonds, these instruments are as close as possible to risk-free in the economy. The interest rate on treasury bills or bonds can normally be found in the sections of the newspaper carrying financial information.

The market-risk premium, given by $R_M - R_F$, can be estimated by determining the historical differences between the market returns and the bond returns, or can be estimated by projecting future market returns. Most practitioners use historical returns and assume that in the future the market-risk premium will remain the same.

The company's beta represents the risk of investing in the company's stock relative to the risk of investing in the market as a whole. Historical values for beta can be estimated from a plot of the data for the company's return against the company's return. The slope of the straight line through this data provides an estimate of beta.

12.2.2 Calculating the Cost of Equity from the Growth Model

The cash flows to an investor in the company come in the form of dividends. If the investor holds the shares in the company forever, eventually she will receive all the possible equity value created by the company as dividends. This means that the present value of the company is the stream of dividends from the company discounted at the cost of equity. This notion can be expressed in the following form:

$$P = \frac{D_1}{(1+R_E)^1} + \frac{D_2}{(1+R_E)^2} + \frac{D_3}{(1+R_E)^3} + \frac{D_4}{(1+R_E)^4} + \dots \qquad (12.5)$$

where P is the present value of the company and D_n represents the dividend in year n. If the dividends are expected to grow at a constant rate, then the dividends can be expressed by the following equation:

$$D_n = D_1(1+g)^n \qquad (12.6)$$

where g is the growth rate of the dividends.

Thus the present value of the company is given by the following expression:

$$P = \sum_{n=1}^{\infty} \frac{D_1(1+g)^n}{(1+R_E)^n} = \sum_{n=1}^{\infty} \frac{D_1}{(R_E-g)^n} = \frac{D_1}{R_E-g} \qquad (12.7)$$

The rearrangement of this expression yields a method for determining the cost of equity:

$$R_E = \frac{D_1}{P} + g \qquad (12.8)$$

This is known as the dividend growth model. The first term on the right-hand side is the current dividend yield and the second term is the expected growth rate in dividends.

12.2.3 Calculating the Cost of Equity from the Historical Returns

The historical average of the returns on the firm's stock or the returns on a stock of a comparable company can be used as an approximation for the cost of equity capital.

The calculation of the cost of equity is demonstrated in the following example.

Example 12.1: Cost of equity.

A company has a beta value of 1.2. The return on the market is 9% and the risk-free rate is 4%. The company's historical returns have been 10%. The dividend yield is 3% and the dividends have been growing a rate of 5%. Determine the cost of equity capital.

Solution:

(i)   CAPM

$$R_E = R_F + \beta(R_M - R_F) = 4\% + 1.2(9\% - 4\%) = 10\%$$

(ii)   Growth model

$$R_E = \frac{D_1}{P} + g = 3\% + 5\% = 8\%$$

(iii)   Historical returns

$$R_E = 10\%$$

The average of the three methods is 9.3%, which can be used as the cost of equity capital. However, most top companies use the CAPM, so following current best practice, a value of 10% is recommended.

12.3 Interest Rates and the Cost of Debt

The after-tax cost of debt is the interest rate paid by the company, R_D, on its loans adjusted for the fact that interest is tax-deductible. Thus the after-tax cost of debt is usually estimated by the following expression:

$$R_D = R_{DBT}(1 - T) \tag{12.9}$$

where R_{DBT} is the interest rate on the company's debts before tax, and T is the tax rate.

The interest rate that a lender charges for a loan reflects the risk that the lender is taking. These risks are the following: (i) the possibility that the borrower defaults on the loan and fails to repay the lender; (ii) the length or term of the loan, which will affect the lender's outlook since a short-term loan will be regarded as less risky than a long-term loan; and (iii) the ease or difficulty with which the lender can sell the loan to someone else. These risks affect the interest so that it is possible to express the interest rate on the company's debt, R, in the following form:

$$R = r_F + IP + DRP + LP + MRP \tag{12.10}$$

where r_F is the real risk-free rate, IP is the inflation premium, DRP is the default risk premium, LP is the liquidity premium and MRP is the maturity risk premium. The inflation premium accounts for the expected change in inflation rate over the loan period, the default risk premium accounts for the risk of the borrower not repaying the loan, the liquidity premium accounts for the possibility that there may not be other buyers for the loan if the lender wishes subsequently to sell it, and the maturity risk premium accounts for the length of the loan.

Not all of the terms in Equation 12.10 are relevant to all situations. As a result, the comparison of different bonds at any one point can yield values for the different risk premiums. For example, compare a short-term Treasury bill with a long-term Treasury bond. The short-term Treasury bill has no default risk, no liquidity risk and no maturity risk because the government backs it, it is in high demand and it is written for a short term. A long-term Treasury bond has the same features, except for the term of the bond, which will alter the maturity risk premium. Therefore the interest rate in these two instruments can be expressed as follows:

Short-term bill: $R = r_F + IP = R_F$

Long-term bond: $R = r_F + IP + MRP = R_F + MRP$

A comparison of the yields for these two debt instruments leads to a value of the maturity-risk premium of the bond.

On the other hand, short-term bills issued by a corporation would have a default risk premium and a liquidity premium. Some companies are more likely to default on their loans than others, giving rise to different levels of the default risk premium. Some corporate bonds may be more difficult to resell to other investors, than others, giving rise to different levels of the liquidity premium. Companies can default and go bankrupt whereas governments can default but cannot go bankrupt. If the country's government defaults on a loan there are widespread and generally long-term consequences that are detrimental to the economy. Long-term corporate bonds will have default risk, liquidity risk and maturity risk.

The interest rate on the short-term and long-term corporate instruments can be written in the following form:

Short-term corporate bill: $R = r_F + IP + DRP + LP = R_F + DRP + LP$

Long-term corporate bond: $R = r_F + IP + DRP + LP + MRP$

$$= R_F + DRP + LP + MRP$$

The default risk of bonds is assessed by ratings agencies such as Moody's and Standard and Poor's. The bonds are divided broadly into investment grade and speculative grade bonds, and further categorised in terms of the issuer's ability to service the debt, that is, to pay the interest and repay the principal. The rating of bonds is also discussed in Chapter 16.

The yield on the bond is the interest rate that the borrower is paying to the lender. The yield is determined by calculating the *IRR* of the cash flows arising from the issuance of the bond. This calculation is similar to the methods used in Chapter 6 to determine the *IRR* of a project. The following example illustrates the calculation of the bond yield.

Example 12.2: Bond yield.

A company issues a bond repayable after ten years. The bond is issued with a face value of $100,000 and carries a 7% coupon. The lender provides the cash to the issuer by buying a bond certificate. Over the years, the lender receives coupon payments (called coupons because the certificate contained coupons for each payment which the lender would have to submit in order to redeem the payment). The coupon is a fixed proportion of the principal, in this case 7%. At the end of the period, the lender is repaid the total face value of the bond by the issuer. The company issued bonds worth $20,000,000. The investment banking costs for arranging the bond amounted to 2.4% of the amount raised. The bond was offered at a discount 1% to the face value to attract investors. The company's marginal tax rate is 40%. What is the cost of debt?

Solution:

The cash flow for a bond to the lender is shown in Figure 12.1.

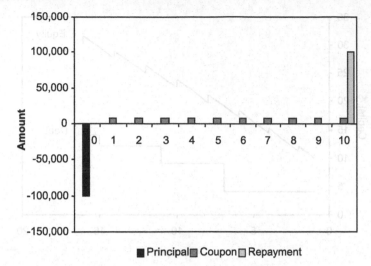

Figure 12.1 Cash flows for the bond from the lender's perspective

Table 12.1 Cash flow to the bond issuer for a 7% coupon bond with a ten-year maturity

Year	0	1	2	3	4	5	6	7	8	9	10
Principal	100										
Investment banking fees	−2.4										
Discount offered	−1.0										
Net cash received	96.6										
Coupons		−7.0	−7.0	−7.0	−7.0	−7.0	−7.0	−7.0	−7.0	−7.0	−7.0
Principal repayment											−100.0
Bond cash flow	96.6	−7.0	−7.0	−7.0	−7.0	−7.0	−7.0	−7.0	−7.0	−7.0	−107.0

The cash flow to the company as a result of issuing the bond is shown in Table 12.1 (amounts in thousands). The company raises $100,000 less the discount of 1% for each bond sold. In addition, the company has to pay the investment bank $2,400 of this amount in fees. Thus, instead of getting $100,000 for each bond, the company gets $96,600. Each year, the company pays $7,000 as a coupon payment, and at the end of the ten-year term, the company pays the full face value of the bond certificate back, that is, $100,000.

The interest rate on this loan can be obtained from the internal rate of return of the bond cash flows. The calculation of the *IRR* by the methods of Chapters 6 and 7 yields a value of 7.5%. This is the cost of debt to the company as a result of raising debt capital with a bond structured in this manner. The after-tax cost of debt to the company is $7.5(1.0 − 0.40) = 4.5\%$.

12.4 Pooling of Funds

The company continually raises equity funding by retaining some of the earnings or profit each year. Depending on the perceptions in the market of the company's investment projects, and the major projects that the company intends to invest in, the company can retain more of the earnings (profits) in order to increase the equity

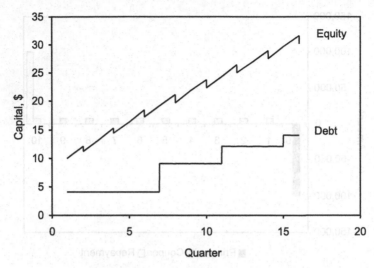

Figure 12.2 Financing of the company's capital requirements through retained earnings and debt

capital of the company. Issuing more stock can also raise additional equity capital; however, management generally views this as expensive due to the costs involved and due to the possible discount required in order to attract investors.

On the other hand, debt is raised discontinuously by the business to ensure that the required capital structure (the proportion of debt and equity) is met.

This raising of funds through retained earnings and the issuance of debt is illustrated in Figure 12.2. The equity of the company rises during every half-year as the company declares profits and drops as the company declares a semi-annual dividend payment. The company's financial management has a targeted debt-equity ratio, and increases the amount of debt in the company at different intervals in order to approximate this value.

The funds that are raised by the company, either as debt or equity, are available for use throughout the company for the company's capital projects as the need arises. The planning of the capital needs of the company and the allocation of the available or anticipated resources is part of the capital budgeting process, which, as discussed in Chapter 2, is a responsibility of the chief financial officer. The actual decisions for investment in projects are decentralised by this process. Central control is limited to specifying the method of analysis, the discount rate and the approval levels. The company's treasury provides the monies required for the project when needed against the planned capital expenditure programme.

The company, even a multinational company, pools its financial resources. The raising of new debt or equity may be justified to the investment community on the basis of a new project that the company is about to embark upon. Even so, there is no matching of specific funding with specific projects. If there was matching, it would be easier to justify projects that are funded with debt than those with equity because debt is generally cheaper than equity. At the same time the company needs to maintain the balance of debt and equity to achieve the targeted capital structure, so the timing of the raising of capital may be out of step with the best projects. All in all, such matching would lead to inconsistent decisions and poor results.

The calculation of the cost of capital from the cost of equity and the cost of debt is discussed in the following section.

12.5 Weighted Average Cost of Capital

The cost of equity and the cost of debt are weighted by the proportion of equity and debt utilized. The weighted average cost of capital, WACC, is given by:

$$WACC = \left(\frac{E}{E+D}\right)R_E + \left(\frac{D}{E+D}\right)(1-T)R_{DBT} \qquad (12.11)$$

where E is the amount of equity, D is the amount of debt, R_E is the cost of equity, R_{DBT} is the before-tax cost of debt and T is the tax rate.

Debt is usually cheaper than equity for a profitable company because R_{DBT} is less than R_E. As a result, there is an incentive to borrow in order to maximise value for shareholders.

An examination of Equation 12.11 could lead to the conclusion that the optimal solution would be to fund the company entirely from debt sources. However, increasing debt makes the company more risky due to the increased fixed costs of servicing the debt. Investors are often more reluctant to invest in companies that have a high proportion of debt. As a result, the cost of equity will increase for companies with an abnormally high proportion of debt. This effect is not incorporated into Equation 12.11. This is discussed further in Section 12.7.

The weighted average cost of capital is the cost of capital for the company as a whole. The marginal cost of capital is the cost of raising more capital in addition to the capital already raised. It is the marginal cost of capital that is of interest in assessing the company's new capital projects. As a result, new projects are assessed against the cost of raising new capital.

The weighting factors $E/(E+D)$ and $D/(E+D)$ can be determined from their values on the balance sheet, that is, their book values, from the market value of the company's equity and debt or from the market value of the targeted values for equity and debt. It is worth returning to the subject of the perspective from which the analysis is performed before further discussing which of these values is appropriate in determining the weighting factors. In Chapter 1, it was suggested that the analysis of a project is from the entity perspective because this enabled the separation of investment and financing decisions. In Chapter 10, two of the case studies examined the cash flow to the equity parties. In the next section, the entity and equity bases are examined in more detail.

12.6 Entity Versus Equity Basis

The analysis of the cash flow has been on the basis of the project, called the entity basis. It has been assumed that the *NPV* calculated on this basis is the value that is added to the equity owner's share of the company. As a result of the choice of the entity basis, the discount rate, as determined by the WACC, accounted for the operating risk of the company, the financing of the company (the debt/equity position) and the reduction in tax due to interest. It is worthwile examining the correspondence between the entity and equity bases. This is discussed next.

The value of the company from the shareholders point of view is determined for both the equity and entity bases in the following two examples.

Example 12.3: Equity basis.

Consider a company in which the shareholders invest $500 and raise a loan for a further $500. The tax rate is 30%, the interest rate is 10% and the cost of equity capital is 30%. The owners invest the $1,000 of capital in equipment that earns $1,300 a year with operating costs of $600. The equipment is depreciated over five years on a straight-line basis. At the end of five years the owners sell the company at market value.

Determine the value of the company using the equity basis.

Solution:

The financials for this situation can be represented as the income statement and the cash flow statement in Table 12.2.

The determination of the cash flows has been discussed in detail in Chapter 3, and the analysis of the figures shown in Table 12.2 follows along those standard lines. At the commencement of operation (year 0), the company raises $500 from the bank and $500 from its owners, and purchases equipment for $1,000. At the end of five years, the owners sell the company for $2,217 and repay the debt of $500. The net present value of the resulting cash stream, *discounted at the cost of equity*, is $1,716. The present value added to the owner's

Table 12.2 Income statement and cash statement

Year	0	1	2	3	4	5
Income statement						
Revenue		1,300	1,300	1,300	1,300	1,300
Costs		600	600	600	600	600
Operating income		700	700	700	700	700
Depreciation		200	200	200	200	200
Interest		50	50	50	50	50
Income before tax		450	450	450	450	450
Tax		135	135	135	135	135
Income after tax		315	315	315	315	315
Cash flow statement						
Cash flow from operations		515	515	515	515	515
Cash flow from investment	−1,000					2,217
Cash flow from financing	1,000					
Equity financing	500					
Debt financing	500					−500
Net cash flow	0	515	515	515	515	2,232

wealth is $1,716 less the equity stake of $500 that the owners invested. Thus, the present value of the amount added to owners' wealth is $1,216.

Why has this been discounted at the cost of equity? The cash flows and the free cash flows are prepared from the equity perspective. The capital is raised from debt and equity. The returns to the debt-holders are specifically accounted for in terms of the interest payments. Thus, what remains as net cash flows in the table above accrues to the owners. The present value of these cash flows must then be evaluated at the return that the investors in this company expect, that is, at the cost of equity.

The calculation of the value of the company in Example 12.3 was performed on the *equity basis*. This method is perfectly acceptable, except that it mixes the investment and financing decisions. This means that the analyst needs to know the amount of debt contributed to the investment in order to calculate the interest charges and the free cash flow, from which the *NPV* is calculated. In practice, this may not be that big a problem: the company's debt is known, and generally the company will have a target debt/equity ratio, so this might be used to determine the contribution of interest to a specific project.

However, the method advocated in Chapter 1 for the evaluation of projects is the entity basis, in which the cash flows are calculated without the effect of interest. The cost of interest is accounted for in the weighted average cost of capital. This is demonstrated in the following example.

Example 12.4: Entity basis.

An alternative method to the equity basis of Example 12.3 is the *entity basis*. Determine the net present value of the company that was described in Example 12.4.

Solution:

The calculation of the free cash flows for the company are shown in Table 12.3. When the entity basis is used, the cash flows are prepared without the contributions of interest and debt. This means that the annual cash flows are slightly higher than those calculated in the case of the full income and cash flow statements given in Table 12.2 in Example 12.3.

The differences between the treatment in Example 12.3 and this are the following: (i) no interest payments, which affects the tax calculation; (ii) no equity or debt in-flows, so that free cash flow in year 0 is the total amount invested; and (iii) no settlement of debt at the end. The free cash flow is discounted at WACC, which takes account of the debt and equity contributions and the effect of interest on the tax. The WACC was given in Equation 12.11 and can be expressed as follows:

$$WACC = \left(\frac{E}{E+D}\right)R_E + \left(\frac{D}{E+D}\right)(1-T)R_{DBT} \qquad (12.12)$$

All of the values for the variables in Equation 12.12 are known, so the value of the discount rate can be calculated as follows:

$$WACC = \left(\frac{500}{500+500}\right)30\% + \left(\frac{500}{500+500}\right)(1-30\%)(10\%) = 18.5\% \qquad (12.13)$$

The *NPV* calculated using a value of 18.5% for the discount rate is $1,649. There is an inconsistency with this value and the equity value.

Table 12.3 The calculation of the free cash flow using the entity basis

Year	0	1	2	3	4	5
Revenue		1,300	1,300	1,300	1,300	1,300
Costs		600	600	600	600	600
Operating income		700	700	700	700	700
Depreciation		200	200	200	200	200
Income before tax		500	500	500	500	500
Tax		150	150	150	150	150
Income after tax		350	350	350	350	350
Cash flow		550	550	550	550	550
Investment	−1,000					2,217
Free cash flow	−1,000	550	550	550	550	2,767

On the other hand, if the value of the equity used in Equation 12.12 was $1,716, the WACC would be 24.8%, and the *NPV* would be $1,216, which is precisely the present value added to the owner's wealth. From where did this value of $1,716 for the equity come? It is the wealth created by the project for the shareholder, that is, the *NPV* (from the equity basis) less the equity investment. This is a circular argument: in order to get the correct answer, the answer must be known!

(It is left as an exercise to show that the condition for the equivalence of the two cases is that the debt at time t, D_t, is given by $D_t = wV_t$ where w is a constant and V_t is the value at t of all the future cash flows.)

These two examples suggest that the values for debt and equity used in the calculation of the WACC must be *market values* (not historical book values) and they must be the *targeted values* at the end of the project, not the current values. This circular argument creates difficulties for the entity basis. However, in a large company with multiple projects the difference between the current debt and equity values and the targeted values is not as dramatic as has been illustrated here. This is because in a profitable company the debt and equity positions are generally more stable than has been illustrated here.

Despite these difficulties, the entity perspective is useful for a number of reasons. The most important is that it separates the analysis of the scenarios for the two decisions that need to be made. The entity perspective is used to decide whether to invest in the project or not without the need of knowing how the project is to be funded or even how the company as a whole is funded. The chief financial officer needs to specify only a discount rate. Surveys of current best practice indicate that the discount rate used is based on WACC.

Projects generally do not only originate from senior management. They arise from different parts of the organisation. The entity perspective allows the economics of the project to be analysed on a stand-alone basis without having a detailed analysis of the entire company's financials and economics with and without the project. Such a requirement would be difficult to manage. In order to allow this simplification, some approximations have to be made that enable practical management of capital budgeting.

As has been mentioned before, the advantage of the entity basis is that the investment and the financing decisions are separated. This is considered a significant advantage since it reflects the separation of the functional roles of capital budgeting and financing within the organisation of the company. The pattern of cash flows for the contribution of debt to the project is not required in the entity approach, resulting in fewer inputs and a more transparent model. The objective is to assess an opportunity for investment in its own right on the same basis as other projects are assessed, and this can be adequately achieved by the weighed average cost of capital because of the pooling of funds. However, reconciliation between the equity *NPV* and the entity *NPV* at the financing stage can lead to some confusion if the difference between the entity and equity basis is not clearly understood.

The practices of the top companies in the Fortune 500 in calculating the discount rate or WACC are reviewed in the next section.

12.7 Practices in the Calculation of WACC

Bruner *et al.* (1998) surveyed the practices of 27 companies in the Fortune 500 with regard to their estimation of the cost of capital. The survey established that the cost of capital was estimated mainly by the weighted average cost of capital. The survey also established how the financial departments in the top companies obtained values for each of the inputs to the weighted average cost of capital for their own use. Each of these factors is discussed in turn.

12.7.1 Cost of Equity Capital

The survey by Bruner and colleagues (1998) indicated that the top companies in the Fortune 500 use the CAPM to estimate the cost of equity. Estimates for expected values for the risk-free rate, the market-risk premium and the company's beta are needed in order to calculate the cost of equity using the CAPM.

12.7.2 Risk-free Rate

The risk-free rate is typically taken as the short-term Treasury bill or a long-term Treasury bond. The question is what term of bond to use. Most textbooks, but not all, suggest the 90-day Treasury bill. Most practitioners use the ten-year or longer Treasury bond. There is a significant difference in rate between bills and ten-year bonds. The reason advanced by practitioners for the use of longer-term bonds is to match the time horizon of the risk-free rate with the term of the investments undertaken by the company.

12.7.3 Market-risk Premium

The market-risk premium, $R_M - R_F$, can be calculated from historical values of the returns on the market and from the risk-free rate. The returns on the market can be estimated using a market index, such as the S&P 500 or the FTSE 100. Ideally, the index used should be calculated based on weighting factors that represent the market capitalization of each of the stocks in the index. The Dow Jones Industrial Average is calculated using the weighting of each stock in the index by the price of the shares, not the total market value of the company. The Nikkei 225, the market indicator for the Japanese market, is calculated in a similar manner to the Dow Jones. In addition, the Dow Jones Industrial Average only includes the top 30 of the largest companies. These are not suitable measures for determining R_M, which is the total return on the market as whole. The S&P 500 does not include the dividends that companies declare. It measures the price appreciation or capital gain, and not the total return to investors.

Over a 60-year period, stocks in the US have returned 9.5% more than treasury bills, and 7.8% more than long-term bonds. However, the market-risk premium has changed over time. Data for the last 40 years yields a market-risk premium in the range of 5 to 6%. The returns to the stock market can be calculated as an arithmetic mean or a geometric mean. There is a significant difference between these two methods. For example, Bruner and colleagues report values for the market-risk premiums given in Table 12.4.

Table 12.4 The market-risk premium, $R_M - R_F$, for the US

	Arithmetic mean	Geometric mean
Stocks over T-bills	8.5%	6.5%
Stocks over T-bonds	7.0%	5.4%

Several financial services firms provide data on the stock market and bond market returns and hence on the market-risk premium.

12.7.4 Beta

The calculation of beta was discussed in Chapter 11. It can be found from a linear regression of the return on the stock against the market return, that is, from the straight line given as follows:

$$R_i = a + \beta_i R_M \tag{12.14}$$

The following example illustrates the calculation of beta in this manner.

Example 12.5: Calculation of beta.

The price of a stock and the index for the market for a 12-month period are given in Table 12.5. Determine the stock's beta.

Table 12.5 Market index and stock price over 12 months

Month	Market index	Stock price
0	9,056	53
1	9,418	55
2	9,324	55
3	9,977	56
4	10,476	60
5	10,057	59
6	10,962	61
7	11,290	63
8	12,420	67
9	11,923	66
10	12,757	69
11	12,502	69
12	12,877	71

Table 12.6 Calculation of the return on the market and return on the share

Month	Market index	Stock price	Market return, %	Return on share, %
0	9,056	53		
1	9,418	55	4	3
2	9,324	55	−1	0
3	9,977	56	7	3
4	10,476	60	5	6
5	10,057	59	−4	−1
6	10,962	61	9	4
7	11,290	63	3	3
8	12,420	67	10	6
9	11,923	66	−4	−2
10	12,757	69	7	5
11	12,502	69	−2	0
12	12,877	71	3	3

Solution:

The stock's beta can be obtained from the slope of the stock's return against the market return, as shown in Equation 12.14. The stock's return and the market return are calculated using Equation 11.1. The calculation of the stock's return and the market return are shown in Table 12.6.

The linear regression of the stock's return against the market return is shown in Figure 12.3. The slope of the line is 0.946, which is the value of beta.

There are three choices that need to be made in order to calculate the beta for a company's stock. These are the period of historical data used (two years, five years), the interval for the collection of data during this period (daily, weekly, monthly) and the stock market index used as a proxy for R_M. Various financial services companies calculate the betas for stocks. Bruner reported that one com-

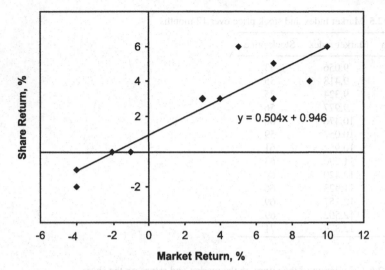

$$y = 0.504x + 0.946$$

Figure 12.3 The linear regression of the stock's return against the market return

pany, for example, uses weekly data over five years and uses the New York Stock Exchange composite for the calculation of the market return. Another financial services company uses monthly data over five years and the S&P 500 as the proxy for the market. Most of the companies that participated in the survey by Bruner and colleagues used betas calculated by financial services companies.

12.7.5 Weighting Factors in WACC

Most of the companies surveyed by Bruner and colleagues used the targeted market values in their calculation of the weighting factors in the calculation of the weighted average cost of capital.

12.7.6 Tax Rate

Most companies surveyed used the marginal or statutory tax rate, while a significant minority (37%) used the historical average tax rate for their company as the tax rate in the calculation of the after-tax cost of debt.

The calculation of the weighted average cost of capital is illustrated in the following example.

Example 12.6: Calculation of the weighted average cost of capital.

The chief financial officer of Black and Decker has obtained three estimates for the company's beta. These are 1.06 from Bloomberg, 1.65 from Value Line and 1.78 from S&P.

Table 12.7 The effect of different inputs estimates on the calculation of WACC

		Bloomberg	Value line	S&P
Beta		1.06	1.65	1.78
R_E	R_F from T-bills	14.4%	19.4%	20.5%
	R_F from T-bonds	15.3%	20.3%	21.4%
WACC	R_F from T-bills	9.7%	12.3%	12.8%
	R_F from T-bonds	10.2%	12.7%	13.3%

Another financial services company has estimated the market-risk premium to be 8.5%. The yield on T-bills was 5.36% and on 30 year T-bonds was 6.26% over the same period that was used for the estimation of the market-risk premium. The marginal tax rate is 38% and debt represents 49% of the capital of Black and Decker. The current interest rate on the debt is estimated to be 7.8% based on the debt rating of the company.

Calculate the cost of capital for Black and Decker.

Solution:

The results for the possible combinations for the calculation of the cost of equity based on the CAPM and for the WACC are given in Table 12.7.

Clearly there is a large variation in values. The largest variation is caused by the variation in the value of beta that is used. The differences in the values for beta obtained from the different agencies are a result of the different methods for calculating the beta. Value Line and S&P use different indexes for the market, but the difference between the results is not as great as with Bloomberg. The principal difference between Bloomberg and the others is the choice of period and the frequency of sample during the period. The chief financial officer would need to further investigate the cause of these differences. Ultimately the chief financial officer will need to use his or her judgement in setting the WACC for the company based on the range of values calculated.

12.7.7 Review Period

The estimates of the cost of capital are normally reviewed on an annual basis. The respondents to survey conducted by Bruner also indicated that they might review the cost of capital in the case of a high impact event. It is worth quoting directly from Bruner: "Firms also recognize a certain ambiguity in any cost number and are willing to live with approximations. While the bond market reacts to minute basis point changes in investor return requirements, investments in real assets, where the decision process itself is time consuming and often decentralized, involve much less precision. To paraphrase one of our sample companies, we use capital costs as a rough yardstick rather than the last word in project evaluation."

12.8 WACC, Leverage and Debt Financing

Any engineer or scientist worth his or her salt would look at Equation 12.12 and say that, if debt was cheaper than equity, the optimal solution is to have as much debt as possible. Another view is that of a manager in a company who examines the equation for the cost of capital and says that it can be lowered if new debt was raised specifically for his (or her) project. This argument is often raised in a multinational corporation, where the division in a different country believes that cheaper financing can be obtained locally and hence the cost of capital for his or her project should reflect this.

Unfortunately, the optimisation of the WACC by indiscriminately increasing debt and the use of local debt are both fallacious arguments. These two issues concerning debt financing are discussed in this section.

Leverage refers to the amount of debt the company has. Debt enables the company to do more with a "lighter weight," that is, more can be done with the same equity capital. As the level of debt increases from zero, the firm is able to gain significantly from leverage. This is the subject of the following example.

Example 12.7: Financial leverage and financial risk.

A company has sales of $500,000. The fixed costs are $50,000, and the variable costs are 30% of sales. The company has $2,000,000 in capital. Consider two scenarios: one in which the company is funded by 80% debt and the other in which the company is funded by 20% debt. The interest is 10%. Determine the earnings and the earnings per share (EPS) if the equity capital was raised at a price of $1 per share.

Solution:

The income statements for the two scenarios are shown in Table 12.8. The different amounts of debt affect the interest payments and the number of shares that the company sold in order to raise its equity capital.

The table clearly indicates the leverage effect of debt. With more debt, the earnings per share have doubled. The sensitivity of the earnings per share to the amount of debt was determined. The effect of leverage is shown in Figure 12.4. These results emphasise the dramatic effect of leverage.

However, increasing debt increases the fixed costs of the business, and higher fixed costs give rise to the possibility of financial distress when sales fall in harder economic times. The effect of the sales on the earnings for the two scenarios is given in Figure 12.5. The break-even point for the low debt scenario is sales of $128,000, whereas the break-even for the high debt scenario is sales of $300,000. The high debt scenario is more exposed in times of trouble and as a result is less robust than the low-debt scenario.

Table 12.8 Income statement for high debt and low debt scenarios

	High debt	Low debt
Sales	500,000	500,000
Fixed costs	50,000	50,000
Variable	150,000	150,000
EBIT	300,000	300,000
Interest	160,000	40,000
EBT	140,000	260,000
Number of shares	400,000	1,600,000
EPS, c/share	35	16.25

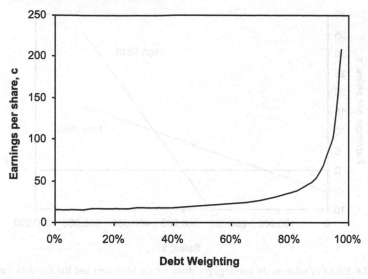

Figure 12.4 Effect of leverage on the earnings per share

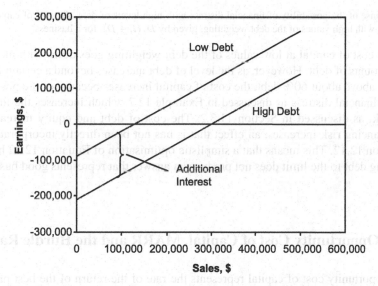

Figure 12.5 Effect of sales on the earnings for the high debt and the low debt scenarios

The effect of sales on the earnings per share is shown in Figure 12.6. These results indicate that the earnings per share for the two scenarios actually reverses as the sales figures fall. Below a sales level of $357,000, the earnings per share for the low debt scenario are higher than those of the high debt scenario and the effect of financial leverage no longer benefits the shareholder. Instead, it places them under stress.

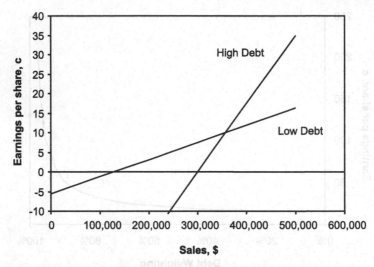

Figure 12.6 Effect of sales on the earnings per share for the high debt and the low debt scenarios

Because of the possibility of financial distress with high levels of debt, the cost of capital rises with high values of the debt weighting, given by $D/(E+D)$, for a business.

The cost of capital at low values of the debt weighting goes down with increasing amounts of debt. However, as the level of debt increases beyond a certain level, maybe above about 60% debt, the cost of capital increases because of the possibility of financial distress as discussed in Example 12.7, which increases the financing risk, as discussed in Section 11.3.2. The cost of debt and equity increase as the financing risk increases, an effect that is has not been directly incorporated in Equation 12.12. This means that a simplistic optimisation of Equation 12.12 by increasing debt to the limit does not produce an answer that represents good business sense.

12.9 Opportunity Cost of Capital, MARR and the Hurdle Rate

The opportunity cost of capital represents the rate of the return of the best project that cannot be pursued due to budget constraints. The opportunity cost of capital is also called the minimum attractive rate of return. This concept was discussed in Chapter 8. While this concept has some attraction, the opportunity cost of capital is not the method for determining the discount rate that is used by the companies surveyed by Bruner.

If *IRR* is the method of analysis that is specified by the company for the assessment of projects, the discount rate is sometimes called a hurdle rate. The hurdle rate can be based on the MARR or on the WACC. However, in practice, the calculation of the hurdle rate is often based on the financial cost of capital, represented by the

Table 12.9 Required rates of return for various business activities (PricewaterhouseCoopers, 2000)

Investment stage	Min *IRR*	Max *IRR*	Average *IRR*
Seed	30	400	105
Start-up	12	200	67
Early expansion	20	100	47
Expansion or development	25	80	35
Mezzanine debt	15	40	23
Management buyout	15	50	29
Turnaround	15	40	26
Overall	25	60	34

weighted average cost of capital. For example, the hurdle rate might be determined as WACC multiplied by 1.3.

The concept of *IRR* is widely used in business. PricewaterhouseCoopers interviewed fund managers in Australia to determine typical values of the required discount rate or hurdle rate that investors at various stages of the establishment of a business would demand from a single investment. The results of this survey are summarised in Table 12.9.

The results of this survey provide an indication of the types of hurdle rates or discount rates that the investment managers use in assessing investments higher risk investments. The seed stage of a business is the earliest investment required to investigate a business proposition and commence the earliest of activities. The start-up stage is the launch of the business's activities or products. Expansion involves the development of the business into a sustainable enterprise. Mezzanine debt is the provision of debt to a business, rather than an equity investment. This debt is subordinated to the senior debt, in other words, in the event of default on the terms of repayment, the senior debt has a higher claim on the assets of the business than the mezzanine debt. Mezzanine debt is also called subordinated debt or junior debt. Management buyout is the activity in which the management of the business purchase the owner's interests in the business from them. Business turnaround is the activity in which an investor purchases an ailing business in order to rectify the causes of the poor performance.

12.10 Summary

A company's main source of funding is from investments in the equity and debt of the company. The cost of capital is the return that investors in the company's equity and debt expect. This return must be sufficient to attract investors to place their money with the company. If the company does not provide the returns, the value of their investment in the company will diminish, making it more expensive for the company to raise funds in the future. As a result, the company's managers have the responsibility of ensuring that the company provides the expected returns, and the projects that the company embarks on are evaluated against the expected returns as the benchmark. Since it is in the setting of the discount rate that financial managers in the company exercise control over the choice of projects, it is important that there is a method for determining the cost of capital.

The cost of capital is composed of two main components representing the main sources of funds. These are the cost of equity and the cost of debt. The cost of debt is easier to determine. Since the company's lenders require their returns in the form of interest, the cost of debt is the interest rate paid by the company to debt-holders.

The cost of equity is more difficult to determine. Investors in the company's equity, that is the shareholders, expect their returns in two forms: as a regular dividend, and as an increase in the value of their share of the company (capital gain). One method of determining the cost of equity is to assume that the company will last forever and the investment will be invested forever. If this is the case, the investor will receive all her investment as dividend in perpetuity. The dividend growth model then asserts that the required rate of return on an equity investment is the current dividend yield plus the expected rate of growth of dividends. A second method for calculating the return on equity is to determine the company's historical return to shareholders.

The most common method of determining the company's cost of equity is to employ the capital asset pricing model, CAPM. This model relates the return required by investors to the risk-free rate obtained on government bills or bonds, the return on the market, and the risk of the company relative to the risk of the market. The risk that the CAPM accounts for is not the total risk, but a measure of the exposure to the market that the investor takes on when investing in the company's equity.

The cost of equity and the cost of debt are combined by a simple weighted average into the cost of capital, known as the weighted average cost of capital. Because the project financials are prepared on the basis of the entity approach in which no cash flow to interest is presented, and interest is tax deductible, the cost of debt is reduced by the tax rate in the calculation of the average cost of capital.

The correspondence between the entity and equity bases was explored. This showed that the weighting factors in the calculation of the weighted average cost of capital depend on the targeted market values of debt and equity, not on book values.

The company does not raise specific funds for specific projects. Rather funds are pooled centrally. The weighted average cost of capital represents the pooling of funds by the company, and as a result is the appropriate discount rate to use in the evaluation of projects. The opportunity cost of capital, also called the minimum attractive rate of return, is the economic cost of capital. In the practice of capital budgeting within firms, the cost of capital is determined as the financial cost of capital (represented by the WACC), not the opportunity cost of capital (represented by the MARR).

12.11 Review Questions

1. What are the sources of capital for a company?
2. What are the components of the cost of capital?
3. What are the assumptions of the dividend growth model?
4. What risk does the CAPM account for in determining the cost of equity?
5. Discuss how you would optimise the amount of debt that a company has?
6. Is there a match between the funds raised and the use of those funds? Explain your answer.
7. What is MARR, and how does it differ from WACC?
8. What considerations should be taken into account in sourcing data for the calculation of WACC?
9. Why is the cost of debt adjusted for tax?
10. Should a division in a foreign country argue for a lower discount rate because debt can be raised locally at a cheaper rate? Why? List some pros and cons.
11. What does "targeted capital structure" mean?

12.12 Exercises

1. A company has a targeted capital structure of 60% equity and 40% debt. The cost of debt is 10% and the cost of equity is 19%. Determine the weighted average cost of capital if the tax rate is 40%.
2. A company's beta is 1.6, the risk-free rate is 6%, and the return on the market index is 11%. What is the company's cost of equity?
3. A company has long-term debt of $25 million, which has average interest rate of 11%. The company's market capitalization is $70 million. If the tax rate is 45% and the cost of equity is 16%, determine the weighted average cost of capital.
4. The company's return on equity is 12% and its beta is 1.2. If the risk-free rate is 4%, what is the market-risk premium?
5. The company has a dividend yield of 15% and is expected to grow at a rate of 12%. Determine the cost of equity.
6. A company has a capital structure comprising 40% debt. New debt can be raised at an interest rate of 12%. The company's dividend yield is 9% and it is expected to grow at 5%. If the tax rate is 48%, determine the weighted average cost of capital.
7. The cost of equity capital is expected to rise as a result of excessive debt. If the cost of equity as a function of debt is given in the table below, determine

the optimal capital structure. Assume that the cost of debt is 10% and the tax rate is 35%.

Percentage debt	Cost of equity capital
0	10.0
10	10.1
20	10.2
30	10.3
40	10.5
50	11.0
60	11.5
70	13.5
80	16.0

8.   The share price and the market index have the following values:

Period	Market index	Share price
1	12	50
2	11	40
3	14	55
4	15	65
5	14	60
6	15	65
7	16	60
8	17	75
9	15	78
10	14	70
11	13	75

Determine the company's beta.

Chapter 13
Risk in Engineering Projects

13.1 Introduction

The cornerstones of finance are risk and return. The returns to the project are a result of the cash flows that the project will generate. The risk is the variability is those anticipated cash flows. These concepts of financial risk and return are applied in this chapter to the evaluation of projects under conditions of uncertainty.

A number of different approaches, such as the certainty equivalent approach, the real options approach and decision tree analysis can be used to account for the variability in returns. These methods are discussed in this and the following two chapters. Each approach has merit, and the discussion is instructive in deepening our appreciation of the issues in economic evaluation and decision-making. In this chapter, the application of the certainty equivalent, the risk-adjusted discount rate, the probability method and the Monte Carlo simulation method are discussed.

13.2 Sources of Uncertainty

The discounted cash flow techniques, as presented in Chapters 5 to 10, implicitly assumed that all the elements of the cash flows are known with certainty. In reality, although these forecasts are made with the best available knowledge and techniques, they are nonetheless uncertain. Although cost engineering estimates of the fixed capital costs are made with increasing accuracy as the project progresses, the approval for the decision to proceed is made with acceptable but usually significant limits of error on the estimate. Predictions of prices and costs into the future become less certain with distance into the future.

Risk is the deviation in the value of a business quantity from its expected value. If the business's management anticipates that the price of gold will be $500 an ounce, and this anticipated value represents the mean of the probable outcomes, then the risk is the chance of the gold price being below this anticipated value.

The variability in a business's results can be viewed either on the basis of past performance or on the basis of future performance. As a result, the variability is either historical or anticipated. The notion of risk concerns the future performance; there is no risk with historical results, they are known with certainty. However, the past can be a good predictor of the future, and it is usually assumed that the variability of the results in the recent past is a good predictor of the risk in the future.

Company-level risks and project level risks are discussed in the following two sections.

13.2.1 Company-level Risks

Some of the risks that a business faces are the following:

(i)   *Business risk* is the variability of the company's anticipated earnings from normal operations. This is comprised of two factors: the variability in the sales and in the operating costs. The price that a business receives for its products can be a major uncertainty. The costs to the business are also subject to uncertainty as a result of raw materials and labour prices. The company's earnings from operations are the difference between these two factors, which may or may not be correlated with each other.

(ii)   *Investment risk* is the variability in the value of the company's investment projects. It could also be viewed as the variability in the initial capital outlay for a project.

(iii)   *Financial risk* is the variability in the profit as a result of changes in the interest rate that affect the company's ability to meet its debt commitments.

13.2.2 Project-level Risks

Both the business risk and the investment risk are important in considering a project. The financial risk is not considered here because of the choice of the basis of the evaluation. The chosen basis is the entity basis, which means that the project is evaluated as if the project is completely funded from the company's equity resources. The factors that affect the input costs and the output price are important for a project, and the variability in the initial investment in the project is a prime variable in the assessment of a project's economic viability. In assessing the risk of the project, all of these factors need to be addressed.

Three levels of risk can be identified for a project. These are the stand-alone risk, the corporate risk and the market risk.

The *stand-alone risk* is the project's variability in its expected returns. It ignores any portfolio effects arising from its grouping with other projects within the company, or the effect that it has on the well-diversified investor. It is the risk as if the project were located in a single project company that was owned by shareholders who only owned stock in this company.

The *corporate* risk is the risk that the project represents for the company. Some of the stand-alone risk of the project will be removed by diversification with the other projects of the company. The corporate risk is the effect of the project increasing the uncertainty of the company's earnings.

The *market* risk is the risk of the project changing the company's beta. This is the risk seen by a well-diversified investor. This should be the only risk with which diversified investors are concerned. However, corporate risk is of interest to undiversified shareholders, to small businesses and to other stakeholders in the business, such as the managers and employees, and the suppliers and the creditors.

Four methods of incorporating risk into the analysis of a project's value have been advocated. Two of these approaches directly alter the net present value calculation to account for risk while the other two views the risk separately. In the next three sections these three methods are considered. They are the probability method, the certainty equivalent method, the risk-adjusted discount rate method and Monte Carlo simulation. The risk-adjusted method is the most commonly used method; however, Monte Carlo simulation is becoming increasingly popular. These four methods are discussed in the remainder of this chapter. The probability method is discussed first so that some of the necessary results from general probability theory that are required for other methods are presented at the same time.

13.3 Probability Method

The probability method estimates the risk of a venture by determining the cash flows for the project in different scenarios. As a result of the variability in the market value of the project's output and the costs of the materials and labour in the production process, the project's free cash flows are variable. The probability of expected outcomes can be established from the analysis of the performance of similar projects and from historical data on prices and costs. Risk is associated with the variability in the free cash flows produced by the project, in much the same way in which risk was associated with the variability of the value of an asset in Chapter 11. This means that the risk can be measured by the standard deviation or variance of the value of the project.

This risk is the *stand-alone risk* for the project. It is the risk as if the project is owned by shareholders who only invested in this project. This is in contrast to the risk that the project exposes the investors to if the investors already owned a well-diversified portfolio of assets. This risk is referred to as the *market risk*, or the *"beta" risk*.

Both the *NPV* of the project and the risk associated with it, as measured by the standard deviation of the *NPV*, are used to make a recommendation on the project using the probability method.

The *NPV* was given by Equation 6.1 and is repeated as the following equation.

$$NPV = \sum_{t=0}^{n} \frac{CF_t}{(1+k)^t} \tag{13.1}$$

where CF_t is the free cash flow in year t, and k is the discount rate.

The calculation of the expected value and the standard deviation are required. The expected value of a distribution of values, V_i, is the weighted average of the values. The weighting factor is the probability of occurrence, given by p_i. Thus the expected value is given by the following expression:

$$Expected\ value = \bar{V} = \sum_{i=1}^{N} p_i V_i \qquad (13.2)$$

The expected value can be written as $\bar{V}$, or as $E[V]$.

The standard deviation, σ, is one measure of the variability of the values about the mean and is given by the following expression:

$$Standard\ Deviation = \sigma = \sqrt{\sum_{i=1}^{N} p_i (V_i - \bar{V})^2} \qquad (13.3)$$

The variance is given by σ^2. The coefficient of variation, which is a useful measure to compare the variance of projects and cash flows of different sizes, is given by $\sigma / \bar{V}$. The calculation of the expected value and the risk are illustrated in the following example.

Example 13.1: Expected value and risk.

There are five anticipated outcomes for a project, shown in Table 13.1, each with its probability of occurence. Determine the expected value and the standard deviation.

Solution:

The calculation of the expected value, using Equation 13.2, is shown in Table 13.2.

The calculation of the standard deviation, using Equation 13.3, is shown in Table 13.3.

Thus the expected value is 1,550, and the standard deviation is 355.6.

Table 13.1 Anticipated outcomes for a project

Outcome	Probability	Value, V_i
1	0.1	1,000
2	0.2	1,200
3	0.4	1,500
4	0.2	2,000
5	0.1	2,100

Table 13.2 Calculation of the expected value

Outcome	Probability, p_i	Value, V_i	$p_i V_i$
1	0.1	1,000	100
2	0.2	1,200	240
3	0.4	1,500	600
4	0.2	2,000	400
5	0.1	2,100	210
Expected value			1,550

Table 13.3 Calculation of the standard deviation

Outcome	Probability, p_i	V_i	$p_i(V_i - \bar{V})^2$
1	0.1	1,000	30,250
2	0.2	1,200	24,500
3	0.4	1,500	1,000
4	0.2	2,000	40,500
5	0.1	2,100	30,250
Standard Deviation			355.6

The difficulty in calculating the project's risk arises from the possibility that the cash flows in different years may be correlated with one another. Recall that this was the cause of the non-linearity in the calculation of the risk in portfolios of investments discussed in Chapter 11. The equation for the standard deviation of a combination or portfolio of assets was provided as Equation 11.6. This equation is relevant to the discussion here. The difference between that treatment and the treatment required here is that the previous discussion focused on the returns in one period for different projects, whereas in this discussion the focus is on the free cash flows in different years from the same project. However, the formula remains the same. The variance of the project's cash flows, σ^2, is given by the following expression:

$$\sigma^2 = \sum_i^N X_i^2 \sigma_i^2 + \sum_{i=1}^{N} \sum_{\substack{j=1 \\ j \neq i}}^{N} X_i X_j \sigma_{ij} \tag{13.4}$$

where σ_i^2 represents the variance of the cash flows in year i, and σ_{ij} represents the covariance between the cash flows in year i and those in year j. The factors X_i and X_j represent the weighting factors. For example, in Equation 11.6, the factors X_i and X_j represented that proportion of stock i or j that was held in the portfolio. In this discussion, they represent the proportion of the contribution of year i or year j cash flow to the present value of the project. In other words, they are the discount factors for each year. This means that Equation 13.4 can be rewritten as follows:

$$\sigma^2 = \sum_i^N \left(\frac{1}{(1+k)^i} \right)^2 \sigma_i^2 + \sum_{i=1}^{N} \sum_{\substack{j=1 \\ j \neq i}}^{N} \frac{1}{(1+k)^i} \frac{1}{(1+k)^j} \sigma_{ij} \tag{13.5}$$

where k is the discount rate. The covariance, σ_{ij}, can be expressed as $\sigma_i \sigma_j \rho_{ij}$, where ρ_{ij} is the correlation coefficient. Thus Equation 13.5 can be expressed as follows:

$$\sigma^2 = \sum_i^N \frac{1}{(1+k)^{2i}} \sigma_i^2 + \sum_{i=1}^{N} \sum_{\substack{j=1 \\ j \neq i}}^{N} \frac{1}{(1+k)^i} \frac{1}{(1+k)^j} \sigma_i \sigma_j \rho_{ij} \tag{13.6}$$

It is useful to examine three limiting cases of Equation 13.6. These are the following: (i) the cash flows in each year are uncorrelated, that is, they are independent; (ii) the cash flows are perfectly correlated; and finally, (iii) the cash flows are negatively correlated. Each of these cases will be discussed in turn.

(i)   *Independent cash flows*. If the changes in the cash flows in each year are independent of one another, then the correlation coefficient is equal to zero, that is, $\rho_{ij} = 0$, and Equation 13.6 reduces to the following:

$$\sigma^2 = \sum_{n}^{N} \frac{1}{(1+k)^{2n}} \sigma_n^2 \qquad (13.7)$$

The calculation of the project's risk for the situation in which the cash flows in each year are independent of one another is demonstrated in the following example.

Example 13.2: Independent cash flows.

Determine the project's standard deviation on the basis of the assumption that the cash flows in each year are independent of each other. The cash flow profile of the project is given in Table 13.4. The symbol CF_t represents the cash flow in year t. The discount rate is 10%.

Solution:

The first step is to calculate the mean and variance of the cash flows for each year. The second step is to apply Equation 13.5 to determine the project's variance or standard deviation.

The calculation of the mean and variance of the cash flows for each year is performed by applying Equations 13.2 and 13.3 to the cash flows for each year. The calculation of the mean is shown in Table 13.5.

The expected *NPV* is the *NPV* of the expected cash flows in each year. This means that the expected *NPV*, $E[NPV]$ is calculated as follows:

$$E[NPV] = \frac{CF_0}{(1+k)^0} + \frac{CF_1}{(1+k)^1} + \frac{CF_2}{(1+k)^2} + \frac{CF_3}{(1+k)^3}$$

$$E[NPV] = \frac{-3,500}{(1+0.1)^0} + \frac{1,275}{(1+0.1)^1} + \frac{1,093}{(1+0.1)^2} + \frac{2,258}{(1+0.1)^3} = 259$$

The calculation of the variance is given in Table 13.6.

The calculation of the project's variance is obtained as follows:

$$\sigma^2 = \frac{\sigma_0^2}{(1+k)^{2(0)}} + \frac{\sigma_1^2}{(1+k)^{2(1)}} + \frac{\sigma_2^2}{(1+k)^{2(2)}} + \frac{\sigma_3^2}{(1+k)^{2(3)}}$$

$$\sigma^2 = \frac{0}{(1+0.1)^0} + \frac{1145}{(1+0.1)^2} + \frac{241}{(1+0.1)^4} + \frac{3776}{(1+0.1)^6} = 3,242$$

Table 13.4 Cash flow profile for a project with different outcomes

Outcome	Probability	Capital cost	CF_1	CF_2	CF_3
1	0.1	−3,500	1,200	1,100	2,400
2	0.2	−3,500	1,240	1,120	2,300
3	0.4	−3,500	1,290	1,090	2,200
4	0.2	−3,500	1,300	1,080	2,250
5	0.1	−3,500	1,310	1,070	2,280

Table 13.5 Calculation of the mean of the project's possible outcomes

Outcome, i	Probability, p_i	CF_0	CF_1	CF_2	CF_3	$p_i CF_{0,i}$	$p_i CF_{1,i}$	$p_i CF_{2,i}$	$p_i CF_{3,i}$
1	0.1	−3,500	1,200	1,100	2,400	−350	120	110	240
2	0.2	−3,500	1,240	1,120	2,300	−700	248	224	460
3	0.4	−3,500	1,290	1,090	2,200	−1,400	516	436	880
4	0.2	−3,500	1,300	1,080	2,250	−700	260	216	450
5	0.1	−3,500	1,310	1,070	2,280	−350	131	107	228
Mean						−3,500	1,275	1,093	2,258

Table 13.6 Calculation of the variance of the project's outcomes

Outcome, i	Probability, p_i	$p_i(C_{1,i} - \overline{CF_1})^2$	$p_i(C_{2,i} - \overline{CF_2})^2$	$p_i(C_{1,i} - \overline{CF_3})^2$
1	0.1	562.5	4.9	2,016.4
2	0.2	245	145.8	352.8
3	0.4	90	3.6	1,345.6
4	0.2	125	33.8	12.8
5	0.1	122.5	52.9	48.4
σ_n^2		1,145	241	3,776

The standard deviation, σ, is 56.9. The expected value of the *NPV* is equal to 259. The coefficient of variance, given by the standard deviation divided by the expected value, is 0.21.

(ii) *Perfectly correlated cash flows.* If the cash flows are perfectly correlated, then the correlation coefficient is equal to one, that is, $\rho_{ij} = 1$, and Equation 13.6 reduces to the following:

$$\sigma^2 = \sum_{i}^{N} \frac{1}{(1+k)^{2i}} \sigma_i^2 + \sum_{i=1}^{N} \sum_{\substack{j=1 \\ j \neq i}}^{N} \frac{1}{(1+k)^i} \frac{1}{(1+k)^j} \sigma_i \sigma_j \qquad (13.8)$$

(iii) *Negatively correlated cash flows.* If the cash flows are negatively correlated, then the correlation coefficient is equal to minus one, that is, $\rho_{ij} = -1$, and Equation 13.6 reduces to the following:

$$\sigma^2 = \sum_{i}^{N} \frac{1}{(1+k)^{2i}} \sigma_i^2 - \sum_{i=1}^{N} \sum_{\substack{j=1 \\ j \neq i}}^{N} \frac{1}{(1+k)^i} \frac{1}{(1+k)^j} \sigma_i \sigma_j \qquad (13.9)$$

The probability method defines the risk as the standard deviation of the project values. It is therefore the project risk on a stand-alone basis. This is important, because this risk does not account for any of the possible portfolio effects, that is, as part of a company, or as part of the market. As mentioned at the beginning of this section, it is the risk to the owner of the project as if that owner had only one investment.

The two methods that are discussed next adjust the *NPV* directly to account for risk. The risk-adjusted discount rate method scales the discount rate depending on the assessment of the risk of the project. The certainty equivalent method adjusts the cash flows to the certainty equivalents and the *NPV* is evaluated using these certainty equivalent cash flows discounted at the risk-free rate. The adjustment of the discount rate based on the risk assessment for the project is discussed in the next section.

13.4 Risk-adjusted Discount Rate

The discount rate may be altered to account for risk that has not already been included in its calculation. The risk-adjusted discount rate, *RADR*, is given by the following expression:

$$RADR = Discount\ Rate + Project\ Risk\ Premium \qquad (13.10)$$

The components of the right-hand side of Equation 13.10 can be separated into their components. Consider a project that has no risk where all the cash flows in the future are known with certainty. In a project such as this, the discount rate needs to account for only one factor: the value of time. Since the project's free cash flows do not occur now, the investor does not have access to them, and because of this timing they are less valuable now to him or her. The discussion leading to Equations 11.13 and 11.14 in Chapter 11 indicated that the appropriate discount rate to use if the project has no risk is the risk-free rate. In practice, the closest possible investment to a risk-free investment is a loan to the government on a short-term basis. These loans, usually for a period of three months, are known as treasury bills, and the interest offered by the government on these loans is taken as representing the risk-free rate.

For a normal project in the company, that is, a project that has the same risk profile as the company as a whole, the discount rate should be the risk-free rate plus a premium for the company's risk. Usually the company's discount rate is the company's weighted average cost of capital and the company's risk is incorporated in the WACC. For projects more risky than the company's normal risk, a risk premium needs to be added to this to account for the additional risk not accounted for by the company risk. The additional premium is called the project-risk premium. In this case, the risk-adjusted discount rate can be written as the following equation:

$$RADR = WACC + Project\ Risk\ Premium \qquad (13.11)$$

The net present value of the project is then obtained by discounting the project's cash flows at the risk-adjusted discount rate. This is expressed as follows:

$$NPV = \sum_{t=0}^{n} \frac{CF_t}{(1 + RADR)^t} \qquad (13.12)$$

If the *NPV* is greater than zero, then the project is recommended for approval.

The company-risk and project-risk premiums are discussed in the following sections.

13.4.1 Company-risk Premium

The discussion leading to Equation 13.10 indicated that the risk-adjustment to the discount rate could account for the company risk and the project risk. In order to examine the company risk in further detail, assume that the company is entirely funded by equity. In this case the weighted average cost of capital is the cost of equity, which can be described by the CAPM. This means that under these conditions the risk-adjusted discount rate is the same as the cost of equity described by CAPM, and is given by the following expression:

$$RADR = R_E = R_F + \beta(R_M - R_F) \tag{13.13}$$

where R_M is the return on the market as a whole, and β is the company's beta.

Equation 13.13 means that the risk premium added to the risk-free rate for company risk is a result of beta risk. Beta risk is the systematic risk, the risk with which a well-diversified investor is concerned. (Recall from Chapter 11 that the beta is a measure of the undiversifiable risk.) Other risks, from the viewpoint of the investor, are diversifiable, and as a result are of less concern. The main concern for a well-diversified investor is exposure to market risk, measured by beta. This means that, from an investor's point of view, there is no need to adjust for project risk unless the project exposes the investor to increased market risk. In other words, adjustment for the project is justified only if the project risk is significantly different from the average risk for the company's other projects.

13.4.2 Project-risk Premium

If a project has a risk that is significantly different from the risk of the "average" project in the company, it may be justified to adjust the discount rate to reflect this increased risk. Two ways of adjusting the discount rate for projects are discussed. The first method involves assessing the project in terms of project risk categories. The second is to determine a "project beta."

(i)   Project risk categories

Different categories of projects may have different risk premiums associated with them. The variability in the cash flows of projects can be used to categorize and rank projects. The coefficient of variation, which is a relative measure given by the standard deviation divided by the mean, is an appropriate measure because projects are often of different sizes. From these categories, a standard template for the risk premium of these different categories can then be established.

The application of this method is illustrated in the following example.

Example 13.3: Risk-adjusted discount rate.

New investments for the company have been grouped into three categories depending on their coefficient of variation. These categories are shown in the following Table 13.7.

The company is currently investigating a project with the cash flow profiles for five different scenarios shown in Table 13.8. (The symbol CF_t represents the cash flow in year t.) Determine the risk-adjusted value of the project and make a recommendation as to whether it should be accepted or not if the risk-free rate is 5% and the company risk premium is 5%.

Solution:

The coefficient of variation for the cash flows is calculated using Equation 13.3 and the identity $v = \sigma/\bar{V}$, where v is the coefficient of variation. The calculation of the coefficient of variation is shown in Table 13.9. The project's coefficient of variation falls in the second category of projects, so the risk premium associated with the project is the company risk premium plus 2%. If the risk free rate is 5%, and the company's risk premium is 5%, then the *RADR* for this project is 12%. The calculation of the *NPV* using a *RADR* of 12% is given in Table 13.10.

$NPV_i(12\%)$ is the net present value for outcome i evaluated using a discount rate or *RADR* of 12%. The expected risk-adjusted value is calculated from the probability of each outcome using Equation 13.1. The *NPV* is 473.6, and the project should be recommended for approval because it is positive.

An alternative, but equivalent, method of calculating the expected *NPV* is to determine the expected cash flow for each year, and then to use these values in Equation 13.6. This alternative is shown in Table 13.11.

The *NPV* is obtained from the cash flows given in Table 13.11 by the following calculation:

$$NPV = \frac{-1500}{(1+0.12)^0} + \frac{547.5}{(1+0.12)^1} + \frac{547.5}{(1+0.12)^2} + \frac{547.5}{(1+0.12)^3} + \frac{547.5}{(1+0.12)^4} + \frac{547.5}{(1+0.12)^5}$$

$$= 473.6$$

The *NPV* is the same as that calculated before. Since the *NPV* is greater than zero, the project is recommended for approval.

Table 13.7 Categories of investment

Investment category	Coefficient of variation, v	Risk premium
1	$v < 0.1$	1%
2	$0.1 < v < 0.2$	2%
3	$v > 0.2$	4%

Table 13.8 Cash flow profiles for a project under five different scenarios

Outcome, i	Probability, p_i	Capital cost	CF_1	CF_2	CF_3	CF_4	CF_5
1	0.1	−1,500	425	425	425	425	425
2	0.2	−1,500	475	475	475	475	475
3	0.4	−1,500	550	550	550	550	550
4	0.2	−1,500	610	610	610	610	610
5	0.1	−1,500	680	680	680	680	680

Table 13.9 Calculation of the coefficient of variation for the project

Outcome, i	Probability, p_i	CF_t	$p_i CF_{t,i}$	$p_i (CF_{t,i} - \overline{CF_t})^2$
1	0.1	425	42.5	1,500
2	0.2	475	95	1,051
3	0.4	550	220	2.5
4	0.2	610	122	781
5	0.1	680	68	1,755
Mean, $\overline{CF_t}$			547.5	
σ				71.3
$v = \sigma / \overline{CF_t}$				0.13

Table 13.10 Calculation of the *NPV* with a *RADR* of 12%

Outcome, i	Probability, p_i	Capital cost	CF_1	CF_2	CF_3	CF_4	CF_5	NPV_i (12%)	$p_i NPV_i$
1	0.1	−1,500	425	425	425	425	425	32	3.2
2	0.2	−1,500	475	475	475	475	475	212	42.4
3	0.4	−1,500	550	550	550	550	550	483	193.0
4	0.2	−1,500	610	610	610	610	610	699	139.7
5	0.1	−1,500	680	680	680	680	680	951	95.1
NPV									473.6

Table 13.11 Calculation of the expected value of the *NPV*

Outcome, i	Probability, p_i	Capital cost	CF_1	CF_2	CF_3	CF_4	CF_5
1	0.1	−1,500	42.5	42.5	42.5	42.5	42.5
2	0.2	−1,500	95	95	95	95	95
3	0.4	−1,500	220	220	220	220	220
4	0.2	−1,500	122	122	122	122	122
5	0.1	−1,500	68	68	68	68	68
Expected value		−1,500	547.5	547.5	547.5	547.5	547.5

(ii) Project beta and divisional cost of capital

If a company was to consider investing in a project or a new line of business that could have a significant impact on the company as a whole, then the effect of the new project on the discount rate must be determined. For example, consider a chemical manufacturer who has developed a biotechnology that will produce a pharmaceutical product. The company wishes to invest 30% of its capital in this new line of business. The stock's beta is 0.75, the market risk premium is 4.85% and the risk-free rate is 5%. The security market line for the company is illustrated in Figure 13.1.

Assume that this company has only equity capital. The return on this equity capital is calculated as follows:

$$R_E = R_F + \beta(R_M - R_F) = 5 + 0.75(4.85) = 8.6\%$$

Since the company is funded only by equity, the cost of capital is 8.6%. This value is the horizontal line labelled WACC on Figure 13.1. Projects with an *NPV* that is

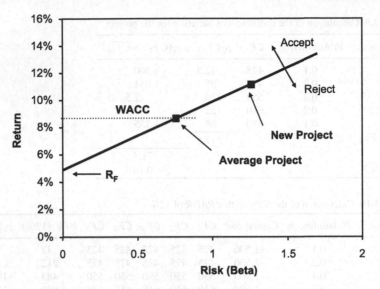

Figure 13.1 Adjusting the discount rate for project risk

positive when the value of WACC is used as the discount rate are recommended for acceptance provided that the risk is the same as that for an average project undertaken by the company. Alternatively, the project is recommended if the *IRR* is greater than the hurdle rate, provided that the risk is the same as the average risk for the company.

In this case, however, the risk of a pharmaceutical company is significantly different to the chemical company.

Suppose that the beta for pharmaceutical companies is higher than that for the chemical company as a whole. It is about 1.3 and labelled on Figure 13.1 as "New Project." The significant investment in the new pharmaceutical business will alter the company's beta from 0.75 to 0.86 calculated as a weighted average $(0.915 = (0.7)0.75 + (0.3)1.3)$. At this value of beta, the company's cost of capital is higher, and can be calculated as follows:

$$R_E = R_F + \beta(R_M - R_F) = 5 + 0.915(4.85) = 9.4\%$$

The average cost of capital for the company after the investment will be 9.4%. After the investment the company will have 70% of its assets in the chemical business and 30% in the pharmaceutical business. The cost of capital for the current business and the new business are related to the average cost of capital after investment by the following expression:

$$0.7(8.6) + 0.3(x) = 9.4\%$$

where x is the return for the pharmaceutical business.

The solution of this expression is that x is equal to 11.3%. This means that the required return for the investment in the new business is not the current cost of capital but significantly higher due to the increased risk. If the *NPV* of the pharmaceutical business is positive with a discount rate of 11.3%, the company should invest in this business. The project risk premium is therefore 11.3% less the company risk of 8.6%, that is, it is equal to 2.7%

It was argued in Chapter 11 that the basis for the decision on the inclusion of an asset in the portfolio was the "excess return to beta." In a nutshell, the excess return to beta determines whether the asset's returns are above or below the security market line (SML). If an asset is above the SML, the asset is included in the portfolio. If it is below the SML, then it is excluded. This analysis led to a method for the selection of the assets to be included in the portfolio and to the calculation of the weight that they should have in the portfolio. While not quite as extensive in its calculations, the method outlined above is in accordance with the use of the security market line as a selection criterion. The method outlined attempts to ensure that only those projects that are above the security market line are selected, in the same manner that the "excess return to beta" method does for portfolio selection.

The adjustment of the cost of capital for the project to account for the increased risk ensures that the project will only be selected if it lies above the security market line. This point is important: it is not only the return that is important in the selection of a project by a company, but also the risk. The method outlined above illustrates how the project return is to be adjusted depending on the risk of the project. Of course, if the beta is lower than the average risk for the company, the discount rate for the project should be lowered accordingly.

In this example, the average cost of capital has increased. However, does this mean that the projects in the chemical business must be evaluated at this new cost of capital? Not necessarily. The two lines of business have different risks associated with them, and the projects in these different businesses should be assessed with discount rates that are appropriate to their risk. This gives rise to the idea of *divisional cost of capital*.

The method outlined can also be used to calculate the cost of capital for different divisions of a company that are in different lines of business. The beta for each division can be obtained by comparison with other companies whose only line of business is the same as that of the division. This is called the "pure-play" approach because these other companies are companies with a single line of business. The calculation of the divisional cost of capital is then performed in the same manner as discussed above.

This section has discussed the methods for adjusting the discount rate to account for risk. In the next section a completely different approach is adopted, that is, adjusting the cash flows to account for risk. This is the certainty equivalent method.

13.5 Certainty Equivalent Method

The certainty equivalent method adjusts the cash flows for a project so that they represent a value that is known "for certain." The certainty equivalent value of each of the cash flows of the project can be assessed by posing the following question: at what price would management be willing to sell risky cash flows? The price received would be the certainty equivalent.

Two mathematical forms of the certainty equivalent value are discussed. In the first form, the cash flows are adjusted using multiplication factors, known as the certainty equivalent coefficients, and in the second form the cash flows are adjusted using subtraction factors, known as certainty equivalent risk premiums. The two methods are discussed in the following two sections.

13.5.1 Certainty Equivalent Coefficients

The certainty equivalent value of the project, CEV, is the NPV of the project cash flows adjusted so that these project cash flows can be regarded as certain. The certainty equivalent value can be expressed in the following form:

$$CEV = \sum_{t=0}^{n} \frac{\alpha_t CF_t}{(1 + R_F)^t} \tag{13.14}$$

where α_t is the factor that converts the risky cash flow in year t, CF_t, to a certain cash flow in year t, and R_F is the risk-free rate. The quantity $\alpha_t CF_t$ is the certainty equivalent of CF_t.

One way to think of the certainty equivalent is that $\alpha_t CF_t$ represents the minimum selling price of the cash flow CF_t. If the decision maker were offered more than $\alpha_t CF_t$, then she would sell it. If she was offered less than $\alpha_t CF_t$, she would hold it, that is, not sell it. If she were offered exactly $\alpha_t CF_t$, she would be indifferent to the decision to either hold or to sell. The cash flows that are used in the certainty equivalent method are not really values that will occur with certainty in future, because those are unknown until that time; rather, they are the values of the cash flows expected at the time that the owner is indifferent to either holding or selling. Thus the term "certainty equivalent" means at the point of indifference.

The certainty equivalent method attempts to separate the two issues that are bundled together as one in the NPV method. The discount rate in the NPV and the risk-adjusted NPV methods accounts for both the price of time and the risk of the venture (refer to Equation 11.13 and its interpretation, Equation 11.14). If the cash flows are known for certain, the discounting need only account for the loss of utility as a result of time. Since the price of time is given by the risk-free rate, certain cash flows should be discounted at the risk-free rate.

The decision rule for the certainty equivalent method is the same as that for the NPV rule: a project is recommended if the CEV is greater than zero.

The application of the certainty equivalent method is illustrated in the following example.

Table 13.12 Certainty equivalent coefficients

Investment category	Coefficient of variation	α_1	α_2	α_3	α_4	α_5
1	$v < 0.1$	0.95	0.92	0.89	0.85	0.80
2	$0.1 < v < 0.2$	0.90	0.85	0.82	0.75	0.70
3	$v > 0.2$	0.88	0.85	0.75	0.70	0.65

Table 13.13 Cash flow profiles for different scenarios for the project

Outcome, i	Probability, p_i	Capital cost	CF_1	CF_2	CF_3	CF_4	CF_5
1	0.1	$-1,500$	425	425	425	425	425
2	0.2	$-1,500$	475	475	475	475	475
3	0.4	$-1,500$	550	550	550	550	550
4	0.2	$-1,500$	610	610	610	610	610
5	0.1	$-1,500$	680	680	680	680	680

Example 13.4: Certainty equivalent coefficients.

A company is investigating a proposal for a new investment. The company has surveyed previous investments of this nature and categorized them on the basis of the coefficient of variation. These categories are shown in Table 13.12.

The value of α_t represents the certainty equivalent coefficient for year t. The project has the cash flows for different scenarios given in Table 13.13.

Determine the certainty equivalent value for this project and make a recommendation on whether it should be accepted if the risk-free rate is 7%.

Solution:

The solution to this problem proceeds in two steps. The first step is to determine the risk category for the project, and the second step is to calculate the certainty equivalent value.

The coefficient of variation is obtained from Equation 13.3 and the identity $v = \sigma / \bar{V}$, where v is the coefficient of variation, σ is the standard deviation and $\bar{V}$ is the expected value. The calculation of the coefficient of variation is demonstrated in Table 13.14 using the cash flows for the first year.

This means that the project falls into the second category. The values of the certainty equivalent coefficients, α_t, are those given in the table in the second category. The expected cash

Table 13.14 Calculation of the coefficient of variation

Outcome	Probability	CF_t	$p_i CF_{t,i}$	$p_i (CF_{t,i} - \overline{CF_t})^2$
1	0.1	425	42.5	1500
2	0.2	475	95	1051
3	0.4	550	220	2.5
4	0.2	610	122	781
5	0.1	680	68	1755
Mean, $\overline{CF_t}$			547.5	
σ				71.3
$v = \sigma / \overline{CF_t}$				0.13

Table 13.15 Calculation of the expected cash flow

	Capital cost	CF_1	CF_2	CF_3	CF_4	CF_5
Expected value	−1,500	547.5	547.5	547.5	547.5	547.5
Certainty equivalent coefficients		0.9	0.85	0.82	0.75	0.7

flows are calculated from the probabilities and the outcomes for each scenario. These values are given in Table 13.15.

The certainty equivalent value is calculated from Equation 13.14 given the certainty equivalent coefficients given in Table 13.15 as follows:

$$CEV = \frac{-1500}{(1+0.07)^0} + \frac{(0.9)547.5}{(1+0.07)^1} + \frac{(0.85)547.5}{(1+0.07)^2} + \frac{(0.82)547.5}{(1+0.07)^3} + \frac{(0.75)547.5}{(1+0.07)^4}$$
$$+ \frac{(0.7)547.5}{(1+0.07)^5} = 320.0$$

The *CEV* is equal to 320. Since the *CEV* is greater than zero, the project is recommended for approval.

13.5.2 Certainty Equivalent Risk Premiums

The cash flows were adjusted to their certainty equivalent values in the previous section using a multiplication factor called the certainty equivalent coefficient. The second form of the certainty equivalent method adjusts the cash flows by subtracting a certainty equivalent risk premium from the cash flow. This can be expressed as follows:

$$NPV = \sum_{t=0}^{n} \frac{CE_t}{(1+R_F)^t} = \sum_{t=0}^{n} \frac{CF_t - RP_t}{(1+R_F)^t} \qquad (13.15)$$

where CE_t is the certainty equivalent of the cash flow at year t, and RP_t is the risk premium associated with the cash flow at year t.

The relationship between the two methods for the certainty equivalent can be uncovered by equating terms at time t. This is given as follows:

$$\alpha_t CF_t = CF_t - RP_t \Rightarrow \alpha_t = 1 - \frac{RP_t}{CF_t} \qquad (13.16)$$

This means that because there is a unique relationship between the certainty equivalent coefficient and the certainty equivalent risk premium, the two methods are identical.

The certainty equivalent method as discussed here is more difficult to implement than the *NPV* method because it has introduced n new parameters in the form of the certainty equivalent factors or certainty equivalent risk premiums required for each year of the project's life. An alternative formulation is to determine a utility function and a risk tolerance. The utility theory can be use to determine the certainty equivalents without introducing n new parameters to the model formulation. Utility theory is discussed in Chapter 14 in the context of decision tree analysis.

The NPV discounts the estimated cash flows at the company's discount rate. The risk-adjusted discount rate changes the discount rate to account for the increased risk of the project compared to the risk of other projects with a discount rate of k. The certainty equivalent method adjusts the cash flows so that they are considered certain and discounts these cash flows at the risk-free rate.

It is worthwhile examining the relationship between the risk-adjusted discount rate and the certainty equivalent. This is discussed in the next section.

13.6 Relationship Between the RADR and the CE Methods

For a simple project that requires an investment I_0 now and produces a cash flow at the end of period one, the NPV for the project using the certainty equivalent and the risk-adjusted discount rate methods can be written as follows:

$$NPV = \frac{CE_1}{1+R_F} - I_0 = \frac{CF_1}{1+RADR} - I_0 \qquad (13.17)$$

where CE_1 represents the certainty equivalent of the expected cash flow at year 1, and CF_1 represents the anticipated free cash flow at year 1.

The rearrangement of this expression yields the following expression for the certainty equivalent value at year 1:

$$CE_1 = \frac{CF_1(1+R_F)}{(1+RADR)} = \frac{CF_1(1+R_F)}{(1+R_F+RP)} \qquad (13.18)$$

where the $RADR$ has been decomposed into a risk-free rate R_F and a risk premium RP.

The risk premium RP associated with the risk-adjusted discount rate can be obtained from Equation 13.18 and expressed in the following form:

$$RP = \left(\frac{CF_1}{CE_1} - 1\right)(1+R_F) = \left(\frac{1}{\alpha_t} - 1\right)(1-R_F) \qquad (13.19)$$

Equation 13.18 or Equation 13.19 allow the conversion between the risk-adjusted discount rate method and the certainty equivalent method. Equation 13.19 means that the risk-adjusted discount rate and the certainty equivalent methods are closely related to one another. It also means that the NPV calculated using the risk-adjusted discount rate and the CEV calculated using the risk-free rate should be equivalent.

It is important to emphasize that the calculation of the two methods should not be confused: the net present value is calculated either from the certainty equivalents discounted at the risk-free rate, or the *NPV* is calculated from the expected cash flows discounted at a risk-adjusted discount rate.

Although they represent two sides of the same coin, the risk-adjusted discount rate is the method that is more predominant of the two in practice. The methods that are popular in companies are discussed next.

13.7 Risk Adjustment Practices

Most companies use the weighted average cost of capital as the discount factor in their determination of the net present value. Bruner and colleagues (1998) surveyed the practices of 27 of the top companies, as well as the practices of a number of analysts and advisors. They report that most companies do not adjust for the risk of individual projects. The reason for this may be that the management of the capital allocation is decentralized and standardized practices need to be adopted to prevent abuse.

However, companies do determine different costs of capital for different divisions. In addition, companies also assess a risk premium for each country and adjust the cost of capital accordingly.

Interestingly, some advisors responded that they prefer to make adjustments to the cash flows than to the discount rate. In other words, rather than changing the discount rate for a particular project, the cash flows are altered to account for increased or decreased levels of risk. This emphasizes that the forecasting of cash flows and the assessment of risk is not an exact science, and the exercise of management's judgment is required.

The fourth and last method for the assessment of risk examined in this chapter is the Monte Carlo simulation method, which is discussed next.

13.8 Monte Carlo Simulation

Monte Carlo simulation is a general class of methods used to solve problems that are difficult to solve using other techniques. Monte Carlo simulation is a stochastic technique, which means that it relies on the generation of random numbers and probability theory to derive a solution to the problem.

The Monte Carlo method, when applied to discounted cash flow techniques, is similar to the probability method with the scenarios described in terms of the probability distributions of their inputs. This means that, instead of a discrete and limited number of outcomes as discussed in Section 13.3, the calculation is repeated for

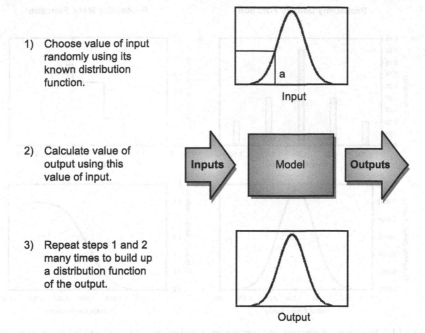

1) Choose value of input randomly using its known distribution function.

2) Calculate value of output using this value of input.

3) Repeat steps 1 and 2 many times to build up a distribution function of the output.

Figure 13.2 The calculation of the distribution of output values by the Monte Carlo method

a much larger number of scenarios. The scenarios are randomly chosen, hence the name "Monte Carlo," after the famous casino.

The Monte Carlo simulation method is illustrated by example. A number of scenarios were suggested in Section 13.3, and the probabilities of their outcomes, p_i, were stipulated. From this data, the overall mean value and the overall standard deviation were determined. An alternative method would have been to describe each cash flow by its mean and standard deviation, and from this data calculate the probability of the outcome. An implementation of this alternative is Monte Carlo simulation. The simulation method draws random samples from the probability distributions of the inputs to the model, shown as step 1 in Figure 13.2, and uses this random sample in the calculation of the output to the model, which is step 2 of Figure 13.2. This is repeated numerous times to determine the distribution of the output from the model. This is shown as step 3 in Figure 13.2. The distribution of the outcome provides the stand-alone risk of the project.

The methodology is illustrated in a number of examples that explain each of the steps in the following sections.

13.8.1 Discrete and Continuous Distributions

The discussion of the probability method earlier assumed that there were a discrete number of outcomes, each of which had a particular value. In other words, there

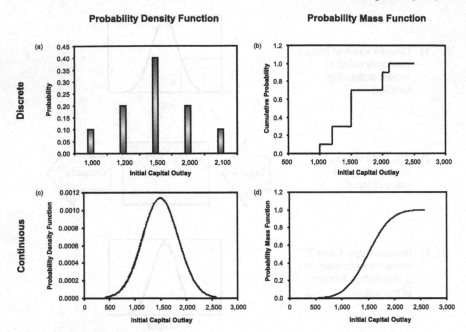

Figure 13.3 Examples of discrete and continuous distributions. Parts (**a**) and (**c**) are probability density functions, while (**b**) and (**d**) are probability mass functions.

are a countable number of outcomes. The probabilities of these discrete outcomes can be plotted against the value of the outcome in two different ways: either as the probability or as the cumulative probability. Such a plot of the probability is a plot of the probability density function, while that for the cumulative probability is a plot of the probability mass function. These two discrete functions are shown in Figure 13.3a and 13.3b.

The probability distributions can also be represented by continuous functions. The continuous probability density function $p(x)$ has the property that the area under the curve is equal to one.

The continuous probability mass function $P(x)$ represents the probability that a particular sample of a random variable falls below value of x. In other words, $P(x)$ is equal to the probability that X is less than x, where X is a random variable and x is the value of interest. The probability mass function has the property that it is zero at low values of x and is one at high values of x. The probability functions for continuous variables are illustrated in Figure 13.3c and 13.3d.

The relationship between the mass function, $P(x)$, and the density function, $p(x)$, are given in the equation:

$$P(x) = \int_{-\infty}^{x} p(y)\mathrm{d}y \tag{13.20}$$

This means that the probability mass function is the area under the probability density between the lowest possible value and the value of interest, x.

The probability that a random variable has a value between the values of a and b, $P(a < X < b)$, is given as follows:

$$P(a < X < b) = \int_{a}^{b} p(x)\mathrm{d}x = P(b) - P(a) \qquad (13.21)$$

The following example demonstrates continuous probability functions.

Example 13.5: Normal distribution.

The capital cost of a project is expected to be \$3,500 with a standard deviation of 10% of the mean. The distribution of the capital cost about the mean is described by a normal distribution. Draw the probability density and probability mass functions if the capital cost is normally distributed.

Solution:

The set of all possible outcomes is called the *sample set*, and an outcome is called a *sample*. The probability density function, $p(x)$, is given by the following expression if it is described by the normal distribution:

$$p(x) = \frac{1}{\sigma\sqrt{2\pi}} \exp\left(-\frac{(x-\mu)^2}{2\sigma^2}\right) \qquad (13.22)$$

where μ is the expected value or mean and σ is the standard deviation of the distribution. The substitution of the values into Equation 13.22 yields the following result for the probability at a capital cost of 3,000:

$$p(3,000) = \frac{1}{350\sqrt{2\pi}} \exp\left(-\frac{(3,000-3,500)^2}{2(350^2)}\right) = 0.000411$$

This calculation is repeated at different values of the capital cost, that is, x, in order to be able to plot the distribution. The normal distribution of these values is shown in Figures

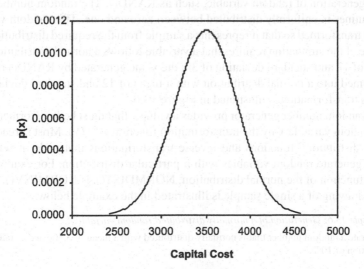

Figure 13.4 Probability density function of the capital cost

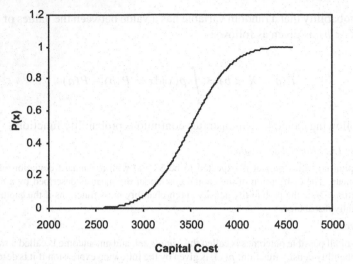

Figure 13.5 Probability mass function of the capital cost

13.4 and 13.5. The results in Figure 13.4 are a direct application of Equation 13.22, and illustrate the "bell-curve" by which the normal distribution is popularly known.

13.8.2 Drawing a Random Sample

Generating random numbers is a science of its own, particularly using a computer that by definition is a deterministic machine. Most spreadsheets supply a routine for the generation of random variable, such as RAND(). The random number from such routines is uniformly distributed between zero and one. This random variable must be transformed so that it represents a sample from the required distribution. For example, if the simulation requires that a variable follows a normal distribution with a mean of 12 and standard deviation of 30, the value generated by RAND() must be transformed into a normal distribution with a mean of 12 and standard deviation of 30. This transformation is illustrated in Figure 13.6.

The random number generator provides a value y that must be transformed to another random variable x by the transformation function $F^{-1}(y)$. Most spreadsheets provide distribution functions and inverse transformations that make it relatively easy to generate random variables with a particular distribution. For example, the inverse function of the normal distribution, NORMDIST(), is NORMINV().

The drawing of a single sample is illustrated in the example below.

Example 13.6: Generation of a normally distributed random variable.

Generate a random number that is normally distributed with a mean of 3,500 and a standard deviation of 10%.

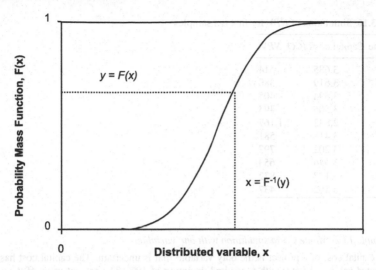

Figure 13.6 The transformation of the uniformly distributed variable y into the variable x that has a distribution $F(x)$

Solution:

The solution is obtained in the following steps:

(i)   Generate a random number between 0 and 1. For the discussion here, assume that RAND() provides a value of 0.419986.

(ii)   Substitute the random number and Equation 13.22 into Equation 13.20. This is given as follows:

$$0.419986 = \int_{-\infty}^{x} \frac{1}{350\sqrt{2\pi}} \exp\left(-\frac{(z-3500)^2}{2(350^2)}\right) dz \qquad (13.23)$$

(iii)   Solve Equation 13.23 for x. The spreadsheet function NORMINV() is used to do this. The result is a value of 3,429.32.

The random value that is generated is 3,429.32.

13.8.3 Monte Carlo Simulation of a Project with One Source of Uncertainty

In Monte Carlo simulation, the value of an uncertain variable is sampled randomly and repeatedly to create a large set of values. The required output, in this case the *NPV*, is calculated for each of this set of values, resulting in a mean (or expected value) and standard deviation for the required output. This procedure is illustrated in the following example.

Table 13.16 Simulation of *NPV* for first ten samples

Outcome	Capital cost, *ICO*	*NPV*
1	3,685	314
2	3,619	380
3	3,594	405
4	3,506	493
5	2,832	1,167
6	3,418	581
7	3,202	797
8	3,346	653
9	3,127	872
10	3,562	437

Example 13.7: Monte Carlo simulation with one variable.

The capital cost of a project is the only variable that is uncertain. The capital cost has an expected value of $3,500 with a standard deviation of 10%. The present value of the cash flows for all future years is $4,000. Determine the expected value and the standard deviation for the project.

Solution:

The *NPV* for the project is given by the following expression:

$$NPV = 4,000 - ICO \tag{13.24}$$

where *ICO* represents the initial capital outlay. In this case, *ICO* is the subject of some uncertainty, and is treated as a random variable. Since *ICO* is a random variable, the *NPV* is a random variable.

A random estimate for *ICO* is calculated by the same method as that illustrated in Example 13.6. In that example, the random number generator yielded a value of 0.419986, which was transformed into a normally distributed random variable with a value of 3429.32.

The *NPV* can be calculated from Equation 13.24 using this estimate of the *ICO*. Thus for this estimate, the *NPV* is 570.6. Now in order to perform a *Monte Carlo simulation*, this entire process is repeated for a large number of times. Table 13.16 shows the results for a simulation of the *NPV* for the first ten samples of the simulation.

The main point of the simulation is to obtain an estimate of the risk of the project, that is, the standard deviation of the *NPV*. This can be obtained by using Equation 13.3. The result, after 1000 simulations, is that the mean *NPV* (or expected *NPV*, $E[NPV]$) is 496.5 and the standard deviation is 347.53. The values are close to the mean of 500 and the standard deviation of 350. The *NPV* from the mean of the *ICO* gives a value of 500. The standard deviation of the *NPV* is attributable to the variance in the *ICO*, and since Equation 13.24 is additive, there should be no change in the standard deviation.

This project risk can be illustrated by plotting a frequency diagram. The range of values for the *NPV* is divided into classes, and the number of times a particular value of the *NPV* falls into a class is its frequency. The frequency of occurrence within each class for the *ICO* and the *NPV* are shown in Figures 13.7 and 13.8.

The coefficient of variance is a useful measure to quantify the risk. For this example, the coefficient of variance, $v = \sigma / \overline{NPV}$, is 0.699. The variation of the results about the mean is sufficiently large that in adverse conditions the *NPV* may even be negative.

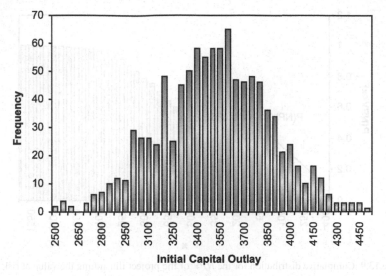

Figure 13.7 Frequency plot for the initial capital outlay, *ICO*

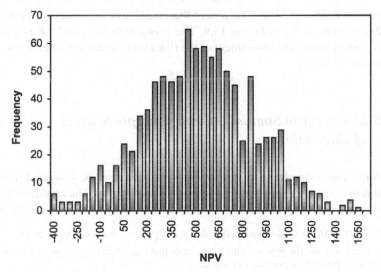

Figure 13.8 Frequency plot for the *NPV*

13.8.4 Value at Risk (VaR)

An alternative way of looking at the risk is to assess the chances of the project failing. This is the probability that the *NPV* is less than zero, that is, *P(NPV < 0)*. This probability can be determined from the results for the simulation. The cumulative frequency plot shows the sum of all the frequencies in all the classes with values below the current value. The cumulative frequency is the probability mass function for a discrete variable. It therefore represents the probability that a random variable *X* will be less than a certain value, that is *P(X < x)*. This is exactly what is required

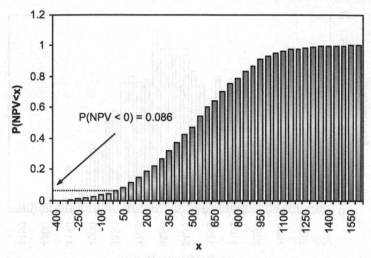

Figure 13.9 Cumulative distribution for the *NPV* of the project illustrating the value at risk

for the assessment of the risk. The plot of the cumulative frequency (or probability mass function) is shown in Figure 13.9. The probability that the *NPV* is less than zero is 0.086 or 8.6%. The assessment of the risk in this manner is sometimes called *value at risk* or *VaR*.

13.8.5 *Monte Carlo Simulation with Multiple Sources of Uncertainty*

The Monte Carlo simulation of the net present value can be extended to model situations in which there are multiple sources of uncertainty. The application of the method with two sources of uncertainty is illustrated in the following example.

Example 13.8: Monte Carlo simulation with two independent sources of uncertainty.

The capital cost and the present value of the cash flows generated by the project are both uncertain. If the *NPV* is expressed as follows:

$$NPV = PV(CF) - ICO \qquad (13.25)$$

then both the *ICO* and the *PV(CF)* are random variables, which makes the *NPV* a random variable.

If the capital cost has a mean of 3,500 and a 5% standard deviation, and the present value of the cash flows generated by the project, *PV(CF)*, has a mean of 4,000 and a standard deviation of 2.5%, determine the risk of the project.

Solution:

Since both sources of uncertainty are independent, the procedure outlined in the previous example can be repeated for both variables. The calculation for the first ten outcomes is shown in Table 13.17. The value for *p(IOC)* and *p(CF)* are the random numbers generated

Table 13.17 The first ten outcomes of the simulation

Outcome	p(ICO)	ICO	p(CF)	PV(CF)	NPV
1	0.5345	3,515	0.8667	4,111	596
2	0.5734	3,394	0.6129	4,029	635
3	0.7736	3,721	0.0760	3,857	136
4	0.1966	3,372	0.4612	3,990	618
5	0.9627	3,523	0.9982	4,291	768
6	0.2624	3,129	0.4448	3,986	857
7	0.6930	3,734	0.6771	4,046	312
8	0.7592	3,512	0.7992	4,084	572
9	0.9339	3,528	0.2280	3,925	398
10	0.7837	3,706	0.0912	3,867	161
Mean, $\overline{NPV}$		3,508.6		4,000.3	491.7
σ		171.3		100.7	197.4
$v = \sigma/\overline{NPV}$		0.049		0.025	0.402

by a numerical random number generator, such as the function RAND() found in most spreadsheets, and the values for *ICO* and *PV(CF)* are the calculations of the estimates for the capital cost and the present value of the cash flows, respectively, calculated in the same manner as steps (ii) and (iii) in the previous example. The calculations are repeated for 1,000 outcomes.

The mean and standard deviation for each of the random variables are shown in Table 13.16. As expected the mean and the standard deviation for the model variables are close to those specified. However, the standard deviation for the *NPV* has increased significantly. This is expected from Equation 11.6. For independent processes, the covariance between variables is zero, so that the variance for two independent variables is described by the following expression:

$$\sigma_{NPV}^2 = \sigma_{ICO}^2 + \sigma_{PV(CF)}^2 \qquad (13.26)$$

As a result, the coefficient of variation for the project indicates that the project risk is significant compared with its mean value. Overall, this project can be regarded as risky. A frequency plot of the simulation outcomes is shown in Figure 13.10. From this plot, it can be seen that even though the coefficient of variation is high, the risk of project failure, that is, of the project having a negative *NPV* is low.

The application of Monte Carlo simulation can be extended to multiple independent sources of uncertainty. This is illustrated in the following example.

Example 13.9: Monte Carlo simulation with multiple independent sources of uncertainty.
Consider the project financials shown in Table 13.18.

(i) If it is assumed that the revenues, costs and investments are independent random variables that are distributed normally, and that the revenues have a 10% standard deviation, while the costs and the investments have a 2.5% standard deviation, determine the standard deviation of the project.

(ii) If the same assumptions are used as those in part (i), except that the costs are 40% of the revenue, determine the standard deviation of the project.

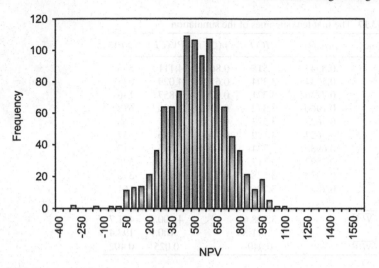

Figure 13.10 Frequency plot for the project

Table 13.18 Project financials for an opportunity

Year	0	1	2	3	4
Revenue	0	1,700	1,870	2,057	2,263
Cost	0	680	748	823	905
Cash flow	0	1,020	1,122	1,234	1,358
Investment	−3,500	0	0	0	0
Free cash flow	−3,500	780	1,122	1,234	1,358

(iii)   Determine if an equation could have been used that would have given the same solution.

The company's discount rate is 10%.

Solution:

(i) The methodology that is adopted is the same as that outlined in the previous two examples. The free cash flow in year 0 is dependent only on the investment made in that year, and is estimated directly. For years 1 to 4, the revenue and costs are treated as random variables and are estimated based on their mean values and their standard deviations.

The results of the first ten outcomes of the simulation are shown in Table 13.19. The symbols *FCF*, *R* and *C* represent the free cash flow, revenue and costs, respectively. The *FCF* for each of the years is simply given by $FCF = R - C - I$, where I represents the investment. The *NPV* can be calculated for each outcome from the following equation:

$$NPV = \sum_{t=0}^{n} \frac{FCF_t}{(1+k)^t} = \sum_{t=0}^{n} \frac{R_t - C_t - I_t}{(1+k)^t} \qquad (13.27)$$

The *NPV* corresponding to the first ten outcomes is given in Table 13.20. This table shows that the project has a simulated mean of 202 and a standard deviation of 185.

Table 13.19 First ten simulations of the cash flow profile for the project

Year	0	1			2			3			4		
Outcome	FCF	R	C	FCF	R	C	FCF	R	C	FCF	R	C	FCF
1	−3,491	1,544	663	881	1,917	757	1,159	1,948	797	1,151	2,176	936	1,240
2	−3,561	1,726	670	1,056	1,797	775	1,022	2,121	848	1,273	2,311	895	1,416
3	−3,610	1,702	712	990	1,782	772	1,010	1,862	824	1,039	2,286	916	1,369
4	−3,573	1,660	682	978	1,762	773	989	2,086	834	1,252	2,305	880	1,425
5	−3,506	1,735	664	1,071	1,968	758	1,210	1,980	814	1,166	2,261	912	1,348
6	−3,409	1,704	642	1,062	1,939	760	1,180	2,007	801	1,206	2,196	914	1,283
7	−3,643	1,589	682	907	1,797	755	1,042	2,041	819	1,221	2,122	916	1,206
8	−3,487	1,640	678	962	1,946	761	1,184	2,228	807	1,421	2,318	925	1,394
9	−3,482	1,691	674	1,017	1,833	741	1,092	2,045	833	1,212	2,253	899	1,354
10	−3,488	1,489	681	809	1,761	768	993	2,059	791	1,269	2,333	915	1,418
Mean	−3,505	1,697	680	1,017	1,869	748	1,121	2,058	823	1,235	2,263	905	1,358
σ	90	85	17	87	94	19	96	102	21	104	112	22	114

Table 13.20 First ten outcomes of the simulation of the NPV

Year	0	1	2	3	4	
Outcome		FCF	FCF	FCF	FCF	NPV
1	−3,491	881	1,159	1,151	1,240	−19
2	−3,561	1,056	1,022	1,273	1,416	168
3	−3,610	990	1,010	1,039	1,369	−160
4	−3,573	978	989	1,252	1,425	47
5	−3,506	1,071	1,210	1,166	1,348	265
6	−3,409	1,062	1,180	1,206	1,283	313
7	−3,643	907	1,042	1,221	1,206	−215
8	−3,487	962	1,184	1,421	1,394	385
9	−3,482	1,017	1,092	1,212	1,354	180
10	−3,488	809	993	1,269	1,418	−10
Mean	−3,505	1,017	1,121	1,235	1,358	202
σ	90	87	96	104	114	185

(ii) The assumption that the costs are 40% of the revenue means that the costs are no longer an independent variable, but are dependent on the revenue in a deterministic manner. This creates a new random variable, say $R - C$, which has a standard deviation of $(1-0.4)(5\%)$.

The calculations are repeated in the same manner as described above, except that the costs are not included as a random variable. The results for the first ten outcomes of the simulation are shown in Table 13.21.

These results indicate that there has been a significant reduction in the standard deviation of the NPV as a result of the assumption that the costs are a fixed percentage of the revenue.

(iii) Could these results have been obtained by the methods of Section 13.3? For part (i) Equation 13.7 can be used to determine the standard deviation:

$$\sigma^2 = \sum_{t}^{n} \frac{1}{(1+k)^{2t}} \sigma_{FCF,t}^2 \tag{13.28}$$

Table 13.21 First ten simulations of the *NPV*

Year	0	1	2	3	4	
Outcome		FCF	FCF	FCF	FCF	NPV
1	−3,615	975	1,171	1,244	1,428	148
2	−3,430	1,017	1,060	1,318	1,429	337
3	−3,644	1,117	1,167	1,176	1,437	201
4	−3,571	997	1,101	1,311	1,285	109
5	−3,503	975	1,116	1,178	1,314	89
6	−3,382	1,001	1,140	1,152	1,316	235
7	−3,583	998	1,098	1,221	1,342	66
8	−3,579	1,009	1,151	1,373	1,454	314
9	−3,590	1,057	997	1,299	1,373	108
10	−3,537	914	1,176	1,331	1,425	240
Mean	−3,497	1,021	1,122	1,234	1,358	214
σ	90	52	56	62	67	131

The standard deviation for the free cash flow for each year, $\sigma_{FCF,t}$, is composed of independent contributions from the revenues, costs and investments, which means that the annual variance, σ_t^2, is given by the following expression:

$$\sigma_{FCF,t}^2 = \sigma_{R,t}^2 + \sigma_{C,t}^2 + \sigma_{I,t}^2 \qquad (13.29)$$

where each of the terms on the right-hand side of Equation 13.29 represents the variances in year t for the revenues, costs and investments. For example, for year 1, the variance is calculated as follows:

$$\sigma_{FCF,1}^2 = (1700 \cdot 0.05)^2 + (680 \cdot 0.025)^2 = 7,514$$

The results of these calculations are summarized in Table 13.22.

The summation of the last line yields a standard deviation, σ, with a value of 180.26, which is the same as that provided by the Monte Carlo simulation.

The standard deviation for part (ii) is calculated in the same manner. For example, the variance for the first year is given by the following expression:

$$\sigma_{FCF,1}^2 = ((1700 - 680) \cdot 0.05)^2 = 2,601$$

The results of these calculations are summarised in Table 13.23.

The summation of the last line yields a standard deviation, σ, with a value of 127.5, which, as in the previous case, is the same as that provided by the Monte Carlo simulation. In effect,

Table 13.22 Calculation of the project's variance for part (i)

Year	0	1	2	3	4
Discount factor	1.000	0.909	0.826	0.751	0.683
Variance, $\sigma_{FCF,t}^2$	7,656	7,514	9,092	11,001	13,312
$\sigma_{FCF,t}^2/(1+k)^{2t}$	7,656	6,209	6,209	6,209	6,209

Table 13.23 Calculation of the project's variance for part (ii)

Year	0	1	2	3	4
Discount factor	1.00	0.909	0.826	0.751	0.683
Variance, $\sigma^2_{FCF,n}$	7,656	2,601	3,147	3,808	4,608
$\sigma^2_{FCF,n}/(1+k)^{2n}$	7,656	2,149	2,149	2,149	2,149

the Monte Carlo method as applied to these types of models has provided no additional insight that could not have been gained easily from the probability method.

13.8.6 Review of Assumptions

The results obtained from the use of Monte Carlo simulation, under the assumption that the variables are independent random variables, have not yielded any additional information that could not have been obtained by the probability method discussed in Section 13.3. The assumption that the variables are independent is part of the usual set of assumptions used in the application of Monte Carlo simulation to the evaluation of capital projects. Although this method has been popularised by software manufacturers, it is not clear that any advantage is gained other than the slight advantage of not having to use formulae.

It should be emphasised that this is not a criticism of Monte Carlo simulation, but of the assumption that the variables are independent. The variables are not independent, and often they are correlated or auto-correlated. Auto-correlation means that the value of a variable is dependent on its value in the previous period. These factors can be accounted for, but because they are not part of the set of assumptions usually used in this context, their incorporation into the Monte Carlo simulation has not been discussed here.

It must also be emphasised that under the usual assumption that variables are independent it is relatively easy to set up a spreadsheet to execute the Monte Carlo simulation. Specialised financial software is not required in this case.

13.9 Summary

Risk is associated with the variability or volatility of results. The sources of risk are the factors whose variability can influence the outcome. Thus, the variability in costs or sales price causes the earnings for a company to be variable, resulting in business risk.

As was seen in Chapter 11, the risk can be reduced or altered by combination with other risks. This is the basis for portfolio selection. Thus risk of a project can be viewed on a stand-alone basis, within the portfolio of other projects undertaken by the company, or from the view of the investor who has his or her own portfolio of assets that diversify the project's risk. Theoretically, it is the investor's viewpoint that is more important.

Four methods of accounting for risk were discussed in this chapter. They are the probability method, the risk-adjusted discount rate, the certainty equivalent method and Monte Carlo simulation. It was argued that Monte Carlo simulation is essentially the same as the probability method with some simplifying assumptions. It was also shown that the risk-adjusted discount rate and the certainty equivalent methods are related to one another.

13.10 Looking Ahead

The assumption has tacitly been made that management has no flexibility to alter the cause of events once the decision has been made to invest. This is not true. Management will move to minimise loss and capitalize on promising opportunities. There is a wide range of management actions that can impact on the value of project. In the next two chapters, the influence of management action and flexibility is discussed.

13.11 Review Questions

1. Should the discount rate be increased for periods in the future to account for the fact that cash flows in the more distant future are known with less certainty than ones expected closer to the present? Explain your answer.
2. Is the risk associated with beta different from project risk? Explain your answer.
3. What are the sources of project risk?
4. What factors can be adjusted to account for risk?
5. What is diversifiable and undiversifiable risk? What is market risk?
6. What are "diversified investors"?
7. Which risk is the most important to diversified investors? Explain your answer.
8. Discuss the following statement: "Corporate risk is the mid-point between stand-alone risk and beta risk."
9. Do the company's projects carry a risk that is different to the company as a whole?
10. Discuss the differences and similarities between the Monte Carlo method and the probability method.
11. Why determine a project beta?
12. How is the cost of capital for a division of a company to be determined?
13. Discuss the differences and similarities of the probability method and the Monte Carlo method.

14. What risk does the Monte Carlo method measure?
15. What is meant by "the usual assumption set of Monte Carlo analysis"?
16. Does a project's discount rate need to be adjusted if it is assessed to have a normal amount of risk compared to the company as a whole? Explain your answer.

13.12 Exercises

1. The anticipated outcomes for a project are shown in the table below. Determine the expected value and the standard deviation for the project.

Outcome	Probability	Value
1	0.1	100
2	0.3	150
3	0.3	162
4	0.2	200
5	0.1	110

2. The *NPV* for a project under different scenarios is shown in the table. Determine the expected *NPV* and the risk of the project.

Outcome	Probability	*NPV*
1	0.2	9
2	0.6	11
3	0.2	7

3. Why do the probabilities in Questions 1 and 2 add up to one?
4. Determine the project's standard deviation if the cash flow profile of the project is given in the table below. Assume that the cash flows are independent of each other.

Outcome	Probability	Capital cost	CF_1	CF_2	CF_3
1	0.1	−3,000	1,200	1,400	2,400
2	0.3	−3,500	1,240	1,600	2,300
3	0.3	−4,500	1,250	1,290	2,200
4	0.2	−3,500	1,300	1,380	2,250
5	0.1	−3,000	1,250	1,370	2,280

The discount rate is 10%.
5. The company's WACC is 10% and the CFO suggests that the country risk premium for a particular project in a foreign country is 3%. Determine the discount rate.
6. The company is considering a project for investment. The cash flow profile for the projects is given in the table below:

Year	Expected cash flow ($ millions)	Standard deviation ($ millions)
0	−55	−5
1	15	1.5
2	17	2.0
3	19	2.1
4	20	2.3

(i)   If the discount rate is 10%, determine the expected *NPV*.

(ii) If the risk free rate is 5%, determine the *NPV* using the certainty equivalent approach if the following certainty equivalent coefficients are applicable:

Year	0	1	2	3	4
Certainty equivalent coefficient	0.95	0.86	0.79	0.76	0.72

7. The company is currently investigating a project with the cash flow profiles for different scenarios shown in the table below.

Outcome	Probability	Capital cost	CF_1	CF_2	CF_3	CF_4	CF_5
1	0.1	−1,250	450	450	450	450	450
2	0.8	−1,500	500	500	500	500	500
3	0.1	−1,750	500	500	500	500	500

(i)   New investments for the company have been grouped into three categories depending on their coefficient of variation. These categories are shown in the table. Determine the risk-adjusted discount rate if the risk free rate is 4% and the company risk premium is 5%.

Investment category	Coefficient of variation, v	Risk premium
1	$v < 0.1$	1%
2	$0.1 < v < 0.2$	2%
3	$v > 0.2$	4%

(i)   Determine the expected *NPV* of the project.

8. A company that has no debt is considering investing in a new division.

(i)   If the risk free rate is 4%, the market risk premium is 5% and the company's beta is 1.3. Determine the return on equity.

(ii) The company wishes to invest 20% of its current value in a business with a beta of 0.8. Determine the company's return on equity.

(iii) Determine the return required on the new division.

9.   A company without debt has a division whose divisional discount rate is 12%.

   (i)   If the division constitutes 30% of the company's value, and the company's overall discount rate is 9%, determine the discount rate for the other divisions.

   (ii)   If the company's beta is 1.2 and the division's beta is 0.8, determine the beta for the other divisions.

10.   The cash flow profile for a project is shown in the table below. Also shown are certainty equivalent coefficients. If the risk-free rate is 5%, determine the expected *NPV*.

	Capital cost	CF_1	CF_2	CF_3	CF_4	CF_5
Expected value	−1,800	600	600	600	600	600
Certainty equivalent coefficients		0.9	0.85	0.82	0.75	0.7

11.   Determine the certainty equivalent risk premium for each of the cash flows given in the table in Question 11.

12.   Determine the risk premium added to the risk free rate in order to account for the risk of each of the cash flows given in Question 11.

13.   The capital cost estimate of $3 million is quoted as ±30%. Determine the probability of the cost being $3.5 million.

14.   Create a spreadsheet to implement a Monte Carlo simulation where the only uncertainty is the capital cost (initial capital outlay).

15.   Determine the project's standard deviation if the cash flow profile of the project is given in the table below. Assume that the cash flows are independent of each other.

Outcome	Probability	Capital cost	CF_1	CF_2	CF_3	CF_3
1	0.1	−10,000	6,200	6,400	6,400	6,200
2	0.3	−15,500	7,200	7,600	7,300	7,200
3	0.3	−14,500	6,200	6,200	6,200	5,000
4	0.2	−19,000	8,300	8,300	8,300	8,300
5	0.1	−20,000	9,300	8,300	5,200	5,200

The discount rate is 19%.

16.   A company has determined that the cash flows for a project correspond to the certainty values given in the table below:

Cash flow	−10,000	2,000	4,000	6,000	8,000	10,000
Certainty equivalent	−5,000	1,500	3,600	5,200	6,800	8,500

Determine certainty equivalent factor and the risk premium for each cash flow.

17. A company determines that the cash flows in the range up to $5 million are to be adjusted according the following table:

Amount, $ millions	Risk premium, $ millions
0	0
1	0.07
2	0.25
3	0.55
4	0.93
5	1.42

If the cash flow profile is given in the table below, and the risk free rate is 4% pa, determine the *NPV* of the project based on certainty equivalents.

Year	0	1	2	3	4	5
Free cash flow, $ millions	−12	3	4	5	5	5

Chapter 14
Decision Tree Analysis and Utility Theory

14.1 Introduction

An important factor in the assessment of a capital project is that managers will not set a course of action and then, come what may, stick to it. Depending on the circumstances or how events transpire, managers will react. For example, if the price of the business's product drops, managers can act to curtail costs. In all the methods previously discussed, it has been assumed that there is only one decision, the decision to invest or not. Once this has happened, the role of management is to wait patiently and passively for the returns to materialize. Of course, this is not the case; once the investment decision has been made, there are many management decisions that can be made and actions that can be taken by managers that may enhance the value of the project.

Investment and operational decisions that are dependent on the circumstances will be made at the appropriate point. However, some of the decisions that managers can make can be anticipated in advance. For example, a factory may be built with excess capacity that, if circumstances warrant, will allow the managers to increase production. The evaluation should include the possibility of increasing production if the demand for the factory's products is good.

A project may be executed in separate stages to allow for the possibility of changing circumstances. A staged investment might consist of a research project in which an initial commitment is made, and, if the results are favourable, the construction of a factory to make the product. The decision to build the factory will only be made once the results of the research are known. The decisions that are made after some uncertainty has been resolved are called contingent decisions.

Contingent decisions are those that are made after an event has transpired. The choice that is made depends on the outcome of the event. For example, if the sales are good, the managers can decide to increase production. If sales are bad, managers can reduce production. Contingent decisions provide managers with the flexibility to choose between different courses of action when different circumstances present themselves.

427

There are two ways to account for this flexibility in making decisions concerning capital investment projects. The first is the use of decision trees, and the second is the application of real option analysis. This chapter is devoted to the use of decision trees, while the next chapter explores the applications of real options analysis. Decision tree analysis can be used to analyse all types of decisions, not only contingent decisions. An important part of decision analysis using decision tree is to incorporate the decision maker's preference for risk. In this chapter, utility theory is used to describe this. The utility theory is used with decision tree analysis to determine the preferred choice or set of choices.

Decision tree analysis is examined in the first part of the chapter, after which utility theory will be discussed.

14.2 Decision Tree Analysis

14.2.1 Decision, Event and Terminal Nodes

A decision tree is a method for structuring and sequencing the decisions and their possible consequences. The decisions and events are arranged in the order or sequence in which they are expected to occur. The decision tree allows the decision problem to be more clearly visualized. It is a branching graph or diagram consisting of decision nodes, events nodes and terminal nodes that are arranged in the sequence in which the decisions and events occur. The decisions are denoted here by squares, the events by circles and the terminal nodes by triangles. Each of these nodes is discussed prior to discussing their combination as a decision tree.

The decision node represents a possible action by the decision maker to choose between various possibilities. The decision node in Figure 14.1 represents a choice between two possibilities: investing and not investing. The outcome of the decision made is the choice. A branch leading from the decision node represents each possible choice.

The event node, shown in Figure 14.2, represents the occurrence of a possible event. The event illustrated is the sales achieved by the company. The possible out-

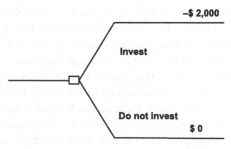

Figure 14.1 A decision node with a choice between two different options, the option to invest $2,000 and the option not to invest

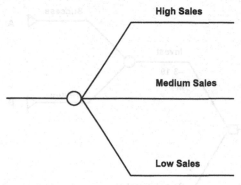

Figure 14.2 An event node with three outcomes

Figure 14.3 A terminal node

comes are high, medium and low sales. A branch leading from the event node represents each of these outcomes.

The first node on the decision tree is regarded as the root node. The final node of a particular branch is the terminal node. The terminal node is shown in Figure 14.3.

14.2.2 Basic Decision Trees

The decision tree is constructed by arranging decision, event and terminal nodes in the sequence in which it is anticipated that they will occur. The consequences of each event and each decision are clearly described in this manner. The basic techniques for the implementation of decision tree analysis are illustrated in the following two examples:

Example 14.1: Investment decision tree.

The company has an opportunity to invest in a project at a cost of $10 million. There are two possible outcomes: either the project is successful, and has a net present value (*NPV*) of $10 million, or it is a failure and has a net present value of –$10 million.

Represent this sequence of decisions and events as a decision tree.

Solution:

The sequence is that a decision to invest is made before an event occurs that determines the outcome. The decision tree for this investment opportunity is given in Figure 14.4.

The structure of the decision problem, that is, the sequence of events and the consequences of decisions, is made clear from the decision tree diagram. The company must decide to invest. If they decide to invest, there are two possible outcomes, either a successful outcome or a failure outcome.

The sequence of decisions and events can be combined repeatedly in order to represent the decision problem. A multiple decision and event problem is illustrated in the following example.

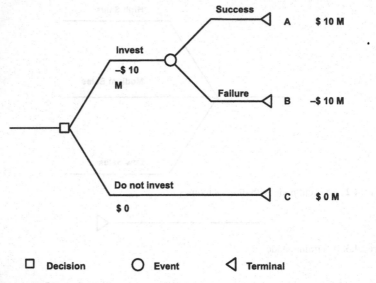

Figure 14.4 Investment decision tree

Example 14.2: Decisions in a licensing opportunity.

A company has the opportunity to invest $2 million in a research and development project. If the project is successful, the company may apply for a patent. If the patent is granted, the company may decide to build a factory at a cost of $10 million or they can license the technology. If the factory project has high sales, it will have an *NPV* of $20 million. If it has low sales, it will have an *NPV* of $2 million. The *NPV* of licensing the technology is estimated to be $10 million.

Represent this sequence of decisions and events as a decision tree.

Solution:

The decision tree for this set of decisions is shown in Figure 14.5.

The decisions are separated by an event in each case. The outcome of the event influences the choice of the next decision. There are seven ultimate outcomes, labeled A through G in Figure 14.5. The values of these final outcomes are also represented.

Although the decision tree provides a structured manner to view the sequence of decisions, it is still not apparent which of the choices, or sets of choices, is preferred. In order to provide an answer to that problem, the probabilities of the different outcomes are required. This is discussed in the next section.

14.2.3 Events and Probabilities

The outcome of an event node in a decision tree has a certain probability of occurring. For example, the probability of the sales being high for the event shown in Figure 14.2 is 10%. If the outcomes shown in an event are exhaustive, then, by the definition of probability, the probabilities for all the branches should add up to one. The event node of Figure 14.2 is shown in Figure 14.6 with the probability of occurrence for each outcome. The symbol p is used for probability. Thus, there are

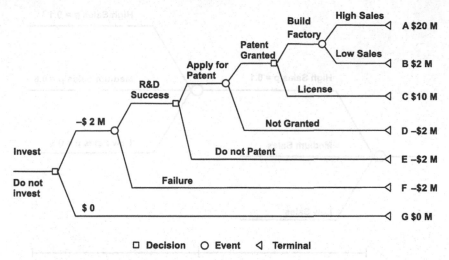

Figure 14.5 Contingent decisions for the licensing decision

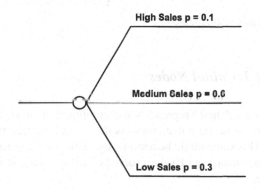

Figure 14.6 An event with the probability of occurrence

three outcomes with probabilities of 0.1, 0.6 and 0.3. The sum of the probabilities for these three outcomes adds up to one.

It is important to understand that the probability that is assigned to a particular branch of an event node is the probability of the occurrence of that particular outcome given that all the information prior to the event is known. Technically, these are known as conditional probabilities. For example, the probability of the sale being high, as shown in Figure 14.6, is 10% given that all the events and decisions that can lead to this event have occurred.

For example, consider the event nodes in Figure 14.7. The probability of the medium sales occurring in period two is 60% conditional on the sales being high in period one. The probability that all the events along a particular path occur is the joint probability. For example, the joint probability of sales being high in both periods is 0.01, which is the probability of the first event multiplied by the probability of the second event. The joint probability will be discussed in more detail later.

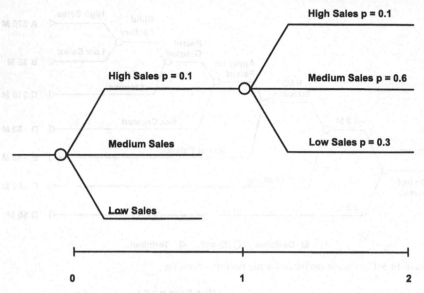

Figure 14.7 The conditional probability

14.2.4 Value of Terminal Nodes

The value of a terminal node represents the combination of all the rewards and expenses along the unique path that connects the root of the tree with the terminal node of that path. This value might be a monetary value, such as profit or net present value, or some other measure of value, such as "utility." This is illustrated in the following example.

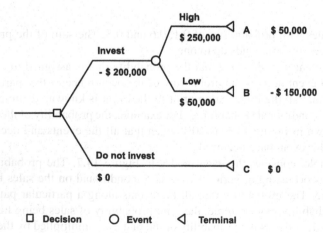

Figure 14.8 Calculation of terminal values

Example 14.3: Terminal values of a decision tree.

The company is considering investing in a new apparatus. The cost of the apparatus is $200,000. The value of the apparatus derived from its use can be either $250,000 or $50,000 depending on the future state of the economy. Draw the decision tree and determine the terminal values.

Solution:

The decision tree for this scenario is shown in Figure 14.8 below, indicating the terminal values for each of the possible paths through the decision tree.

The terminal value of path A is the value of the apparatus from its use in the case of the optimistic circumstances occurring less the cost of the investment. The terminal value of path B is the value of the apparatus derived from its use in the case of the pessimistic scenario less the cost of the investment. The terminal value of path C is zero because no revenues are received or costs incurred if the investment is not made.

14.2.5 Expected Value and Decision Trees

The expected value of the different options is used to determine which of them might be preferred. The concept of expected value was discussed in Chapters 13, and is calculated using Equation 13.2. If each of the outcomes has equal probability of occurrence, then the expected value is the same as the arithmetic average. In general, some of the outcomes are more probable than others, and the use of the expected value properly accounts for this.

The expected value of all of the scenarios represented in a decision tree can be obtained by calculating the value of the root node from the values of the terminal nodes. The expected value of the root node, and the paths that dominate, determine which decisions are favoured. The calculation of the root node that starts at the terminal node and moves successively through the tree to the root is sometimes referred to as "rolling back the decision tree." There are three rules associated with this process, one for each of the types of nodes. These rules are as follows:

(i)   Event node

The value of an event node is the expected value of all the outcomes. The value is obtained by the summation of the probability multipied by the value of each outcome. The expected value, $E[V]$, is given as follows:

$$E[V] = \sum_{i=1}^{N} p_i V_i \tag{14.1}$$

where p_i is the probability of the outcome with value V_i. This means that if there are two outcomes that have values of 20 and 40, and their respective probabilities of occurrence are 0.6 and 0.4, then the value of the event node is equal to $28 = 20(0.6) + 40(0.4)$.

(ii)   Decision node

The decision-maker will choose the option that has the highest value. Thus, the value of a decision node is the maximum value of the branches at that decision node.

(iii)   Terminal node

The value of a terminal node is the sum of all the rewards and expenses along the unique path that connects the root of the tree with the terminal node. This was discussed in Section 14.2.4 above.

The application of these rules to determine the value of the root node is illustrated in the following example.

Example 14.4: Rolling back the decision tree.

A company has an opportunity to invest in a project at a cost of $10 million. There are two possible outcomes if the company decides to invest: either the project is successful with a probability of 0.3, and has a net present value of $10 million, or it is a failure and has a net present value of –$3 million with a probability of 0.7. Calculate the expected value for the decision tree.

Solution:

The decision tree for the investment opportunity is given in Figure 14.9.

From the problem description, the terminal nodes at A, B and C have values of $10 million, –$3 million and $0, respectively. Consider the node at point 2. There are two outcomes from the event node, so the value at this node is the expected value of the outcomes. The calculation of the expected value using Equation 13.2 to the events at point 2 is given as follows:

$$E[V] = p_1 V_1 + p_2 V_2 = 0.3(10) + 0.7(-3) = \$0.9 \text{ million}$$

Now, consider the decision node at point 1. The value of the "invest" branch is $0.9 million, while that of the "do not invest" branch is zero. Therefore, the invest branch dominates, so that the value at point 1 is $0.9 million. The preferred choice at node 1 is to invest because the value of the "invest" option is higher than the value of the "do not invest" option. The preferred choice can be shown on the decision tree as the negation lines on the "do not invest" branch.

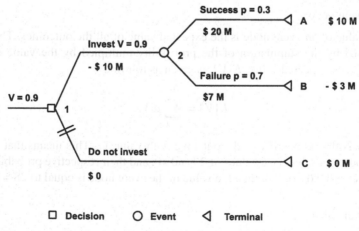

Figure 14.9 The decision tree for the investment opportunity

14.2.6 Net Present Value

The net present value is easily incorporated into the decision tree analysis. The net present value is determined at each terminal node and then the method of "rolling back" the decision tree is performed in terms of these terminal values. Consider the following example.

Example 14.5: Decision tree with three time periods.

A company can invest $1,500,000 in a new product. After the short implementation phase, the company will manufacture and sell the product. In the first year of operation it is estimated that there is a 50% chance of high sales yielding an after tax cash flow of $1,000,000 and a 50% chance of low sales, yielding an after tax cash flow of $200,000. In the second year of operation, there are four possibilities. If the sales were high in the first period, the sales in the second period may be high, yielding $1,300,000 with a probability of 80%, or low, yielding $600,000 with a probability of 20%. If the sales in the first period were low, the sales in the second period may be high, yielding a cash flow of $500,000 with a probability of 30%, or they may be low, yielding a cash flow of $200,000 with a probability of 70%.

Determine the expected value of the project if the discount rate is 12%.

Solution:

The decision tree for the project is illustrated in Figure 14.10 below.

Each of the possible routes to the possible final outcome is labeled A through D. The calculation of the net present value in each of these terminal nodes is shown in Table 14.1.

In order to determine the value of the decision tree and the optimal strategy, we need to roll back the decision tree. Consider node 3. The expected value at node 3 is equal to $0.32 = 0.8(0.429) + 0.2(-0.128)$. Similarly, the node at 4 has a value of $-1.09 = 0.3(-0.922) + 0.7(-1.162)$.

Rolling back to node 2, the value is equal to $-0.39 = 0.5 \times 0.32 + 0.5 \times (-1.09)$. Node 1 is a decision node, which means the choice is between not investing, or investing in a project

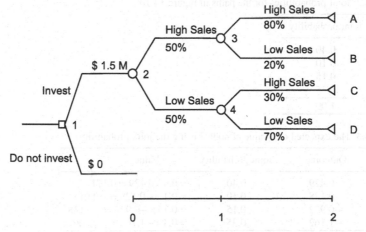

Figure 14.10 Decision tree for an investment with returns in different periods

Table 14.1 Calculation of the *NPV* for each of the terminal nodes

Path	Calculation	NPV
A	$-1.5 + 1/(1+0.12)^1 + 1.3/(1+0.12)^2$	0.429
B	$-1.5 + 1/(1+0.12)^1 + 0.6/(1+0.12)^2$	−0.128
C	$-1.5 + 0.2/(1+0.12)^1 + 0.5/(1+0.12)^2$	−0.922
D	$-1.5 + 0.2/(1+0.12)^1 + 0.2/(1+0.12)^2$	−1.162

whose expected net present value ($E[NPV]$) is –$0.39 million. The decision, in this case, is to not invest in this opportunity because the expected value of the *NPV* is negative. Therefore, the value at node 1 is zero.

14.2.7 *Joint Probability*

The joint probability is the probability of the outcome at the terminal node. For example, what is the probability that the path D represents the terminal outcome for the decision tree shown in Figure 14.10? It is the multiple of the probabilities along the length of the path. Thus, it is $0.5(0.7) = 0.35$. Since all of the outcomes are depicted in the decision diagram, the joint probabilities should add up to one. The joint probabilities for each of the paths in Figure 14.10 are given in Table 14.2 below.

The joint probability can be used to find the value at a node from which all the paths in consideration originate. For example, the node 2 in Figure 14.10 has four paths, labelled A to D, leading from it. The value of the node is the expected value of the outcome evaluated using the probability of each path. The probability of each path is the joint probability between the origin of the path and the end of

Table 14.2 Joint probabilities for the paths in Figure 14.10

Path	Joint probability
A	0.40
B	0.10
C	0.15
D	0.35
Total	1.00

Table 14.3 The expected of a value of node 2 using the joint probability

Path	Outcome	Joint probability	Value
A	0.429	0.40	$= 0.4 \times 0.429 = 0.171$
B	−0.128	0.10	$= 0.1 \times -0.128 = -0.013$
C	−0.922	0.15	$= 0.15 \times -0.922 = -0.138$
D	−1.162	0.35	$= 0.4 \times -1.162 = -0.406$
Expected value, $E[NPV]$			−0.386

the path. The calculation of the expected value from the joint probability is shown in Table 14.3.

The expected value of the *NPV* is −$0.386 million, as obtained before in Example 14.5.

14.2.8 Short-cut Notation

A decision tree can rapidly get very dense as the number of decisions and events increases. Detractors refer to it as "decision bush" analysis rather than decision tree analysis. Much of the power of the graph is lost because it becomes so complicated and visually unattractive. As a result, decision analysts use a short hand notation rather than drawing out the tree in full. This notation consists of arranging the nodes in sequence without linking them explicitly. An example of a decision followed by an event, another decision and another event is shown in Figure 14.11. The total number of paths in this fairly simple set of options and events would be 36, emphasising the need for a short-cut notation.

In the next section, the method of representing a decision in the form of a decision tree is used to analyse decision opportunities.

14.3 Decision Analysis

A decision strategy is a coherent set of choices that are made as events unfold. The decision tree framework is an ideal tool for representing the various strategies involving sequential decisions that might be adopted. The decision tree and the associated decision strategies can be integrated with other decision-making tools, such as the strategy table discussed in Chapters 2, and modelling tools, such as the influence diagram discussed in Chapters 9. The application of these tools together with decision trees is illustrated in the following two detailed examples. The first example provides the decision frame, and the second example uses this decision frame to determine the preferred strategy.

Example 14.6: Framing the decision for a mining development opportunity.

The senior management of Johnson Mining are considering their options to develop a mining property. The property, wholly owned by the company, is at Lyon, Western Australia, and the company's current processing facilities are in Queensland, on the other side of Australia, a country the size of the USA. The deposit in Queensland was exhausted years

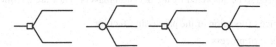

Figure 14.11 Short-cut notation for decision trees for a decision-event-decision-event sequence

ago, and since then the company has imported nickel ores from a variety of mines in the Philippines.

The deposit at Lyon is a nickel laterite. Laterites form a major source of nickel, but their processing can be difficult. Two processing routes are traditionally used: smelting and roasting. The Queensland operations use the roasting technology with downstream purification. Queensland's downstream purification is very efficient.

The smelting route has been excluded because tests have shown that this material from Lyon responds poorly.

In recent years, a new process has been developed, called high pressure leaching. The operating costs of this new technology are much lower than either of the other processes. Unfortunately, other operations using high pressure leaching, coincidently all of them in Western Australia, have failed for a variety of reasons. The main reason for these failures is downstream of the actual high pressure-leaching unit.

Engineers at Johnson have suggested a hybrid process, in which high pressure leaching is used, but, rather than processing this product further as the other unsuccessful operations have done, produce an impure nickel salt, which can be transported to Queensland for purification. This hybrid process will reduce the transport costs, while using what the engineers consider to be the best of the company's current operations and the best of the new technology.

The average grade of the nickel in the deposit is estimated to be between 1 and 2.5%. The extent of the ore body is estimated to be 200 million tonnes. Extensive sampling of the deposit has not been performed due to the costs involved. The mine capacity and the possible plant capacity have not been finalised. There are two options being debated internally, an option to produce 30,000 tons per annum (tpa) of nickel and a 60,000 tpa option. The efficiency of the conversion of nickel in the ore to nickel finally produced is about 70%.

The demand for nickel is about 1 million tonnes a year, increasing at about 4% per annum. However, the price is rather volatile. It has increased sharply during the last five years to about $20 per pound, but the long-term average has a range of between $2 and $5 per pound.

The company uses a discount rate based on its WACC and all decisions are based on *NPV*. The company values all mining investments, including operating plants, based on the mine having a life of twenty years. The financing of the operations is by using internally generated funds and by raising debt in the form of corporate bonds.

Determine the decisions to be made, and classify them using the decision hierarchy. Determine the key questions that need focus now. Determine the strategies that the company can adopt. Use the decision framing techniques outlined in Chapters 2.

Solution:

The first part of the framing exercise outlined in Chapters 2 is to unpack the issues into facts, decisions and uncertainties. Each sentence is examined as if it were an issue in an "issue-raising" meeting. The facts, uncertainties and decisions are listed in the Table 14.4.

The eight decisions listed in Table 14.4 can further be divided into categories of policy, strategic and tactical decisions. This division is shown in Figure 14.12.

The policy decisions are those decisions that have already been made. The strategic decisions are those that are the current centre of focus, while the tactical decisions are those for later implementation. The strategic decisions in the decision hierarchy represent the key questions that need to be answered by the decision makers. They are as follows:

(i)   The choice of location of the processing facility

(ii)   The choice of process

(iii)   The choice of capacity

Table 14.4 Facts, uncertainties and decisions for the nickel laterite project

	Item
Facts	1. Johnson Mining owns a nickel laterite property in Lyon, WA.
	2. The company owes a laterite processing facility in Queensland.
	3. The processing of laterites can be difficult.
	4. There are two major processing routes for laterites: smelting and roasting.
	5. A third processing route, "high pressure leaching," has been recently introduced.
	6. The new technology has had a number of failures.
Uncertainties	1. The capital and operating costs of each processing route.
	2. The nickel grade of the ore body.
	3. The size of the ore body.
	4. The conversion of nickel in the ore to nickel product.
	5. The market demand for nickel.
	6. The price of nickel.
Decisions	1. The process route.
	2. The processing location.
	3. The capacity of the mine and the processing plant.
	4. A new facility will not use the roasting process.
	5. WACC is used as the discount rate.
	6. The economic evaluation is based on *NPV*.
	7. The life of the project is twenty years.
	8. The form of financing for the project.

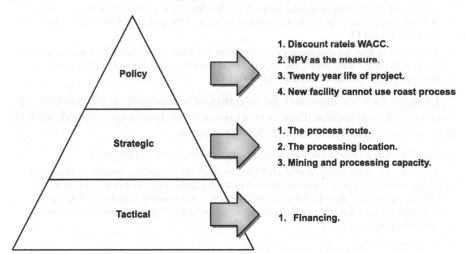

Figure 14.12 Decision hierarchy for the framing of the Johnson Mining laterite opportunity

In order to answer these questions, the company's analysts and engineers need to understand the key uncertainties and drivers for this opportunity. The three choices listed are not completely independent. If the company decides to use the Queensland facility, the process choice is governed by the process in place in Queensland, that is, the roasting process. If the

Table 14.5 Strategy table for strategic choices

Mining capacity	Processing route	Processing location
30,000 tpa	High pressure leaching Hybrid technology	Western Australia
60,000 tpa	Roasting	Queensland

Table 14.6 Four scenarios for the nickel laterite opportunity

Theme	Mining capacity	Processing route	Processing location	Rationale
Conservative	60,000 tpa	Roasting	Queensland	Use new deposit as feed to current plant – minimises capital and technology risk
Innovative	60,000 tpa	Hybrid	Western Australia	Minimizes transport costs
Baby steps	30,000 tpa	Hybrid	Western Australia	Minimizes risk
Business	60,000 tpa	High pressure leaching	Western Australia	Minimum capital and operating expenses

process choice is the new technology, choice of location must be Western Australia, since it would be illogical to transport highly unrefined material all the way to a new facility in Queensland.

The strategic decisions can be formulated into strategies using a strategy table. The strategic decisions form the headings of the columns of the strategy table. The items in each of the columns are the options available for each of the strategic decisions. A decision strategy is a coherent choice from the items available in each of the columns. The strategy table is shown in Table 14.5.

The choice of four decision strategies is shown in the Table 14.6. Each decision strategy is a coherent choice from among the options of presented in each column. The rationale for the strategy as a whole is also shown in Table 14.6.

Example 14.6 has illustrated the selection of scenarios from the available options in a coherent fashion. These scenarios can be analysed using methods such as decision trees. This is done in the following example.

Example 14.7: Evaluating the decision for a mining development opportunity.

The next step in the decision analysis of the mining and processing investment discussed in the previous example is the construction of an influence diagram which links the uncertainties and the decisions to the decision criterion. The influence diagram will be used to construct the decision tree, which links the uncertainties and the decisions in the order in which they will occur.

Solution:

The development of the influence diagram for this decision begins with the decision criterion, which has been specified by the company as net present value (*NPV*). The influence diagram is developed by asking what is needed in order to determine this factor. For example, the answer to this question is that the free cash flow and the discount rate are required for the calculation of the *NPV*. The company has specified WACC as the discount rate, so that the limb of the diagram ends with that policy decision. The components of the free cash flow are required, and these are the cash flow and the capital. The cash flow is determined from income and depreciation. The income in turn requires the revenue, the operating costs

(Opex) and the tax. The revenue requires the nickel production and the nickel price. The nickel price requires knowledge of the nickel market. This activity is continued until one of the uncertainties or a decision is reached. Some of the uncertainties are influenced by decisions. For example, the choice of location and process will influence, but not necessarily resolve, the uncertainties in the capital and operating costs. The completed influence diagram is shown in Figure 14.13.

The five uncertainties that are depicted in the influence diagram and are mentioned in Table 14.4 above are the nickel price, the nickel grade, the nickel market, the recovery, and the capital and operating costs. The influence diagram has broken the uncertainty of operating costs into its components, that is, the variable and fixed operating costs.

The market uncertainty affects the nickel price and it is assumed that this variable is the only one to consider. Nickel is a well-established commodity metal that is used mainly in the manufacture of another commodity, namely, stainless steel. The metal is traded on international exchanges, and traders and merchants temporarily store any excess production. The dynamics of this market are described by supply-demand economics, so that the impact on the company of any global shortfall or excess in production is registered in the price rather than in the sales volume. The company will always be able to sell its entire production; however, it is unsure of the price that it will receive.

One of the uncertainties mentioned in Table 14.4 is not present on the influence diagram. This is the size of the ore body. The reason for omitting it is the following. Management have determined that the life of the mine should be evaluated on a twenty-year basis. If the lowest estimated grade is used with the highest production rate over a twenty-year life, only 172 million tons will be mined. The size of the ore body exceeds this, which means that the policy of evaluation based on a twenty-year life implies that the size of the ore body does not affect the decisions on how to exploit this resource.

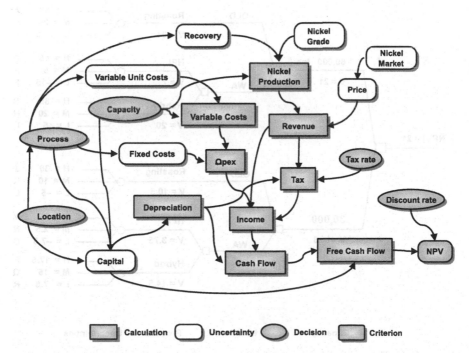

Figure 14.13 Influence diagram for the mining development opportunity

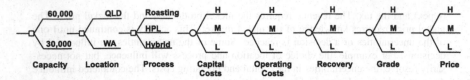

Figure 14.14 The decision tree for the mining development opportunity

The influence diagram can be used to construct a transparent model for the calculation of the decision criteria, in this case, the *NPV*. The influence diagram is the same for all the cases in this decision analysis. There may be cases in which the influence diagram may differ between scenarios.

The influence diagram can also be used to construct the decision tree. The order of the decisions and the uncertainties must be logically consistent with the problem and the decision frame. Once this order is established, the decision tree can be drawn. The decision tree is shown in Figure 14.14 in the short-cut notation discussed before.

The total number of possible paths is 2,916, some of which are redundant. For example, if the location is Queensland, only the roasting process is available. Removing this from the set of possibilities results in 1,215 valid paths. The probabilities that are assigned to each event are conditional probabilities given that the decisions and uncertainties before have been resolved. For example, the probability of the capital costs being higher than anticipated is conditional on the path that is currently being assessed, that is, given that the

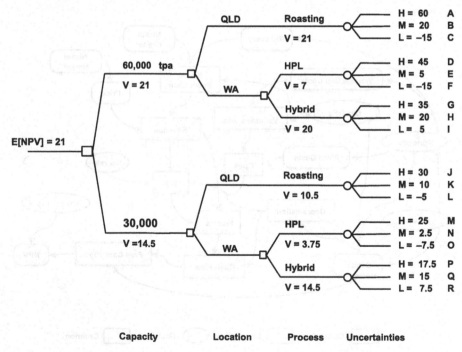

Figure 14.15 The modified decision tree for the mining development opportunity

Table 14.7 Joint probability for each path and the value for each path

Path	Probability	NPV	$p_i NPV_i$	Expected value at node
A	0.1	60.0	6.0	
B	0.8	20.0	16.0	
C	0.1	−10.0	−1.0	21
D	0.1	45.0	4.5	
E	0.8	5.0	4.0	
F	0.1	−15.0	−1.5	7
G	0.1	35.0	3.5	
H	0.8	20.0	16.0	
I	0.1	5.0	0.5	20
J	0.1	30.0	3.0	
K	0.8	10.0	8.0	
L	0.1	−5.0	−0.5	10.5
M	0.1	25.0	2.5	
N	0.8	2.5	2.0	
O	0.1	−7.5	−0.75	3.75
P	0.1	17.5	1.75	
Q	0.8	15.0	12.0	
R	0.1	7.5	0.75	14.5

hybrid process was chosen, given that the chosen location is Western Australia, and given that the chosen capacity is 60,000 tpa of nickel. Because of the large number of options and the extent of the data required, some assumptions need to be made in order to simply the calculations. It is commonly assumed that each event is independent of the others. This simplifies the assessment of the probabilities at each event node, and the calculations of the terminal values, the joint probabilities and the expected value of the NPV are performed as before. However, it must be borne in mind that this simplification may not be justified.

For the assessment of this problem, the event uncertainties have been collected into a single event uncertainty that affects the value of the project. This simplification is shown as the decision tree in Figure 14.15. The net present value for each of the uncertainty events has been evaluated, and the results are given in Figure 14.15 and in Table 14.7.

The method of the expected value at each of the six event nodes shown in Figure 14.15 is calculated in Table 14.7. These expected values are rolled back through the three decisions that must be made, that is, the choice of process, location and capacity. At decision nodes, the branch with the highest value dominates. The highest value branch is the choice of a capacity of 60,000 tpa, with operations in Queensland using the roasting process. The expected value of the NPV is $21 million. This is the option that is recommended.

The method outlined in the previous two examples has illustrated the steps in framing and analysing a decision problem. These same steps can be applied generally to complex decisions. As a result, the tools of decision analysis, that is, the decision hierarchy, the decision strategy table, the influence diagram and the decision tree are powerful methods that are applicable to any difficult decision situation, including capital projects. Another powerful concept in economics and decision analysis is that of utility theory, which is discussed in the next section.

14.4 Utility Theory and Risk

14.4.1 Utility

A decision maker needs to define a set of opportunities that exist and a measure of preference in order to make a decision. The set of opportunities are the alternatives that are available between which a choice is made, while the measure of preference is a rule for ranking the opportunities in terms of their desirability. It has been assumed that the decision maker prefers more wealth to less wealth, and a number of measures of "more wealth," such as *NPV* and *IRR*, have been presented. Two measures of preference were discussed in Chapters 11. These were wealth and risk, and the selection of an optimal portfolio from all the opportunities, called the "feasible set," was discussed. These two measures of preference led to the selection of a portfolio that was on the "efficient frontier."

Utility expresses a person's preference for the value and risk of a payoff. The payoff is a particular outcome or consequence. The utility is dependent on the payoff, but not necessarily in a linear fashion. It can be established by interviewing a decision maker about his or her choices for different levels of risk. This can be done using the options of different levels of risk, and is discussed below in the section on lotteries.

The utility can be illustrated by the following examples. Consider the following fair bet. If the value of a random number generator is greater than 0.5, I will pay you $5,000. If it is less than 0.5, you will pay me $5,000. The game will be played only once. It is a fair game with an expected value of zero. Would you play? Consider the following game. I will pay you $5,000 if the random number is greater than 0.5. If it is less than 0.5, you win nothing. The fair value of this game, calculated from its expected value, is $2,500. How much would you be willing to pay to play this game? These and similar games reveal a person's preferences and risk tolerance. These preferences can be expressed in mathematical form as a utility function.

14.4.2 Utility Function

The *utility function* is a mathematical function that transforms the payoff into a value of the utility. In other words, if the outcome or payoff is x, then the utility of that outcome is $U(x)$. Utility is a means of quantifying a person's preference for the payoff of one event over that of another event. As a result, it can incorporate not only quantifiable aspects, such as monetary value, but also aspects that are less quantifiable, such as attitudes towards risk.

The first property of utility is such that if a decision maker prefers outcome x_1 to outcome x_2, then the utility of x_1, given by $U(x_1)$, is greater than the utility of x_2, that is, $U(x_1) > U(x_2)$. Such a preference expresses the decision-maker's attitude towards risk.

The absolute values of the utilities are not important because they are only used to compare the utilities of opportunities for the same decision maker, and the decision maker only wants to know if the utility for one opportunity is greater than the utility for another opportunity. More formally, utilities are *ordinal* rather than *cardinal*. This means an arbitrary scale for utility can be chosen, and in this case, let it be between 0 and 1. It is assumed that the decision maker prefers more wealth to less wealth. As a result, the utility for the maximum outcome can be assigned a value of one, and the utility for the minimum outcome is zero.

A method to determine the values of the utility in between the two extremities is required. In order to do this, the certainty equivalent value of a lottery will be used.

14.4.3 Lotteries and Certainty Equivalents

A lottery is a gamble of the following form: the gambler is offered the chance of winning \$1,000 with a probability of 0.5, or winning nothing with a probability of 0.5. Instead of accepting the gamble, the gambler could also give up the opportunity of the lottery and accept an amount Z instead. The value of Z must lie between \$0 and \$1,000 because these are the extremities of the outcomes of the gamble. It cannot have a value greater than the maximum payout, and because the expected value of the lottery is \$500, it must be worth something greater than or equal to zero.

The interesting value of Z is the amount that makes the gambler *indifferent* to either playing the lottery or receiving the cash amount. This value of Z is called the certainty equivalent of the gamble. The certainty equivalent can be viewed as the minimum selling price of the lottery. If the gambler owned this lottery, for what amount of cash would the gambler be willing to sell the lottery? For example, the gambler may be willing to sell this lottery for \$400. In his view, the lottery has an expected value of \$500, which is risky, so accepting a value less than this makes sense to him. Another gambler may be less *risk averse*, and she would only be willing to part with the gamble for \$500.

The certainty equivalent, Z, is dependent on the gambler. A "conservative" gambler will assign a lower value to the gamble than a less conservative gambler. This means that the certainty equivalent, like the utility, expresses the preference of an investor (or gambler) for risk.

14.4.4 Expected Utility

Consider a more general lottery than those that have been discussed in the previous section. This general lottery has a number of payoffs $X_0, X_1, \ldots X_n$ that have the probabilities $p_0, p_1, \ldots p_n$ of occurring. The expected value of this lottery, $E[X]$, is given by the following expression:

$$E[X] = p_0 X_0 + p_1 X_1 + \dots + p_n X_n = \sum_{i=1}^{n} p_i X_i \qquad (14.2)$$

The utility of each of the outcomes can be expressed as $U(X_0), U(X_1) \dots U(X_n)$. The first property of the utility that is of interest is the expected utility. The expected value of the utility is defined in a similar manner to the expected value of the outcomes:

$$E[U] = p_0 U(X_0) + p_1 U(X_1) + \dots + p_n U(X_n) = \sum_{i=1}^{n} p_i U(X_i) \qquad (14.3)$$

The second property of the utility that is of interest in this discussion is the following: the expected value of the utility is equal to the utility of the certainty equivalent. Thus, if Z represents the value of the certainty equivalent, and $U(Z)$ represents the utility of the certainty equivalent, then the utility of the certainty equivalent is given by the following expression:

$$U(Z) = E[U] \qquad (14.4)$$

This is a powerful property that enables the evaluation of the utility function. In order to determine the utility of the values in the interior of the range, that is, between utilities of zero and one, the certainty equivalents are determined from lotteries using known outcomes with known utilities, and the utility is determined from Equation 14.4.

The following example demonstrates the use of Equation 14.4 and lotteries to determine the utility function for a decision maker.

Example 14.8: Decision-making using the utility function.

A decision that consists of a choice between two investments must be made. There are three possible outcomes once the choice has been made, each with equal probability of occurring. The value of the outcomes is shown in Table 14.8.

The expected value for each of these investments is given by:

$$E[A] = (1/3)100 + (1/3)50 + (1/3)5 = 51.6$$
$$E[B] = (1/3)90 + (1/3)70 + (1/3)6 = 55.3$$

Table 14.8 Value of two alternative investments with three different outcomes

Outcome	Investment A	Investment B
1	100	90
2	50	70
3	5	6

Based on the expected value of the outcomes, investment B is preferred because the utility of B is greater that A. However, this choice can also be made based on utility. The utilities for these outcomes are discussed next.

The utility of the extreme points in the range are known: the $U(100)$ is equal to one, and $U(5)$ is equal to zero. The values of the utility between these two extremities are required. They can be uncovered by offering the investor a lottery. The lottery is constructed from these known points. For example, the first lottery is one that is constructed from the extremities, that is, one that pays 100 with a probability of 0.5, and 5 with a probability of 0.5. The aim of interviewing the investor using the lottery is to determine the investor's certainty equivalent for this lottery. The investor responds with a value of 30. This means that the utility of an outcome of 30 can be determined, as follows:

$$U(30) = E[U] = p_5 U_5 + p_{100} U_{100} = 0.5(0) + 0.5(1) = 0.5$$

This means that the utility of an outcome of 30 is 0.5. The next interview question concerns the certainty equivalent of a lottery that pays 5 with a probability of 0.5, and 30 with a probability of 0.5. The investor responds with a value of 15. The utility of 15 is calculated as follows:

$$U(15) = E[U] = p_5 U_5 + p_{30} U_{30} = 0.5(0) + 0.5(0.5) = 0.25$$

Similarly, the investor indicates that the certainty equivalent for a lottery that pays 30 with a probability of 0.5 and 100 with a probability of 0.5 is 60, so that the utility of 60 is found by the following calculation:

$$U(60) = E[U] = p_{30} U_{30} + p_{100} U_{100} = 0.5(0.5) + 0.5(1) = 0.75$$

The utility function for an investor considering this type of investment is shown in Figure 14.16.

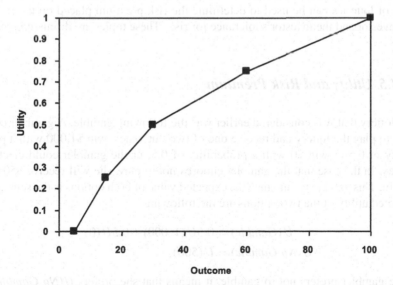

Figure 14.16 Utility function for the investor

Table 14.9 Utilities associated with outcomes from each investment

Outcome	Investment A	Investment B	Utility A	Utility B
1	100	90	1.00	0.94
2	50	70	0.66	0.81
3	5	6	0.00	0.025

The optimal choice between the two alternatives is made on the basis of the alternative that maximizes the utility. In other words, the alternatives are ranked by expected utility, and the alternative with the highest utility is the recommended choice.

The utilities for each of the outcomes of the two choices given in Table 14.9 were calculated from the results shown in Figure 14.16. These calculated utilities are given in Table 14.9. The utilities at points between the values shown on Figure 14.16 were evaluated by linear interpolation between the points.

The expected utility for investment A is calculated as follows:

$$E[U] = (1/3)U(100) + (1/3)U(50) + (1/3)U(5) = (1/3)1 + (1/3)0.66 + (1/3)0 = 0.55$$

and that for investment B is calculated as follows:n

$$E[U] = (1/3)U(90) + (1/3)U(70) + (1/3)U(6)$$
$$= (1/3)0.94 + (1/3)0.81 + (1/3)0.025 = 0.59$$

Thus, the investment B is preferred, in line with the choice that would have been made if the expected value of the outcomes had been used.

The ability to explore the investor's preferences in uncertain situations in the form of lotteries can be used to determine the risk premium placed on the risk by the investor and the investor's tolerance for risk. These topics are discussed next.

14.4.5 Utility and Risk Premium

The lottery that was considered earlier was the following gamble. A gambler could elect to play the lottery and receive one of two outcomes: win $1,000 with a probability of 0.5 or win $0 with a probability of 0.5; or the gambler could elect not to play. In the case that the gambler chooses not to play, she will receive $500 for certain. This gamble is fair since the expected value of both options is the same. The expected utility of the two options are the following:

$$U(Gamble) = 0.5U(1,000) + 0.5U(0)$$
$$U(No\ Gamble) = U(500)$$

If the gambler prefers not to gamble, it means that she prefers $U(No\ Gamble)$ to $U(Gamble)$, that is, $U(No\ Gamble) > U(Gamble)$. This can be expressed as follows:

$$U(500) > 0.5U(1,000) + 0.5U(0)$$

This expression can be rearranged as follows:

$$U(500) - U(0) > U(1,000) - U(500)$$

This implies that the change in utility over the interval [0, 500] is preferred to the change in utility over the interval [500, 1000]. The shape of the utility function is *concave*, similar to that found in Figure 14.16. The more that is at stake, the less likely she is to gamble. This means that she is risk averse, and hence risk aversion is described by a concave utility function.

The expected value of this lottery is given by the following expression:

$$E[X] = pX_0 + (1 - p)X_1 \tag{14.5}$$

where p is the probability of payoff X_0.

The expected value of the utility of the lottery is the utility of the certainty equivalent, given by the following expression:

$$E[U] = pU(X_0) - (1 - p)U(X_1) = U(CE) \tag{14.6}$$

A risk premium can be defined as the difference between the expected value of the lottery and the certainty equivalent in the following manner:

$$Risk\ Premium = E(X) - CE \tag{14.7}$$

If the risk premium is zero, the gambler is regarded as *risk neutral*, and the following expression holds:

$$U(E[X]) = U(CE) \tag{14.8}$$

The substitution of Equations 14.6 and 14.7 into Equation 14.8 yields the following expression:

$$U(pX_0 + (1 - p)X_1) = pU(X_0) + (1 - p)U(X_1) \tag{14.9}$$

The only way in which Equation 14.9 can be true is if the utility function, U, is a linear function of X. This is an important result: the utility function of a risk-neutral investor is a straight line.

If the risk premium is positive, the certainty equivalent is the expected value less the risk premium, that is, $CE = E[X] - RP$, where RP is the risk premium. Since the utility of $E[X]$ must be greater than the utility of $E[X] - RP$, the following relationship holds:

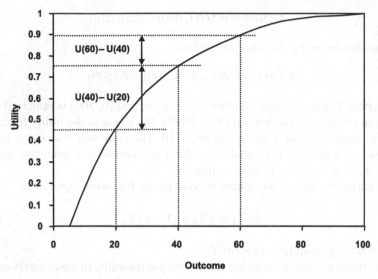

Figure 14.17 Marginal changes in utility for a risk-averse investor

$$U(E[X]) > U(E[X] - RP) \qquad (14.10)$$

This relationship in turn means that the utility of the expected value of X is greater than the utility of the certainty equivalent, since $CE = E[X] - RP$. This result is given in the following expression:

$$U(E[X]) > U(CE) \qquad (14.11)$$

The risk-averse investor or gambler will exchange the utility of $E[X]$ for the utility of CE in order to avoid the risk of the gamble.

Another way to look at risk-averse behaviour is to examine the marginal changes in utility. This is illustrated in Figure 14.17. The difference between $U(40)$ and $U(20)$ is greater than the difference between $U(60)$ and $U(40)$. This means that the loss of utility for a loss in wealth of 20 is greater than the gain in utility for a gain in wealth of 20. The prevention of loss is more important than the possibility of a risky gain for a risk-averse investor.

14.4.6 Exponential Utility Function

Any mathematical function that has a concave curvature can be used to describe the utility function. A form that is common is the exponential equation, which is given as follows:

$$U(x) = 1 - \alpha e^{-x/RT} \qquad (14.12)$$

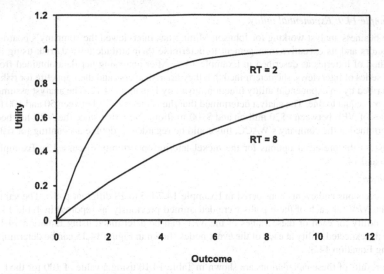

Figure 14.18 The effect of the risk tolerance on the utility curve with α

where α is a proportionality constant and RT is the risk tolerance. The risk tolerance measures the degree of risk to which the investor is willing to be exposed.

The function given in Equation 14.12 is plotted in Figure 14.18 for two values of the risk tolerance. As the risk tolerance increases, the utility function becomes more flat, consistent with the result obtained previously that the utility function for a risk neutral investor is a straight line.

The certainty equivalent can be obtained from the utility by solving Equation 14.12 for x. The certainty equivalent value, CE, for that utility is given by the following expression:

$$CE = -RT\ln\left(\frac{1 - U(x)}{\alpha}\right) \qquad (14.13)$$

The investor or decision-maker is indifferent to a lottery that has a utility of $U(x)$ and the certainty equivalent given by CE.

14.4.7 Using Utility to Account for Risk in NPV Calculations

The utility function can be used to determine the certainty equivalent value of a capital project using Equation 14.13. This method determines the net present value (discounted at the risk-free rate), and determines the certainty equivalent from the utility function. From a practical point of view, the method of utility has reduced the number of parameters required to implement the certainty equivalent method from n, one for each year in the study, to two, that is, α and RT.

This is illustrated in the following example.

Example 14.9: Exponential utility.

The business analyst working for Johnson Mining has interviewed the company's board of directors and its executive management to determine their attitude towards risk using the method of lotteries as described in Example 14.8. After analysing her data obtained from this series of interviews, she determined that they are risk averse and their appetite for risk is described by an exponential utility function, given by Equation 14.12. The analyst assumed that α is equal to one. The analyst determined that the value of RT is between 50 and 100 for values of NPV between −$20 million and $100 million. (Note that since the NPV has been determined at the company's WACC, this could be regarded as double accounting for risk.)

Reassess the preferred options for the nickel laterite opportunity discussed in Examples 14.6 and 14.7.

Solution:

The decisions and events considered in Example 14.7 led to 18 different paths. The values of the NPV for each of these paths were determined previously, as reported in Table 14.7. The utility for each of these values of the NPV can be determined using Equation 14.12, and the expected utility at each of the event nodes shown in Figure 14.15 can be determined using Equation 14.3.

The results of these calculations are shown in Table 14.10 using a value of 100 for the risk tolerance. This is the upper limit of the risk tolerance for the company. The decision remains unchanged from that recommended in Example 14.7, that is, the company should operate the mine at a capacity equivalent to 60,000 tpa of nickel and should ship the ore to the Queensland operation for processing using the current process (roasting). This is because the expected utility and the certainty equivalent at the node that leads to paths A, B, and C is greater than those values of any of the other paths.

The results for these calculations at the lower limit of the risk tolerance, that is, at a value of 50 for the risk tolerance, gives a different result. The results of the same set of calculations as those shown in Table 14.10, but with a risk tolerance of 50, are given in Table 14.11.

Table 14.10 Calculation of the expected utility and certainty equivalent at each of the event nodes using $RT = 100$

Path	Probability	NPV	Utility	$p_i U_i$	Expected U at node	Certainty equivalent
A	0.1	60	0.451	0.045		
B	0.8	20	0.181	0.145		
C	0.1	−10	−0.105	−0.011	0.1796	19.80
D	0.1	45	0.362	0.036		
E	0.8	5	0.049	0.039		
F	0.1	−15	−0.162	−0.016	0.0591	6.09
G	0.1	35	0.295	0.030		
H	0.8	20	0.181	0.145		
I	0.1	5	0.049	0.005	0.1794	19.77
J	0.1	30	0.259	0.026		
K	0.8	10	0.095	0.076		
L	0.1	−5	−0.051	−0.005	0.0969	10.19
M	0.1	25	0.221	0.022		
N	0.8	2.5	0.025	0.020		
O	0.1	−7.5	−0.078	−0.008	0.0341	3.47
P	0.1	17.5	0.161	0.016		
Q	0.8	15	0.139	0.111		
R	0.1	7.5	0.072	0.007	0.1347	14.47

Table 14.11 Calculation of the expected utility and certainty equivalent at each of the event nodes using $RT = 50$

Path	Probability	NPV	Utility	$p_i U_i$	Expected U at node	Certainty equivalent
A	0.1	60	0.699	0.070		
B	0.8	20	0.330	0.264		
C	0.1	-10	-0.221	-0.022	0.3115	18.66
D	0.1	45	0.593	0.059		
E	0.8	5	0.095	0.076		
F	0.1	-15	-0.350	-0.035	0.1005	5.30
G	0.1	35	0.503	0.050		
H	0.8	20	0.330	0.264		
I	0.1	5	0.095	0.010	0.3236	19.55
J	0.1	30	0.451	0.045		
K	0.8	10	0.181	0.145		
L	0.1	-5	-0.105	-0.011	0.1796	9.90
M	0.1	25	0.393	0.039		
N	0.8	2.5	0.049	0.039		
O	0.1	-7.5	-0.162	-0.016	0.0622	3.21
P	0.1	17.5	0.295	0.030		
Q	0.8	15	0.259	0.207		
R	0.1	7.5	0.139	0.014	0.2508	14.44

The expected utility and the certainty equivalent for the node leading to paths G, H, and I is the largest. This means that with a lower risk tolerance, the decisions leading to the third event node in Figure 14.15 dominate. This suggests that if the company's decision makers are more risk averse, the preferred option is to mine at a capacity of 60,000 tpa of nickel equivalent, to build an upfront pressure leaching operation in Western Australia, and to ship the impure nickel salt to Queensland for purification.

14.5 Summary

A decision tree is a graphical representation of the sequence of decisions, events and their anticipated outcomes. The graph consists of decision, event and terminal nodes linked by branches indicating either the choice of a decision or the outcome of an event node. This graphical representation allows decisions to be deconstructed into their components so that the analyst can easily view the options and the outcomes. The probability of each outcome is used to evaluate the preferred decision.

Utility theory is a way of accounting for a decision maker's risk tolerance. The utility function describes the utility of an outcome at the point of indifference, that is, the point at which the decision maker is indifferent to the risky option or to the certain option. The value of an outcome is transformed into a utility by the utility function. The preferred option is that which maximizes the utility.

14.6 Review Questions

1. What are the elements of a decision tree?
2. To what type of decisions is decision tree analysis applicable?
3. Under what conditions do the probabilities of occurrence sum to one?
4. Can decision trees be used with *NPV* analysis?
5. What does "expected value" mean?
6. What is the difference between conditional probability and joint probability?
7. What is "utility"?
8. Explain the difference between "ordinal" and "cardinal."
9. What does "indifference" mean in relation to a choice?
10. Explain the difference between "the expected value of an outcome" and "the expected utility of an outcome."
11. Can the notion of utility be used to account for a decision maker's risk tolerance?
12. Explain how lotteries can be used to determine the certainty equivalent.
13. Explain the terms "risk neutral" and "risk averse."
14. Describe how the exponential utility function can be used to account for the decision maker's preferences.

14.7 Exercises

1. An event has three outcomes: high; medium; and low. The value of the high outcome is 10, that for the medium outcome is 8, and that for the low outcome is 6. If the probability of each of these outcomes is 0.1, 0.65 and 0.25, respectively, draw the event node and determine the expected value of the event.
2. A company has the opportunity to invest in a pilot study. There is an equal chance of the pilot study being good or bad. If the outcome of the pilot study is good, the company may decide to build a full-scale operation. The outcome of the full-scale operation depends on the state of the economy. If economic conditions are good, the value of the operation will be 10. If the economic conditions are poor, the value of the operation will be –5. There is an equal chance of the opportunities being either good or poor. Draw the decision tree for this example. Determine the value of the opportunity to invest in the pilot plant, if the pilot plant costs 0.1 and the full scale plant costs 1.
3. A company is contemplating the investment in a factory to produce garden equipment. The company has the choice of building two different sized factories, a small one and a large one. The demand for the garden equipment is anticipated to be high or low. If the large factory is built, and the demand is high, the profit will be $10 million. If the demand is low, then the profit is expected to be $2 million. If a small factory is built, the profit will be either $4 million or $1 million, depending on whether the demand is high or low. Draw

the decision tree for this option. If the probability of the demand being high is 0.6, determine which choice should be made.

4. An oil production company is considering the opportunity to drill for oil. The company needs to purchase an exploration lease for $2 million for a certain concession area from the authorities that will give it the right to drill. The drilling program will take six months and cost $2 million. The chances of finding oil are summarized in the table below:

Size of oil field	Probability
None	0.4
High	0.3
Low	0.3

If oil is discovered, the company can sell the oil field or develop it itself. If the company develops the oil field there are two possible outcomes. In the case of a small field, these outcomes are that the oil production facility has a value of $60 million if the oil price is high, or $10 million if it is low. If the oil field is large, the value of the production facility is $120 million if the price is high and $80 million if it is low. If the company sells the oil field without developing it, it will receive a price that is 10% of the expected value of the production facility. There is an equal probability that prices are high or low.

(i)   Draw the decision tree for this situation.
(ii) Determine the optimal choice.

5. Pharmaceutical products go through three stages of clinical trials and two stages prior to clinical trials. The outcome for each stage may be favourable, or unfavourable. If it is favour, the pharmaceutical company has the choice of entering the next stage of trials. The success rate for moving through the stages is given in the table below:

Stage	Number of trials per successful drug launch	Estimated cost per stage
Screening	5000.0	150
Preclinical	1019.3	4,000
Phase 1	3.7	30,000,000
Phase 2	2.6	30,000,000
Phase 3	1.2	30,000,000
Product launch	1.0	

(i)   Given success in a stage, determine the probability of success in the next stage.
(ii) Determine the expected cost of developing a successful drug.

6. Two investments have the values for three different economic scenarios:

(i)   Determine the expected values of the two investments.

Outcome	Investment A	Investment B	Utility A	Utility B
1	100	90	1.00	0.84
2	75	65	0.76	0.71
3	12	23	0.00	0.05

(ii)   Determine the expected utilities of the two investments.

(iii)   Suggest which option is preferred.

Chapter 15
Real Options Analysis

15.1 Introduction

Decision tree analysis enables engineers and managers to understand the decisions that can be made as events unfold. Managers can make decisions, called contingent decisions, based on what happens. The decision tree analysis specifically incorporates the complexity and flexibility of decision-making that makes management situations interesting. Such flexibility is valuable. For example, an operation that switches between products at minimal cost depending on the market demand provides a high degree of managerial flexibility. The management of this operation has options that they can exercise, when required. These options are known as real options, the subject of this chapter.

Real option analysis is a radical departure from the methods presented in earlier chapters. It specifically accounts for the uncertainty in cash flows arising from the investment, which is a drawback of the traditional *NPV* method, and it discounts the uncertain cash flows at the correct rate, which is a drawback of decision tree analysis. Real options analysis draws on an analogy with financial options, which has found widespread acceptance. Financial options are discussed in the next section, after which examples of real options are developed.

15.2 Financial Options

15.2.1 Options Contracts

An options contract provides the owner of the contract with the right to buy or sell within a period at a known price. For example, an option contract may allow its owner the right to purchase 100 shares of Johnson Engineering at a price of $19 per share in three months from the date of the contract. If the price of the shares at the end of the period is $21, the owner of the option may *exercise* the option and make a profit equal to $21 less the $19, that is, $2. The phrase "exercise the option"

457

means that the owner of the option will take up her rights to purchase the shares at $19 from the person from whom she has purchased the option. If the shares are lower than $19, the owner could buy them in the market for less, so the owner would not exercise the option. There is no obligation to exercise the option. This means that the owner of the option cannot make a loss.

The contracted price is called the *strike price* and the length of the option is called the *maturity*. If the asset price is greater than the strike price, the option is *in the money*, and if not, it is *out of the money*. The asset that the option is written on is called the *underlying asset*. The option to buy shares in Johnson Engineering had a strike price of $19, a maturity of three months, and the underlying asset was 100 shares of Johnson Engineering. If the current price of Johnson Engineering is $21 per share, the option is in the money.

The option on the shares of Johnson Engineering is called a *call option*, which is the right to buy an asset. An option that gives the owner the right to sell an asset is called a *put option*.

The owner of an option is said to have a *long position*, while the writer has a *short position*. The investor purchased the call option on Johnson Engineering from a bank. The investor has a long position while the bank has a short position. If the investor makes a profit on the option, the bank will make a loss.

The final term that is of interest concerns the timing of the exercise of the option. If the option can only be exercised at maturity, it is called a *European option*. If it can be exercised at any time before the option expires, it is called an *American option*. The option to purchase the shares of Johnson Engineering could only be exercised at the end of the contract period, and hence it was a European option. If it could have been exercised at any time during the period, it would have been an American option. American options are more valuable, because they offer more flexibility to the owner of the option.

The investor purchases the option from the writer. If the price of the shares drops below the strike price before maturity, the owner does not exercise and so only looses the amount paid for the purchase of the option. The sequence of events and decisions can be represented by a decision tree, shown in Figure 15.1. The investor decides to purchase the option for $1. At the maturity of the option, the share price is either above or below the strike price of $19. For this example, assume that it is either $23 or $16. The investor will make a profit on the purchase of the options if the price is higher than the strike and the investor exercises the option (Path A). The investor may fail to exercise, even if the option is in the money (Path B). However, this is unlikely to be intentional. The investor will not exercise the option if the price is less than the strike, so Path C will never be followed. Overall, the investor stands to gain an overall profit of $3 from an initial outlay of $1 if the share price at the end of the period is $23.

The outcome of the price of the underlying asset is unknown, and hence the owner of the option does not know whether the option will be in the money (or not) at the maturity of the option. This uncertainty is the source of financial risk, as discussed in previous chapters. Such risk was not regarded favourably, and techniques, such as portfolio management, were discussed as a means to reduce the risk.

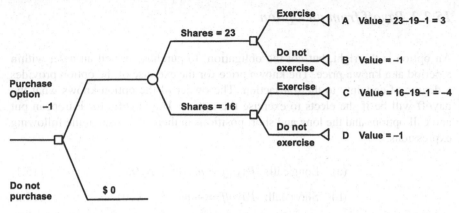

Figure 15.1 Decision tree for the purchase and exercise of a call option with a strike of $19 and a purchase price of $1

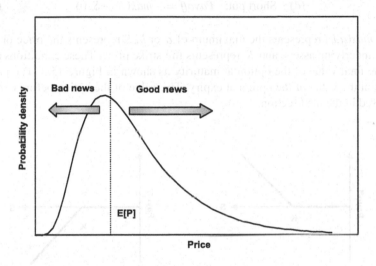

Figure 15.2 The two sides of uncertainty

The attitude of the owner of an option to this same financial risk is different. The most that the owner of an option can loose is the price she initially paid for the option. If the uncertainty is higher, the possibility of a higher outcome, and hence profit, is also higher. This means that for the owner of an option, uncertainty represents an opportunity. The attitude towards uncertainty is turned from a negative one to a positive one, as illustrated in Figure 15.2, because the effect of negative events is limited. The option provides "downside protection," and turns uncertainty into opportunity. This change in how uncertainty is viewed makes the options approach to financial markets radically different from that of an investor in other financial assets.

15.2.2 Payoff from an Option

An option is the right, but not the obligation, to purchase or sell an asset within a period at a known price. The known price for the exercise of the option provides a degree of certainty to the transaction. The owner of the option knows what the payoff will be if she elects to exercise the option. The payoffs for European put and call options and the long and short positions in them are given in the following expressions:

$$\text{(a)} \quad \text{Long call:} \quad Payoff = max(S - K, 0) \qquad (15.1)$$

$$\text{(b)} \quad \text{Short call:} \quad Payoff = -max(S - K, 0) \qquad (15.2)$$

$$\text{(c)} \quad \text{Long put:} \quad Payoff = max(K - S, 0) \qquad (15.3)$$

$$\text{(d)} \quad \text{Short put:} \quad Payoff = -max(K - S, 0) \qquad (15.4)$$

where $max(a,b)$ represents the maximum of a or b, S represents the price or value of the underlying assets, and K represents the strike price. These conditions represent the final value of the option at maturity as shown in Figure 15.3. The payoffs represent the value of the option at expiry. The value of the option before expiry is discussed in the next section.

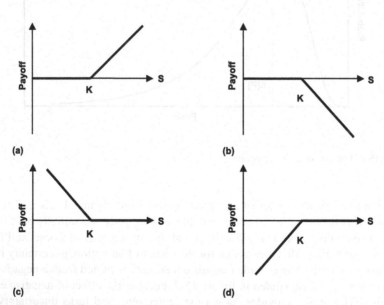

Figure 15.3 Payoff from various option positions. (a) Owner of call option, (b) Writer of a call option, (c) Owner of a put option, (d) Writer of a put option. S represents the price of the underlying asset, and K represents the strike price.

15.2.3 Price of an Option

The payoff of the option is the value of the option at maturity. Before maturity, the option has additional value because there is still time for the price of the underlying asset to increase, and hence for the possible profits from the option to increase. This means that the value of the option is higher than the payoff value before maturity. The price of a European call option is given in Figure 15.4 as a function of the price of the underlying asset, S. The option has no value if the price of the asset is much less than the strike price; it is said to be "deep out-of-the-money." As the price of the underlying asset increases above that of the strike price, the option value increases. For a European option, the value of the option is always higher than the payoff before maturity since it cannot be exercised early. For an American option, this may not be the case.

There are five variables that affect the price of the option: the share price, S, the strike price, K, the time to maturity, T, the volatility, σ, and the risk-free interest rate, r. The affect of each of these variables will be discussed.

(i)   Price of underlying asset

The price of the underlying asset is the prime variable affecting the price of the option. The value of the option increases as the value of the underlying asset increases. Options are written on a wide range of assets, such as stocks, futures, bonds and market indexes. The asset does not need to be a traded asset for an option to be written on it or for the option pricing theory to be valid.

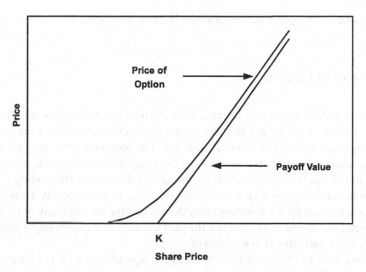

Figure 15.4 Price of a European call option

(ii)   Strike price

The strike price sets the purchase price. The value of the option decreases as the level of the strike price increases.

(iii)   Time to maturity

The value of the option increases as the time to maturity increases. Traded options are typically about 3 months. However, longer term options, called LEAPS (Long-term Equity Anticipation Security), that have a maturity of two and half years are available.

(iv)   Volatility

The volatility is the standard deviation of the price changes of the underlying asset. It is the anticipated volatility between the current date and the maturity of the option. This is the only variable that cannot be observed in the market, since it is the anticipated value rather than the historical value. It can be inferred from historical volatilities, or it can be implied from the price of the option. As a result, option trading is sometimes referred to as trading on volatility.

The changes in price, which result in volatility, are often thought of as being caused by new information about the asset. However, empirical research has shown that volatility is due mainly to trading itself rather than by the arrival of information.

The value of the option increases as the volatility increases. This means that the higher the uncertainty, the higher the value of the option. Risk, uncertainty and volatility increase the value of an option because the investor is not exposed to the effects of negative events.

(v)   Risk-free interest rate

The value of the option increases as the interest rate increases.

15.2.4 Use of Options

Options are called derivatives because they derive their value from another security. As a result, they are not primary securities, such as shares, but are traded as secondary securities on the financial markets. The options written on a company's stock represent no gain for the company. If the company issues shares, it gains from the amount of equity capital raised from the sale of the shares. The trading of shares between investors represents the changing ownership of the company. However, options are not issued by the company; they do not involve the company or its owners. Options can be viewed as side bets on the performance of the company's share price made by other participants in the market.

Options are used by speculators and hedgers. Speculators wish to take a position in the market. They take a view on the movement in price of the asset and gamble on that belief. For example, if a speculator took the view that the oil price was going to increase, he could purchase call options written on the crude oil price. The same speculator might also believe that the stock market was going to decline, and hence buy put options on the stock market index. On the other hand, hedgers wish to purchase options in order to reduce risk. For example, a put option increases as the price moves down, providing a form of insurance against the poor performance of an asset. The combination of the put option and the asset means that the hedger has no exposure to the price if it declines below the option strike. This provides a floor value to the portfolio of assets held by the hedger. Hedgers combine assets and options in numerous different trading strategies like this one. These strategies are all designed to limit exposure to risk.

In the next section, options on assets that are not traded in the market are discussed. These options are known as real options.

15.3 Options on Non-financial Assets: Real Options

A call option is the right to purchase an asset at a given price. If the owner chooses to exercise the option, the owner forfeits the option but gains the asset. The purchase price for the asset is the strike price. Many investment opportunities have a similar structure. For example, a company that has spent money on an engineering design has the right, but not the obligation, to implement the design by building the factory. By performing the design work, the company has earned an option that is not available to all the participants in the market. If the company decides to invest in the factory, it is exercising this option. It forfeits the option, but gains the asset. The strike price for this asset is the cost of building the factory. This type of option is called a *real option*.

The options that the manager of a company has arise because of different activities and opportunities. They are derived mainly from the flexibility that a company has to make decisions based on the information available to it. Inherent options are those decisions to stop, start, delay or increase production. Created options are opportunities that are created due to the firm's anticipation of the future, such as an oil company with an investment in research in alternative energy sources. This type of option may provide the company with flexibility in case of changes in market forces. Other options are switching options, such as switching feed stock from oil to natural gas, and platform options, which allow the company to enter new businesses.

The ability to delay the implementation of a project is a prime example of managerial flexibility. The delaying of the investment allows time for more information to be gathered that may clarify the desirability of the investment. In addition, it may allow more time for better economic conditions. The value of the option comes from both the value of the underlying asset, and flexibility of being able to choose the best time to make the investment.

The present value of the underlying asset is defined as $V(P)$, where P is the variable that is the key uncertainty in the valuation of the project. For example, the price of the company's products may be the key uncertainty. The investment in

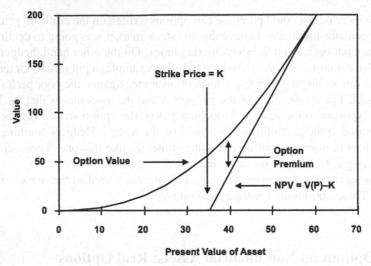

Figure 15.5 The value of the project as an option to invest

Table 15.1 Analogy between a financial call option and a real option on a capital investment

	Financial call option	Real option to invest in a project
Underlying	Share or asset, S	Present value of future cash flows, V
Strike price	Strike price, K	Capital investment, K
Maturity	Contract maturity, T	Opportunity expires, T
Uncertainty	Share price uncertain	Project value uncertain
Volatility	Stock volatility, σ	Project volatility, σ
Discount rate	Risk-free interest rate, r	Risk-free interest rate, r
Exercise	When call $= S - K$	Invest when $F(P) = V(P) - K$
Intrinsic value	Payoff $= S - K$	$NPV = V(P) - K$

the project is K, which is the initial capital outlay or capital expenditure for the construction of the project. The net present value of the project, NPV, is given by the following expression:

$$NPV = V(P) - K \tag{15.5}$$

This is the payoff from exercising the option. The payoff, or the NPV, is the value that is added to the firm because of the project. These values are shown in Figure 15.5. The payoff has the same profile as a call option, shown in Figure 15.3a, because the company will not invest in a project with negative NPV. If $V(P)$ is less than K, the project is "out of the money." If $V(P)$ is greater than K, then the project is "in the money."

The treatment of the investment opportunities as an option has been called the *real options approach* or *real options analysis*. The analogy between a financial call option and a real option (to invest in a project) is made in Table 15.1.

The value of the option prior to exercise is the payoff value plus the additional value created, for example, by the managerial flexibility of being able to time the decision. The value of the flexibility is the option premium. The option value is sometimes expressed as a *Strategic NPV*, given by the following definition:

$$Strategic\ NPV = NPV + Value\ of\ Flexibility \qquad (15.6)$$

The sum of the *NPV* and the value of flexibility has been called "strategic" to emphasise the role of the decisions that management can make to optimise the value of the operations. The *NPV* is the same as that which has been discussed in previous chapters. The calculation of the option premium or flexibility value is discussed in a later section of this chapter. In the next section, some examples of real options are discussed.

15.4 Examples of Real Options

Real options are applicable to many of the situations where management has the ability to decide on a course of action. Examples are the option to delay, to expand, to contract, to shutdown, to restart, to abandon, to switch between products, suppliers and raw materials and many others.

15.4.1 Option to Invest (or Deferral Option)

A company that has the opportunity to invest in a project does not face a now-or-never decision. The decision can be delayed or deferred. This option to defer is a major constituent of investment decisions. An example of a deferral option is the following. An oil company has purchased lease rights on a property that contains oil. The company does not have to produce oil from the field immediately. The management of the company can wait until the price justifies the development of the oil field. This means that the company has the right but not the obligation to develop the oil field. This option will expire when the lease expires. The rights to some resource leases might never expire. If this is the case, they may be held for decades by the owner until the price justifies the cost of the investment.

The decision to delay an investment might provide the company's competition with an opportunity to fill a gap in the market. The value of the option to delay is derived from the exclusion of competitors by a proprietary position. If the company owns proprietary technology, patents, oil leases, or mineral rights, the deferral option has value and it will continue to maintain value because competitive action is stunted.

15.4.2 Time-to-built Options

The development of an engineering project in stages was discussed in Chapter 4. Each stage of the project involved a series of capital outlays before the major investment, for the construction of the facility, was required. Each of these stages can be viewed as an option to invest in the next stage. These options are illustrated in Figure 15.6.

   A similar example is that of the development and approval of pharmaceuticals. The development of a pharmaceutical drug goes through specific stages as it gets regulatory approval. Each stage of the research and development process is an option to invest in the next stage.

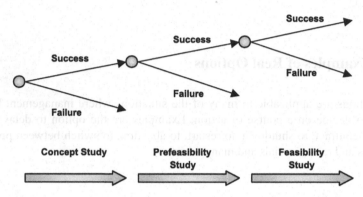

Figure 15.6 The development of an engineering design as a series of staged options

15.4.3 Growth Options

Growth options are options that a company possesses to expand its operations. The options arise from the company's position in the market and its current knowledge and skills. Because of these options, the company can increase its market share or develop new markets. For example, Amazon.com developed a new route to market for books through online marketing. The investment that Amazon.com made in software, distribution and marketing gave Amazon.com a customer base that allowed it to expand into other products that customers preferred to purchase online. For example, it expanded into a new market by selling music, then into another new market by selling movie DVDs. The initial investment made by Amazon.com in delivery systems for the marketing and distribution of books gave the company growth options to invest in the distribution of other products.

15.4.4 Abandonment Options

Factories and operations can be closed or abandoned prematurely. The abandonment option of an operation is a put option on the value of the future cash flows from the operation with a strike price of the salvage value. These may not necessarily be the abandonment of whole operations, but rather a line of business, an unprofitable customer, or an order. In this form, the abandonment scenario does not seem as drastic. For example, the cancelling of a particular contract that loses money is the exercise of an abandonment option.

15.4.5 Switching Options

A company may be able to switch between alternatives. For example, the manufacturing process may be sufficiently flexible to switch between energy sources such as oil or coal. The process may be able to produce a variety of products and management can choose the most profitable at any particular time. For options to be valuable to the company, they should be owned only by the company. If the company's competitors have the same flexibility, then these real options are of little value. For example, an oil refinery that can change the product streams does not really own an option, since most oil refineries can also do this. However, a multi-purpose batch plant that can produce a wide variety of chemicals using the same equipment has utility that few competitors have, and therefore the switching options have value.

The methods for the calculation of the value of an option are presented in the next section.

15.5 The Valuation of Financial Options

The calculation of the value of an option is discussed in this section. Two main methods are used. The first is the binomial lattice using risk-neutral probabilities, and the second is the Black–Scholes equation. The binomial lattice is a numerical approximation of the Black–Scholes equation. The concepts necessary for the development of both models are discussed. The concepts are those of the risk-free portfolio, and the risk-neutral probability.

15.5.1 Risk-free Portfolio

The calculation of the value of options relies on establishing a risk-free portfolio. If the portfolio is risk-free, then its value is not affected by the movements in price. As a result, its value can be calculated. This method of valuation of an option is

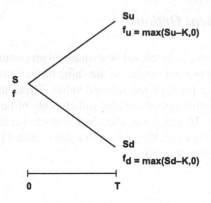

Figure 15.7 Share and option prices over a single period

illustrated in this section. A single time step is used to simplify the calculations so that the method is clear.

Consider the price of the underlying asset, S, and the price of the option, f, over a single period. This is shown in Figure 15.7. In one time step, the price of the asset can move up to a value of Su where u represents a factor that represents the increase in price. The price of the asset could also move down to Sd, where d represents the decrease in price. The price of the option in the event of an increase in asset price is f_u, while that in the event of a decrease in price is f_d.

The objective is to create a portfolio of the asset and the option that is riskless. This can be achieved by purchasing m shares of the asset and selling an option on the asset. The initial value of this portfolio, given by V_0, is the following:

$$V_0 = mS - f \tag{15.7}$$

The number of shares is chosen so that the portfolio is riskless. This means that the portfolio will have constant value independent of whether the asset price increases or decreases. If this is the case, then the following relationship holds:

$$mSu - f_u = mSd - f_d \tag{15.8}$$

The solution of Equation 15.8 gives an expression for m, the number of shares in the portfolio.

$$m = \frac{f_u - f_d}{Su - Sd} \tag{15.9}$$

The present value of this risk-free portfolio at the maturity T is the value of the portfolio discounted at the risk-free rate. This is given as the following expression:

$$PV(V_T) = (mSu - f_u)\frac{1}{1+r} \tag{15.10}$$

where r represents the risk-free rate and $PV()$ represents the present value function. The present value of the portfolio must be the same as the initial value of the portfolio so that the following relationship holds:

$$mS - f = (mSu - f_u)\frac{1}{1+r} \tag{15.11}$$

The substitution of the expression for m, given in Equation 15.9, into Equation 15.11 yields the following expression for the value of the option:

$$f = \frac{1}{1+r}(pf_u + (1-p)f_d) \tag{15.12}$$

where the factor p is given as follows:

$$p = \frac{(1+r) - d}{u - d} \tag{15.13}$$

The following example illustrates the calculation of the value of the option using the formulation above.

Example 15.1: Simplified option calculation.

If the asset price is initially \$20, and goes up or down by 10% in the period and the strike is \$21, determine the value of the call option. The risk-free rate is 8% over the period.

Solution:

The factor p is calculated as follows:

$$p = \frac{1 + 0.08 - 0.9}{1.1 - 0.9} = 0.9$$

The value of the asset at the end of the period in the up position at the end of the period is $Su = 20(1.1) = 22$ and the value in the down position is $Sd = 20(0.9) = 18$. The value of the option is $max(22 - 21, 0) = 1$ in the up position, and $max(18 - 21, 0) = 0$ in the down position. The substitution of these values into Equation 15.12 gives the value of the option. This calculation is given as follows:

$$f = \frac{1}{1+r}(pf_u + (1-p)f_d) = \frac{1}{1.08}(0.9(1) - 0.1(0)) = 0.833$$

The value of the call option is 0.833.

The value of p in Equation 15.12 can be interpreted as a probability. This interpretation is discussed next.

15.5.2 Risk-neutral Probability

The calculation of the option using the risk-free portfolio over one time step lead to Equation 15.12. The use of a single time step means that it is not an accurate approximation for the value of the option. It is useful to examine Equation 15.12 more thoroughly before developing a more accurate method of calculation.

The value of p in Equation 15.12 can be interpreted as a probability, that is, the probability of an increase in the price of the share over the period. In this case, the expected value of the option is given by the following expression:

$$E[f] = pf_u + (1-p)f_d \qquad (15.14)$$

and the expected value of the stock price at maturity is given by the following equation:

$$E[S_T] = pSu + (1-p)Sd \qquad (15.15)$$

The substitution of the expression for p, given by Equation 15.13, into Equation 15.15 yields the following expression:

$$E[S_T] = S(1+r) \qquad (15.16)$$

Equation 15.16 says that the expected value of the share price is the future value of the current share price using a compound rate of r. This means that the share price grows at the risk-free rate. The probability of an increase in price at each time step is given by p, the risk-neutral probability. These prices and price movements are those calculated in the risk-neutral world.

The risk-neutral world is not the objective world, or real world. The growth is not the real world growth, μ, and the probability of an increase in price is not the real world probability, q. The real world growth rate of the share price is irrelevant in the calculation of the option value using risk-neutral valuation. The expected return from the risk-free portfolio of assets is the risk-free rate, and future cash flows can be valued by discounting at the risk-free rate in the risk-neutral world.

The risk-neutral world invoked for option valuation is analogous to the certainty equivalent method discussed in Chapter 13. An alternative to the certainty equivalent method, the risk-adjusted discount rate method, was also examined in that chapter. A method for the calculation of the value of an option that is analogous to the risk-adjusted discount rate method and that uses real-world probabilities and risk-adjusted rates rather than risk-neutral probabilities and the risk-free rate has also been proposed. The calculation of the option value using real-world probabilities and the risk-adjusted discount rate can be shown to be mathematically the same as risk-neutral valuation. The method relies not on a risk-free portfolio, but on a replicating portfolio. However, this method is computationally more intense, and will not be discussed further.

15.5.3 Binomial Lattice

The binomial lattice is a calculation that relies on the risk-free portfolio and the risk-neutral probabilities to determine the price of an option. The method is developed and illustrated in this section.

The values of the up and down factors, u and d, and the risk-neutral probability, p, must be chosen so that Equations 15.15 and 15.16 hold. This means that the expected value of the share at the end of the period is given by the following equation:

$$Se^{rT} = pSu + (1-p)Sd \qquad (15.17)$$

where the exponential factor, e^{rT}, represents compounding at a continuous rate. The standard deviation of the share price movements must also be equal to the volatility of the share price. This means that the variance of the change in share price over the period T is $S^2\sigma^2 T$, where σ is the standard deviation of the share price. Since the variance of a variable X is given by $E[X^2] - E[X]^2$, the variance of the share price is given by the following expression:

$$S^2\sigma^2 T = pS^2u^2 + (1-p)S^2d^2 - S^2(pu + (1-p)d)^2 \qquad (15.18)$$

Equations 15.17 and 15.18 represent two equations in three unknowns. A third condition, that $u = 1/d$, is made so that the lattice recombines, which is illustrated later. If these conditions are used, the expressions for p, u and d are given by the following expressions:

$$p = \frac{e^{rT} - d}{u - d} \qquad (15.19)$$

$$u = e^{\sigma\sqrt{T}} \qquad (15.20)$$

$$d = e^{-\sigma\sqrt{T}} \qquad (15.21)$$

This set of conditions can be used to calculate the movement of the share price in the risk-neutral world. The difference between Equations 15.19 and 15.12 is that continuous discounting is used in Equation 15.19 while annual compounding was used in Equation 15.12.

The movements of the price can be represented as a lattice of price movements called the binomial lattice or binomial tree. This is considered in the following example.

Example 15.2: Binomial lattice of share price movements in a risk-neutral world.

A share price of a company is currently $20. Draw the lattice over five months for the share prices if the volatility is 20% using a period of one month between price movements.

Solution:

The lattice for the share price movements is shown in Figure 15.8. The factors given by Equations 15.20 and 15.21 are designed so that the multiple of the up and down factors, ud, is equal to one, which means that the lattice is recombining. In other words, with each new time step, only one extra node is added to the lattice. The share price at any node is equal to the original share price multiplied by the up and down factors required to get to that node. The share price is shown in Figure 15.4.

The up and down factors are calculated from Equations 15.20 and 15.21 as follows:

$$u = e^{\sigma\sqrt{T}} = \exp(0.2\sqrt{1/12}) = 1.0594$$
$$d = e^{-\sigma\sqrt{T}} = \exp(-0.2\sqrt{1/12}) = 0.9439$$

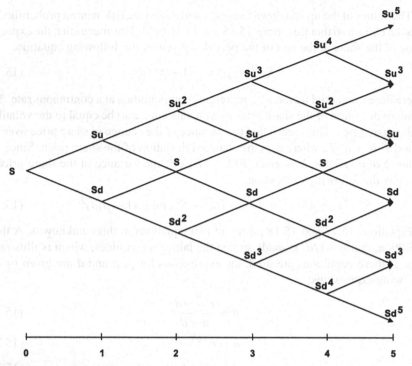

Figure 15.8 Binomial lattice for the share price. Note that the lattice is recombining because $ud = 1$

These values are used to calculate the binomial lattice of asset price movements.

The values of the share price are calculated according to this lattice, and shown in Table 15.2.

The first row represents the upper line of the lattice, and the values in each column represent the nodes at each time step.

The value of the option is calculated from the binomial lattice of the share prices by starting at maturity where the value of the option is known, and moving backwards to the beginning. At maturity, the value of the option is known, and is given

Table 15.2 Binomial lattice for the share prices

	0	1	2	3	4	5
0	20.00	21.19	22.45	23.78	25.20	26.69
1		18.88	20.00	21.19	22.45	23.78
2			17.82	18.88	20.00	21.19
3				16.82	17.82	18.88
4					15.88	16.82
5						14.99

by one of the payoff conditions discussed in Section 15.2.2. For example, the value of a call option with a strike of K is given by $max(S_T - K, 0)$, where S_T is the value of the node at the maturity T. As a result, the value of the option corresponding to the share price in the last row of binomial lattice can be calculated.

The value of the option at the time steps prior to the maturity date is calculated from the expected value of the option using risk-neutral probabilities. This is given by the following expression:

$$f = e^{-r\Delta t}(pf_u + (1-p)f_d) \tag{15.22}$$

where Δt represents the time step, that is, $n\Delta t = T$, where n is the number of time steps. As before, the variable f represents the value of the option at the current time, while the variables f_u and f_d represent the value of the option in the up and down state in the next time step. The method is to start at maturity and work back through the binomial lattice to find the option value at time zero.

The calculation of the value of the option in this manner is illustrated in the example given below.

Example 15.3: Using the binomial lattice to determine the option value.

Determine the value of a European call option with a strike price of $20 from the binomial lattice given in Example 15.2. Assume a risk-free rate of 10%.

Solution:

At maturity, the call option has a value of $max(S_T - K, 0)$. These values constitute the boundary conditions for the problem and are shown in Figure 15.9.

The values of the option at the boundary are shown in the final column of Table 15.3.

The risk-neutral probabilities are required in order to calculate the value of the option. The risk-neutral probability, p, is calculated as follows:

$$p = \frac{e^{r\Delta T} - d}{u - d} = \frac{\exp(0.1(1/12)) - 0.9835}{1.0168 - 0.9835} = 0.558$$

The value of the option at one time step back from the maturity date, say at point A in Figure 15.9, is calculated from the expected value of the option under risk-neutral probabilities. This calculation is as follows:

$$f = e^{-r\Delta T}(pf_u + (1-p)f_d) = \exp(-0.1/12)(0.558(6.69) + (1 - 0.558)3.78) = 5.36$$

Similarly, the value of the option at point B is calculated as follows:

$$f = \exp(-0.1/12)(0.558(3.78) + (1 - 0.558)1.19) = 2.61$$

Table 15.3 Binomial lattice for the option prices

	0	1	2	3	4	5
0	1.51	2.17	3.04	4.11	5.36	6.69
1		0.71	1.12	1.73	2.61	3.78
2			0.20	0.36	0.66	1.19
3				0.00	0.00	0.00
4					0.00	0.00
5						0.00

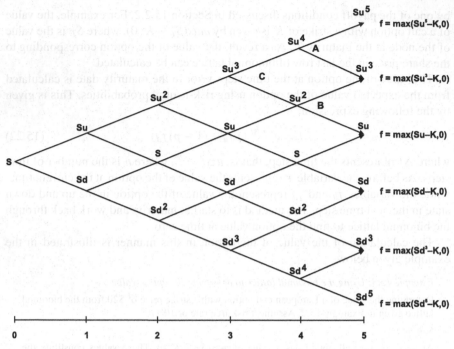

Figure 15.9 Calculation of the option values at maturity

These two values contribute to the value of the option at point C, which is calculated in a similar manner as follows:

$$f = \exp(-0.1/12)(0.558(5.36) + (1 - 0.558)2.61) = 4.11$$

The calculation is extended back until the origin, in a similar manner, to obtain the value of the option. The values for these calculations at each node are shown in Table 15.3. The value of the option is 1.51.

If the calculation in Examples 15.2 and 15.3 was performed for the same option but with a smaller time step, say a week instead of months, the value of the option would change slightly. As the time step was reduced, it would eventually converge to a result. This convergence is a result of the fact that the binomial lattice method provides an approximate solution to the Black–Scholes differential equation. The binomial lattice is a good method for the approximation of the value of an option. The Black–Scholes formula is discussed in the next section.

15.5.4 Black–Scholes Option Pricing Formula

The up and down movements depicted in the binomial lattice approximate the random movements of the share price with time. At any node, the price may move up or down. The path travelled through the lattice is called a *random walk* and the process is called a *stochastic process*. The stochastic process that drives stock prices is assumed to follow a Markov process, in which all the information of the past is incorporated in the present value. For example, the current share price is based on all the history of the company and all information about the share price. This information is contained in the current value of the share price. The only variable affecting the future value is the current value. The stochastic process is also assumed to follow a Wiener process, also called Brownian motion. This means that movements in the price of the share in different periods are independent of each other, and the change in price is normally distributed with a mean of zero and standard deviation of $\sqrt{\Delta t}$, where Δt represents the time step.

The change in price of the share, ΔS, over the time step Δt can be represented by the equation of geometric Brownian motion given as the following expression:

$$\frac{\Delta S}{S} = \mu \Delta t + \sigma \varepsilon \sqrt{\Delta t} \tag{15.23}$$

where μ is the growth rate of the share price, σ is the volatility of the share price, and ε is a random number that is normally distributed with a mean of zero and a standard deviation of 1.0.

Black, Scholes and Merton showed that the geometric Brownian motion for a risk-free portfolio of shares and options led to the following partial differential equation:

$$\frac{\partial f}{\partial t} + rS\frac{\partial f}{\partial S} + \frac{1}{2}\sigma^2 S^2 \frac{\partial^2 f}{\partial S^2} = rf \tag{15.24}$$

This equation describes the value of an option with time as a function of the share price. The boundary condition for the solution of the differential equation is dependent on the type of option. These boundary conditions were discussed in Section 15.2.2. Black and Scholes provided solutions to the differential equation and these are given as follows for the price of a European call, *call*, and a European put, *put*:

$$call = SN(d_1) - Ke^{-r(T-t)}N(d_2) \tag{15.25}$$

$$put = Ke^{-r(T-t)}N(-d_2) - SN(-d_1) \tag{15.26}$$

where the function $N()$ is the normal distribution, and d_1 and d_2 are defined as follows:

$$d_1 = \frac{\ln(S/X) + (r + \sigma^2/2)(T-t)}{\sigma\sqrt{T-t}} \tag{15.27}$$

$$d_2 = d_1 - \sigma\sqrt{T-t} \tag{15.28}$$

The calculation of the option price using the Black–Scholes equation is illustrated in the following example.

Example 15.4: Black–Scholes option pricing model.

Use the Black–Scholes equation to determine the value of a European call option with a strike price of $40 on a share whose current price is $42. The volatility is 40%, the risk-free interest rate is 10%, and the option has 6 months to maturity.

Solution:

The values for the parameters of Equation 15.19 are calculated as follows:

$$d_1 = \frac{\ln(42/40) + (0.1 + 0.4^2/2)(0.5)}{0.4\sqrt{0.5}} = 0.490697$$

$$d_2 = 0.490697 - 0.4\sqrt{0.5} = 0.207855$$

$$N(d_1) = 0.68818$$

$$N(d_2) = 0.582329$$

The value of the call option is calculated from Equation 15.25 as follows:

$$call = (42)0.68818 - 40e^{-0.1(0.5)}0.58233 = 6.74642$$

Thus, the value of the call option is $6.74.

If the company on whose share the option is written issues dividends, then the option calculation should account for this. The reason is that the dividends represent a transfer of value from the company to the shareholders. The holders of options on the share are not entitled to this value. The Black–Scholes equation was modified by Merton to account for dividends, and the result is given in the following expression for a European call option:

$$call = Se^{-\delta(T-t)}N(d_1) - Ke^{-r(T-t)}N(d_2) \tag{15.29}$$

where δ represents the continuous dividend rate. The other variables have their same definition as before, except that the factor d_1 is given by the following expression:

$$d_1 = \frac{\ln(S/X) + ((r - \delta) + \sigma^2/2))(T - t)}{\sigma\sqrt{T - t}} \tag{15.30}$$

There are six variables that affect the price of the option: the share price, S, the strike price, K, the time to maturity, T, the volatility, σ, the dividend yield, δ, and the risk-free interest rate, r. Interestingly, the growth rate of the share price does not affect the price. The reason for this is the same as that discussed earlier in the context of binomial lattices. The option is valued in the same risk-neutral world discussed earlier where growth is at the risk-free rate.

15.6 Valuation of Real Options

15.6.1 Option to Invest

The company faced with an opportunity to invest does not have to execute immediately. It can wait and execute later. As a result, the option to invest is also known as the deferral option. The option value is the cost of ownership of this flexibility. The option may arise from a proprietary position in technology, or the market, or access to raw materials, amongst others. The following example illustrates that calculation of the option value.

Example 15.5: Option to defer investment.

A company is considering the purchase of an oil lease, which would allow it to produce oil. The expected present value of the positive cash flows from the lease, calculated using the probabilities of finding oil and other success factors, is $120 million. The present value of the costs of oil exploration and production are $90 million. The lease expires in three years. This means that if the company has not produced oil from the lease, it looses it. What price should the company bid for the lease? The risk-free rate is taken as 5.5%.

Solution:

The lease can be described as an option to explore for and produce oil. The value of the underlying asset is the total benefits from oil production from the lease, and the strike price is the total costs. The time to expiry is three years. This option is a call option, with a payoff of $max(V - K, 0)$. The volatility of the value of an oil project is 20%. (Note that this is not necessarily the same as the volatility of the oil price.) The option can be exercised at any point during the three years, that is, oil can be produced as soon as it is found and production facilities are erected. As a result, the option is an American call option.

The value of the option is calculated using the binomial lattice method with ten time steps over the three years. The calculation of the up and down factors is shown as follow:

$$u = e^{\sigma\sqrt{T}} = \exp(0.2\sqrt{0.3}) = 1.115770$$

$$d = e^{-\sigma\sqrt{T}} = \exp(-0.2\sqrt{0.3}) = 0.896242$$

The risk-neutral probability is calculated as follows:

$$p = \frac{e^{r\Delta T} - d}{u - d} = \frac{\exp(0.055(0.3)) - 0.896242}{1.11577 - 0.896242} = 0.54843$$

The binomial lattice for the value of the underlying asset is calculated in the same manner as described in Exercise 15.2. The results of this calculation yield the binomial lattice for the asset price shown in Table 15.4. The binomial lattice in Table 15.4 gives the risk-neutral asset prices at intervals of 0.3 years.

Table 15.4 Binomial lattice of the value of the underlying asset, $V(P)$

0	0.3	0.6	0.9	1.2	1.5	1.8	2.1	2.4	2.7	3
120.00	133.89	149.39	166.69	185.99	207.52	231.54	258.35	288.26	321.63	358.86
	107.55	120.00	133.89	149.39	166.69	185.99	207.52	231.54	258.35	288.26
		96.39	107.55	120.00	133.89	149.39	166.69	185.99	207.52	231.54
			86.39	96.39	107.55	120.00	133.89	149.39	166.69	185.99
				77.43	86.39	96.39	107.55	120.00	133.89	149.39
					69.39	77.43	86.39	96.39	107.55	120.00
						62.19	69.39	77.43	86.39	96.39
							55.74	62.19	69.39	77.43
								49.96	55.74	62.19
									44.77	49.96
										40.13

Table 15.5 Binomial lattice for the value of the option

0	0.3	0.6	0.9	1.2	1.5	1.8	2.1	2.4	2.7	3
45.21	56.99	70.75	86.55	104.47	124.64	147.29	172.69	201.18	233.10	268.86
	32.56	42.39	54.17	67.97	83.82	101.73	121.86	144.46	169.82	198.26
		21.84	29.64	39.40	51.24	65.14	81.03	98.91	118.99	141.54
			13.17	18.87	26.48	36.24	48.24	62.31	78.16	95.99
				6.72	10.33	15.60	23.00	32.92	45.37	59.39
					2.57	4.32	7.19	11.79	19.02	30.00
						0.54	1.00	1.86	3.45	6.39
							0.00	0.00	0.00	0.00
								0.00	0.00	0.00
									0.00	0.00
										0.00

The option price is calculated by working back the column representing the option maturity, which establishes the boundary. The boundary condition of the option at maturity is equal to $max(V - K, 0)$, that is $max(V - 90, 0)$. This boundary condition is shown as the final column of the option values in Table 15.5.

The value of the call at one time step prior to maturity, that is, in the column at 2.7 years, is calculated based on Equation 15.22. However, since this is an American option there is the possibility of early exercise, so Equation 15.22 is modified to account for this possibility. This modification is expressed as follows:

$$f = max\left(e^{-r\Delta t}\left(pf_u + (1-p)f_d\right), V - K\right)$$

where V is the value of the underlying asset at the appropriate point in the asset value lattice. The calculation proceeds in the same manner as that outlined in Exercise 15.3. As shown in the first column of Table 15.5, the value of the option is \$45.21 million.

Thus, the company should bid \$45.21 million for this lease.

Comments:

An American option will not be exercised early unless the underlying asset pays dividends. Dividends mean that the option holder is not a beneficiary of the total return, but only of the capital gain. The capital gain does not increase at the growth rate μ, which is the rate that investors are expecting, but at $\mu - D$, where D is the dividend rate. The effect of dividends is accounted for in the asset lattice by a continuous loss in value due to dividend payments. In the absence of dividends, the American call and European call options are identical. A continuous loss of value for holders of commodities derivatives that is analogous to dividends is also found. The rate of this loss of value is called the convenience yield.

15.6.2 Option to Abandon

Factories and facilities can be abandoned in exchange for the salvage value or sale price. The CEO of a corporation reviews the poor performance of a division or an operation with a view to "stem the bleeding" by either selling it or closing it down. The abandonment is a put option to place the asset on the market for its salvage value. The sale price or the salvage value for the asset is the strike. The put option is calculated in exactly the same manner as the call option discussed in Example 15.2 and 15.3, except that the payoff at the boundary is equal to $max(K - V_T,0)$. The following example illustrates that calculation of the option value for abandonment.

Example 15.6: Option to abandon for the salvage value.

A line of business within the company has a current value of $8.5 million. This value was calculated as the present value of the future projections of the free cash flows. The company could abandon this business for the salvage value of its equipment and other fixed assets amounting to $8 million within the next year. If the risk-free rate is 5.5%, and the volatility is 20%, determine the value of the option to abandon.

Solution:

The option to abandon is an American put option with the strike price of $8 million. The value of the option is calculated in the same manner as outlined in Examples 15.2 and 15.3, except that (i) the payoff is that of a long put; and (ii) Equation 15.22 is replaced with the following expression:

$$f = max\left(e^{-r\Delta t}\left(pf_u + (1-p)f_d\right), K - V\right)$$

The value of the put option as a function of the value of the underlying business is shown in Figure 15.10. The value of the option is $315,000. The total value of the business can be viewed as the value without flexibility plus the value of the option to abandon. This gives a total value of $8.815 million.

Comments:

It is optimal to exercise an option when the option value is equal to the payoff. This establishes a different investment rule or decision criterion to that of the *NPV* rule. The option to abandon is not exercised immediately when the value decreases below its salvage value. There is some inertia because holding the option is more valuable. This is clearly shown in Figure 15.10, where the option value is greater than the intrinsic value. The real options approach indicates that this option should not be exercised until the option value is equal to the intrinsic value. The results of Figure 15.10 show that the value of the underlying business would have to drop to $6.5 million for this option to be exercised.

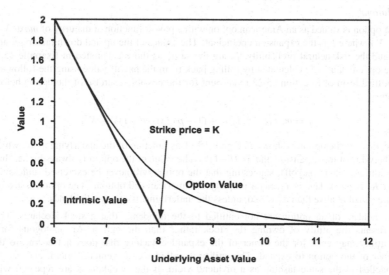

Figure 15.10 Value of the option to abandon

15.6.3 Option to Temporarily Close Operations

The option that has been valued in Example 15.6 is the option to abandon the business for its salvage value. Another option that management has is to simply halt operations if the costs exceed the revenues. The payoff from this option is $max(P - C,0)$, where P represents the unit revenues and C represents the unit expenses. This views the option as a claim on P by paying an exercise price of C. If the revenues in any year are less than the variable costs of the operation, management has the option to temporarily shut down operations. If operations are closed, management has the right to restart them.

15.6.4 Option to Expand or Contract

If the economic conditions warrant it, the management of company can elect to expand the operations. The option to expand is the option to increase the value of the project, V, by a factor b, where $b > 1$, at a cost of K'. Thus, the payoff from the option to expand is $max(bV - K', V)$. This expresses the condition that if the cost of the expansion is too great, or the gain from the expansion is too low, management will not elect to exercise the option. The following example illustrates that calculation of the option value for expansion.

Example 15.7: Option to expand.

Determine the value of the option to expand the business whose asset value, V, is described by the binomial lattice given in Table 15.4 (Example 15.5). Management would like to double the value of the business. The volatility is 20% and the risk-free rate is 5.5%. Evaluate this option as if it expires in three years.

Solution:

The option is valued as an American option with a payoff function at maturity of $max(bV - K', V)$, where b is the expansion coefficient. The values of the up and down factors, u and d, and the risk-neutral probability, p, are the same as those calculated in Example 15.5. The option values are calculated by rolling back from the payoff value using the following modified form of Equation 15.22 to account for the possible exercise of the option before maturity:

$$f = max \left(e^{-r\Delta t} \left(pf_u + (1 - p) f_d \right), (b - 1)V - K' \right)$$

The value of the option is shown in Figure 15.11 as function of the underlying asset, which is the value of the expansion, that is, $(b-1)V$. The value of the option is always higher than the intrinsic value (payoff), suggesting that the option will never be exercised (unless the cost K' is zero). This is a general property of American call options. The option is always more valuable alive than exercised, unless the underlying asset pays dividends.

The holders of an option are not entitled to the dividends that a stock declares. This represents the utility of owning the stock rather than the option. An analogous form of utility may exist for the owner of the expanded factory that does not accrue to the owner of the option to expand. This utility is called the convenience yield, and can be modelled in the same fashion as a dividend yield. If the calculations are repeated with a convenience yield of 5% (see Figure 15.12), it is optimal to exercise the option if the

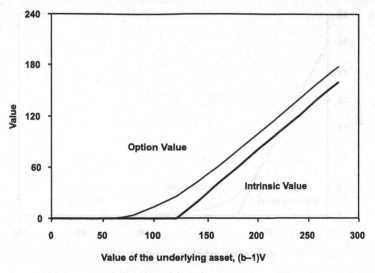

Figure 15.11 Value of the option to double capacity at a cost of $K' = 120$

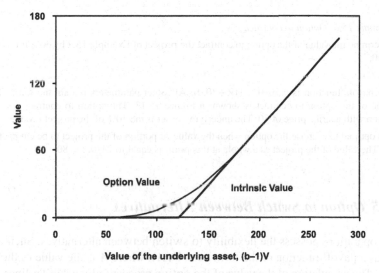

Figure 15.12 The value of the option to expand with a convenience yield of 5%

underlying asset has a value greater than 180 (with $K' = 120$). This means that management should double operations at a cost of 120 if the value of the operation, V, reaches 180.

The option to contract is analysed in a similar manner as that of the option to expand. The payoff of the options is $max((b-1)V + K', 0)$, where b is the contraction coefficient, which has a value that is less than one.

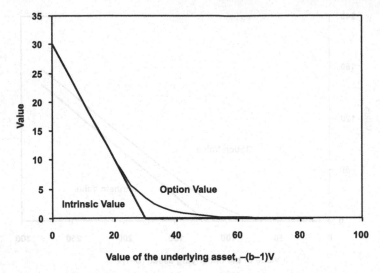

Figure 15.13 The value of the option to shrink the project by 30% at a cost of 30

Example 15.8: Option to contract.

Determine the value of the option to contract the project of Example 15.7 by 30% for a cost of 30.

Solution:

The payoff function is $max((0.7-1)V+30,0)$. All other parameters remain the same. The value of the option to contract is shown in Figure 15.13. The option to contract is a put option with a strike price of 30. The underlying asset is the 30% of the project's value.

It is optimal to exercise the option when the value of portion of the project to be cut back is 24. The value of the project as a whole at this point is equal to $24/0.3 = 80$.

15.6.5 Option to Switch Between Alternatives

Many operations possess the flexibility to switch between alternatives. Such an option has a payoff function of the form $max(V, A)$, where A is the value of the alternative. The calculation of the value of the option proceeds along similar lines as the calculations discussed before.

15.7 Decision-making Process

A general procedure for making decisions in a normative manner was presented in Chapter 2. The process consisted of three phases: frame, evaluate and decide. The decision-making process is not altered by real options; however, all three phases need to bear the options in mind.

(i)   Decision framing
        During the framing process, it is important to identify the real options that
        may be associated with the decision:
        1.   Determine the structure of the option decision, such as that shown in
              Figure 15.1, which gives rise to management flexibility.
        2.   Identify the uncertainty that is the source of volatility for the option.
        3.   Classify the type of option.

(ii)   Decision evaluation
        Calculate the value of the options by considering the following factors:
        1.   Identify the payoff of the option.
        2.   Quantify the uncertainty by determining the volatility.
        3.   Get data from the financial and commodities markets for the parameters
              that are observable in the market.
        4.   Determine the "trigger" points that would signal investment both in the
              option and in the exercise of the option.
        5.   Create a decision rule that incorporates the real option.

(iii)   Decision and review
        Ensure that the value of the option is transparent and can be clearly explained.
        It is also worth broadening the frame by asking questions about how the in-
        vestment can be restructured to increase value, or how options can be incor-
        porated into the investment proposal by staging the investment or by creating
        module units in the project that allow for expansion at a later stage.

15.8 Real Options Analysis is Not Decision Tree Analysis

Decision tree analysis is well established and accepted in the analysis of decisions.
Some confusion has been created amongst practitioners who call their decision trees
"real options." While there are some similarities, the two approaches are fundamen-
tally different. Although many decisions do have the structure of an option (shown
in Figure 15.1), not all decisions are options. As a result, real options is not a general
decision-making tool.

   The second major difference is that real options analysis uses the mathematical
framework established by Black, Scholes and Merton that is an economically sound
method for valuing the option. There are two methods for doing this: the replicating
portfolio approach and the risk-neutral approach. Both approaches are equivalent to
each other. The risk-neutral approach alters the growth rate and the probabilities of
growth to a risk-neutral position so that the value can be determined by discounting at the risk-free rate. The replicating portfolio approach adjusts the discount rate
to account for the change in risk. These two approaches are analogous to the certainty equivalent and risk adjusted approaches discussed in Chapter 13. Both these
approaches create a portfolio of assets, a hedge, that excludes the possibility of making risk-free profits that could be made if the combination of assets that replicates

the option is not priced correctly. For example, a combination of shares and bonds can be combined to yield the payoff of an option. Since they have the same payoff, they should have the same price. If they do not have the same price, an investor could create a money machine by continuously purchasing the cheaper and selling the more expensive. Such profits are called arbitrage profits. The derivation of the option value using the Black–Scholes and binomial lattice excludes the possibility of arbitrage or risk-free profits.

The application of decision trees as commonly undertaken does not account for this possibility of arbitrage, and is not economically sound for the evaluation of option-like decisions. In most application of decision trees with discount cash flows, the "law of one price" is violated, resulting in the possibility of arbitrage profits. Option pricing is economically and mathematically sound. Decision trees do not correctly approximate the option pricing method.

Both decision tree analysis and real options are valuable tools in the assessment of capital projects. However, it serves no purpose to confuse the two.

15.9 Strategic Thinking and Real Options

Real options changes the way risk is viewed since the option limits the downside effects while the option is held. Option-like arrangements are common, particularly in investment situations. Since investments signify the implementation of the company's strategy, real options can add to the development of that strategy. A few examples of how this is done are discussed in this section.

An example of real options at work in the field of strategy is a large company taking an equity stake in a smaller company while the smaller company develops its technology or products. This allows it to monitor the performance of the smaller company and to have first option should events turn out better than expected. Pharmaceutical companies are dependent on a pipeline of new discoveries of potential drugs. A strategy that is pursued by some pharmaceutical companies is to assist start-up companies by funding them. In return, the larger corporation offers a route to market for the start-up's products. If the product is better than expected, the corporation may even exercise an option to acquire the start-up.

A similar example to that of the pharmaceutical company mentioned above is in the field of geological exploration. Junior explorers are known to be better at discovering new ore deposits than the larger "majors." By investing in juniors at an early stage, a major gains the option to develop a mine from any deposits found by the junior.

Another form of the development of a project in stages is to create modules that can be repeated once the first module has been tested. Thus, the first module is an option to develop the others. A franchising concept for a chain of retail outlets is tested by investing in the first outlet. If this is successful, the outlets are rolled out into every shopping mall in the country.

There are two sides to real options. The first is the recognition of the existence of the real option, which adds to the value of the opportunity arising from the value of flexibility. The other side of real options is that options are not exercised easily. It is only optimal to exercise options that are deep in-the-money. This means that the conditions for the decision to invest are more stringent than the *NPV* rule, that is, that the *NPV* must be greater than zero. This is clearly demonstrated in the first of the two cases studies that follow this section.

15.10 Case Study: Phased Expansion of Gas-to-Liquids Operation

15.10.1 Introduction

Natural gas is an abundant source of energy. Unfortunately, the largest sources of natural gas tend to be far from the markets for energy. The natural gas can be piped to markets, it can be refrigerated and transported in tankers as liquefied natural gas (LNG), or it can be converted to a liquid fuel, such as diesoline, and distributed to the market. Qatar, a small emirate in the Arabian or Persian Gulf has abundant supplies of natural gas. A number of major oil companies are investing in the various processes to extract the natural gas and transport it to Western consumers.

An oil producer is investigating the construction of a gas-to-liquids (GTL) operation. For a number of reasons, most of them related to negotiation strategy rather than economics, the project will be developed in two stages. An initial smaller plant will be installed. Once this plant is producing liquid fuels, the second plant will be approved. The project financials are given in Table 15.6.

The terminal value of the operation is calculated using a price to earning multiple of 11, which is currently the average P/E ratio for oil companies. The capital allowance (tax depreciation) is calculated using a 25% declining balance, which is the same as that used in the UK.

Table 15.6 Project financials for the GTL investment

Year	0	1	2	3	4	5	6	7	8
Revenues	0	200	255	280	1,080	1,195	1,255	1,256	1,257
Cost of goods sold	0	90	115	126	486	538	565	565	566
SG&A expenses	0	44	56	62	238	263	276	276	277
Operating profit (EBIT)	0	66	84	92	356	394	414	414	415
Capital allowance	0	138	103	77	708	531	398	299	224
Tax	0	0	0	5	0	0	5	39	65
Cash flow	0	66	84	87	356	394	409	375	350
Capital expenditure	550	0	0	2,600	0	0	0	0	0
Increase in working capital	127	0	0	598	0	0	0	0	0
Free cash flow	−677	66	84	−3,111	356	394	409	375	350
Terminal value									3,849

Table 15.7 *NPV* calculation for the GTL investment

Present values	0	1	2	3	4	5	6	7	8
CF+TV	−677	66	84	−3,111	356	394	409	375	4,199
Discount factor (12%)	1.000	0.893	0.797	0.712	0.636	0.567	0.507	0.452	0.404
PV(CF+TV)	−677	59	67	−2,214	226	224	207	170	1,696
NPV	−242								

The net present value is calculated in Table 15.7. Clearly, the investment in this opportunity does not qualify for approval because the *NPV* is less than zero.

The investment in a GTL plant is a major investment. The company believes that the technology has greater potential than competing LNG operations, which are energy inefficient and costly. There are only a few companies with GTL technology, and this investment represents an opportunity to get further ahead of other oil and gas companies. The senior management of the GTL division requested the business analyst to re-examine this investment.

15.10.2 Staging the Investment Decision

The investment decision that is required now concerns the smaller plant. The project financials can be split into two to represent the two decisions: the first decision to invest in the smaller plant, and the second decision to invest in the larger operation. Splitting the project financials into the two stages is shown in Table 15.8.

The analysis of these two stages separately is shown in Table 15.9.

Unfortunately, the *NPV* for both stages is much less than zero. Neither of these investment stages meets the minimum requirements for approval. It could also be argued that the capital requirement for the second stage, $2,600 and $568 million is known now, and should not be discounted at the risk-adjusted discount rate of 12% but at the risk-free rate of 5.5%, which is the current yield on 10-year government bonds. This would make the *NPV* for the second stage even more negative than the value of −$219.6 million shown in Table 15.9.

15.10.3 Real Options Analysis

The investment in the first stage is the prerequisite for the investment in the second stage. The two investments are not independent of each other. In the real options way of thinking, the first investment includes an option to invest in the second stage. The decision tree for this representation of the decision is shown in Figure 15.14. A comparison of this decision tree and that of an option shown in Figure 15.1 indicates that this investment decision follows the same pattern.

Table 15.8 Project financials split into two phases

Year	0	1	2	3	4	5	6	7	8
Phase 1									
Revenues	0	200	255	280	280	280	280	280	280
COGS	0	90	115	126	126	126	126	126	126
SG&A expenses	0	44	56	62	62	62	62	62	62
Operating profit	0	66	84	92	92	92	92	92	92
Capital allowance	0	100	75	56	42	32	24	18	13
Tax	0	0	3	12	17	21	23	25	27
Cash flow	0	66	81	80	75	72	69	67	66
Capital expenditure	550	0	0	0	0	0	0	0	0
Increase working capital	127	0	0	0	0	0	0	0	0
Free cash flow	−677	66	81	80	75	72	69	67	66
Terminal value									721
Phase 2									
Revenues				0	800	915	975	976	977
COGS				0	360	412	439	439	440
SG&A expenses				0	176	201	215	215	215
Operating profit				0	264	302	322	322	322
Capital allowance				0	650	488	366	274	206
Tax				−7	−17	−21	−18	14	38
Cash flow				7	281	323	340	308	284
Capital expenditure				2,600	0	0	0	0	0
Increase working capital				598	0	0	0	0	0
Free cash flow				−3,191	281	323	340	308	284
Terminal value									3,129

Table 15.9 Calculation of the *NPV* for the two investment stages

Year	0	1	2	3	4	5	6	7	8
Discount factor (12%)	1.000	0.893	0.797	0.712	0.636	0.567	0.507	0.452	0.404
PV(CF)	−676.5	58.9	64.6	57.0	47.9	40.7	35.0	30.3	317.6
NPV (phase 1)	−24.5								
PV(CF)	0.0	0.0	0.0	−2271.2	178.6	183.1	172.1	139.4	1378.5
NPV (phase 2)	−219.6								

The *NPV* analysis presented previously was based on the assumption that both the first stage and the second stage decisions need to be made now. In splitting the stages, it was subtly hoped that either of the stages had a positive *NPV*, and some justification could be made for investment in the first stage.

Investment in the first stage means that the owners not only acquire the smaller plant but also the right but not the obligation to build the larger plant. Thus, the purchase price for the smaller plant inludes the price of the option to build the bigger operation in three years from now.

The second stage can be evaluated not as an investment decision that must be made now, but as an option to invest in three years time. This investment can only

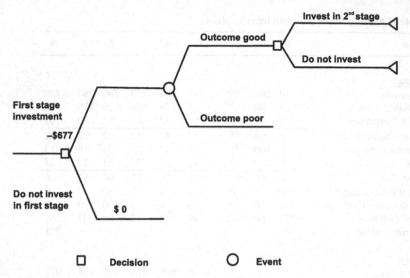

Figure 15.14 Decision tree for two-stage investment

be made in three years, not earlier, so it is a European style option. This means that the Black–Scholes equation can be used to evaluate this option.

The values for the input into the Black–Scholes equation are given in Table 15.10.

$$d_1 = \frac{\ln(2{,}052/3{,}191) + (0.055 + 0.2^2/2)(3)}{0.2\sqrt{3}} = -0.638$$

$$d_2 = 0.490697 - 0.2\sqrt{3} = -0.984$$

$$N(d_1) = 0.262$$

$$N(d_2) = 0.162$$

The value of the call option is calculated from Equation 15.19 as follows:

$$call = (2{,}052)0.262 - 3{,}191e^{-0.055(3)}(0.162) = 95.5$$

Thus, the value of the option to invest in the larger plant is worth $95 million. The total *NPV* is the sum of the *NPV* for the first stage and the value of the option. This is expressed as follows:

Table 15.10 Inputs to the Black–Scholes equation

Parameter	Meaning	Value	Units
S	Value of underlying asset	2,052 = 178.6+183.1+172.1+139.4+1378.5	$ million
K	Investment	3,191 = 7+2,600+598	$ million
T	Time to investment decision	3	Years
r	Risk-free rate	5.5%	Per year
σ	Volatility	20%	(Per year)$^{1/2}$

$$NPV = NPV_{Stage1} + call\ option\ on\ larger\ plant = -24.5 + 95.5 = 71.0$$

The total *NPV* is greater than zero, and investment in the first stage of the project is recommended.

This analysis was presented to the senior management of the GTL division and accepted by them. It will be presented to the corporate investment committee before requesting approval from the board of directors.

15.10.4 Concluding Comments

The value of the option is derived from the possibility of improved economic conditions at the time that the decision for the second stage is required. At the end of the three years, the company does not have to make a now-or-never decision for the second stage. It can further delay that investment. This means that the option does not have to be exercised at the end of the three-year period, but it can wait for a lot longer. The option is probably better represented by a European call option for the first three years, and an American call option after that period. This second option has been ignored in this analysis.

On the surface, it appears that the real options approach has recognised value from the flexibility and made it easier to justify the investment. This interpretation should be viewed as incorrect. The real options approach does encourage the purchase of options, because options that are out of the money can have value. This means that marginal investments on an *NPV* basis can be justified by the real options approach. However, the conditions for the exercise of the option, that is, the actual investment decision, are much more stringent than the *NPV* analysis. An option is not exercised if it simply is in-the-money. It must be deep in-the-money in order to justify the exercise of the option and hence invest. This same condition applies to the real options analysis. The real options approach encourages the purchase of options but discourages the exercise of them unless they are deep in-the-money. The exercise price, that is, the condition for investment, is discussed in the next case study.

15.11 Case Study: Value of the Joint Venture Contract for Cuprum

15.11.1 Introduction

The production of copper proceeds in several steps and by a variety of processing routes. The ore is mined, crushed and ground, and then upgraded by flotation. Flotation is a process in which ore particles that stick to soap bubbles are separated from waste rock by scooping the foam from the top of the flotation tank. The flotation product is called a concentrate. The concentrate is then processed by smelting, in which the concentrate is melted and reacts at a high temperature to separate into copper metal and the waste products. The copper is further processed by refining to yield a high-grade product that is sold on the international metals exchanges.

Cuprum has invented a new process for the production of copper. The process relies on a proprietary biotechnology. The input to the process is the copper concentrate and the product from the process is high-grade copper. Cuprum's marketing material states that their process is going to revolutionize the industry. Cuprum is considering entering into a joint venture agreement with a mining company to build a demonstration plant. Cuprum, a company with limited resources, would like to know the value of the joint venture agreement so that it can negotiate a fair contract for itself. It would also like to know under what conditions the company should build the plant.

15.11.2 Market for Copper Concentrates

The choices for the treatment of the copper concentrate are shown in Figure 15.15. Either the concentrate is treated by the mine in an integrated operation or it is sold to a custom smelter. There are also merchants that operate in the market for copper concentrates, purchasing from the mine and selling to custom smelters. This means that the raw materials to Cuprum's operations are a marketable commodity.

The sale of concentrates to a smelting operation is based on a toll treatment contract called treatment costs and refining charges, or TC/RC, contracts. (Toll treatments were discussed in Chapter 4.) They are usually negotiated for a year in advance between a mine and a smelter or between a merchant and a smelter. Consequently, the TC/RC spot price is uncorrelated with the spot copper price. However, if the copper price increases, the TC/RC price will generally increase about a year later.

The participation of merchants in this market has grown dramatically. The merchant business is around 50% of the total amount of concentrate produced, while twenty years ago it was close to 15%. These merchants have substantial financial strength and expertise in risk management. They hedge their risks in the concentrate market with positions in the copper spot market, and often take speculative

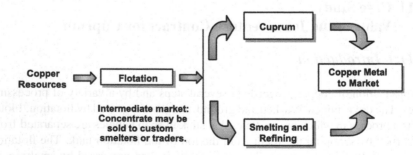

Figure 15.15 Treatment options and markets for copper concentrates and copper metal

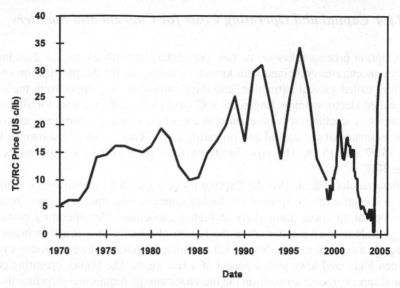

Figure 15.16 TC/RC prices for copper concentrate over the last 30 years

positions in the concentrate market. This has resulted in greater financial and pricing flexibility, and greater volatility in the TC/RC prices.

The treatment charge is a fixed amount based on the amount of material treated, and the refining charge is based on the metal content. The TC/RC contract allows for some participation by the smelter in the movements of the copper spot price. However, this is a very weak function, so that the seller of the concentrate absorbs most of the risk of the variation in the copper spot price. This means that the revenue for the smelter, and hence the profitability of the smelter, is almost entirely determined by the TC/RC price. The TC/RC prices over thirty-five years are shown in Figure 15.16. It is clear that the volatility of the TC/RC price has increased consistently over the period.

15.11.3 Revenues

Since Cuprum is competing directly with smelters and not with mines that produce the copper concentrates, the evaluation of Cuprum's value proposition should be based on a direct comparison with the economics of smelting.

The TC/RC contract is a tolling contract, which was discussed in Chapter 4. Although the copper concentrate is a raw material to their operations, the revenue to the smelter is based on the quantity of material to be treated and the TC/RC price, which is typical for a tolling agreement. The economics of the smelting operation is not dependent on the price of copper, but on the TC/RC price. The mines, through this contract with smelters, have retained exposure to the copper price. Thus, the TC/RC price sets the revenue for the smelters.

15.11.4 Capital and Operating Costs for Cuprum and Smelters

The Cuprum process relies on its new proprietary technology for the dissolution of the concentrates combined with known technologies for the purification of the solution, called solvent extraction, and the recovery of the copper from the solution, called electrowinning. Engineers at Cuprum have collected data for all the installations of smelting, solvent extraction and electrowinning operations, and have made estimates of the capital and operating costs. These costs are shown in Figures 15.17 and 15.18. The error limits on the capital costs are regarded as being ± 30%.

These results indicate that the Cuprum process may have an advantage with regard to capital costs. In spite of the higher capital costs, smelters generally have lower operating costs, particularly at higher capacities. The operating costs for a large-scale modern smelter are in the region of 13 c/lb. This is a significant advantage in a cyclical commodity market in which the price of the commodity cycles between highs and lows with a period of a few years. The higher operating costs of the Cuprum process arises from the increased energy requirements, primarily because of the need to recover the copper from solution by electrolysis. The electrical costs for electrowinning cannot be significantly reduced because of thermodynamic limitations.

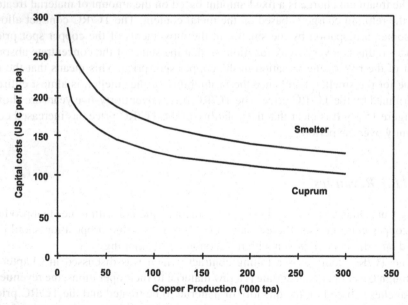

Figure 15.17 Capital costs for the Cuprum and smelting process as a function of production scale

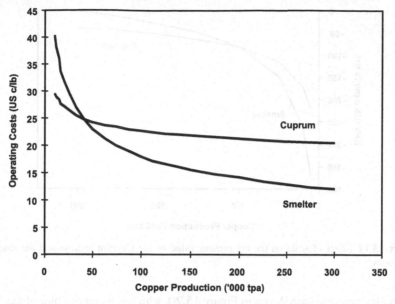

Figure 15.18 Operating costs for the Cuprum and smelting process as a function of production scale

15.11.5 NPV of Copper Smelting and Cuprum's Process

Since the revenues and costs for the two alternatives have been established, it is now possible to compare the net present value for these operations. The cash flows for each year, π_i, are determined from the after-tax profits, determined by the following formula representing the income statement:

$$\pi = (R - C)(1 - T) - TdK \tag{15.31}$$

where R is the revenue, that is the TC/RC price, in US c/lb, C are the costs in US c/lb, T is the tax rate, d is the depreciation rate and K is the capital cost in US c/lb pa. The substitution of Equation 15.31 into the equation for the net present value gives the following expression:

$$NPV = \sum_{t=1}^{n} \frac{(R_i - C_i)(1 - T) + TdK}{(1 + k)^t} - K \tag{15.32}$$

where k is the discount rate. A risk-adjusted discount rate of 12% was used. The net present value for a project with an economic life of twenty years using a TC/RC price of 29 c/lb is shown in Figure 15.19. The effects of tax were neglected, that is $T = 0$.

These results indicate that neither smelting nor the Cuprum alternative meets the decision criteria that the *NPV* should be greater than zero even in the optimistic situation of no taxes and the TC/RC price being constant for twenty years. The

494

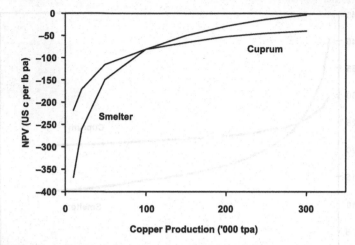

Figure 15.19 Effect of scale on the net present value of the Cuprum process and the smelting operations

alternative processes are shown in Figure 15.20, which is a contour plot of the *NPV* against the operating and capital costs. This plot reveals that the higher capital of the smelters is compensated by their lower operating cost, which removes most of Cuprum's advantage.

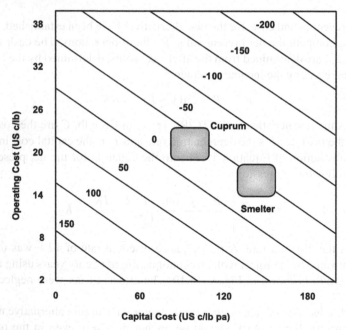

Figure 15.20 Contour plot of the *NPV* as a function of operating and capital costs, showing the trade-off between these costs for the two options

The capital costs are expressed in terms of annual production. The production from a mine is typically in the range of 100,000 tons per annum (tpa), while smelters typically have capacities of 200,000 tpa. The total capital cost for 100,000 tpa facility with a cost of 100 US c/lb pa is calculated as follows:

$$Capital\ cost = 100,000\,tpa\,(2,204\ lb/t)\,100\ c/lbpa\frac{1}{100c/\$} = \$220.4\ million$$

The conclusion that is drawn from the *NPV* analysis is that, although smelters have an advantage over the Cuprum competitor, there is no economic incentive to invest in either operation to treat copper concentrates. Neither option is economically viable unless government subsidies are received or the capital costs are dramatically lowered. The prices paid for the acquisition of smelters in the last ten years have varied from 27% to 58% of their replacement costs. This implies, as the *NPV* analysis indicates, that the revenues provided by the smelting sector as a whole provides little or no return on the invested capital.

The *NPV* analysis and the comparison of the economics of the process with that of smelters presented above led the design team to set a target for the capital cost of 90 US c/lb pa and the operating cost of 17 US c/lb. The *NPV* for this scenario is –0.4 c/lb pa, which translates to –$881,600 for a 100,000 tpa facility. Based on the uncertainties of the costs, they believe that this can be achieved, and that the company's executives will approve this investment.

15.11.6 The Value of the Technology

Given the bleak picture of the economics outlined in the previous section, it would come as no surprise if the answer to the question "what is the value of the technology?" were zero. In fact, this is the view taken by many executives in the industry. However, they are wrong because even though the TC/RC price has been low, there is much volatility in the market, and the future price is far from a known constant. The low TC/RC prices have resulted in under investment in smelting capacity, and a squeeze on capacity may easily lead to a sustained rise in prices. The technology represents an option on the TC/RC prices.

There is another source of uncertainty, the technical uncertainty of the new process. While the engineers at Cuprum have made their best possible estimates of the capital and operating costs, the only way to determine these is to build the operation. This type of uncertainty is endogenous, meaning that it is internal to the investment, and the only way in which this uncertainty is resolved is by investing in the operation. In other words, with endogenous uncertainty, an incentive exists to start the business, and "to learn by doing" even though the *NPV* is less than zero. Such investments are sometimes justified as "strategic."

Although the technical uncertainty is important, the value of the technology will be determined by considering the price uncertainty. The price uncertainty is exogenous, that is, it is external to the investment, and there may be value in waiting to see how the price evolves. This option on the price provides an incentive to delaying the investment decision. Real option analysis is used to determine the value of the technology as an option to delay the investment, and to determine the price at which investment should be made in the Cuprum process. The value of this option is derived from the ability of the owners to wait for the right time to invest.

If the price is uncertain, then both the value of the project and the option to invest in the project are uncertain. As a result, the real option approach must first calculate the value of the project, and then calculate the value of the option to invest. A similar procedure was followed in the valuation of an option using the binomial lattice: first, the value of the asset was determined, and then the value of the option, based on the asset lattice, was calculated.

(i)   Real options model

The model follows that of McDonald and Siegel (1986). The firm must decide to invest in a single project at a cost, K. The value of K is known, and fixed and is irreversible, that is, the investment represents a sunk cost. The value of the project, V, is uncertain, but analogous to stock prices, it is assumed that it follows geometric Brownian motion. The value of the project is determined by the price, P, of the commodity produced by the project. The returns on P, given by $\Delta P/P$, are normally distributed with mean α and standard deviation σ. The spot price is determined competitively, and follows the stochastic Brownian motion given by:

$$dP = \alpha P dt + \sigma P dz \tag{15.33}$$

where dz is the increment to a Brownian motion, α is the drift in the spot price, and σ is the standard deviation of the spot price.

A flow of benefit or services accrues to the owner of a commodity that does not accrue to the owner of a futures contract for delivery of the commodity. The owner of the commodity can store the commodity or liquidate inventories of the commodity, transport it, or benefit from local shortages of the commodity. This continuous flow of benefit is analogous to a dividend yield on stocks. From the point of view of a holder of an option on a stock, the stock continuously losses value due to dividend payments. The convenience yield, δ is the difference in the expected change in the commodity price α, and the expected return μ. The expected rate of return that investors would require is given by the capital asset pricing model. Thus, convenience yield is given by the following expression:

$$\delta = \mu - \alpha \tag{15.34}$$

Following McDonald and Siegel, the value of the project is given by the following differential equation:

$$\frac{\partial V}{\partial t} + \frac{1}{2}\sigma^2 P^2 \frac{\partial^2 V}{\partial P^2} + P(r - \delta)\frac{\partial V}{\partial P} = rV - \pi$$

This is similar to the Black–Scholes equation for a dividend paying stock. The general solution to this differential equation for a perpetual project is given by the following expression:

$$V = P/\delta - C/r \tag{15.35}$$

The term P/δ is the expected present value of the revenue stream discounted at the risk-adjusted rate μ. This is shown as follows. The expected value at time t of the revenue stream P that follows a geometric Brownian motion with a drift rate of α is $Pe^{\alpha t}$. The present value of this revenue at t is therefore $Pe^{\alpha t}e^{-\mu t}$. Thus, the total present value of this stream over the life of the project is given by the following expression:

$$\int_0^\infty Pe^{\alpha t}e^{-\mu t}dt = P/(\mu - \alpha) = P/\delta \tag{15.36}$$

Similarly, the term C/r is the present value of the constant cost stream, discounted at the risk-free rate, r. It is discounted at the risk-free rate because it is known with certainty. Therefore, the value of the project is calculated with discount factors that are appropriate to the level of risk of the cash flows. The risky cash stream is discounted at $\delta = \mu - \alpha$, while the sure constant cash stream is discounted at the risk-free rate, r.

The value of the option to invest is described by the following differential equation:

$$\frac{\partial F}{\partial t} + \tfrac{1}{2}\sigma^2 F^2 \frac{\partial^2 F}{\partial P^2} + F(r - \delta)\frac{\partial F}{\partial P} = rF \tag{15.37}$$

McDonald and Siegel showed that if the option never expires, that is, it is perpetual, then the solution of the differential equation is given by the following expression:

$$F = (P^*/\delta - C/r - K)\left(\frac{P}{P^*}\right)^{\beta_1} \tag{15.38}$$

The parameters P^*, and β_1 are given by the following equations:

$$P^* = \frac{\delta\beta_1}{\beta_1 - 1}(C/r + K) \tag{15.39}$$

$$\beta_1 = \tfrac{1}{2} - (r - \delta)/\sigma^2 + \sqrt{((r - \delta)/\sigma^2 - \tfrac{1}{2})^2 + 2r/\sigma^2} \tag{15.40}$$

In this analysis, the project and the option on the project have been considered to last forever, that is, to be perpetual, and the costs have been considered to be constant. These simplifactions have been made so that analytical solutions to the differential equations can be obtained.

The result from the model is that it is optimal to invest, that is, to exercise the option when $P = P^*$. This is an important result, because it establishes a decision point for management action.

(ii)   Results

The calculation of the value of the technology is based on the set of parameters shown in Table 15.11.

The value of the cash flows from the project is given by $V(P)$ which has a value of 240 US c/lb pa. The NPV for the opportunity is given by $max(V(P) - K, 0)$, and has a value of 150 US c/lb pa. This is significantly higher than the discounted cash flow (DCF) analysis.

The effect of the price of the TC/RC contract is on the value of the project and the option to invest is shown in Figure 15.21. As the TC/RC price increases, so does the value of the project and the option to invest. The intrinsic value of the option is the NPV of the project calculated using the risk-neutral method, that is, given by Equation 15.35.

The value of the option is derived from the ability to delay investment decisions. The proprietary technology of Cuprum allows the company to delay the decision without the fear that the opportunity for investment will be lost by competitive action. This means that the value of the technology is the value of the option. For the

Table 15.11 Parameter set for the calculation of the value of the technology

Parameter	Symbol	Value	Unit
Current concentrate price	P	29	US c/lb
Current operating costs	C	17	US c/lb
Convenience yield	δ	5%	pa
Risk-free rate	r	5%	pa
Investment	K	90	US c/lb pa
Volatility	σ	20%	$(pa)^{1/2}$

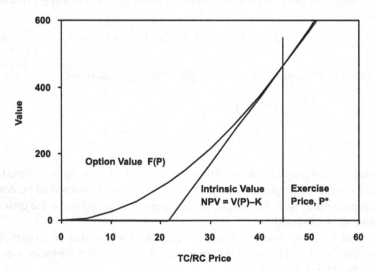

Figure 15.21 Effect of the TC/RC price on the value of the project and the option

set of parameters used, this means that the technology is worth 202 US c/lb pa, or for a 100,000 tpa facility, it is worth $440 million.

The optimal exercise price is given by P^*. The value of P^* for the parameters used is shown in Figure 15.21. The value of P^* is 45 US c/lb, which is significantly higher than the current price of 29 US c/lb, and is more than double the price at which the *NPV* is zero. The real options approach recommends waiting until the TC/RC price reaches a value of 45 US c/lb.

A major difference between the DCF analysis and the real options analysis is in the discounting procedure. The DCF analysis discounted all cash flows at 12%. The real options analysis recognised that the costs were known and constant, and are, therefore, discounted at the risk-free rate. The price is subject to drift described by Brownian motion. This means that the revenue price will increase upwards at the growth rate of the price. The present value of the revenues is obtained by discounting the revenues at the risk-adjusted return required by investors. It was shown in Equation 15.36 that the term P/δ is the expected present value of all the revenues discounted at the risk-adjusted rate μ, which is described by the capital asset pricing model (CAPM). The convenience yield, δ, is the difference between the growth rate, α, which represents the increase in the average price with time, and the risk-adjusted rate, μ, which is the investors' required rate of return.

The solution is sensitive to the value of the convenience yield. The effect of the choice of the value of the convenience yield and the optimal investment price is shown in Figure 15.22. These results indicate that as the convenience yield increases, the *NPV* decreases and the optimal exercise price increases. The value of the convenience yield chosen, 5%, is close to the estimated long-term average for copper.

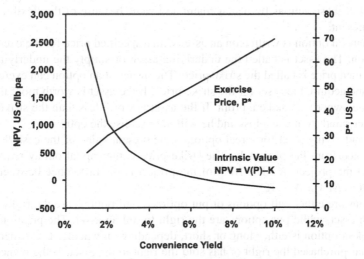

Figure 15.22 The effect of the convenience yield on the *NPV* and P^*

15.11.7 Concluding Comments

The analysis of the economics of the new technology and its competition has led to specific targets for the capital and operating costs. The real options analysis has provided a price at which it is optimal to invest, and, importantly, it has led to a clear idea of the value of the technology, so that Cuprum's principle negotiators have a clear understanding of their value.

The *NPV* calculated using the real options approach, which has similarities with the certainty equivalent approach, gave values of the *NPV* that were significantly higher than the DCF value for the *NPV* using a discount rate of 12%. The differences between the two calculations were that the real options approach discounted the risky cash flows at the convenience yield and the certain costs at the risk-free rate.

15.12 Summary

An option is the right to purchase an asset some time in the future for a price that is settled now. The owner of the option does not have any obligation to do so. Many investment decisions have this same structure: a company has the right to invest in the project, but it has no obligation to do so. By analogy with options traded on the financial markets, this option to invest has value. The real option approach recognizes that the options that a company has, such as the option to delay an investment, has value. Other forms of management flexibility also have value. For example, management can abandon operations, can switch between products or suppliers, can grow the business, can develop a project in stages and abandon it at any stage to limit loss. All of these actions by management add value because of the flexibility that they represent.

A financial option is written on an asset with a specified price for the exercise of the option. The asset is called the underlying asset, or simply the underlying, and the specified price is called the strike price. The owner of an option can exercise the option and acquire the asset for the strike price. If the asset is worth more than the strike, the owner will make a profit. If the asset is worth less than the strike price, the owner stands to make a loss and he will not exercise the option.

The underlying asset for a real option is the present value of the cash flow that is generated from the project V. The strike price is the capital outlay required to construct the project, K. The *NPV* of the project is the difference between these two, that is, $NPV = V - K$.

Options are either call options or put options. Call options are the right to purchase an asset, while put options are the right to sell an asset. The position of the owner of an option is either long or short, depending on whether the owner of the option has purchased the right or has sold the right to the asset. If the owner of an option can exercise the option early, he has an American option. If the option cannot be exercised until the end of the option period, it is a European option.

Options are valued based on a differential equation that has some similarities with molecular transport by flow and diffusion. The solution to European options is known as the Black–Scholes equation. There is an approximate solution to the Black–Scholes equation, called the binomial lattice. This is a technique for the numerical solution of the differential equation. It is suitable for the calculation of other options in addition to European options. Both the binomial lattice and the Black–Scholes solution rely on valuing the option in a risk-neutral world. This is conceptually similar to the certainty equivalent discussed in Chapter 13.

The real options approach provides a method for the valuation of the option, of the underlying asset and the optimal point to exercise the decision. The options approach has implications for the way in which capital projects are viewed, and strategy is formulated. The real options approach is sometimes confused with decision tree analysis, presented in Chapter 14. Both approaches are valuable, and because this confusion simply lowers the value of both approaches, it is not encouraged.

15.13 Looking Ahead

In the next chapter, the sources of capital are examined. The markets for raising equity and debt capital are discussed. The financial environment is generally discussed. Different types of loans are examined.

15.14 Review Questions

1. What is an option?
2. What is the difference between purchasing an option on shares and purchasing the shares?
3. Describe the way in which investment decisions are similar to financial options.
4. What is management flexibility?
5. Does management flexibility have value? Explain your answer.
6. What is the difference between an option and a bet?
7. What is the difference between a call option and a put option?
8. What does "underlying asset" mean?
9. What is "risk-neutral valuation?"
10. What is "volatility?"
11. What is a "real option?"
12. What is the underlying asset in a real option?
13. What is the "intrinsic value" of an option?
14. What are the differences between American options and European options?

15. What does real options analysis provide that discounted cash flow analysis does not provide?

16. Discuss the similarities and differences between real option analysis and decision tree analysis.

17. What is the binomial lattice?

18. Discuss the meaning of the parameters of the Black–Scholes equation.

19. Does real options analysis make *NPV* redundant?

15.15 Exercises

1. An asset has a price of $100. The asset price can increase or decrease by 10% over one interval. The risk-free rate is 5%.

(i) Determine the probability p of an increase in price.

(ii) Determine the value of the asset if the price increases and if it decreases.

(iii) Determine the payoff of an option with a strike price of $80 that matures at the end of the interval.

(iv) Determine the price of the option.

2. Consider the following European call option: the price is currently $100 and the strike is $95. The volatility is 20% pa$^{1/2}$. The option matures in a year.

(i) Create the binomial lattice for the value of the underlying asset if two time steps are used.

(ii) If the risk-free rate is 5%, calculate the value of the option.

3. Repeat question 2 using three time steps. Is the answer the same? Explain your results.

4. Determine the value of the following American put option using both the binomial lattice and the Black–Scholes equation: asset price: $8; strike price: $10; risk-free rate: 5%; volatility: 40%; time to maturity: 5 months. For the binomial lattice use steps of 1 month.

(i) Determine the value of the option if the volatility increases to 60% pa$^{1/2}$.

(ii) Determine the value of the option if the risk-free rate increases to 12%.

(iii) Comment on the results.

5. A company has a patent on a new technology that allows it to manufacture a new product. Market surveys are positive, suggesting that there is a latent demand for the product. The project requires an investment of $80 million and the present value of the returns is $120 million. The patent expires in 8 years. If the risk-free rate is 5% and the volatility of the underlying business is assumed to be 30% pa, determine whether the company should invest now or defer the investment.

6. A company has an oil lease that allows its owner to explore for oil and if found produce oil from the discovery. The probability of the discovery of oil is 10%,

and the *NPV* of the production from such as discovery is expected to be $100 million, with an investment of $30 million in wells and pipelines. If no oil is found the cost of the lease and drilling amounts to $5 million. The company wishes to sell this lease, which expires in three years, for $10 million. If the volatility of the *NPV* is 20%, determine whether the company should sell this lease. The risk-free rate is 5%.

7. A company is assessing its options on a division that is under-performing. The company estimates that the division's assets are worth $20 million. Interest in the division from a private equity company suggests that it can sell the business for $30 million if a deal is struck within the next six months. If the risk-free rate is 5% and the volatility of the business is 30%, determine whether the company should sell.

8. Within the next two years, a company can expand its operations by 150% for a capital cost of $25 million. The net present value of the current business is $20 million and the company pays a 5% dividend. If the risk-free rate is 10% and the volatility is 15%, what is the value of the option to expand? Should the company expand now, or delay the decision until later?

9. A business can contract its operations by 30% at a cost of $2 million. If the current value of the business is $10 million, the risk-free rate is 5% and the volatility is 50%, determine the value of the option to contract within the next six months.

The financial resources for capital investments come from two main sources. These are equity and debt sources. Equity refers to the contribution of the owner and debt refers to the contribution of a lender. A company funds its requirements for capital from both these sources. The different ways of raising finance for the company and the financial markets in which this is done are discussed in Chapter 16. The financial markets include a variety of instruments that are designed to mitigate financial risk. These instruments, such as futures and options, and their use in hedging risk are also examined.

The providers of debt often require security, or collateral, for the debt. An alternative to this is for the debt providers to rely on the anticipated cash flows. In other words, the lenders must be satisfied that the cash flows can pay back the debt. This type of funding is called project finance. A public–private partnership is an application of project finance to fund public infrastructure projects. The details of project finance and public–private partnerships are discussed in Chapter 17.

Chapter 16
Sources of Finance

16.1 Introduction

Projects, businesses, and companies need capital to conduct their business. This capital is used to invest in fixed capital and in working capital, so that the business can make its products, meet the market needs, pay its taxes and provide a return to the investors. In return for the capital, the company issues financial instruments, such as share and bond certificates. These financial instruments or securities must be attractive to investors in amount, timing and risk of the return that they promise.

There are two main types of investors, or providers of capital. They are the providers of either debt capital or equity capital. The cost of debt and equity financing to the company was discussed in Chapter 12. The different forms of debt and equity securities, the markets for these securities, and the choice of the mix of debt and equity financing are discussed in this chapter.

16.2 Lenders, Borrowers and Financial Institutions

The financial system links the people and organizations that wish to loan money with the people and organizations that wish to borrow money. The lenders and the borrowers both come from the same economic sectors, such as the household sector and the corporate sector. The difference between them is that the lenders have surplus cash while the borrowers have a deficit of cash. Lenders are also called savers or investors. Borrowers wish to gain access to funds that the lenders have, that is, they wish to raise capital from investors. Lenders wish to earn a return on their savings. In return for the cash provided by lenders, borrowers issue securities, which are contracts that obligate the borrower to pay the lenders their money back. The financial system is shown schematically in Figure 16.1.

Lenders and borrowers can make financial arrangements directly with each other or through a financial institution or intermediary. Direct transactions may be ar-

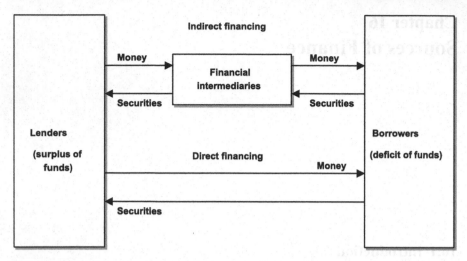

Figure 16.1 The financial system

ranged by a broker who collects a commission, or by an investment bank that serves as an underwriter. The investment bank assists in the design of the borrower's securities so that they are attractive to investors, purchases the entire issue of these securities from the borrowers, and then resells them to investors.

A financial intermediary collects money from lenders and issues its own securities, and provides money to borrowers in exchange for the borrowers' securities. The function of the financial intermediary, such as a bank, is to provide lenders with the financial securities they require and to accept the securities borrowers prefer. Typically, the terms of securities that borrowers provide and lenders demand differ in size, maturity, and liquidity, amongst others. The financial intermediary resolves this difference by creating new forms of capital more suited to the needs of lenders. In other words, it creates new securities that are more in demand.

The financial intermediary creates two markets, one for the lenders and one for the borrowers. A financial intermediary thus acts more than a broker between lenders and borrowers. They aggregate smaller funds into larger loans that are more suitable for borrowers. They diversify risk for the lenders or savers by holding a portfolio of securities from a wide range of borrowers. In this manner, they facilitate the flow of funds between lenders and borrowers.

The types of financial intermediaries can be divided into deposit-taking and non-deposit-taking institutions. Deposit-taking intermediaries are the banks, mutual banks and credit unions. Examples of non-deposit-taking intermediaries are life insurance and short-term insurance companies, mutual funds, unit trusts and pension funds.

Banks offer a wide range of services to lenders and borrowers in all sectors, that is, household, corporate, government and foreign sectors. They take deposits from savers and loan this money to borrowers. They also interact with other financial intermediaries, taking loans from other intermediaries and holding the securities of others. Life insurance companies accept premiums in exchange for a life policy. These premiums are invested in a diversified portfolio of stocks, bonds, real estate and mortgages. The insurer makes payments to the beneficiaries of the policies. Pension funds provide retirement plans. The funds collected are aggregated and invested in the same classes of assets in which life insurers invest. Mutual funds and unit trusts accept savings from investors, pool these savings, and invest them in a manner defined by the founding statements of the fund.

16.3 Financial Securities

There are two broad categories of financial securities. These are equity and debt instruments, reflecting the two major sources of funds. The type and characteristics of these instruments are discussed in the following sections.

16.3.1 Equity Instruments

Equities are the source of finance that the owners provide to the business. There are two forms of equity in a business. These are ordinary shares (common stock) and preference shares (preferred stock).

(i)   Ordinary shares

The ordinary shares or common stock in the business represent part or shared ownership of the business. The investors provide the share capital for the business in exchange for this partial ownership. From the viewpoint of the company, the company issues shares in order to raise the capital that is needed to participate in business. The return to the holders of common stock is directly related to the performance of the company. This return is in two forms, the growth in the value of the company and the dividends paid by the company. As discussed in Chapter 3, the claim of owners against the business is represented on the balance sheet as the equity, comprised of the share capital and the retained earnings.

The holders of ordinary shares are exposed to the entire risk of the business. If the business performs well, the value of the company increases and dividend payments increase; if the business performs poorly, the value of the business performs poorly, and dividend payments decrease or cease. The holders of ordinary shares have a right to a share of the profits only once the company has made a profit, and the board of directors declares a dividend.

The company raises equity capital by issuing new shares that the investors purchase. Consider the funding of the development of Amazon.com. The new share issues and the equity capital raised are shown in Figure 16.2. The company was started in 1994 with the founder investing $10,000 at a share price of $0.001. After

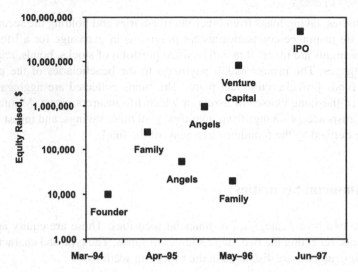

Figure 16.2 Equity raised by Amazon.com and the source of the funds

about a year, the founder's parents invested $245,000 at a share price of $0.17. In the next three rounds, angel investors provided $54,000 and $937,000 and family members provided $20,000 at a share price of $0.33. Two venture capital funds provided $8 million in June 1996 at a share price of $2.34. The company offered its shares to the public in the initial public offering (IPO) in May 1997. The IPO raised $49.1 million at a share price of $18.

The sale of additional shares in each new round of financing is at the expense of the current owners, since their portion of ownership of the company is diluted. However, the value of their investment usually increases as the value of the business and the share price increases.

The sale of new shares by a listed company, that is, subsequent to the initial public offer, can be in the form of a rights offer or a follow-on offer. The shareholders usually have a pre-emptive right to subscribe to new shares that the company may offer. This allows the shareholders to maintain their proportional share of the ownership of the company. Current shareholders may also waive this right. If the company sells to existing shareholders, it is called a rights offer. If the current shareholders waive their rights, the company can raise equity capital by selling new shares to new shareholders for cash. This is called a follow-on equity offer. In both cases, the share offer is based on market prices, although at a slight discount. The amount of the discount depends on the demand for the shares.

Part of the return to ordinary shareholders is a dividend, which is a cash payment to the shareholders. Companies can retain more of the profits or earnings by paying a smaller dividend. This increases the equity capital of the company without selling additional shares. The company may offer additional shares instead of a cash dividend. This practice, known as a scrip dividend or a stock dividend, allows the company to retain more of the earnings and thereby to increase the equity capital.

(ii)   Preference shares

Preference shares, or preferred stock, have the characteristic of both equity and debt. The dividend offered by preference shares is fixed, a percent of the face value of the shares. The dividends are more predictable than those from ordinary shares. These payments are similar to the interest payments on debt. However, if there are insufficient funds, the dividends are passed, that is, the company does not pay them. Unlike debt, failure on the part of the company to pay the preference dividend does not throw the company into bankruptcy. Usually, these unpaid dividends accumulate and holders of preference shares are paid in arrears. Preference shares usually have no maturity date. Unlike debt, the dividend payment from preference shares is not deductible for tax purposes. In the case of liquidation of the company, the holders of preference shares have a claim to the assets prior to that of the ordinary shareholders.

There are three additional forms of preference shares. Participating preference shares have a fixed dividend but also share in the profit, like ordinary shares, according to a predetermined formula. Redeemable preference shares give the company the option to redeem the shares at a specified price during a specified period. Convertible preference shares grant the holder the right to exchange them for ordinary shares according to prearranged terms.

16.3.2 Debt Instruments

Debt securities are the promises made by the issuer to the holders to repay the principal and to make payments of interest according to an agreed schedule. Some of the main features are the term of the loan, the security offered for the loan and the interest charged.

The term of the debt required by a company may be short or long-term. Short-term debt is for less than a year, and is usually required to fund changes in the working capital. The business of many companies is seasonal, and the increased working capital requirements can be funded with short-term debt. Long-term debt is used to fund the general activities of the business. It creates leverage for the owners of the business.

Debt can be secured or unsecured. Secured debt is secured over one or more of the assets of the company, which means that in the case of liquidation, the proceeds from the sale of the secured assets would be used to first meet the claims of the creditors or debt-holders. The secured assets are called collateral. In the case of liquidation, there is a ranking of the claims on the assets of the company. Senior debt has first claim and subordinated debt has a subsequent claim. Unsecured debt does not lay claim to any of the assets.

Debt can also be based on a fixed interest or floating interest. Fixed interest debt is that in which the interest rate is fixed for the period of the loan. Bonds are often fixed interest instruments. Variable interest or floating interest debt is that in which the interest rate is not fixed but varies according to the market rate. Mortgages are a common form of floating interest debt.

Various forms of loans are discussed in the next section. The features of long-term and short-term debt are presented in the sections following that.

16.3.3 Types of Loans

The different types of loans are regular loans, discount loans, instalment loans, amortized loans and bonds. A primary difference between these loans is the repayment schedules, which can have a dramatic effect on the effective interest rate that is paid. The effective interest rate represents the cost of the loan to the borrower, and the return of the loan to the lender. Because of the different returns to the lenders of these different types of loans and the risk associated with a particular financing situation, lenders are not flexible with the form of the loan.

The different forms of loans are discussed in terms of the repayment schedule. The techniques of discounted cash-flow analysis are used to determine the effective rate, which determines the cost of the debt, for each of the different types of debt.

(i)   Regular loan

A regular loan is one in which the interest is calculated at a flat rate. The interest is paid at regular periods during the term of the loan and the principal is paid at the end. The following two examples illustrate the regular loan.

Example 16.1: Regular loan with interest paid annually.

Consider a regular loan of $10,000 over a year at an interest rate of 13%. Interest is charge annually. Determine the interest repayment, the schedule repayment and the effective interest rate for this loan.

Solution:

The interest is charged at a flat rate. This means that the interest is calculated by the following expression:

$$Interest = (Days\ per\ period)(Interest\ rate\ per\ day)(Loan\ principal)$$
$$Interest = 365(0.13/365)(10,000) = \$1,300$$

The loan repayment schedule is shown as a time line in Figure 16.3.

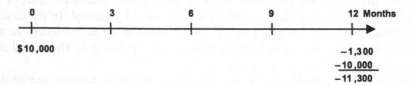

Figure 16.3 Repayment schedule for a regular loan of $10,000 at an interest rate of 13% with interest paid annually

The effective rate is given by Equation 5.6. In this case, it is straightforward to calculate, as shown below:

$$e = \left(1 + \frac{i}{m}\right)^m - 1 = (1 + \frac{0.13}{1})^1 - 1 = 0.13$$

The interest on the regular loan can be paid at more regular periods than annually. Consider the quarterly payment of interest in the following example.

Example 16.2: Regular loan with quarterly interest payments.

Consider the loan of Example 16.2. Calculate the effective interest rate if the interest is paid quarterly instead of the annual payment made in Example 16.1.

Solution:

The interest is calculated as follows:

$$Interest = (Days\ per\ period)(Interest\ rate\ per\ day)(Loan\ principal)$$
$$Interest = 91(0.13/365)(10,000) = \$324.11$$

The loan repayment schedule is shown in Figure 16.4. One of the quarters has an extra day so that the total number of days in four quarters sums to 365 days. Sometimes, this convention is not used. Instead, the loan is calculated as if there were 360 days in a year, which means monthly and quarterly payments are equivalent.

The rate per quarter is the internal rate of return for the cash flows. The effective annual rate is calculated as follows:

$$e = \left(1 + \frac{i}{m}\right)^m - 1 = \left(1 + \frac{0.13}{4}\right)^4 - 1 = 0.1365$$

The effective rate is 13.65%. The annual percentage rate, *APR*, is calculated as follows:

$$APR = (Rate\ per\ period)(Periods\ per\ year) = 0.0325(4) = 13\%$$

The earlier payment of the interest amounts has lead to an increase in the effective interest rate from the stated 13% to 13.65%.

The result obtained in Example 16.2 demonstrated the effect of the timing of the cash flows on their value. From a borrower's point of view, the early payment of the interest has increased the cost of the loan. From the lender's point of view, this has increased the lender's return on the loan.

(ii)   Discount loan

The discount loan is structured so that the entire interest amount is paid in advance. The borrower receives the loan amount less the interest payment. At the end

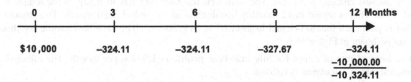

Figure 16.4 Repayment schedule for a regular loan of $10,000 at an interest rate of 13% with interest paid each quarter

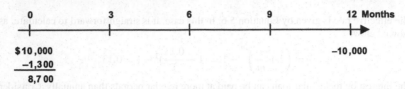

Figure 16.5 Cash-flow profile of a discount loan of $10,000 that carries an interest rate of 13%

of the loan period, the borrower repays the principal amount. The effect that this has on the effective rate is illustrated in the following example.

Example 16.3: Discount loan.

A borrower applies for a loan of $10,000 at a rate 13% for a year. The lender offers the loan on a discount basis. Determine the amount the borrower initially receives and the effective rate.

Solution:

The interest charged is 13% of $10,000, which is equal to $1,300. This is deducted from the principal, so that the borrower receives $8,700. The cash-flow profile for this loan is shown in Figure 16.5.

The effective rate for this loan is the internal rate of return of this cash flow profile, which is equal to 14.94%.

Sometimes, a bank will require that the borrower retain an amount, called the compensating balance, in the bank account. The compensating balance has a significant effect on the effective rate. For example, if the compensating balance is 10%, the borrower in Example 16.3 will receive $7,700 at the beginning of the year as a loan, with $1,000 remaining in the account. At the end of the year, the borrower repays $9,000, since $1,000 is still with the bank. The effective rate for this cash flow is 16.88%, significantly higher than the *APR* of 13%.

(iii)   Instalment loan

An instalment loan, often used for vehicle finance, is an "add-on" loan. The interest charge is added onto the value of the loan. This total amount is divided by the number of repayments to determine the repayment schedule. A typical instalment loan is illustrated in Example 16.4.

Example 16.4: Instalment loan.

A borrower applies for vehicle finance of $10,000 for a year. The finance charge is 13%. Determine the repayment schedule and the effective rate if monthly repayments are required.

Solution:

The interest charged is $1,300. The total that the borrower has to repay to the lender is $11,300. This is repaid in 12 monthly instalments at the end of each month. The amount that is paid each month is equal to $941.67 = $11,300/12. The cash-flow profile for this loan is shown in Figure 16.6.

The internal rate of return for this cash-flow profile is 1.932% per month. The effective interest rate is calculated as follows:

$$e = \left(1 + \frac{i}{m}\right)^m - 1 = (1 + 0.01932)^{12} - 1 = 0.2582$$

Figure 16.6 Cash-flow profile of an instalment loan with monthly repayments

The effective rate is 25.82%, close to double the quoted rate of 13%. The *APR* is 23.19% = 1.923%(12).

(iv)   Amortized loan

A loan in which part of the principal is repaid together with the interest is called an amortized loan. The instalment loan discussed above is a form of amortized loan. Another form of amortized loan is that in which the interest is not calculated on the entire loan amount, but on the outstanding balance. Mortgages are loans in which the interest charge is calculated in this fashion. This type of loan is illustrated in the following example.

Example 16.5: Interest charged on outstanding balance.

A borrower applies for a loan of $10,000 over a year. The interest is calculated at a 13% on the outstanding balance. Determine the repayment schedule and the effective interest rate.

Solution:

The interest and the principal on the loan is to be repaid in equal instalments each month. The instalments can be calculated from the annuity formula, given by Equation 6 of Table 5.3. This formula is given as follows:

$$A = P(A|PV, i, n) = Pi(1+i)^n/[(1+i)^n - 1] = \frac{10{,}000(1.13/12)^{12}}{(1.13/12)^{12} - 1} = 893.17$$

The opening and closing balances for each of the 12 months of the loan are given in Table 16.1. The closing balance is the opening balance less the monthly instalment plus the interest charged. The interest charged is calculated on the outstanding amount. For example, the interest charged for the first month is equal to $108.33 = 10,000(0.13/12)$, and in the second month it is equal to $99.83 = 9,215(0.13/12)$.

The cash-flow profile for this loan is shown in Figure 16.7.

The internal rate of return for the cash-flow profile shown in Figure 16.7 is 1.083% per month. The effective interest rate for this loan is calculated as follows:

$$e = \left(1 + \frac{i}{m}\right)^m - 1 = (1 + 0.01083)^{12} - 1 = 0.1380$$

The effective rate is 13.8% and the *APR* is 13%.

Figure 16.7 Cash-flow profile of an amortized loan with interest charged on outstanding balance

Table 16.1 Opening and closing balances for each month of the amortized loan

Month	Opening amount	Monthly payment	Interest charged	Closing amount
1	10,000	893.17	108.33	9,215
2	9,215	893.17	99.83	8,422
3	8,422	893.17	91.24	7,620
4	7,620	893.17	82.55	6,809
5	6,809	893.17	73.77	5,990
6	5,990	893.17	64.89	5,162
7	5,162	893.17	55.92	4,324
8	4,324	893.17	46.85	3,478
9	3,478	893.17	37.68	2,622
10	2,622	893.17	28.41	1,758
11	1,758	893.17	19.04	884
12	884	893.17	9.57	0.00

(v)   Bonds

A bond is a long-term loan agreement. The bond sets terms for the principal, the coupon rate, the coupon frequency and the maturity. The principal is the face value of the loan agreement; it is the amount that the borrower or issuer of the bond will repay to the owner of the bond at end of the life of bond. The coupon rate is the interest portion that the borrower promises to pay the bondholder for the duration of the bond period, called the maturity or term of the bond. The coupon rate may be a fixed or variable percentage of the principal. In the case of variable rates, the coupon rate is indexed to a specific rate, such as the prime rate, or in the case of inflation-indexed bonds, to the consumer price index. The coupons or interest payments are usually paid semi-annually or annually, which is the coupon frequency.

A bond has the same cash-flow profile as a regular loan. The borrower gets the buying price, and repays the interest at specified dates, and finally, the principal is paid at the maturity of the bond.

Bonds may be issued at a value different from the face value. If the bond is issued at the face value, it is issued at par. The bond may be offered at a discount to the face value, that is, it is issued below par, and if it is issued at a price higher than the face value, it is issued above par. The effect of issuing the bond at a discount to the face value is illustrated in the following example.

Example 16.6: Bonds.

A bond is issued with a face value of $10,000 at a coupon rate of 13% for 5 years.

Determine the effective rate under the following conditions:

(i)   Coupons are paid annually.

(ii)   The bond is issued at a discount of 2%.

(iii)   Coupons are paid semi-annually.

Solution:

(i) The coupon is the coupon rate multiplied by the face value, that is, the coupon is equal to $1,300 = 0.13($10,000)$. The cash-flow profile for the bond is shown in Figure 16.8.

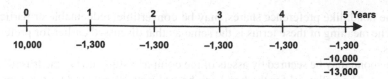

Figure 16.8 Cash-flow profile for a five-year bond carrying a 13% annual coupon

The internal rate of return for the cash-flow profile shown in Figure 16.8 is 13% per annum (since payments are at annual intervals). Thus, the effective interest rate for this bond is 13%.

(ii) If the bond is issued at a discount of 2%, the issuing price is \$9,800. The *IRR* for the cash-flow profile with this issuing price is 13.58%. The effective rate has increased from 13% to 13.58% due to the discount offered by the borrower (issuer).

(iii) If the coupon is paid semi-annually, the payment is half the annual payment. That is, the coupon amount is \$650 every six months. The *IRR* for this new cash-flow profile is 6.50% per half year. The effective rate is calculated as follows:

$$e = (1 + 0.065)^2 - 1 = 0.1342$$

The effective rate for the bond with semi-annual payments is 13.42%. This means that semi-annual coupon payments increase the cost to the borrower compared with annual payments.

16.3.4 Long-term Debt

Long-term funding is used to finance the general operation of the company. It lowers the equity capital requirement and increases the leverage. A company's debt may be issued publicly or placed privately. A public issue is offered openly to the public, while private placement is a direct transaction between the borrower and the lender. A bank loan or debt placed with an insurance company is an example of a private placement. Bonds listed on a bond exchange are public issues.

Debt securities are called bonds, debentures and commercial paper. There is no standard definition of these terms; also, they have different meanings in different countries. Strictly, a bond is secured debt, whereas a debenture is unsecured debt. The term "bond" is used here to refer to long-term debt and commercial paper is regarded here as shorter-term debt than bonds.

(i)   Bonds

The main features of the cash flow of a bond were discussed in the previous section. The borrower promises to repay the lender interest payments according to a specified schedule and the principal at maturity.

The bond issue is governed by a trust deed or a prospectus that specifies the terms of the agreement. The ownership of the bonds may be recorded, and coupons and principal are only paid to the bondholders in the register on a specified date prior to the date of payment. If no register is held, the bonds are said to be bearer bonds. The certificate holder detaches a coupon from the certificate in order to claim payment.

The bond, like preference shares, may be convertible, redeemable or participating. The meaning of these terms is the same as that discussed earlier for preference shares.

The bond may be secured by assets of the company stipulated in the trust deed of the issue. The collateral for the bond can be ordinary shares held by the company. Bonds secured by real property are referred to as mortgage bonds, and are discussed in the next section.

(ii)   Mortgage bonds

A mortgage bond is a long-term loan secured usually by real property (real estate and buildings). The mortgage is generally structured as an amortized loan with interest on the outstanding balance.

(iii)   Leases

A lease provides long-term funding by allowing the company to use the leased assets without incurring the initial cost of purchase.

(iv)   Loans or corporate debt

Loans are a private arrangement between the borrower and the lender. The arrangers of the debt generally rely on the business's record of performance, rather than an assessment of the business prospects in the future. The record of the business performance is assessed on the financial ratios such as the interest coverage ratio, which is discussed in Chapter 17. The lender may impose several restrictions in order to protect the debt. These include the distributions to shareholders, and the operation of the business as measured by some financial ratios, such as the debt-to-equity ratio, the interest coverage ratio and the EBITDA-to-debt ratio.

16.3.5 Short-term Debt

Short-term finance generally has a term of less than a year. It is mainly used to fund the changes in the working capital of the company. The company must stock-up if it anticipates an increase in sales. This increase in inventory increases the working capital. This is particularly the case for those businesses that have peaks and troughs of activity.

(i)   Line of credit and revolving credit

A line of credit is a flexible form of short-term financing that allows the borrower to access funds up to a specified limit. The interest charge depends on the borrower's financial standing. The borrower may need to sign a promissory note to access the funds, which are credited to the checking account of the company.

A revolving credit facility is a formal line of credit that allows the company to draw down funds as needed. The company may be required to pay a charge on the funds that remain on the unused balance in addition to the interest charged on the used funds.

(ii)   Banker's acceptance

Bills of exchange allow a seller of goods to access funds sooner than payment by the buyer. The seller creates a bill that commits the buyer to payment by a certain date. The buyer endorses, or accepts, the bill. The seller can then exchange the bill with a third party directly for cash. A banker's acceptance is a bill of exchange that the company sells to a bank. The company promises to settle the bill at a specified date, usually 3 months later. Banker's acceptances are of the form of discount loans, which were discussed earlier.

(iii)   Commercial paper

Commercial paper is an unsecured short-term loan issued by companies of high standing. As a result, the interest rate for commercial paper is favourable.

16.3.6 Public Issue and Private Placement of Financial Securities

A company can raise funds in two broad ways. The first is through private arrangements with investors, and the second is through public arrangements with investors. For example, the company may be able to place its debt securities with banks or insurance companies as a private issue. The company on the other hand may sell its common stock (shares) to the public after registering the issue with the authorities, such as the Securities and Exchange Commission (SEC) in the US and the UK Listing Authority. Generally, private placement of equity is used for smaller businesses, while the private placement of debt is a significant portion of the corporate debt market.

A number of debt issues in the US are designed to take advantage of an amendment to the rule that prohibited the resale of privately issued securities. This prohibition made these securities illiquid. An amendment, known as Rule 144A, allows qualified institutional buyers to trade unregistered debt and equity securities. Debt securities that take advantage of Rule 144A can be underwritten, and are rated in the same way by rating agents such as Standard & Poor's and Moody's as publicly issued debt. The main purchasers of privately placed debt securities are insurance companies. Rule 144A has created a quasi-public market for these private securities.

16.4 Financial Markets

The markets for financial commodities bring the people with surplus funds together with those with deficit funds. For example, investors with surplus funds who want to invest in equities meet the sellers of equities at the stock exchange. Markets are either for physical assets, such as copper and maize, or for financial assets, such as shares and bonds. The financial markets deal in financial instruments, which are contractual agreements between parties that grant rights to physical assets. The financial markets can be divided into the money market and the bond market, the equity market, the foreign currency market and the commodities market. The bond and money markets constitute the interest-bearing market, while the bond and equity markets constitute the capital market. This classification is illustrated in Figure 16.9.

Terms that are used to describe the financial markets are spot and futures markets, capital and money markets, and primary and secondary markets.

Spot markets are those in which payment and delivery occur within a few days, whereas futures markets are for the delivery and settlement at a date in the future (at a price agreed today).

Money markets are for short-term debt, usually of a duration of less than a year. The capital funds of a company are raised on the capital markets, consisting of the bond market and the stock market. The bond market is for long-term debt.

The primary market is that in which a company raises new capital. For example, the sales of the shares of the stock and the initial public offering of Amazon.com that were discussed earlier represent primary market transactions. The purpose of the primary market is for companies to raise capital. The secondary markets are those in which existing securities are traded between investors. The company whose securities are traded on the secondary market does not receive any funds from the transactions between investors.

Markets are organized in different ways. Broker markets involve a broker who acts as an agent for the investor. Investors traded with another through the agency of brokers. In a dealer market, the investor purchases from a dealer who holds securities in his own inventory. Dealers set bid and ask prices. The bid price is the price the dealer will pay for the security, and the ask price is the price at which the dealer will sell the share. The dealer sets these prices so that the size of the inventory remains

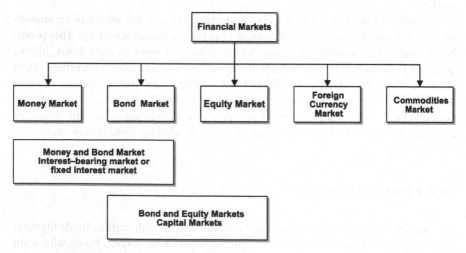

Figure 16.9 Classification of the financial markets

constant. The bid-ask spread is the dealer's profit margin. Of course, dealers can make or loose money on the value of their inventory. Dealer markets are also called over-the-counter (OTC) markets.

16.4.1 Equity Markets

The stock exchanges are organized exchanges that represent the primary and secondary markets for the exchange of equity securities. Examples of stock exchanges are the London Stock Exchange (LSE), the New York Stock Exchange (NYSE), and the Toronto Stock Exchange (TSX). For a stock to be traded on an exchange it must be listed and meet the requirements for listing. The requirements for listing differ from country to country; generally, these requirements relate to the size and profit record of the company. The advantage of listing is that it increases the liquidity of the share and significantly increases the profile of the company. The disadvantages of listing are the costs, both initially and reporting and disclosure. Shares in unlisted companies are traded in the over-the-counter market, in which a dealer buys from and sells to investors.

16.4.2 Bond Markets

Although some bonds are listed on exchanges such as the NYSE, most bond trades are in the OTC market. Treasury issues are the bonds with the most liquidity. Corporate bonds are much less liquid.

Bonds may be traded on price or on yield. Settlement, of course, is in price, so there must be a clear understanding of the method of determining price for those trades that are made on yield. The price of the bond is the present value of the bond. This is expressed as follows:

$$Price = \sum_{t=0}^{n} \frac{C_t}{(1+y)^t} + \frac{P}{(1+y)^n} \qquad (16.1)$$

where C_t is the coupon at the time t, P is the principal or face value, and y is the bond yield. The values of the coupons and the principal are known. The unknown variables are the price and the yield to maturity, y. The choice of one sets the other. For example, if the bond was sold at a yield of 11.4389%, the price can be determined precisely.

The calculation of the price, described by Equation 16.1, must be modified to account for the accrued interest between the payments of coupons. For example, if the bond pays coupons semi-annually on 30 June and 31 December, and the trade is made on 30 November, the seller is entitled to the portion of the coupon between 1 July and 30 November, and the buyer to the portion between 1 December and 31 December.

The yield to maturity, or more simply the yield, is actually the interest rate that the lender earns and the borrower pays on the bond. For bonds of different times to maturity, there are different yields to maturity. A plot of the yield to maturity as a function of the remaining term of bond for different bonds gives the term structure of interest rates. This term structure provides the market's outlook on future interest rates. For government bonds, the term structure of interest rates provides the market's estimation of future inflation.

Bonds are rated by different rating agencies. Broadly, bonds are divided into those of investment quality and those of speculative quality. The lower quality bonds are also called high-yield bonds or junk bonds. Although the rating symbols and categories differ slightly between the various rating agencies, those of the highest quality are rated AAA, and those in default or with a poor chance of meeting their obligations, are rated D. Investment grade bonds are those with a rating higher than BBB, while speculative grade bonds have a rating below that. Junk bonds have a rating that falls into the speculative bond category.

16.4.3 Futures and Derivatives Markets

The three main forms of derivatives are futures, options and swaps. Futures are agreements to purchase an asset at a date in the future at a price that is agreed now. A swap is an agreement to exchange the rights to one asset for the rights to another. Options were discussed in the previous chapter.

The origins of derivatives lie in the agricultural markets. Farmers and merchants locked in prices before the growing season by selling forward an agreement that is similar to a futures contract. This provided the farmer with secure knowledge of the price he would get for his crop, allowing him to determine how much to invest in growing the crop. On the other hand, a purchaser of agricultural produce might need to limit the maximum price paid, which has the elements of an option contract. These contracts altered or limited the party's exposure to risk. Derivative instruments provide a form of insurance for the parties.

There are three types of participants in the derivatives markets: hedgers, speculators and arbitrageurs.

Hedgers purchase derivative instruments in order to alter their risk exposure or achieve more certainty. The use of futures contracts to hedge the output price for a mining operation is illustrated in the following example.

Example 16.7: Gold hedge.

A gold mining company is exposed to the price of gold. In order to mitigate this risk, it can sell gold futures. The company expects to produce 100,000 troy ounces for sale in six months. The spot price is $400 per ounce and the six-month futures price is $380 per ounce. Demonstrate a hedging strategy for the company.

Solution:

The futures contract to purchase gold at a price of $380 in six months from now. The hedging strategy would be to sell futures now with a six-month maturity for 100,000 ounces, and then close out this position in six months time. (Positions are usually closed by taking the opposite position. So, this position is closed by buying the same futures in the market.)

If the price in six months is $420 per ounce, the company sells the gold output for $420 per ounce and losses $40 = (420 − 380) on the futures contracts. Thus, the effective price that the company receives is $380 per ounce.

If the price in six months is $350 per ounce, the company sells the gold production for $350 per ounce and gains $30 = (380 − 350) from the futures contracts. This means that the effective price that the company receives is $380 per ounce.

Therefore, if the price goes up or down, the company will receive the same price for its output, $380 per ounce.

This hedging strategy alleviates the miner's fears of a price collapse.

Speculators wish to be exposed to a risk; they take a position in the market. For example, a speculator may believe that the British pound will strengthen against the Canadian dollar within a month. The speculator may purchase currency futures, and wait for the month until he closes out his position. The current exchange rate is 1.7492, and the currency future is trading for 1.7410. If the speculator is willing to back his beliefs to the amount of £500,000, and the futures price in a month is 1.7819, he will make a profit equal to $20,450 = (1.7819 − 1.7410)500,000. Of course, if the futures price is trading for 1.73, the speculator will make a loss. Why should the speculator do this in the futures market rather than purchasing the pounds in the currency market? The main reason is the mechanism of the futures market. The speculator need only deposit a margin, and settle the profit and loss each day, called marking to market. This is discussed in more detail later. With a relatively small outlay, the speculator can take a large position. For this reason, the margin and marking to market features of the futures market are regarded as highly geared. In this context, gearing means the increased exposure to the price that the futures market provides compared to the size of the initial investment.

Arbitrageurs seek to make risk-less profit, by trading the same instrument in different markets. Arbitrage opportunities exist whenever assets are not correctly priced in the markets. Consider a share that is trading at $250 on the New York Stock Exchange and at £160 on the London Stock Exchange. If the exchange rate is 1.6 $/£, the arbitrageur can purchase the share in New York for $250 and sell it in London for $256 (= 160 × 1.6) and make $6 without taking any risk whatsoever. Arbitrage opportunities like this do not last. They represent a deviation from equilibrium, and the arbitrageur's actions cause a return to equilibrium.

Standardized futures and options contracts are traded publicly on exchanges, such as the Chicago Mercantile Exchange, the Chicago Board of Trade, the London International Financial Futures Exchange and the Tokyo International Financial Futures Exchange.

Trading on a futures exchange is organized slightly differently to other markets. An investor who wishes to trade in futures contracts must open a margin account with a broker. The initial margin that is deposited with the broker may be as low as 10% of the value of the trade. The margin account is adjusted to reflect the investor's profit or loss at the end of the day's trading. For example, if the futures price had increased by $100 during the day, the investor's margin account would be credited with $100. If the price had decreased by $100, the margin account would be debited by $100. The settlement of the margin account at the end of the day is called "marking to market." The settlement is not between broker and investor, but it is passed on

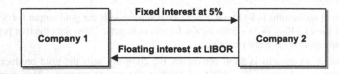

Figure 16.10 An interest rate swap

to investors with the opposite position through the brokers and the exchange. The futures market is "zero-sum gain" on a daily basis.

The margin account creates significant leverage in the futures market. However, the margin calls could present significant cash flow problems to hedgers even if later in the term of the future the price turns to favour the hedger. For example, a gold mining company with a hedge to cover declining prices may find itself with cash flow problems from margin calls if the price increases. It is easy to see how management can get panicky under these circumstances, even if the price returns to its previous level.

A swap is an agreement between two parties to exchange streams of cash flows in the future. For example, two companies may decide to enter an interest rate swap in which one company pays interest to the second company at a fixed rate, say 5%, while the second pays interest to the first company at a floating rate, say that defined by the London Interbank Offer Rate (LIBOR). LIBOR is the interest rate offered by a bank for a deposit from another bank. Since the principal amount is the same for both parties, it is not exchanged. This swap is illustrated in Figure 16.10.

Swaps are used to transform a liability. For example, if Company 1 shown in Figure 16.10 had a loan that required it to pay a floating interest rate of LIBOR plus 1.5%, the net result of the loan and the swap is that the company must pay a fixed rate at 6.5%. The swap has transformed the floating rate to a fixed rate, and removed the interest rate risk to the company. The difference in floating interest rate and the base rate, which in this case is LIBOR, is frequently quoted in basis points, which is 100th of a percentage point. For this example, the floating rate is at 150 basis points. The transformation of currency risk is illustrated in the following example.

Example 16.8: Currency swap.

A project company has arranged a loan in one currency at 9% pa, but it receives its revenues in another currency. The exchange rate between the currencies poses a risk to the operations of the project. Demonstrate how a currency swap can transform this risk.

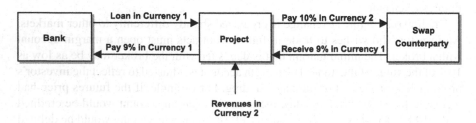

Figure 16.11 Currency swap

Solution:

The project company enters a swap agreement with a counter party, usually a financial intermediary, to pay interest in the same currency as its revenues and to receive interest in the currency of the loan. The loan and the swap are illustrated in Figure 16.11.

Because the project receives exactly the amount in Currency 1 from the swap counterparty that it need to pay to the bank, the project has no exposure to changes in value of Currency 1 compared to those of Currency 2. The currency swap has transformed the liability by removing the exposure to exchange rates.

16.5 Financing Decisions Within a Company

Three major decisions affect the financing of a company. These decisions, or policies, are the following: (i) how much long-term debt should the company have compared with the equity of the company; (ii) how much of the profit must the company retain; and (iii) how much short-term debt does the company need. The first decision is referred to as the capital structure of the company, the second is the company's dividend policy and the third is the company's credit policy.

16.5.1 Capital Structure

The amount of debt in a company affects the discount rate, or the weighted average cost of capital (WACC), used to value the company, and the company's investments in projects. Increasing amounts of debt increase the leverage of the company and decrease the WACC. The effects of financial leverage were discussed in Chapter 12. In the presence of taxes, increasing the debt also increases the value of the company. This increase in value results from the interest payments on debt being tax deductible.

However, as the debt increases, the financial risk of the company increases. This risk is a result of the possibility of the company not being able to meet its interest payments, that is, the company is in financial distress. Therefore, an optimum exists in the amount of debt that the company can accommodate. This is illustrated in Figure 16.12. The effects of financial risk on the company were discussed in Chapter 12.

The optimal amount of debt that the company can accommodate without increasing the cost of capital due to perceptions of increased risk is a matter of judgement. Financial managers need to balance the amount of debt with the perceptions of risk.

16.5.2 Dividend Policy

The returns to the shareholders of the company are in the form of capital gains and dividends. Retaining all the profit should be reflected in increased capital gains.

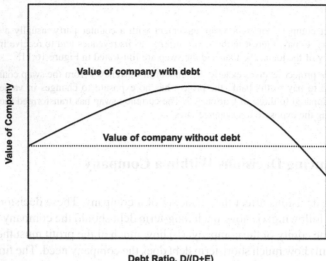

Figure 16.12 Effect of the amount of debt, D, on the value of the company. E refers to the amount of equity

This means that it should not matter to investors whether they receive their returns as capital gains or as dividends. An alternative view is that if an investor holds a company's shares in perpetuity, the only returns are the dividends. This gives rise to the dividend model for the determination of the cost of equity discussed in Chapter 12.

The company can adopt one of a number of different options, called dividend policies, regarding the payment of dividends over time. The company can pay dividends in one of the following predictable manners: (i) constant dividend amount; (ii) constant dividend ratio; (iii) constant dividend plus bonus. The constant dividend policy sets the dividend to a constant absolute value, say $0.25 per share, for a number of years. The amount may be increased once it is clear that the earnings have permanently increased. The constant dividend ratio policy pays a dividend at a fixed percentage of the profit or earnings. For example, the company may pay 45% of the profits as a dividend. The company may also pay a constant dividend, but augment this with a bonus payment if performance has been particularly good. The company need not state their policy; some do, and the policy of others is clear to investors.

The dividend declared by a company sends strong signals to investors about the state of the company. The payment of a dividend indicates that the company is profitable. If the dividend is similar to the previous dividends, this indicates that the company is stable. Financial managers must balance the need to maintain shareholder loyalty, through dividend payments and the need to reinvest the profits of the business to grow the business, and thereby increase the share price or the capital gains to investors.

16.6 Comparison of Equity and Debt Financing

The financing of the company is through equity and debt instruments. The choice between equity and debt financing represents a trade-off between risk and return.

The interest costs are known and predictable. If the company is able to meet the interest charges, the returns to the shareholders are increased by increasing debt. The leverage effect of debt was discussed in Chapter 12. The company can do more with owners' equity because of the debt that it raises. The interest on debt is usually deductible for tax, whereas the dividends are not. From the viewpoint of the company, this makes debt financing significantly cheaper than equity financing.

On the other hand, interest must be paid on debt, whereas dividends can be passed. If the company is not able to make its interest payments, it is commercially insolvent. High levels of debt will increase the financial risk of the company, and significantly reduce the return to the shareholders if the company is under financial stress.

16.7 Summary

The sources and types of securities that are offered by companies to raise capital have been discussed in this chapter. The two major types of securities are debt and equity. Equity securities offer part ownership in the company. Returns are in the form of capital gains and dividends. The risks of equity ownership are the possibility of the loss of investment in the case of bankruptcy. Debt securities offer returns in the form of more predictable interest payments and the repayment of the loan principal. Preference shares are a mixture between the equity and debt, since they provide interest-like returns.

There are number of different forms of debt that are offered by borrowers. The differences are concerned with the repayment schedule, the collateral offered for the loan and the maturity of the loan.

The markets for financial securities concern the capital markets for equity and long-term debt securities, the money market for short-term debt, and the foreign currency and commodities markets. Equities, that is, shares or stock of a company, are traded on the stock exchanges, whereas most bonds are traded in the over-the-counter or dealer markets. There are three main forms of derivatives: futures; options; and swaps. These instruments are used to transform the risk to the holder.

The financial managers of a company must decide on the capital structure of the company and on the dividend policy. The capital structure of the company refers to the proportion of debt and equity capital in the financing of the company. The dividend policy refers to the amount of dividends to be retained each year by the company. Both influence the investors' perception of the company, and alter the form that shareholders receive their returns, either as capital gains or as dividends.

16.8 Review Questions

1. What elements does the financial system link?
2. From what economic sector do lenders and borrowers come?
3. What is a financial intermediary?
4. Provide some examples of financial intermediaries.
5. What does underwriting mean?
6. Is a bank a financial intermediary?
7. What are the two forms of equity ownership in a company?
8. How do investors receive the returns from equity ownership?
9. Why would a company want to sell more of its shares to investors?
10. What is a rights issue?
11. What are preference shares?
12. What are the different forms of debt instruments?
13. What does "secured" mean in the context of "secured loan"?
14. What is senior debt?
15. Describe the differences between a regular loan, a discount loan, an instalment loan, an amortized loan and a bond.
16. What is the difference between a bond and a debenture?
17. What is a mortgage bond?
18. What is the difference between revolving credit and a banker's acceptance?
19. To what does the term "capital markets" refer?
20. What is the difference between a broker market and a dealer market?
21. What is the difference between a futures contract and an option contract?
22. What is hedging?
23. Why is there an optimum in the debt ratio for a company?

16.9 Exercises

1. A bank has proposed a regular loan for a year with monthly interest payments. The loan is at 9% pa. If the principal amount is $1,000,000, determine the monthly payments, the final settlement and the effective rate.
2. What are the values for the *IRR* and the *APR* of the loan in Question 1?
3. A company has been offered a regular loan with an interest rate of 9% and a discount loan for 8%. Which option should the company choose?
4. An engineering company needs a loan of $2,000,000 for a three-month period to bridge cash flows on a construction project. The bank offers a discount loan with an interest rate of 10%. What should the loan principal be? What is the effective rate on this loan?

5. A contractor has purchased vehicles on an instalment loan with monthly payments over a three-year period. If the interest rate on the principal of $250,000 is 9% pa, determine the repayment schedule and the *APR* on the loan.

6. An engineering consultancy has purchased the building in which it has its offices for $450,000. The mortgage has a term of 20 years and carries an interest rate of 7% pa. Determine the monthly repayments on this mortgage.

7. A company's bonds are selling at $91,000. If the face value is $100,000 carrying a coupon of 5% that is paid semi-annually and mature in 4 years, determine the effective interest rate.

8. If the yield on a five-year bond with face value of $1,000,000 carries a 5% pa semi-annual coupon and the yield to maturity is 8%, determine the price of the bond.

9. A company requires short-term finance to cover cash flow shortfalls of $150,000. The company sells a three-month banker's acceptance at a rate of 10%. What principal is required and what is the settlement amount?

10. A copper company produces 100,000 tpa of copper. The spot price is $1.5/lb, and the six month futures price is $1.60/lb. However, the company anticipates a decline in price, and wishes to enter a hedge. Determine the hedge position that the company should take if the futures contracts are for $1000 t of copper.

Chapter 17
Financing Engineering Projects

17.1 Introduction

Three main topics are examined in this chapter. The first is the financing of the construction phase of a capital project. The financing of the construction of the project is dependent on the contractual arrangements between the contractor and the owner. These contracts apportion the construction and performance risks to the different parties, and as a result, they determine the financing that can be provided.

Once the project is built and delivered, the construction loans and financing terminates, and the permanent financing is provided. The second topic of the chapter is the use of project finance to fund the permanent capital requirements of the project. Project finance is the funding of a project as a venture separate from the other activities of the company. This form of funding relies on the forecast profitability of the project, not on the creditworthiness of the company. It is sometimes called non-recourse or limited recourse financing, because the assets of the sponsor are not provided as collateral. This means that in the even of the failure of the project to meet its debt obligations, the lenders do not have recourse to the sponsor's assets.

The application of project finance is not limited to the private sector, but has found application in the financing of public infrastructure projects through public–private partnerships. This forms the third topic of the chapter.

17.2 Financing Engineering Construction of Capital Projects

17.2.1 Project Delivery Systems

The owner of a project can implement a project in a number of ways, depending on the skills and experience of the owner. The three main methods of implementing the project, referred to as delivery systems or procurement methods, are owner-managed projects, lump-sum turnkey projects, and engineering, procurement and construc-

531

tion management projects. The differences between these procurement methods are related to the risks assumed by the different parties during construction. Consequently, they affect the financing during construction. Each of these delivery systems is discussed.

(i)   Owner-managed projects

The owner prepares the design package using either in-house resources or an external engineering consultancy. The design package is generally at an accuracy level of $\pm 15\%$. The design package is also called a basic engineering package, a feasibility study, a bankable feasibility study or the front-end engineering design. Once the design package is complete, the owner places an order with an engineering company, which need not be the same company that provided the design package, for the detailed engineering design. On the basis of the detailed engineering design, the owner procures the equipment and materials, and engages a construction company to build the project.

The owner actively manages the design and construction process. The owner retains responsibility for the performance of the project. The owner has a number of major contracts to coordinate: that for the basic design package; for the detailed design specifications; and for the construction. The owner assumes the business risk associated with both the project itself, and that associated with the management of the consultants and contractors.

This is a traditional form of contract in many industries, including the building, construction and infrastructure industries. The owner appoints an engineer or an architect (either in-house or externally) to develop detailed plans and specifications. The owner then seeks tenders for the construction of the project based on these plans. The tenders often require bids for a fixed price, and as a result, the form of contract for the construction phase is regarded as a lump-sum contract. Many government infrastructure projects are organized in this manner.

(ii)   Lump-sum turnkey projects

In this type of project, which is also called an engineering procurement and construction contract (EPC) contract, the owner develops a conceptual design specification and issues requests for proposal and engages an engineering company as the contractor. The contractor is required to deliver the project on a particular date for a fixed price, hence the term "lump sum." The contractor also needs to guarantee the performance, quality and efficiency of the operations, hence the name "turnkey," which suggests that all the owner has to do is turn the key for operations to commence without any problem.

In the building and construction industry, this form of contract is known as design-build contract. The contractor provides architectural design, engineering, and construction to the owner for an agreed price.

The risk is mostly transferred from the owner, who responsible for setting only the conceptual specifications, to the EPC contractor, who is responsible for delivering the project for the agreed price. As a result, owners view EPC contracts favourably. The disadvantage is that the project may cost more because the contractor will include a charge for risk in the price. Governments preclude this type in a bid to ensure the lowest price for infrastructure development. The acceptance of public–private partnerships and project finance means that EPC contracts have become more common in the development of public infrastructure.

(iii)   Engineering procurement and construction management projects

In the engineering procurement and construction management (EPCM) contract, the owner of the project appoints a contractor to manage the engineering design, the procurement of the equipment and materials, and the management of the construction phase. Construction contractors and equipment suppliers perform the work required to build the project under supervisory management of the EPCM contractor. The EPCM contractor, who is paid on a cost reimbursable basis, acts on behalf of the owner, who retains responsibility for the project. (An EPCM contract should not be confused with an EPC contract. The EPC contract is to deliver the entire project, generally for a lump-sum price, while the EPCM contract is a design and management function, generally rewarded on a cost reimbursable basis.)

A major difference between these contracts is the manner in which the risk is shared between the owner and the contractor. This is discussed next.

17.2.2 Risk in Engineering Contracting

The three main forms of engineering contract share the construction and completion risk between parties differently. The owner retains most of the risk if it decides to manage the project itself or to appoint an EPCM contractor. The contractor accepts most of the risk in an EPC contract. This is illustrated in Table 17.1. The majority of large capital projects are executed as lump-sum turnkey (EPC) contracts.

The risks that are important to a contractor and an owner are those that will affect either the completion date, the performance of the project or the cost of the project.

Table 17.1 Sharing of risks between contractor and owner in a capital project

Risk sharing	Types of contract
Risks retained by project owner	Engineering, procurement, and construction management (EPCM)
	Cost reimbursable
	Project management
	Fee for service
Risks shared with contractor	Alliance
	Performance incentive
	Construction management
Risks accepted by contractor	Lump-sum turnkey
	Fixed price
	Guaranteed maximum price
	Engineering, procurement and construction (EPC)

An EPC contractor is unlikely to accept responsibility for the engineering design package that has been prepared by the owner or on behalf of the owner by another contractor. The providers of such front-end engineering services will limit their liability to a percentage of fees. If it is suspected that there is a fault in the design, it is difficult to determine which of the parties is at fault. As a result, the owner does not divest himself of all risks through a lump-sum turnkey or EPC contract.

17.2.3 Construction Loans

The financing of the projects is often separated into the construction phase and the permanent financing on completion of the construction. The construction of the project is funded by a short-term loan from a commercial bank or from the project owner's internal funds. The permanent funding replaces the construction loan on completion.

The contractor will provide a draw down schedule based on the construction schedule. Draw down refers to the use of finance with time. As the project incurs costs, these costs are met with funds from the construction loan. In other words, the draw down schedule details the funding requirements as the project proceeds. The loan schedule may be based on this, or may allow for flexible draw down. The draw down schedule influences the cost of financing for the construction of the project. The lender may charge interest on the amounts withdrawn and commitment fees on the unused balance. In addition, an arranging fee is frequently charged.

The construction phase from the contractor's point of view is quite different. The timing of expenses and payments is critical to the cash flows of the contractor. In addition, the owner may retain the contractor's profit for the work until the completion of the project. This *retainage* of profits has a significant effect on the contractor's cash flows.

17.2.4 Financial Guarantees

The owner may require guarantees to ensure that the contractor will honour its obligations. There are three main guarantees that may be required: (i) a bid bond; (ii) a performance bond; and (iii) a payment bond. A bid bond guarantees that the contractor will honour its bid and sign the contracts. A performance bond guarantees that the contractor will complete the contract as agreed. It means that the contractor will suffer a financial penalty if he fails to complete the project. A payment bond guarantees that the subcontractors and suppliers will be paid by the contractor.

Banks, insurance companies and surety companies provide the financial guarantees. A contractor that frequently provides guarantees to owners will have a "bonding line," in the same way that companies have a credit line, with a bank. The amount of the performance bonds is usually between 2 and 25% of the contract price, with a value of 10% regarded as the norm. The cost of the guarantee is between 0.125

and 2% of the amount guaranteed per annum. Bank charges are in the lower half of the range, while surety companies charge in the upper part of the range.

A classic example of guarantees in the tender and award of the construction contract is the building of the Hoover Dam. The Bureau of Reclamation issued specifications for tenders in January 1931. Each bid was to be accompanied by a $2 million bid bond, and the winner was expected to post a $5 million performance bond.

The following two sections provide examples of the financing of the construction phase of the project. These examples illustrate the different forms of engineering contract, the different forms of reward to the construction contractor, and the form of the financing during the construction period.

17.2.5 Indiantown Cogeneration Facility

Bechtel Power Corporation was awarded a fixed price turnkey construction contract for the Indiantown Cogeneration facility in Florida. Cogeneration refers to the production of electricity and steam from the same power plant. A fixed price of $438.7 million was agreed on for a guaranteed completion date of January 21, 1996. Three different definitions of completion were defined: (i) mechanical completion meant that all the materials and equipment required for the operation of the facility had been installed and operated safely; (ii) substantial completion meant that the facility met certain environmental requirements and could be operated at 88% of the net electrical output; and (iii) final completion meant that mechanical completion in addition to all performance tests. Substantial completion was to be met by the completion date, and final completion within one year of substantial completion. If final completion was not met by the guaranteed completion date, Bechtel would pay liquidated damages. The liquidated damages were limited to $100 million. The contract limited changes to the scope of the project.

Construction was financed from four sources: (i) a bank loan of $202.6 million; (ii) tax-exempt bonds totalling $113 million; (iii) a $139 million loan from GE capital; and (iv) $100,000 of equity capital.

The project was refinanced before completion of construction due to cost overruns during construction that called for additional capital. Bonds, called first mortgage bonds, were issued for the amount of $505 million. In addition, new tax-exempt bonds amounting to $125 million were issued. These funds were used to repay the bank loans, the earlier tax-exempt bonds, and the original loan from GE Capital. GE Capital was committed to funding $140 million of the construction costs. On completion, this loan would effectively be converted to equity, and the other partners would pay GE Capital for their share of the equity. Thus, the total capital on completion of the project would be $630 million in debt from the mortgage bonds and the tax-exempt bonds, and $140 million in equity.

17.2.6 Eurotunnel

The Eurotunnel links the UK to France by tunnel under the English Channel. A consortium, Transmanche Link, agreed to design, construct and commission the rail tunnel within seven years. The tunnels and underground structures would be constructed based on a contract that paid for the costs and a 12% profit. Transmanche would get 50% of the savings if the costs were less than the target cost. However, if they were higher, they would pay 30% of the increase. The terminals, fixed equipment such as mechanical and electrical installations, were to be provided on a lump-sum basis. Subcontractors would provide the rolling stock and locomotives. Transmanche would receive a 12% profit on the procurement of these items.

A performance bond of 10% of the total contract was required, and 5% of the progress payments due to Transmanche were withheld.

It was estimated that £4.8 billion was required in funding for the project. The construction was financed in stages. The initial shareholders provided £50 million, and 40 banks provided a syndicated loan of £5 billion. Drawdowns on the loan were not permitted until at least £1 billion in equity had been raised, and at least £700 million of the equity had been invested in the project. A further two equity offerings were required to raise the equity requirement. After completion, Eurotunnel would raise cheaper financing through other sources (permanent debt) in order to repay the construction loans.

The expected costs for the project are given in Table 17.2. Construction costs were expected to amount to £2.8 billion, Eurotunnel's management and operational costs to £565 million, financing costs to £1 billion, and inflation to £489 million.

The bank loan of £5 billion and the equity contribution of £1 billion meant that the Eurotunnel project had funding that was 25% in excess of its anticipated requirements. As events turned out, this was insufficient. The contractor underestimated the logistics of boring the tunnel, changes were required for the terminals, and the costs of the rolling stock were seriously underestimated. The tunnel was opened a year behind schedule on May 6, 1994. The original cost estimate of £4.8 billion eventually became £10.5 billion. In addition, Eurotunnel underestimated the competitive response from ferries and the aggressive pricing of passenger airline flights. As a result, the company has struggled through numerous difficulties.

Table 17.2 Expected costs in £ millions for the Eurotunnel project (Finnerty, 1996)

	1986	1987	1988	1989	1990	1991	1992	1993	Total
Construtcion costs	14	168	504	575	671	507	300	22	2,761
Eurotunnel expenses	37	103	81	74	70	66	73	61	565
Inflation	-	3	30	68	118	130	110	30	489
Financing charges	8	49	29	95	160	245	327	111	1,024
Total	59	323	644	812	1,019	948	810	224	4,839

17.3 Project Finance

17.3.1 Overview of Project Finance

The debt financing that has been discussed in the previous chapter has been the context of the direct funding of project through the company's general credit. The debt is issued because of the company's general standing and record of accomplishment. If the debt is secured, some of the assets of the company provide collateral in case of default. The debt is also provided for the general operations of the company, not for a particular project. In other words, the lenders look at all the assets and the projects of the company to service the debt. In addition, all the company's funds are pooled, and activities are funded from this pool.

An alternative form of funding is *project finance*, which is the funding of an economically attractive project from the cash flows of the project. The lenders look to the expected cash flows from the project to service the loans. The security of the loans depends on the anticipated profitability of the project. Projects that are financed in this manner are separated in both legal and accounting senses from the sponsors and equity owners of the project.

Project finance is significantly different from the general funding of a company. For example, if a company approaches lenders for funding, the lender will examine the balance sheet and determine the level of debt already in the company, and examine the income statement to determine the ability of the company to meet interest payments (the interest cover). The lender looks to past performance and capacity to determine the creditworthiness of the company. The lender expects the assets of the company to serve as collateral for the loan. Generally, companies plan to last; the debt can be refinanced, and is therefore regarded as permanent funding. Project finance, on the other hand, examines the future projections of the cash flows to determine the level of debt that the project can sustain. Debt funding is based on past performance; project finance is based on anticipated earnings. The project is finite, so the project financing considers the repayment of both interest and principal.

Project finance has been used to fund large-scale natural resource and infrastructure projects, such as pipelines, oil fields, mines, toll roads and power plants. The Trans Alaska Pipeline System is an 800-mile pipeline that transports crude oil and natural gas from the North Slope to Valdez in Alaska. It was built between 1969 and 1977 at a cost of $7.7 billion, more than the combined cost of all the other pipelines on the continental US. The Eurotunnel project was built between 1984 and 1993 at a cost of over £10 billion. It is a twin bore tunnel joining the rails systems of the UK and France. The Ras Gas project for the production of liquefied natural gas was built between 1996 and 1999 at a cost of $3.4 billion.

17.3.2 Project Structure

The project is financed and executed as a legal entity separate from the company sponsoring the project (the project's sponsor). This separation has several important features. The assets of the sponsor are separated from the effects of the project. The lenders do not have recourse to the sponsor's assets in the event of any failure to meet the payments on either the interest or the principal. In other words, the sponsor's assets are not provided as collateral for the loan. The sponsors may be required to supplement the cash flows of the project under some conditions. In this case, the financing is referred to as limited recourse financing. If the borrowings are funded entirely from the cash flow from the project without any supplementation from the sponsor in the case of cash shortfalls, then the financing is referred to as non-recourse financing. The separation of the sponsor from the project may allow the sponsor to pursue a project that is much larger than it could justify based on the size of its own balance sheet. In addition, depending on the jurisdiction, the project finance structure may be regarded as an off-balance sheet technique for reporting purposes. This means that the debt incurred by the project does not have to be reported as part of the sponsor's debt on the sponsor's balance sheet, which also allows the sponsor to enter a venture that incurs a large amount of debt.

Large projects may generate tax credits during construction and in the early years of operation. Access to these tax credits influences the choice of legal entity for the project. For example, a limited liability partnership provides the limited liability of a corporation for the limited partners, but each partner is taxed as a single entity with their share of the partnership. The requirements for the separation of the risks of the project from the sponsor and the access to tax credits form an important part of the choice of legal structure.

17.3.3 Risks and Risk Mitigation

The risks to the lender in project financing are higher than for the financing of a corporation because the project has no assets and no record of cash flows. The lenders will seek to mitigate these risks through contractual agreement and financial planning.

Three key risks related to the project are the completion risk, the off-take risk, and the production disruption risk. The completion risk is the risk that the project may not be completed because of increased costs, shortages of supplies, delays to construction, or abandonment because cost overruns make the project unprofitable. The off-take risk is that the products or services of the project are no longer in demand. For example, the Eurotunnel project mentioned earlier faced intense competition on prices with ferries and airlines. This meant that it did not meet the projections of revenue that underpinned the financial projections of the project. EuroDisney was another project funded by project finance that did not meet revenue projections. The third project risk is that of disruption to production or service, which means that revenues are disrupted and the project is not able to meet its debt obligations on time.

As a result, three important features to project finance are required to mitigate the project risks to the lenders. These are the following: (i) an agreement between parties to complete the project; (ii) and agreement for the purchase of the project's output or production, called an "off-take" agreement; and (iii) an agreement between parties to meet any shortfalls in the event of a disruption to the operation. The off-take agreement is crucial. The project's products or services must be in such demand that customers would be willing to enter long-term contracts to purchase the products, and they must be of sufficient quality to justify the funding for the construction of the project.

Other risks are financing risks, currency risks, and political risks. *Financing risks* refer to the negative impact that a change in the interest rate can have on the project's ability to meet its debt repayments. Interest rate risk can be mitigated through entering an interest rate cap agreement, which limits the maximum interest rate payable, or through an interest rate swap agreement, which transform the interest repayments from those based on a floating rate to a fixed rate. *Currency risks* occur when the revenues to the project occur in a currency different from the interest payments. A currency swap can be used to transform this risk. *Political risk* refers to the negative impact on a project because of the interference from political authorities. This risk is not only related to foreign destinations. The political authorities in the United States have the ability to make laws that are retrospective, such as environmental legislation, which can negatively impact a project. Project sponsors may have to invest a significant amount of effort in getting the regulatory approvals required and political backing for the project. A new government may reverse this effort, resulting significant loss to the project and its stakeholders.

17.3.4 Assessing Debt Capacity

An important aspect of the financial planning of a project finance opportunity is the debt capacity of the project. This borrowing capacity is the amount of debt that the project can service during the loan repayment period. The loan period depends on the duration of the project and on the lender's requirements and assessment of the risks. The borrowing capacity is determined from an assessment of either the discounted cash flows or the interest coverage ratio.

The maximum amount of debt that the project can carry, D, is related to the net present value of the project, NPV, as follows:

$$D = \frac{NPV}{\alpha} \tag{17.1}$$

where α is a multiple greater than one. The NPV of the project is assessed as in Chapters 5–10. The calculation of the borrowing capacity is illustrated in the following example.

Example 17.1: Borrowing capacity.

The net present value of a project using the interest rate of the debt as the discount rate is \$640 million. The consortium of banks providing debt to the project require the debt coverage ratio to be 1.5. Calculate the maximum amount of debt that the project can sustain.

Solution:

$$D = \frac{NPV}{\alpha} = \frac{640}{1.5} = \$426 \text{ million}$$

The project can sustain a loan of $426 million.

The interest coverage ratio can also used to determine if the project can sustain debt repayments. The interest coverage ratio is defined as follows:

$$Interest\ coverage\ ratio = \frac{EBIT}{Interest} \qquad (17.2)$$

where *EBIT* is the earnings before interest and taxation, and *Interest* is the interest charge on the debt. If the interest coverage ratio is below one, the project cannot sustain the level of debt. The lenders may require an interest coverage ratio of greater than, for example, 1.25 to allow for uncertainties.

17.3.5 Application of Project Finance

Project finance is a method of financing capital projects. In order for a project to be a suitable candidate for project finance, it must meet some requirements. These are as follows:

(i)   The project must be economically independent of and separable from the sponsor's other business interests.

(ii)   The project must be profitable to sustain the levels of debt, since credit is supplied based on the project's forecasted profitability rather than on the sponsor's creditworthiness.

(iii)   It must be possible to mitigate the major risks either by contract or by financial design. For example, a major risk is that customers fail to purchase the project's output at the forecast prices, which puts the project's revenues and cash flows at risk. An agreement for the purchase of the product with a customer provides lenders with the assurance that the forecasted cash flows can be realized.

Project finance offers major advantages to the sponsor of the project. The leverage available to a project financed opportunity is significantly higher than if the sponsor funded it off its own balance sheet (that is, on the general credit of the company). It also expands the amount of debt available to the sponsor without creating concerns about the sponsor's financial risk. The debt in the project can usually be arranged so that it does not appear on the sponsor's balance sheet because the lenders to the project have limited recourse to the sponsors assets in the case that the project defaults. The sponsor can undertake projects that are larger than usual, because of the separation of the sponsor's assets from those of the project and because of the increased debt capacity.

The disadvantages are the complexity of the contracts and arrangements that need to be made. This complexity increases the costs of financing and the time required to arrange the financing. The contracts need to be prepared by legal teams, the tax issues need to be researched and the project needs to be thoroughly assessed. This will also consume a greater amount of management's time than direct financing on the general credit of the company.

The application of project finance to the Pembroke Cracking Company and the Ras Gas project are discussed in the following two sections. These are examples of the practice of project finance. The first example illustrates the use of project finance to fund a project on almost all debt by providing the necessary assurances to the lenders. The second example, that is the Ras Gas project, highlights the contractual arrangements between the parties, including the engineering construction company.

17.3.6 Project Finance for the Pembroke Cracking Company

The Pembroke Cracking Company was formed as a partnership to build and operate a catalytic cracking plant in Pembroke, Wales. The sponsors, Gulf Oil Corporation and Texaco, Inc, each owned 50% of Pembroke through subsidiaries. A company, Pembroke Capital Company, Inc, which was also equally owned by the two sponsors, was formed to raise funds for the project. Pembroke Capital was financed by $1,000 of equity, and $900 million of long term bonds. The security for the debt issue were the following agreements with Gulf and Texaco: (i) A throughput agreement (see Chapter 4) that obligated Gulf and Texaco in the ratio 35:65 to process sufficient crude oil feed so that Pembroke could meet all of its expenses; (ii) an agreement between Gulf and Texaco that ensured the construction of the cracking plant was completed; (iii) if there was a deficiency in cash flow, Gulf and Texaco agreed to advance sufficient cash to Pembroke to make up for the deficiency. Advances of cash would be credited to future payments for the service of the cracking plant. The strength of the throughput agreement and cash deficiency agreement enabled Pembroke to be financed with almost 100% debt.

17.3.7 Project Finance for the Ras Gas Project

The North Field, off the coast of Qatar in the Middle East, is one of the largest deposits of natural gas in the world. It is estimated that there are at least 300 trillion cubic feet of natural gas in the field. The Ras Gas project consists of two process units that liquefy 5.2 million tons per year of natural gas. The project consists of offshore platforms, a pipeline to shore, the onshore process and storage facilities and port facilities. In 1996, the project costs were anticipated to be $3,400 million, of which $2,750 million was for fixed capital and $650 million was for finance and interest during the construction of the facility.

The Qatar General Petroleum Corporation (QPGC) and Exxon Mobil are the owners of the Ras Gas project through the Ras Laffan Liquefied Natural Gas Company. QPGC has a 70% stake in Ras Gas, while Mobil owns 30%. Ras Gas has a 25-year sale and purchase agreement for the delivery of 4.8 million tons per annum of liquefied natural gas (LNG). This contract is on a take-or-pay basis. Ras Gas will also sell condensate, a liquid hydrocarbon, to oil refineries. The sale of condensate on the spot market generates 20 to 25% of the project's income.

Ras Gas entered three engineering, procurement and construction (EPC) contracts for the construction of the onshore operations, the offshore platforms and the pipelines. The project was completed in 1999 at a capital cost of $3.264 billion, slightly less than the anticipated costs.

The financing needs for the Ras Gas project were provided by 75% debt and 25% equity. The debt was provided on a non-recourse basis after the completion of the project in the form of $1.2 billion worth of bonds and $1.35 billion worth of bank finance. Most of the bank finance was supported by export credit agencies. The purchasers of the bonds were insurance companies, mutual funds banks and pension funds.

The construction contractors arranged the bank debt as part of the competitive bidding process. Four groups of contractors bid on the main EPC contract. Each of these groups was expected to include financial proposals from commercial banks on their financial commitments to funding the project. The only way for the banks to participate in the financing of the project was if the contractor that they supported was awarded the construction contract. The main EPC contract for the onshore process plant was awarded to JGC/MW Kellogg and a group of banks led by the Industrial Bank of Japan and Credit Suisse.

17.4 Public–private Partnerships and the Funding of Public Infrastructure

The traditional model for the procurement of infrastructure in the public sector is through a tender process. The government or state owns the project, compiles a tender, advertises for tenders and contractors tender their bids in response. The government department selects the winning bid based on the criteria for the project. All responsibility for the procurement, operation, maintenance and funding of the infrastructure is taken by the public sector. The responsibility of the private sector is limited to the construction of the asset.

In a number of countries, the infrastructure needs have risen faster than the available funding, mainly as a result of cuts in infrastructure spending. The proportion of infrastructure spending has declined as a proportion of GDP as governments have attempted to curb public debt. In response, the public sector has attracted private funding to infrastructure projects through public–private partnerships that allow the private sector to construct and operate public infrastructure. An example of such a project is the 19-mile M1-A1 link in Yorkshire, UK, which was built at a cost of £214 million. It gives the private-sector consortium a 30-year contract to operate the road. The road users do not pay tolls; rather the government pays the private-sector consortium a "shadow" toll based on road usage in three different bands.

A public–private partnership is a joint venture between a government agency and a business to implement a particular project. Both parties aim to construct and operate the project more efficiently than the government agency could do alone. Arguments in favour of public–private partnerships stress the limited funds to government, and the efficiency of the private sector, while arguments against emphasise that because private finance is more expensive public-sector finance, the public sector is not getting value for money. (The public sector is financed through taxes and government bonds.)

The government departments are often blamed for high construction costs, time and budget overruns and inefficient operations. Business efficiency and the discipline of the market are proposed as the cure. However, there is no objective evidence that business or project-financed ventures are more efficient.

The involvement of the private sector in the financing of the project shifts some or all of the risks of the project from the government agency to the business partner. As a result, the returns to the business partner must be in line with returns in the market for the same amount of risk. Since governments can borrow at the lowest rate in the economy, private funding must be more expensive than government funding. The risks in public–private partnerships are discussed next.

17.4.1 Risk Sharing in Public–private Partnerships

There are four main aspects to the sharing of risks and rewards in a public–private partnership involving an infrastructure project. These are the construction, the operation, the ownership and the financing of the project.

(i)   Construction

The traditional model of infrastructure projects is for the construction to be either by government agencies and their employees or by competitive tender. The incentives for government employees to minimize costs and to deliver projects on time is less than for those working for a private contractor appointed by tender. The risks for the construction of the project can be easily shifted to the contractor by penalties for overruns in cost and time. In addition, the contractor may be required to place a performance bond, increasing the risks to the contractor.

The public–private partnership alters the nature of this relationship if the contractor is part of the ownership consortium. Even if the contractor is not part of the ownership structure, the risk of the project not being completed can be addressed in the same manner as is done in project financing. The private partners carry the risk of cost overruns and delays in construction.

(ii)   Operation

Assets within the public sector are operated as a portfolio. Risks are diversified over a very wide range of public sector projects. As a result, not every project is regarded on a stand-alone basis or as self-funding. Profitable activities subsidize unprofitable ones that are regarded as necessary for the public good. On the other hand, management of public facilities may not be as focused on efficient operation as their counterparts in industry.

The operations of an infrastructure project funded by public–private partnership must generate sufficient returns to reward the equity and debt holders for the risks that they carry. The private partners have a large incentive to operate the project efficiently. Surprisingly, there is no objective evidence to suggest that the private sector is inherently more efficient than the public sector.

Significant commercial risks are transferred to the private partners during the operation of the public–private partnership. Many of these projects provide services, such as toll roads, airports and hospitals. The demand for these services is unpredictable, and the public sector agency is unlikely to provide guarantees. For example, the demand for a toll road is uncertain, and as a result the commercial party carries the risk of the anticipated demand not materializing. Risks other than demand risk that are carried by the private partners are that the cost estimates for operating the facility are inadequate, and that the infrastructure requires refurbishment that is costly.

(iii)   Ownership

In the traditional model, the public sector owns the project. It takes all the risks, and bears the responsibility of ownership. For many of the public sector projects, the government is the main source of risk. For example, consider the operation of an airport. The profitability of the airport is greatly affected by government decisions regarding aspects such as environmental noise, and the provision of land transport and links. If government is the main source of risk, it can be argued that they are the natural owners of the asset.

The ownership rights in the public–private partnership vary from a franchise or lease model through to full ownership of the assets. A common arrangement is called BOOT, which is an acronym for build, own, operate and transfer. The private business owns the operation for a period, after which ownership and operations are transferred to the public sector.

(iv)   Financing

The financing of public–private infrastructure projects follows the project finance model, where debt is used to fund a significant portion of the capital requirements. In this model, commercial risks are mitigated through contractual obligations. In the example of the Pembroke Cracker discussed above, the sponsors agreed to pay for the use of the facilities in the form of a throughput agreement, and to make up for any shortages of cash. These types of agreements provided a greater degree of certainty for the lenders.

17.4.2 Structuring of a Public–private Partnership

A number of different structures other than BOOT exist for public–private partnerships. A perpetual franchise transfers all responsibilities to the private partners, while the BOOT model allows for the transfer of ownership to the public sector after a period. In a build, transfer, operate model (BTO) the private partners design and finance the construction of the facility, which is then transferred to the public sector. The private partners then lease the facility from the public sector owners.

The Private Finance Initiative (PFI) was established in the UK in 1992 to encourage private sector investment in infrastructure. The model adopted in this initiative is a concession contract referred to as design, build, finance and operate. The A1-M1 link road is an example of a public–private partnership undertaken as part of the PFI using project finance to fund the venture. This project is discussed in the next section.

17.4.3 The A1-M1 Link Road

The A1-M1 link road east of Leeds completes what was an obvious gap in the road system. A 30-year concession was granted to Yorkshire Link Ltd., a joint venture by Kverner plc and BICC plc to design, build, finance and operate the toll road. It is a 19-mile highway consisting of 2 and 3 lanes in both directions, a viaduct, 37 bridges and 2 tunnels. The major risks included construction risk and demand risk. No data on the anticipated the traffic volumes or road usage were available, and the investors and lenders were expected to assume this demand risk. Although it is a toll road, tolls are not collected from users. Instead, the government agency pays the toll based on the road usage, called a "shadow toll" agreement. These toll payments to the concessionaire are capped to limit the total risk to the Highways Agency.

The debt financing was provided by commercial banks. Senior debt of £300 million, comprising 81% of the funding requirements, was arranged by Lloyds Bank, Da-Ichi Kangyo Bank, Banque Indosuez, Credit Suisse ABN Amro and National Westminster Bank. Subordinated debt, comprising 11% of the funding, was provided by Deutsche Bank, 3i plc, and the Royal Bank of Scotland. The remainder of the finance was provided by shareholders funds.

The agreements governing the relationships between and the responsibilities of the parties are shown in Figure 17.1. The contractor and the operator were equity owners in the concessionaire, in addition to their participation in the construction and the operation of the project. The terms of the debt financing is governed by the loan agreement with the concessionaire, Yorkshire Link.

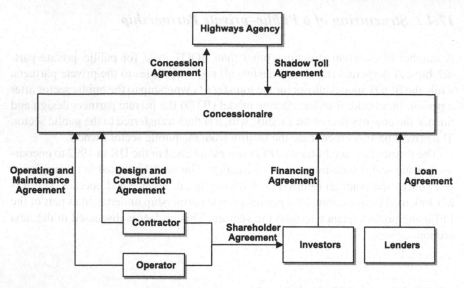

Figure 17.1 Contractual agreements between parties in a Project Finance Initiative toll road project

17.5 Case Study: Project Finance of a Cogeneration Facility

17.5.1 Introduction

A utility company, Johnson Energy, and an engineering company, Gatty Engineering, are both investors in a project to produce electricity and steam. This type of operation is called cogeneration, because two forms of energy are produced. The electricity from the project will be sold to Johnson Energy, while the steam will be sold to a nearby chemical processor, Nitrocity.

The project will be discussed in terms of the legal structure, the contractual arrangements and the financial arrangements.

17.5.2 Legal Structure

The project sponsors are Gatty Engineering and Johnson Energy. Gatty Engineering has limited ability to make equity investments since it does not have large capital resources. It cannot afford to accept any risks of debt default by the project. Like Gatty, Johnson Energy is willing to make an equity investment, but cannot assume responsibility for repayment of debt arising from the project.

An undivided joint interest structure was unsuitable because of the lack of liability protection. Liabilities in the project could have a detrimental effect on the sponsor's business. A company or corporation was also considered unsuitable because

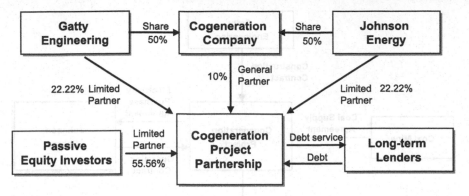

Figure 17.2 Legal structure of the project

the tax losses during the construction and early operating years would be confined to the company. The sponsors chose to enter into a limited partnership that would allow the tax benefits to "flow through" to the sponsors themselves.

The general partners of a partnership are jointly and severally liable for the debts that are incurred in the ordinary course of the business of the partnership. General partners are subject to unlimited liability. This liability can be limited by forming a subsidiary company that acts as the general partner in the project partnership. The liability of the partners can also be contractually limited to the assets of the partnership.

The partnership can also have limited partners, as long as there is at least one general partner. The limited partners are not exposed to unlimited liability. The sponsors act as both general and limited partners through a subsidiary company. Other equity investors, referred to as passive investors, are accommodated as limited partners.

The legal structure of the project is illustrated in Figure 17.2.

17.5.3 Contractual Arrangements for the Project

The contractual arrangements that support the cogeneration project are illustrated in Figure 17.3. The contracts are the construction contract, the coal supply contract, the electricity purchase contract, the steam purchase contract and the operating contact. They will be discussed in turn.

(i)   Construction contract

Gatty Engineering entered an agreement with the partnership to complete the construction of the project on a fixed-cost turnkey basis. The facility will produce 330 MW of electricity and 225 million tons per year of high-pressure steam. It is estimated that the construction and commissioning will take two years to complete at a cost of $300 million. The contract is backed by a performance bond to ensure

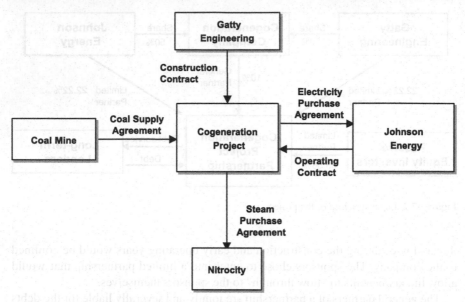

Figure 17.3 Contractual agreements between parties in the cogeneration project

completion of the project. The contract defines the requirements for mechanical completion, substantial completion and final completion. Substantial completion, which requires that the plant meet certain performance tests, has to be achieved by the guaranteed completion date. Final completion requires that the plant is mechanically complete and meets an availability that exceeds 88% over a one month test period.

(ii)   Operating contract

Johnson Energy has experience in operating similar facilities within its portfolio of assets. They will assume all responsibilities for the management of the operation and maintenance of the cogeneration plant under the terms of an operating agreement. The fee for this service is $6 million per year, renewable each year with adjustments for inflation. The plant can be operated and maintained with a staff contingent of less than fifteen people.

(iii)   Coal purchase contract

The partnership has entered a supply agreement with a coalmine, Deel Coal, to provide pulverized bituminous coal for 15 years. The contract specifies that Deel Coal will supply 1.1 million tons of coal each year. The contract also specifies the price, quality and calorific value of the coal that is supplied. A 30-day stockpile will be maintained at the cogeneration premises. In addition, Deel Coal will remove and dispose of the fly ash residue from the plant. The supply agreement also includes a sub-agreement with a trucking company for the transport of the coal and the fly ash. This transportation agreement is the responsibility of Deel Coal.

(iv)   Electricity purchase contract

Johnson Energy has entered a 15-year purchase agreement for the electricity. The terms of the contract obligates Johnson Energy to purchase all of the electricity offered by the partnership to it. The payments for the electricity will be in two forms: a fixed charge based on the cogeneration capacity; and a variable charge based on energy production. The pricing is adjusted for changes in the price of coal so that the partnership carries substantially no price risk for input costs.

(v)   Steam purchase contract

Nitrocity has entered into a 15-year agreement for the purchase of the steam. The company will accept and pay for the steam that is delivered (called a *take-if-offered* contract in Chapter 4).

17.5.4 Project Financials

A project must be economically viable for project financing to be applicable. The purpose of the project financials is to establish the viability of the project, and to determine the debt capacity of the project. The total capital cost is discussed next and then the project financials are presented.

(i)   Total capital costs

The capital items are detailed in Table 17.3. Although Gatty Engineering has committed to the construction of the plant on a turnkey basis, changes in the design or other changes agreed to by the partners will fall outside of the scope of the EPC contract. This accounts for the largest item in Table 17.3, the owner's contingency.

Table 17.3 Capital costs

Item	Amount, $ millions
Engineering, procurement and construction costs	297.84
Electrical, potable water, sewerage connections	4.62
Land acquisition	5.98
Development costs and fees	9.86
Spare parts	7.21
General and administrative costs	8.84
Start-up consumables	2.38
Initial working capital	2.35
Fuel reserve	3.40
Insurance	2.11
Other construction costs	2.72
Owner's contingency	25.16
Total	372.47

Table 17.4 Construction loan draw down and financing costs

End of month	Cumulative construction drawdown	Commitment fees	Interest	Total financing cost	Total construction and financing cost	Cumulative funds used	Unused balance
0	0.00	0	0.000	0	0.00	0.00	420.00
1	0.93	0.175	0.000	0.175	1.10	1.10	418.90
2	1.86	0.175	0.009	0.184	1.11	2.22	417.78
3	3.26	0.174	0.018	0.193	1.59	3.81	416.19
4	4.65	0.173	0.032	0.205	1.60	5.41	414.59
5	6.51	0.173	0.045	0.218	2.08	7.48	412.52
6	8.37	0.172	0.062	0.234	2.09	9.58	410.42
7	12.09	0.171	0.080	0.251	3.97	13.55	406.45
8	17.67	0.169	0.113	0.282	5.86	19.41	400.59
9	25.11	0.167	0.162	0.329	7.77	27.18	392.82
10	32.55	0.164	0.226	0.390	7.83	35.01	384.99
11	43.71	0.160	0.292	0.452	11.61	46.62	373.38
12	58.59	0.156	0.389	0.544	15.42	62.05	357.95
13	77.19	0.149	0.517	0.666	19.27	81.31	338.69
14	99.51	0.141	0.678	0.819	23.14	104.45	315.55
15	125.5	0.131	0.870	1.002	27.04	131.49	288.51
16	155.3	0.120	1.096	1.216	30.98	162.47	257.53
17	192.5	0.107	1.354	1.461	38.66	201.13	218.87
18	233.4	0.091	1.676	1.767	42.69	243.82	176.18
19	270.6	0.073	2.032	2.105	39.31	283.12	136.88
20	303.1	0.057	2.359	2.416	34.97	318.09	101.91
21	329.2	0.042	2.651	2.693	28.73	346.82	73.18
22	351.5	0.030	2.890	2.921	25.24	372.06	47.94
23	364.5	0.020	3.101	3.120	16.14	388.20	31.80
24	372.0	0.013	3.235	3.248	10.69	398.89	21.11
Total		3.005	23.887		398.89		

The construction phase of the project is funded by a commercial loan. The partnership will pay the bank a 1.2% fee on closing of the loan agreement, and will itself incur \$2 million in costs in arranging the finance. This loan will be for \$420 million. The loan includes interest on the drawn amount and commitment fees on the unused balance. The interest rate is 10% pa on the used balance, and the commitment fees are 0.5% of the unused balance.

Gatty Engineering has provided a schedule of the financial requirements for the construction of the project on a monthly basis, which is given in Table 17.4. Also shown in Table 17.4 is the calculation of the financing costs for the construction phase of the project.

The total capital cost with financing is equal to \$398.89 million plus \$5.04 million for arranging fees and \$2 million for costs incurred by the partnership. This means that the total capital cost is \$405.9 million.

(ii) Revenue

The revenues for the cogeneration facility come from two sources, the sale of electricity and steam. The electricity charge is calculated on a capacity charge and an energy charge. The capacity charge is 374 $/kW, and the since the plant produces 330,000 kW, the annual capacity charge is $123,472,000 ($=(330$ MW) (1000 kW/MW)(374 $/kW)). The energy charge is 0.02422 $/kWhr. Since the plant produces 2,553,155,000 kW per year ($=(330$ MW)(1000 kW/MW)(24 hr/day) (365 day/yr)(88.32% dispatch)), the energy charge yields $61,837,414 each year. The price of steam is $0.19 per million pounds. The income from the sale of steam is equal to $94,000 ($=$ (225 million tons per year)(2,204 lbs/t)(0.19 $/Mlb)). All revenues and costs are escalated at an inflation rate of 5%.

(iii) Costs

The cost of the pulverized coal is $21.1 per ton, and 834,717 tons of coal is consumed each year. The cost of disposal of the fly ash is equal to that of the coal purchased. Additional costs are the transportation of the coal and the fly ash, maintenance costs, and other operating expenses that arise.

These revenues and costs are represented in the project financials, which are given in Table 17.5. The project is expected to have a life of 15 years. The project financials are prepared as of the date of delivery of the plant for operation. The capital cost is assumed to be the total capital commitment required for the construction of the plant.

Table 17.5 Cash flow projections for the first five years of the cogeneration partnership. The years 6 to 15 are identical to year 5 escalated to the inflation rate of 5%

Year	0	1	2	3	4	5
Revenues ($000s)						
Capacity		129,646	136,128	142,934	150,081	157,585
Energy		64,929	68,176	71,585	75,164	78,922
Steam		99	104	109	115	120
Total revenues		194,674	204,408	214,628	225,359	236,627
Costs ($000s)						
Coal		23,752	24,939	26,186	27,496	28,871
Fly ash disposal		23,752	24,939	26,186	27,496	28,871
Transportation		15,225	15,986	16,786	17,625	18,506
Maintenance		12,075	12,679	13,313	13,978	14,677
Other operating expenses		12,600	13,230	13,892	14,586	15,315
Total expenses		87,404	91,774	96,363	101,181	106,240
Operating income ($000s)		107,270	112,634	118,265	124,179	130,388
Capital ($000s)	−405,932					
Free cash flow ($000s)	−405,932	107,270	112,634	118,265	124,179	130,388

17.5.5 Assessment

The measures of performance of the project are given in Table 17.6. All of the measures are extremely positive.

Table 17.6 Assessment measure for the cogeneration project

Measure	Value
NPV (9%)	$745,225,000
IRR	30.4%
PI	2.84
Payback period	4 years

17.5.6 Financing

The commercial loan discussed earlier covered the period of the construction of the plant. Permanent financing must be arranged for the partnership, which is sourced from the equity contributions and the debt raised. The debt will be non-recourse to the equity investors, that is, the lenders can only look to the project for security. In addition, the project is of a fixed life, so the cash flow must support both the interest and the repayments of principal. The debt capacity is discussed next and then the equity financing structure.

(i)   Debt finance

The debt is raised in the form of mortgage bonds with the following repayment schedule. Five percent of the principal amount is repaid in the first three years, ten percent in the following four years, and fifteen percent in the last three years. Interest is charged at a rate of 10% on the unpaid principal.

The capital requirement for the partnership is $406 million, and the target capital structure is 75% debt and 25% equity. This means that the partnership would like to raise permanent debt funding of $304.45 million.

The loan repayment schedule is given in Table 17.7.

The debt capacity can be determined in one of three ways: from the interest cover; from the debt service cover; and from the present value multiple. The *NPV* of the project's cash flows for the first ten years of the project is $430.6 million. The lenders require a coverage ratio, given by α, of 1.33. This means that the debt capacity of the partnership is calculated using Equation 17.2 as follows:

$$D = \frac{NPV}{\alpha} = \frac{430.6}{1.33} = \$323 \text{ million}$$

The interest coverage is given by Equation 17.2, while the debt service coverage is defined as follows:

Table 17.7 Permanent loan repayment schedule

Year	1	2	3	4	5	6	7	8	9	10
Opening balance	304.4	289.2	274.0	258.8	228.3	197.9	167.4	137.0	91.3	45.7
Interest charged	30.4	28.9	27.4	25.9	22.8	19.8	16.7	13.7	9.1	4.6
Principal repaid	15.2	15.2	15.2	30.4	30.4	30.4	30.4	45.7	45.7	45.7
Closing balance	289.2	274.0	258.8	228.3	197.9	167.4	137.0	91.3	45.7	0.0

$$Debt\ Service\ Coverage = \frac{EBITDA}{Interest + Principal(1 - T)} \qquad (17.3)$$

where *EBITDA* is the earnings before interest, taxation, depreciation and amortization. *T* represents the tax rate. *EBIT* is in this case is the operating income less the depreciation, which is calculated on a straight-line basis over ten years, and *EBITDA* in this case is the same as the operating income. The results of the calculation of the interest cover and the debt service coverage are presented in Table 17.8.

The debt service coverage drops to a low of 1.7 in year 4, which is above the lender's requirement of 1.5. The lenders agree to loan the money to the partnership on the basis of these cash flow projections.

(ii)   Equity finance

The partners have agreed to fund the cogeneration project with equity capital of $101.48 million. The general partners will fund 10% of this amount, and the limited partners will fund 90% of it. The equity interests of the general and limited partners are also split in the same ratio. The limited partners consist of the sponsors, Johnson Energy and Gatty Engineering, and passive equity partners. Each of the sponsors has contributed 25% of the equity required by the limited partners, while the passive investors have contributed 50% of the equity capital for 55.56% of the ownership. These relationships were presented in Figure 17.2. The proportional contributions are summarised in Table 17.9.

The distribution of the cash flows for the project are in proportion to the capital contribution and the ownership structure until the partnership has paid the limited

Table 17.8 Calculation of the interest coverage and debt coverage

Values in millions	1	2	3	4	5	6	7	8	9	10
Operating income	107.3	112.6	118.3	124.2	130.4	136.9	143.8	150.9	158.5	166.4
Depreciation	40.6	40.6	40.6	40.6	40.6	40.6	40.6	40.6	40.6	40.6
EBIT	66.7	72.0	77.7	83.6	89.8	96.3	103.2	110.3	117.9	125.8
Interest	30.4	28.9	27.4	25.9	22.8	19.8	16.7	13.7	9.1	4.6
Principal repayment	15.2	15.2	15.2	30.4	30.4	30.4	30.4	45.7	45.7	45.7
Amount available for disbursement	61.6	68.5	75.6	67.9	77.1	86.7	96.6	91.6	103.7	116.2
Interest cover	2.2	2.5	2.8	3.2	3.9	4.9	6.2	8.1	12.9	27.6
Debt service coverage	2.0	2.2	2.3	1.7	1.9	2.1	2.3	1.8	2.0	2.2

Table 17.9 Equity ownership in the partnership

Partner	Equity	Percentage	Johnson	Gatty	Passive
General partner	30.44	20%	50.00%	50.00%	0%
Limited partners	71.04	80%	22.22%	22.22%	55.56%
Total	101.48	100.00%	27.78%	27.78%	44.45%

partners their capital back. At that point, called reversion, the split in the distribution of the proceeds between the general partners and the limited partners changes to a 50:50 ratio. The reversion is to reward the general partners for the increased risk that they are exposed to compared with the limited partners. This distribution of the proceeds to the three parties is shown in Table 17.10.

The before and after tax cash flows to both the limited and general partners is shown in Table 17.11.

Table 17.10 Distribution of proceeds after reversion

Distribution after reversion	Johnson	Gatty	Passive
General partner	50.00%	50.00%	
Limited partners	22.22%	22.22%	55.56%
Total	36.11%	36.11%	27.78%

Table 17.11 Before and after-tax cash flows to the partners for the first ten years. All amounts in millions

Year	0	1	2	3	4	5	6	7	8	9	10
Operating income		107.3	112.6	118.3	124.2	130.4	136.9	143.8	150.9	158.5	166.4
Debt service		45.7	44.1	42.6	56.3	53.3	50.2	47.2	59.4	54.8	50.2
Amount for distribution		61.6	68.5	75.6	67.9	77.1	86.7	96.6	91.6	103.7	116.2
Limited partners share		80%	80%	50%	50%	50%	50%	50%	50%	50%	50%
Cash to limited partners		49.3	54.8	37.8	33.9	38.6	43.3	48.3	45.8	51.8	58.1
Cash to general partners		12.3	13.7	37.8	33.9	38.6	43.3	48.3	45.8	51.8	58.1
Tax calculation											
Depreciation		40.6	40.6	40.6	40.6	40.6	40.6	40.6	40.6	40.6	40.6
Interest		30.4	28.9	27.4	25.9	22.8	19.8	16.7	13.7	9.1	4.6
Taxable income		36.2	43.1	50.3	57.7	67.0	76.5	86.4	96.6	108.8	121.3
Tax		12.7	15.1	17.6	20.2	23.4	26.8	30.2	33.8	38.1	42.4
Cash flow after tax (CFAT)		48.9	53.4	58.0	47.7	53.7	59.9	66.3	57.7	65.6	73.7
Limited partners tax		10.1	12.1	8.8	10.1	11.7	13.4	15.1	16.9	19.0	21.2
General partners tax		2.5	3.0	8.8	10.1	11.7	13.4	15.1	16.9	19.0	21.2
CFAT to limited partners	−71.0	39.1	42.7	29.0	23.8	26.8	29.9	33.2	28.9	32.8	36.9
CFAT to general partners	−30.4	9.8	10.7	29.0	23.8	26.8	29.9	33.2	28.9	32.8	36.9
CFAT (passive)	−35.5	21.7	23.7	16.1	13.2	14.9	16.6	18.4	16.0	18.2	20.5
CFAT (sponsors)	−66.0	29.7	41.9	34.4	38.8	43.3	47.9	41.7	47.4	53.3	29.7

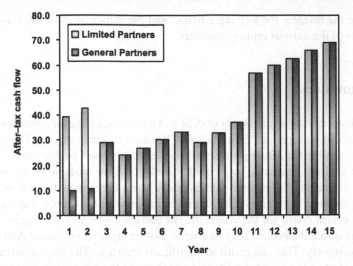

Figure 17.4 After-tax cash flows for the limited and general partners

Table 17.12 *IRR* and *NPV* for the after-tax cash flow to partners, passive investors and sponsors

Party	IRR	NPV
General partner	57.8%	227.7
Limited partners	49.9%	241.0
Passive investors	55.8%	137.8
Sponsors	51.7%	330.8

The after-tax cash flow to the limited and general partners is shown in Figure 17.4. This figure indicates the effect of switching of the proportional payments at reversion, which occurs at year 3.

(iii) Returns

The values for the *IRR* and *NPV* for each of the equity parties are given in Table 17.12, which indicates that this is an excellent investment. It has been assumed throughout that the equity investment is made at the end of construction. The equity investment will probably be made as construction proceeds. If the calculation is performed in this manner, the cash flow profile is affected, which alters the values of the *IRR*.

17.5.7 Concluding Remarks

This detailed example has illustrated the key ingredients of project finance. The legal structure, financial arrangements and financial plan have been presented. A model like this one can be used by the business analyst to determine the best method of arranging the finance, the key ingredients, and the sensitivities of the financing to the values of the various input parameters.

17.6 Summary

Project finance is the arrangement of debt and equity financing for a business opportunity that can be separated from the other business of the company. The providers of funds examine the future profitability, rather than passed performance, to determine the venture's capacity for debt. The risks of these cash flows are mitigated by several agreements between parties, such as an agreement for the sale of the project's products or services. The lending arrangements are more complex than arranging general credit for the company, but there are several advantages. The first advantage is that the project may be able to sustain significantly more debt than the sponsor company. This can result in significant leverage. The second advantage is that the assets of the sponsor company and those of the project are separated from each other. This allows the sponsor company to pursue projects that may be large in comparison with the size of the company itself. The disadvantage is that the complexities in arranging the agreements and finance are significantly greater than for general credit.

17.7 Review Questions

1. Is project finance defined as the "finance for the project?"
2. How does project finance differ from financing a project from the general credit of the company?
3. What agreements are generally required in project finance to mitigate project related risks?

17.8 Exercises

1. A construction company provides a bid bond of $1,000,000 for the competitive tendering on a road project. The evaluation of the bids takes 5 months and the negotiations and execution of agreements takes a further 4 months. If the bank charges a 12% interest rate for this guarantee, what is the cost to the construction company if it wins the bid and if it looses the bid?
2. The project owner requires a performance bond of 10% of the EPCM costs of $40 million. If the project is expected to take three years to complete, and the bank charges the construction company an interest rate of 15% pa, determine the cost of the performance bond.

3. Determine the borrowing capacity on a $3.2 billion project if the debt multiple is 1.75.

4. The construction of a project is financed by a bank loan in which the owner pays 12% interest on the loan amount used, and 1% on the unused balance. The draw down schedule is given in the table below. Determine the total construction and financing charges.

Month	Amount, $ million
0	1
3	2.3
6	4
9	3.1
12	2
15	1
18	0

Appendix A
Equivalent US and UK Terms Used in the Financial Statements

UK Term	US Equivalent
Accounts	Financial statements
Capital allowances	Tax depreciation
Creditors	Accounts payable
Debtors	Accounts receivable
Financial year	Fiscal year
Fixed tangible assets	Property, plant and equipment
Gearing	Leverage
Loan capital	Long-term debt
Net asset value	Book value
Profit	Income (or earnings)
Profit and loss account	Income statement
Profit attributable to ordinary shareholders	Net income
Reserves	Stockholder's equity (other than capital stock)
Share capital	Ordinary shares, capital stock
Share premium account	Additional paid-up capital relating to proceeds of sale of stock in excess of par value
Shares in issue	Shares outstanding
Stocks	Inventory
Turnover	Revenues (or sales)

Appendix B
Answers to Selected Exercises

Chapter 1

1. First calculate the cumulative free cash flow, as shown in the table below:

Year	0	1	2	3	4	5
Free cash flow	−100	35	40	45	50	55
Cumulative free cash flow	−100	−65	−25	20	70	125

The cumulative free cash flow turns positive between the end of year 2 and the end of year 3. Notice that the dates on the years are at the end of the year. The equation for linear interpolation between two points (y_1, x_1) and (y_2, x_2) is given by the following equation:

$$y - y_1 = \frac{y_1 - y_2}{x_1 - x_2}(x - x_1)$$

The two points here are (20,3) and (−25,2) and y is equal to zero. The necessary substitutions yield the following result:

$$0 - (-25) = \frac{45}{2}(x - 2)$$

$x = 2.55$.
This means that the payback period is 2.55 years.

2.   The free cash flows are given in the table below:

Year	0	1	2	3	4	5
Revenue		70.0	70.0	70.0	70.0	70.0
Costs		42.0	42.0	42.0	42.0	42.0
Tax		8.4	8.4	8.4	8.4	8.4
Investment	−100.0					
Working capital	−15.0					15.0
Free cash flow	−115	19.6	19.6	19.6	19.6	34.6

3.   The cumulative free cash flow is given in the table below:

Year	Free cash flow	Cumulative free cash flow
0	−10,000	−10,000
1	−5,000	−15,000
2	3,000	−12,000
3	3,000	−9,000
4	3,000	−6,000
5	3,000	−3,000
6	3,000	0
7	3,000	3,000
8	3,000	6,000
9	3,000	9,000
10	2,000	11,000
11	2,000	13,000
12	2,000	15,000
13	2,000	17,000
14	2,000	19,000
15	3,000	22,000

The free cash flows are zero at the end of the sixth year. The payback period is 6 years.

4.   The cumulative free cash flows for the three options are given in the table below:

Year	Philippines	Malaysia	Thailand	Cumulative Philippines	Cumulative Malaysia	Cumulative Thailand
0	−30,000	−40,000	−38,000	−30,000	−40,000	−38,000
1	−10,000	−6,000	−7,000	−40,000	−46,000	−45,000
2	9,000	8,000	5,000	−31,000	−38,000	−40,000
3	7,800	8,000	7,500	−23,200	−30,000	−32,500
4	7,800	8,000	7,500	−15,400	−22,000	−25,000
5	7,800	8,000	7,500	−7,600	−14,000	−17,500
6	7,800	8,000	7,500	200	−6,000	−10,000
7	7,800	8,000	7,500	8,000	2,000	−2,500
8	7,800	8,000	7,500	15,800	10,000	5,000
9	7,800	8,000	7,500	23,600	18,000	12,500
10	7,800	8,000	7,500	31,400	26,000	20,000

The payback period for the Philippines options is slightly less than the other two, and is the preferred option.

5. The various values are calculated as follows:

(i) $Return\ on\ investment = \dfrac{Annual\ income(profit)}{Investment} = \dfrac{3}{10} = 30\%$

(ii) $Return\ on\ investment = \dfrac{Annual\ income(profit)}{Investment/2} = \dfrac{3}{(10/2)} = 60\%$

(iii) $Return\ on\ investment = \dfrac{Total\ income(profit) - Investment}{(Investment/2)(Project\ life)}$

$$= \dfrac{5(3) - 10}{(10/2)(5)} = 20\%$$

6. The income increases at 5% pa, which means that for each year the income is that of the previous year multiplied by (1+0.05). As a result, the cumulative free cash flow is calculated as shown in the table below:

Year	0	1	2	3	4	5	6	7	8	9
Income		750	788	827	868	912	957	1005	1055	1108
Investment	−8,000									
Free cash flow	−8,000	750	788	827	868	912	957	1005	1055	1108
Cumulative free cash flow	−8,000	−7,250	−6,463	−5,636	−4,767	−3,856	−2,899	−1,893	−838	270

The payback period occurs in the ninth year.

7. Neither the original value nor the book value is of any interest. The only value of interest is the price at which she could sell the equipment for now.

8. In 2.5 years, the project will earn $2.5(10,000) = 25,000$. This is the maximum that can be invested.

Chapter 3

1. The income statement for the company is as follows:

	Amount
Revenue (sales)	1,000,000
COGS	600,000
Gross profit	400,000
Overheads (SGA)	50,000
EBIT (PBIT)	350,000
Interest	50,000
Tax	105,000
Net income (net profit)	195,000

(i) The gross profit, given by sales less operating expenses, is $400,000.

(ii)   The EBIT is the gross profit less the overheads, and is equal to $50,000.
(iii) The tax is given by the expression (EBIT-interest) tax rate, which in this case is equal to 105,000.
(iv)   The net profit is given by the expression (EBIT-interest-tax), and is equal to $195,000.

2.   The transactions can be divided as follows:

	Amount	Cash	Sales
(i) Loan	600,000	600,000	
(ii) Assets	50,000	−50,000	
(iii) Assets	15,000	−15,000	
(iv) Stock	200,000	−200,000	
(v) Accounts receivable	100,000		100,000
Cash		335,000	

This allows the amount of cash to be determined. As a result, the balance sheet is given as follows:

	Amount
Assets	
Fixed	65,000
Current	635,000
Total assets	700,000
Liabilities	
Loan	600,000
Equity	
Earnings (sales)	100,000
Total liabilities and equity	700,000

Since information is given only on sales amount, it is assumed to be all profit and that it is all retained. The current assets are the cash, the inventory and the accounts receivable.

3.   The income statement can be represented as follows:

	Amount
Sales	100,000
COGS	67,000
Gross profit	33,000
Overheads	18,650
Depreciation	1,500
Interest	500
Income (before tax)	12,350

The profit for the business is $12,350. Sales or revenues for the consultancy represent the amount received for the selling of services.

4. The forecast income statement is shown below:

Year	1	2	3	4	5	6
Sales	600,000	720,000	840,000	960,000	1,080,000	1,200,000
Engineers	5	6	7	8	9	10
Expenses	210,000	252,000	294,000	336,000	378,000	420,000
Gross profit	390,000	468,000	546,000	624,000	702,000	780,000
Overheads	195,000	234,000	273,000	312,000	351,000	390,000
Depreciation	20,000	20,000	20,000	20,000	20,000	0
Interest	9,000	9,000	9,000	9,000	9,000	9,000
Profit before tax	166,000	205,000	244,000	283,000	322,000	381,000
Tax	58,100	71,750	85,400	99,050	112,700	133,350
Net profit	107,900	133,250	158,600	183,950	209,300	247,650

6. The calculation of the dividends and the cash flow are shown in the table below:

	Amount
EBIT	3,000,000
Depreciation	300,000
Interest	500,000
Tax	880,000
Net income	1,320,000
Dividends	792,000
Retaining earnings	528,000
Cash flow	1,620,000
Cash retained	828,000
Cash paid in dividends	792,000

10. The income statement, cash flow statement and the balance sheet for the business at the end of the first year are shown in the tables below.

Income statement	Amount
Sales	89,000
COGS	35,000
Gross earnings	54,000
SGA	10,000
Depreciation	24,000
Interest	2,000
Tax	7,200
Earnings	10,800

Cash flow statement	Amount
Cash flow from operating activities	
Operating income	44,000
Increase in working capital	30,000
Tax paid	7,200
Total	6,800
Cash flow from financing activities	
Equity	100,000
Loan	75,000
Interest payment	2,000
Total	173,000
Cash flow used in investment activities	
Investment	120,000
Total	120,000
Net cash flow	59,800

Balance sheet	Amount
Non-current assets	
Equipment	120,000
Deprecation	24,000
Total non-current assets	96,000
Current assets	
Cash	59,800
Accounts receivable	10,000
Inventory	20,000
Total current assets	89,800
Total assets	185,800
Liabilities	
Loan	75,000
Total liabilities	75,000
Equity	
Retained earnings	10,800
Share capital	100,000
Total equity	110,800
Total equity and liabilities	185,800

11. The value of Apex Pumps is equal to $5,000(15) = $75,000$.
12. The balance sheet for the company is given in the table below:

Balance sheet	Amount
Assets	
Non-current	
Equipment	82,000
Buildings and land	92,000
Vehicles	5,000
Office equipment	9,000
Total non-current assets	188,000
Current	
Cash	12,000
Accounts receivable	11,000
Inventory	20,000
Total current assets	43,000
Total assets	231,000
Liabilities	
Bank overdraft	16,000
Creditors	12,000
Total liabilities	28,000
Equity	
Retained earnings	13,000
Share capital	190,000
Total equity	203,000
Total equity and liabilities	231,000

Chapter 4

1. A "class 3 estimate" is an estimate of the capital expenses with an error level of about $\pm 15\%$. It is usually prepared for budget authorization, or for full authorization of the funding of the project. It is also called the "bankable feasibility study."

2. The cost of the reactor is calculated as follows:

$$Cost = 3,000,000\frac{567}{211} = 8,061,611$$

3. The cost of the boiler can be calculated as follows:

$$Cost = 200,000(1+0.047)^5 = 251,631$$

4.   The cost of the compressor is calculated as follows:

$$Cost = 2,000\left(\frac{1,500}{300}\right)^{0.66} = 5,786$$

5.   The cost of the ammonia plant is estimated as follows:

$$Cost = 4,000,000\left(\frac{442}{100}\right) = 17,680,000$$

6.   The costs of producing the item from the input materials are calculated in the table below:

Item	Quantity	Unit price, $	Total cost
12-k Resistor	2	0.11	0.22
20-k Resistor	3	0.15	0.45
150-k Potentiometer	1	0.98	0.98
Diode	4	0.43	1.72
Switch	1	0.66	0.66
Capacitor	2	0.67	1.34
Transistor	1	1.53	1.53
Grand total			6.9

The labour costs are calculated below:

	Time per 100 units	Cost	Cost per 100 units
Machine shop	6	25	150
Assembly	5	25	125
Inspection	2	25	50
Production costs			325
Overhead			325
Total labour and overheads			650
Cost per unit			13.4

7.   The depreciation charge each year is $20,000/10 = 2,000$. The depreciation schedule is given below:

Year	0	1	2	3	4	5	6	7	8	9	10
Asset	20,000										
Depreciation charge		2,000	2,000	2,000	2,000	2,000	2,000	2,000	2,000	2,000	2,000
Book value		18,000	16,000	14,000	12,000	10,000	8,000	6,000	4,000	2,000	0

The dates for each year represent the end of the year.

8.   The calculation of the free cash flow and the cumulative free cash flow (FCF) is shown in the table below:

Year	0	1	2	3	4	5	6	7	8	9	10
Revenue	0	5	5	5	5	5	5	5	5	5	5
Expenses	0	3	3	3	3	3	3	3	3	3	3
Tax calculation	0										
Depreciation	0	1	1	1	1	1	1	1	1	1	1
Taxable income	0	1	1	1	1	1	1	1	1	1	1
Tax	0	0.35	0.35	0.35	0.35	0.35	0.35	0.35	0.35	0.35	0.35
Investment	−10										
Change in working capital	−2.5										
Free cash flow	−12.5	1.65	1.65	1.65	1.65	1.65	1.65	1.65	1.65	1.65	1.65
Cumulative FCF	−12.5	−10.85	−9.2	−7.55	−5.9	−4.25	−2.6	−0.95	0.7	2.35	4

The payback period is in the eighth year. The return on investment can be calculated in a number of different ways (see Chapter 1, Exercise 5). The simplest method is as follows:

$$Return\ on\ investment = \frac{Annual\ income(profit)}{Investment} = \frac{5-3-1-0.35}{10} = 6.5\%$$

9. The cost of the vessel for different size tanks:

Size	Cost
1	4,309
10	20,000
100	92,832
1000	430,887

11. The total costs are given in the table below:

	Amount
Raw materials	655,200
Utilities	405,600
Packaging and delivery	47,736
Total costs	1,108,536

The project financials for the epoxide plant are given in the table below, assuming that the fixed capital can be depreciated over 15 years:

Year	0	1	2	3	4	5
Revenue		2,901,600	2,901,600	2,901,600	2,901,600	2,901,600
Total costs		1,108,536	1,108,536	1,108,536	1,108,536	1,108,536
Tax calculation:						
Depreciation		666,667	666,667	666,667	666,667	666,667
Tax		394,239	394,239	394,239	394,239	394,239
Investment	−10,000,000					
Change in working capital	−2,000,000					
Free cash flow	−12,000,000	1,398,825	1,398,825	1,398,825	1,398,825	1,398,825

Chapter 5

1.   The present value is calculated as follows:

$$Present\ Value = 1{,}000/(1+0.05)^5 = 783.58$$

2.   The future value is calculated as follows:

$$Future\ Value = 1{,}000(1+0.10)^{10} = 2{,}593.74$$

3.   The future value is calculated as follows:

$$Future\ Value = 250(1+3.5\cdot0.10) = 337.50$$

4.   The following two formulae apply:

$$FV = P\left(1+\frac{i}{m}\right)^{mn} \quad \text{and} \quad FV = P(1+e)$$

where P is the principal, m the number of periods per year, n the number of years and e is the effective rate. Equating the two formulae and eliminating P, the effective rate is given by:

$$e = \left(1+\frac{i}{m}\right)^{m} - 1$$

Thus the effective rate is calculated as follows:

$$e = \left(1+\frac{0.06}{12}\right)^{12} - 1 = 6.168\%$$

The effective rate is 6.188%.

5.   If the amount doubles, the future value is twice the present value. Hence, the period is calculated as follows:

$$2 = 1(1+0.08)^n \Rightarrow n = \log(2)/\log(1.08) = 9.006$$

It would take roughly 9 nine years for the amount to double in value.

6.   The annuity formula is given as follows:

$$PV = A(PV|A,i,n) = 250[(1+0.05)^4 - 1]/[0.05(1+0.05)^4] = 886.49$$

7.   The formula is given as follows:

$$PV = A(PV|A,i,n) \Rightarrow 1000 = 300[(1+i)^5 - 1]/[i(1+i)^5]$$

The value of i is found by trial and error. A value for i is guessed, from which PV is calculated. Based on the value of PV, a new guess is made until the calculated PV is equal to 1,000. The value of i is 15.24%.

8.   The value of the annuity is calculated as follows:

$$FV = A(FV|A, i, n) = 5[(1+0.08)^5 - 1]/0.08 = 29.33$$

9.   The value of the deposit is calculated as follows:

$$FV = PV(1+i)^n \Rightarrow 100{,}000 = PV(1+0.11)^5 \Rightarrow PV = 59{,}345.13$$

10.   The interest rate is calculated as follows:

$$FV = PV(1+i)^n \Rightarrow i = 2^{1/n} - 1 \Rightarrow 2^{0.1} - 1 = 7.177\%$$

11.   The time is calculated as follows:

$$FV = PV(1+i)^n \Rightarrow n = \log(45{,}000/15{,}000)/\log(1+0.09) = 12.75$$

12.   The future value of the amount is calculated as follows:

$$FV = PV(1+i)^n = 250(1+0.12)^5 = 384.66$$

13.   The present value of the annuity is calculated as follows:

$$PV = A[(1+i)^n - 1]/[i(1+i)^n]$$
$$= 1{,}200[(1+0.12)^4 - 1]/[0.12(1+0.12)^4] = 3644.82$$

The present value of the annuity is 3,644.82. You would prefer $4,500.

14.   The calculation of the present value and the future value of the deposits is shown in the table below:

Year	Cash flow	Compound factor	Future value	Discount factor	Present value
0	1,000	1.828	1,828.04	1	1,000.00
1		1.677	0.00	0.917431	0
2	2,000	1.539	3,077.25	0.84168	1,683.36
3	−3,000	1.412	−4,234.74	0.772183	−2,316.55
4		1.295	0.00	0.708425	0
5	8,000	1.188	9,504.80	0.649931	5,199.451
6		1.090	0.00	0.596267	0
7	−5,000	1.000	−5,000.00	0.547034	−2,735.17
Total			5,175.34		2,831.09

The value at the end of the period is 5,175.34, which has a present value of 2,831.09.

The future value at each year was calculated as follows:

$$FV = (Deposit\ or\ Withdrawal)(1+0.09)^{7-t}$$

where t is the year in which the deposit or withdrawal is made.

This calculation can be checked by determining the interest on the balance in the bank account. This is shown in the table below.

Year	Opening	Interest	Deposit or withdrawal	Closing
0	1,000.00	0.00	1,000.00	1,000.00
1	1,000.00	90.00		1,090.00
2	1,090.00	98.10	2,000.00	3,188.10
3	3,188.10	286.93	−3,000.00	475.03
4	475.03	42.75		517.78
5	517.78	46.60	8,000.00	8,564.38
6	8,564.38	770.79		9,335.18
7	9,335.18	840.17	−5,000.00	5,175.34

15. The effective rate is calculated as follows:

$$e = \left(1 + \frac{0.09}{12}\right)^{12} - 1 = 9.381\%$$

19. The interest rate is calculated as follows:

$$FV = PV(1+i)^n \Rightarrow 750,000 = 500,000(1+i)^4 \Rightarrow i = 10.67\%$$

20. The present value of the loan is calculated as follows:

$$PV = A(P|A, i, n) = A[(1+i)^n - 1]/[i(1+i)^n]$$
$$PV = 7,000[(1+0.14)^{10} - 1]/[0.14(1+0.14)^{10}] = 36,512.81$$

Chapter 6

1. The present value of the annuity is calculated as follows:

$$PV = A[(1+k)^n - 1]/[k(1+k)^n]$$
$$= 1,200[(1+0.12)^5 - 1]/[0.12(1+0.12)^5] = 4,549$$

The present value of the annuity is $4,549. You would prefer $5,500.

2. The calculation of the present values of each of the cash flow, and the cumulative values are shown in the table below:

Year	Cash flow	Discount factor	Present value	Cumulative cash flow	Cumulative present value
0	−100	1.000	−100.00	−100	−100.00
1	50	0.935	46.73	−450	−53.27
2	60	0.873	52.41	10	−0.86
3	70	0.816	57.14	80	56.28
4	200	0.763	152.58	280	208.86

The decision criteria are given in the table below:

NPV	208.86
IRR	61%
PI	3.09
EAC	707
Payback (years)	1.833
Discount Payback (years)	2.018

3.   The calculation of the NPV and the IRR is given in the table below:

Year	Project A	Project B	Discount factor	PV(A)	PV(B)
0	−20,000	−40,000	1	−20,000	−40,000
1	10,000	20,000	0.900901	9,009	18,018
2	10,000	20,000	0.811622	8,116	16,232
3	10,000	20,000	0.731191	7,312	14,624
NPV	4,437	8,874			
IRR	23%	23%			

4.   The calculation of the NPV and IRR is given in the table below:

Year	Project A	Project B	Discount factor	PV(A)	PV(B)
0	−20,000	−40,000	1	−20,000	−40,000
1	−10,000	−20,000	0.884956	−8,850	−17,699
2	−10,000	−20,000	0.783147	−7,831	−15,663
3	−10,000	−20,000	0.69305	−6,931	−13,861
NPV	−43,612	−87,223			
IRR	#NUM!	#NUM!			

#NUM! indicates that the IRR calculation does not yield a valid result.
5.   The free cash flows and the decision criteria are calculated in the table below:

Year	0	1	2	3	4	5	6	7	8	9	10
Revenue		35.00	35.00	35.00	35.00	35.00	35.00	35.00	35.00	35.00	35.00
Expenses		19.00	19.00	19.00	19.00	19.00	19.00	19.00	19.00	19.00	19.00
Tax calculation:											
Depreciation		5.00	5.00	5.00	5.00	5.00	5.00	5.00	5.00	5.00	5.00
Taxable income		11.00	11.00	11.00	11.00	11.00	11.00	11.00	11.00	11.00	11.00
Tax		3.85	3.85	3.85	3.85	3.85	3.85	3.85	3.85	3.85	3.85
Capital	-50.00										
Working capital	-4.05										
Free cash flow	-54.05	12.15	12.15	12.15	12.15	12.15	12.15	12.15	12.15	12.15	12.15
Discount factor	1.000	0.909	0.826	0.751	0.683	0.621	0.564	0.513	0.467	0.424	0.386
PV	-54.05	11.05	10.04	9.13	8.30	7.54	6.86	6.23	5.67	5.15	4.68
NPV	20.60										
IRR	18.28%										

6. The present value of the cash in-flows required by the company is $41 million. This represents an annual cash flow of $6.24 million, calculated as follows:

$$A = PV(A|P,k,n) = 41[0.15(1+0.15)^{30}]/[(1+0.15)^{30} - 1]$$
$$= 6.24 \text{ million}$$

The costs are $2 million a year, which means that the revenue must be $8.24 million each year (ignoring any taxes). The rentals must then be at least $1099.24 per square metre.

7. The capital recovery is calculated as follows:

$$A = PV(A|P,k,n) = 15[0.08(1+0.08)^{25}]/[(1+0.08)^{25} - 1]$$
$$= 1.405 \text{ million}$$

The maintenance costs are $2 million, which makes the total cost $3.405 million. The fee per round of golf is equal to $56.75.

8. The capital recovery is given as follows:

$$A = P(A|P,k,n) - S(A|FV,k,n)$$
$$A = 100,000[0.2(1+0.2)^5]/[(1+0.2)^5 - 1] - 30,000[0.2/\{(1+0.2)^5 - 1\}]$$
$$= 29,406.58$$

The annual maintenance costs are $20,000, which means that the total costs are 49,406.58.

9. The capital recovery is calculated as follows:

$$A = PV(A|P,k,n) = 75,000[0.15(1+0.15)^{15}]/[(1+0.15)^{15} - 1]$$
$$= 12,826.28$$

The operating profit is \$20,000, which means that the net *EAC* is $7,173.72 = 20,000 - 12,826.28$.

The net present value is given by the following expression:

$$P = A(P|A, k, n) = 7,173[(1+0.15)^{15} - 1]/[0.15(1+0.15)^{15}]$$
$$= 41,947.40$$

The *IRR* can be calculated in the usual fashion (by trial and error) and yields a value of 25.83%.

10. The cost of the endowment is calculated as follows:

$$P = A(P|A, k, n) = 10,000[(1+0.05)^{50} - 1]/[0.05(1+0.05)^{50}]$$
$$= 182,559.25$$

11. The monthly payment is calculated as follows:

$$A = PV(A|P, k, n)$$
$$A = 500,000[0.07/12(1+0.07/12)^{30(12)}]/[(1+0.07/12)^{30(12)} - 1]$$
$$= 3,326.51$$

12. The *EAC* and *NPV* of the two options are calculated as follows (amount is in millions):

	Project A	Project B
Capital	100.00	200.00
Capital recovery	−11.02	−22.03
Maintenance	−7.50	−4.00
Total *EAC*	−18.52	−26.03
NPV	−168.08	−236.31

13. The *IRR* is 18.60%.
16. The *NPV* is 9,077.04.
17. The *NPV* is described by the following expression:

$$NPV = R/k - 1,000 \Rightarrow 0 = R/0.10 - 1,000 \Rightarrow R = 100$$

18. The annual savings must be \$1,764.76

Chapter 7

1. The calculation of the various discounted cash flow measures is shown in the table below:

Year	Project A	Project B	Discount factor	PV(A)	PV(B)
0	−24,000	−24,000	1.0000	−24,000	−24,000
1	12,000	0	0.8929	10,714	0
2	12,000	5,000	0.7972	9,566	3,986
3	12,000	10,000	0.7118	8,541	7,118
4	12,000	25,000	0.6355	7,626	15,888
NPV				12,448	2,992
IRR				35%	16%
EAC				4,098	985
PI				1.519	1.125

Project A is preferred by all the measures.

2. The calculation of the various discounted cash flow measures is shown in the table below:

Year	Defender	Challenger	Discount factor	PV(A)	PV(B)
0	−20,000	−40,000	1.0000	−20,000	−40,000
1	−5,000	−5,000	0.9434	−4,717	−4,717
2	−7,500	−5,000	0.8900	−6,675	−4,450
3	−10,000	−5,000	0.8396	−8,396	−4,198
4	−12,500	−5,000	0.7921	−9,901	−3,960
NPV				−49,689	−57,326
EAC				−14,340	−16,544

The lowest cost alternative is Project A.

3. The cash flows from operations represent an annuity. The net present value is given by the expression for the case without taxes:

$$NPV = (R - C)(A|P, k, n) - I$$

where R, C and I represent the revenues, costs and initial capital investment, respectively.
The annuity is given by the following expression:

$$(A|P, k, n) = [(1 + k)^n - 1]/[k(1 + k)^n]$$

The *EAC*, if required, can be calculated from the following expression:

$$EAC = NPV(P|A, k, n)$$

The factor $(P|A, k, n)$ is the inverse of the factor $(A|P, k, n)$.

The results of these calculations are given in the table below:

	NPV
Alternative 1	46,469.70
Alternative 2	68,879.28
Alternative 3	73,782.87

In the case of taxes, the NPV is given by the following expression:

$$NPV = [(R-C)(1-t)+tD](A|P,k,n) - I$$

where t is the tax rate, and D is the depreciation charge. In this case, the depreciation charge is equal to the investment divided by ten, so that the NPV is given by the following expression:

$$NPV = [(R-C)(1-t)+\frac{tI}{10}](A|P,k,n) - I$$

The results of these calculations are given in the table below:

	NPV
Alternative 1	20,170.96
Alternative 2	31,766.09
Alternative 3	33,474.51

In all cases, the third option is the preferred choice.

4. This is similar to the previous question, except that there are no revenues, and there is a salvage value at the end of the project. As a result, the NPV can be written as follows:

$$NPV =(-C-0.01(L+B))(A|P,k,n) - I + 0.5B(P|F,k,n)$$
$$+0.9L(P|F,k,n)$$

where C, L, B, and I represent the costs, land, buildings and initial capital investment, respectively.

The results of these calculations are given in the tables below:

	Value
$P|A,k,n$	6.8109
$P|F,k,n$	0.1827

Alternative	C	L	B	I	NPV
East Coast	27,000,000	3,000,000	6,000,000	18,000,000	−201,464,950
West Coast	23,000,000	4,000,000	6,500,000	20,500,000	−176,613,555
Midlands	29,000,000	3,500,000	7,000,000	21,500,000	−218,515,281

The lowest cost alternative is the West Coast option.

5.   The calculation of the economic service life and the *EAC* follows the method
        discussed in Example 7.6. The decrease in the market value and the increase
        in the operating costs for Option A are given in the table below. *PV*(Opex)
        is the present value of the operating costs and PV_n(Opex) is the cumulative
        value for the *PV*(Opex).

Year	Market value	Operating costs	Discount factor	PV(Opex)	Cumulative PV_n(Opex)
0	−1,000,000	0	1.000	0	0
1	−850,000	−21,000	0.917	−19,266	−19,266
2	−722,500	−29,400	0.842	−24,745	−44,011
3	−614,125	−41,160	0.772	−31,783	−75,795
4	−522,006	−57,624	0.708	−40,822	−116,617
5	−443,705	−80,674	0.650	−52,432	−169,049
6	−377,150	−112,943	0.596	−67,344	−236,393
7	−320,577	−158,120	0.547	−86,497	−322,891
8	−272,491	−221,368	0.502	−111,097	−433,988

These values are used to calculate the *EAC* as shown in the table below.

Remaining life, n	A|P,k,n	A|FV,k,n	AOC	CR	EAC (total)
1	1.090	1.000	−22,500	−1,090,000	−1,112,500
2	0.568	0.478	−27,883	−568,469	−596,352
3	0.395	0.305	−34,820	−395,055	−429,875
4	0.309	0.219	−43,811	−308,669	−352,480
5	0.257	0.167	−55,524	−257,092	−312,616
6	0.223	0.133	−70,854	−222,920	−293,774
7	0.199	0.109	−91,009	−198,691	−289,700
8	0.181	0.091	−117,615	−180,674	−298,290

The *EAC* for the second option is calculated in the same manner. The results
of these calculations are given in Figure B.1.

(i)   The *ESL* for Option 1 is seven years and it is six years for Option 2.
(ii)   The *EAC* for Option 1 at the ESL is −$289,700 and it is −$261,622 for
        Option 2.
(iii)   The preferred alternative is Option 2.

6.   The *EAC* is given as follows:

$$EAC = CR + AOC$$

The capital recovery, *CR*, is given by the following expression:

$$CR = Investment(A|PV, k, n) - Salvage(A|FV, k, n)$$

and the annual operating charge, *AOC*, is given as follows:

$$AOC = PV_n(Opex)(A|PV, k, n)$$

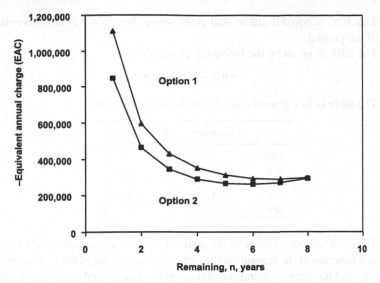

Figure B.1 The equivalent annual charge as a function of the remaining life, n

The cumulative present value of the operating costs is calculated in the table below:

Year	Operating costs	Discount factor	PV(Opex)	Cumulative PV_n(Opex)
0	0	1.000	0	0
1	40,000	0.901	36,036	36,036
2	45,000	0.812	36,523	72,559
3	50,000	0.731	36,560	109,119
4	55,000	0.659	36,230	145,349

The factors $A|PV,k,n$ and $A|FV,k,n$ are equal to 0.322 and 0.212, respectively.

	Defender	Challenger
CR	−35,609	−82,219
AOC	−46,850	−15,000
EAC	−82,459	−97,219

The *EAC* for the defender is lower. Hence it is recommended that it is retained.

7.   The *EAC* is given by the expression:

$$EAC = CR + AOC$$

and the capital recovery, *CR*, is given by the expression:

$$CR = Investment(A|PV, k, n)$$

The *AOC* is equal to the annual costs, since they do not change over the life of the project.

The *NPV* is given by the following expression:

$$NPV = EAC(P|A, k, n)$$

The table below gives the results of these calculations.

	Heat pump	Air conditioning
A|P,k,n	0.110	0.110
AOC	−30,000	−50,000
CR	−55,084	−33,050
EAC	−85,084	−83,050
NPV	−772,311	−753,852

8.   The economic service life of the instrument is the minimum value of the *EAC* as a function of the remaining life. The *EAC* is the sum of the capital recovery, *CR*, and the annual operating charge, *AOC*. The capital recovery is given by the expression:

$$CR = Investment(A|PV, k, n) - Market\ Value(A|FV, k, n)$$

while the *AOC* is constant in this case. The calculation of the *EAC* is given in the table below:

Remaining life, *n*	*A|P,k,n*	*A|FV,k,n*	Market value	*AOC*	*CR*	*EAC*
1	1.120	1.000	−29,750	−3,000	−9,450	−12,450
2	0.592	0.472	−25,288	−3,000	−8,781	−11,781
3	0.416	0.296	−21,494	−3,000	−8,202	−11,202
4	0.329	0.209	−18,270	−3,000	−7,700	−10,700
5	0.277	0.157	−15,530	−3,000	−7,265	−10,265
6	0.243	0.123	−13,200	−3,000	−6,886	−9,886
7	0.219	0.099	−11,220	−3,000	−6,557	−9,557
8	0.201	0.081	−9,537	−3,000	−6,270	−9,270
9	0.188	0.068	−8,107	−3,000	−6,020	−9,020
10	0.177	0.057	−6,891	−3,000	−5,802	−8,802

The economic service life is ten years. The general result is that if the operating costs do not increase, it is best to retain the equipment for as long as possible.

The charge for a sample is calculated as follows:

$$Charge\ per\ sample = \frac{12(5,000) + 8,802}{10,000} = \$6.88$$

9.   The *EAC* of the two options is given by the expression:

$$EAC = P(A|PV, k, n)$$

The price, P, is $0.5 for the cheaper option and $1 for the more expensive option. The EAC of the cheaper option is equal to $0.201. By changing n so that the EAC of the more expensive is equal to $0.201, a value of 7.22 is obtained. This means that the more expensive option must last for 7.22 years. Note that this is more than double the time that the cheaper option should last.

10. The ranking of the project by NPV and PI is given in the table below:

Project	Initial investment	Present value of the cash flows	NPV	PI	Ranking by NPV	Ranking by PI
A	100	120	20	1.20	3	3
B	150	190	40	1.27	1	2
C	200	220	20	1.10	3	4
D	400	320	−80	0.80	4	6
E	10	30	20	3.00	3	1
F	280	305	25	1.09	2	5

If the projects are ranked by PI or NPV, all but project D should be approved.

11. The ranking of the projects by NPV is given in the following table:

Project	Initial investment	IRR	Cumulative investment
D	300	0.14	300
A	200	0.13	500
C	275	0.12	775
B	250	0.11	1025

Only projects A and D will be approved.

12. The calculation of the selected projects is shown in the table below:

Project	Investment	NPV	X	$X_i I_i$	$X_i NPV_i$
A	12,000	17,000	1	12,000	17,000
B	5,000	12,000	1	5,000	12,000
C	8,000	9,000	0	0	0
D	10,000	12,000	0	0	0
F	15,000	25,000	1	15,000	25,000
G	17,000	24,000	1	17,000	24,000
H	12,000	18,500	1	12,000	18,500
I	11,000	19,000	1	11,000	19,000
J	8,000	11,500	1	8,000	11,500
K	8,000	11,000	0	0	0
M	2,000	4,000	1	2,000	4,000
Total				82,000	131,000
Budget				85,000	

The projects whose value of X_i is one are accepted. The total spending is $82,000, which has an NPV of 131,000.

Chapter 8

1.   The nominal rate is given by the expression:

$$r = \frac{1+i}{1+h} - 1 = \frac{1.2}{1.08} - 1 = 11.11\%$$

2.   The values of the real and the nominal NPV are equivalent and equal to $165,852.

3.   The cost is calculated as follows:

$$Cost = 202(1+0.07)^{10} = 397.36$$

4.   The nominal discount rate is calculated as follows:

$$k = (1+r)(1+h) - 1 = 19.84\%$$

5.   The calculation of the NPV in nominal terms is shown in the table below:

Year	0	1	2	3	4	5
Annual cash flow (nominal)		26,250	27,563	28,941	30,388	31,907
Depreciation		10,000	10,000	10,000	10,000	10,000
Tax		5,688	6,147	6,629	7,136	7,667
CFAT		20,563	21,416	22,311	23,252	24,240
PV	−50,000	18,693	17,699	16,763	15,881	15,051
NPV	34,087					

The calculation of the NPV in real terms is shown in the table below:

Year	0	1	2	3	4	5
Annual cash flow (real)		25,000	25,000	25,000	25,000	25,000
Depreciation		10,000	10,000	10,000	10,000	10,000
Tax		5,250	5,250	5,250	5,250	5,250
CFAT		19,750	19,750	19,750	19,750	19,750
PV	−50,000	18,852	17,995	17,177	16,397	15,651
NPV	36,073					

6.   The value in five years is equal to $63,516.

7.   The calculation of the purchase price is shown in the table below:

Year	0	1	2	3
Purchase	32,980			
Property tax		330	330	330
Sale				50,000
Capital gains tax				2,553
CF	−32,980	−330	−330	47,117
PV	−32,980	−294	−263	33,537
NPV	0			

The method used consists of setting up the project financials as shown above, and then search for a value for the purchase price that makes the *NPV* equal to zero.

9. The depreciation schedules using the different methods are given below:

(i) Straight-line depreciation

Year	Annual depreciation charge	Book value
0		100,000
1	20,000	80,000
2	20,000	60,000
3	20,000	40,000
4	20,000	20,000
5	20,000	0
Total	100,000	

(ii) Double declining balance

Year	Annual depreciation charge	Book value
0		100,000
1	40,000	60,000
2	24,000	36,000
3	14,400	21,600
4	8,640	12,960
5	5,184	7,776
Total	92,224	

(iii)   Quarter declining balance

Year	Annual depreciation charge	Book value
0		100,000
1	25,000	75,000
2	18,750	56,250
3	14,063	42,188
4	10,547	31,641
5	7,910	23,730
Total	76,270	

(iv)   MACRS depreciation

Year	Depreciation rate (%)	Annual depreciation charge	Book value
0			100,000
1	20	20,000	80,000
2	32	32,000	48,000
3	19.2	19,200	28,800
4	11.52	11,520	17,280
5	11.52	11,520	5,760
6	5.76	5,760	0
Total		100,000	

Chapter 9

6.   The sensitivities are shown in Figure B.2.

Chapter 11

1.   The monthly returns, the average return and the standard deviation of the returns is given in the table below:

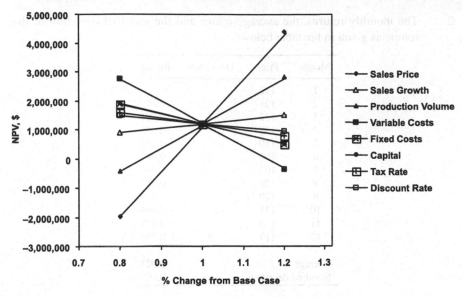

Figure B.2 The sensitivity of the *NPV* to the values of the input variables

Month	Price	Return
1	10	
2	12	0.200
3	9	−0.250
4	13	0.444
5	11	−0.154
6	10	−0.091
7	13	0.300
8	12	−0.077
9	14	0.167
10	11	−0.214
11	10	−0.091
12	9	−0.100
Average		0.012
Standard deviation		0.228

2.   The monthly returns, the average return and the standard deviation of the returns is given in the table below:

Month	Price	Dividends	Return
1	110		
2	121		0.100
3	132		0.091
4	130		−0.015
5	110		−0.154
6	100	5	−0.045
7	103		0.030
8	120		0.165
9	140		0.167
10	131		−0.064
11	130		−0.008
12	119	9	−0.015
Average			0.023
Standard deviation			0.099

4.   The break-even sales price is $167.67. The price should be $192.67 if it is 15% above break-even.

5.   The answers are as follows:

(i)   The break-even number of units in both cases is the same, that is, equal to 20,000 units.

(ii)   The DOL is 5 in both cases.

(iii) The denominator in the DOL is the EBIT. Therefore, an increase of 15% in the DOL means an increase of $5(15\%) = 75\%$ in the EBIT.

6.   The break-even number of units is 8,751.

7.   The return is 43.36%.

8.   The weighted average return on the portfolio is calculated in the table below:

Share	Return	Holding	
A	10%	10%	1.00%
B	21%	30%	6.30%
C	15%	40%	6.00%
D	9%	30%	2.70%
Total			16.00%

The weighted average return is 16%.

10. (i) The returns, the average return and the standard deviation for the portfolio are shown below:

Month	Asset 1	Asset 2	Portfolio
1	17	22	19.5
2	2	3	2.5
3	3	5	4
4	-9	6	-1.5
5	1	-1	0
6	7	0.1	3.55
7	21	6	13.5
8	3	-4	-0.5
9	-14	1.5	-6.25
10	-8	9	0.5
11	-0.7	3	1.15
12	10	1	5.5
Average return	2.69	4.30	3.50
Standard deviation	10.25	6.59	6.90

(ii) The portfolio return can also be calculated from the expression given below:

$$\sigma_p^2 = X_1^2\sigma_1^2 + X_2^2\sigma_2^2 + 2X_1X_2\sigma_{12}$$

The results of these calculations are given in the table below:

Correlation coefficient	Variance	Standard deviation
1	70.97	8.42
0	37.16	6.10
-1	3.35	1.83

(iii) This is obtained by varying X_1 in part (ii) over the range.

(iv) The standard deviation of the portfolio as the percentage of the holding of asset 1 is changed is shown in the table below:

Percentage of holding of asset 1	Standard deviation of portfolio
0	4.30
0.2	3.98
0.4	3.66
0.6	3.34
0.8	3.01
1	2.69

11.   The variance of the portfolio is given by the expression:

$$\sigma_P^2 = \frac{1}{N}(Average\ Variance) + \frac{N-1}{N}(Average\ Covariance)$$

The minimum risk is the average covariance. This means that in order to reduce the risk by 90%, the portfolio risk must be 14. The portfolio must consist of 10 assets to reduce the portfolio variance by 90%.

12.   The expected return given by the CAPM is expressed as follows:

$$R_i = R_F + \beta_i(R_M - R_F) = 0.08 + 0.6(0.14 - 0.08) = 11.6\%$$
$$R_i = R_F + \beta_i(R_M - R_F) \Rightarrow 0.2 = 0.08 + \beta_i(0.14 - 0.08) \Rightarrow \beta_i = 2.0$$

13.   The expected return is given by the expression:

$$R_i = R_F + \beta_i(R_M - R_F) = 0.06 + 0.8(0.07) = 11.6\%$$

14.   The market risk premium is $R_M - R_F$. Solving the two equations for the CAPM to eliminate R_F gives a value for the risk premium of 4.29%. If the shares are correctly priced, the risk premium is the same for both of them. The substitution of the values into the CAPM yields a value for the risk-free rate of 12.14%.

15.   The regression line is Return = 3.4568(Beta) + 11.42. The assets B and D lie above the line, and undervalued, while A and C lie below the line and are overvalued.

Chapter 12

1.   The WACC is 15.4%.
2.   The cost of equity is calculated using CAPM, which gives a value of 14%.
3.   The WACC is 12.45%.
4.   The market risk premium is 6.67%.
5.   The calculation of the cost of equity is shown as follows:

$$R_E = \frac{D_1}{P} + g = 15\% + 12\% = 27\%$$

6.   The values calculated are shown in the table below:

Item	Value
Cost of equity	14.00%
Cost of debt	12.00%
Tax rate	45.00%
WACC	11.04%

7. The results of the calculation of the WACC as a function of the percentage of debt is given in the table below:

Percentage debt, %	Cost of equity capital, %	WACC,%
0	10	10.00
10	10.1	9.74
20	10.2	9.46
30	10.3	9.16
40	10.5	8.90
50	11	8.75
60	11.5	8.50
70	13.5	8.60
80	16	8.40

It is optimal to hold 60% debt.

8. The market return and the share return are shown in the table below:

Period	Market index	Share price	Market return	Share return
1	12	50		
2	11	40	−0.083	−0.200
3	14	55	0.273	0.375
4	15	65	0.071	0.182
5	14	60	−0.067	−0.077
6	15	65	0.071	0.083
7	16	60	0.067	−0.077
8	17	75	0.063	0.250
9	15	78	−0.118	0.040
10	14	70	−0.067	−0.103
11	13	75	−0.071	0.071

A plot of the share return versus the market return has a slope of 1.137, which is the value of beta.

Chapter 13

1. The calculation of the expected value and the standard deviation are given in the table below:

Outcome	Probability	NPV	$p_i NPV_i$	$p_i(NPV_i - \overline{NPV})^2$
1	0.1	100	10.0	298.1
2	0.3	150	45.0	6.3
3	0.3	162	48.6	16.4
4	0.2	200	40.0	412.2
5	0.1	110	11.0	198.9
Expected value			154.6	
Standard deviation				30.5

2.   Expected value is equal to 9.8 and the standard deviation is equal to 1.6.
3.   By definition of probability. If they don't add up to one, then the data does
     not make sense.
4.   The results of the calculation of the NPV, the expected NPV and the standard
     deviation of the NPV are given in the table below:

Outcome	Probability	NPV	$p_i NPV_i$	$p_i(NPV_i - \overline{NPV})^2$
1	0.1	1051.09	105.11	54077.2
2	0.3	677.61	203.28	39290.0
3	0.3	−644.63	−193.39	276679.1
4	0.2	512.77	102.55	7766.1
5	0.1	981.59	98.16	44339.0
Expected value			315.72	
Standard deviation				649.73

5.   The discount rate is 13%.
6.   The results of the calculation of the NPV and the certainty equivalent of the
     NPV are shown in the table below:

Year	Expected cash flow ($ millions)	Certainty equivalent coefficient	Certainty equivalent
0	−55	0.95	−52.25
1	15	0.86	12.90
2	17	0.79	13.43
3	19	0.76	14.44
4	20	0.72	14.40
NPV	0.62		
NPV (CE)			−3.46

7.   The results of the calculation of the coefficient of variation of the cash flows
     is given in the table below:

Outcome	Probability	CF_i	$p_i CF_i$	$p_i(CF_i - \overline{CF})^2$
1	0.1	450	45	250
2	0.8	500	400	0
3	0.1	550	55	250
Mean CF			500	
Standard deviation of CF				22
Coefficient of variation				0.045

8.   (i)   The return on equity for the company is 10.5%.
     (ii) The beta with the new division is 1.2. The new return on equity is 10%.
     (iii) The return for the division is 8%. The reason for the lower return re-
           quired by the division is the lower risk.

9.   (i)   The divisional return is given by the expression:

$$R_E = (1-x)R_{Division} + x R_{OtherDivisions}$$
$$\Rightarrow 9\% = 0.3(12\%) + 0.7 R_{OtherDivisions} \Rightarrow R_{OtherDivisions} = 7.71\%$$

     (ii)   The beta for the other divisions is calculated as follows:

$$\beta_E = (1-x)\beta_{Division} + x\beta_{OtherDivisions}$$
$$\Rightarrow 1.2 = 0.3(0.8) + 0.7\beta_{OtherDivisions} \Rightarrow \beta_{OtherDivisions} = 1.371$$

10. The calculation of the certainty equivalent cash flows and the *NPV* are shown
     in the table below:

	Capital cost	CF_1	CF_2	CF_3	CF_4	CF_5
Expected value	−1,800	600	600	600	600	600
Certainty equivalent coefficients		0.9	0.85	0.82	0.75	0.7
Certainty equivalent cash flows	−1800	540	510	492	450	420
NPV	301.18					

Chapter 14

1.   The expected value of the event is 8.3.
2.   The decision tree is shown in Figure B.3:
3.

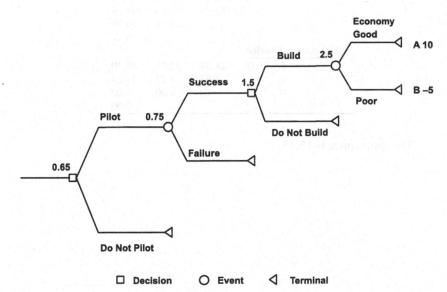

Figure B.3 Decision tree for Exercise 14.2

Chapter 15

1. (i)   Equal to 0.75.
    (ii) Price if it increases is $110, if it decreases it is $90.
    (iii) Payoff if the price increases is $110 - 80 = 30$. If the price decreases it
          is $90 - 80 = 10$.
    (iv) $f = \dfrac{1}{1+r}(pf_u + (1-p)f_d) = \dfrac{1}{1.05}(0.75(30) + 0.25(10)) = 23.809$

2. The stock price lattice and the option price lattice are given in the table below:

	0	1	2
Stock price movements			
	100.00	115.19	132.69
		86.81	100.00
			75.36
European call option			
	13.35	22.54	37.69
		2.70	5.00
			0.00

The option price is 13.35.

3. The stock price lattice and the option price lattice are given in the table below:

	0	1	2	3
Stock price movements				
	100.00	112.24	125.98	141.40
		89.09	100.00	112.24
			79.38	89.09
				70.72
European call option				
	13.73	21.54	32.55	46.40
		4.93	9.22	17.24
			0.00	0.00
				0.00

The option price is 13.73.

4.   The stock price lattice and the option price lattice are given in the table below:

	0	1	2	3	4	5
Stock price movements						
	8.00	8.98	10.08	11.31	12.70	14.25
		7.13	8.00	8.98	10.08	11.31
			6.35	7.13	8.00	8.98
				5.66	6.35	7.13
					5.04	5.66
						4.49
American put option						
	2.16	1.44	0.77	0.26	0.00	0.00
		2.87	2.08	1.27	0.52	0.00
			3.65	2.87	2.00	1.02
				4.34	3.65	2.87
					4.96	4.34
						5.51

The option price is 2.16.

(i)   If the volatility increases to 60%, the option price is $2.49.
(ii)   If the risk-free rate increases to 12%, the option price is $2.05.

5.   The value of the option from the Black–Scholes equation is $72.27 million. The intrinsic value, or NPV, is $40 million. If this were a financial option, it would not be exercised as this point. However, both the option value and the present value of returns will diminish with time as the patent runs out. Since the option is in-the-money, and the underlying asset (the project) will only decrease in value as the patent runs out, it is recommended to invest now.

6.   The expected NPV of the lease is equal $5.5 million $(= 0.1(100) + 0.9(-5))$. The value of the underlying asset is equal to $35.5 million ($V = NPV + K = 5.5 + 30$). The value of the option from the Black–Scholes equation is $10.68 million. This means that a sale price for the lease, which is the option to produce oil, is lower than the option price. It is recommended that the company sell the lease for the option price, that is, $10.68 million.

7.   The option price is $10.53 million and the intrinsic value is $10 million. This option is deep-in-the-money. It is recommended that the company sell.

8.   The option to expand is worth $10.36 million and the intrinsic value is $10 million. The option is deep-in-the-money, and hence the company should expand.

9.   This option, whose value is $0.047 million, is out-of-the-money, and should not be exercised.

Chapter 16

1. The monthly repayments are $7,500 a month, the final settlement is $1,007,500, which includes the principal and the last interest payment. The effective rate is 9.38%.

2. The *APR* is 9%. The *IRR* is 0.75% per month, that is, 9.38% pa $[=(1+0.75\%)^{12}-1]$.

3. The effective rate for the regular loan is 9.38%, and that for the discount loan is 8.70%. Select the discount loan.

4. The loan principal should be $2,051,282.05. The effective rate is 10.6577%.

5. The payments are $8,819.44 per month. The present value of this annuity stream is described by the expression:

$$P = A\left(\frac{1-(1+i)^{12T}}{i}\right) \Rightarrow 250,000 = 8,819.44\left(\frac{1-(1+i)^{36}}{i}\right)$$

Solving this equation for the value of i yields a value of 1.3537% per month, and the effective rate is 17.51% $[= (1+1.3537\%)^{12}-1]$.

6. The monthly repayments are calculated from the following annuity expression:

$$A = P\left(\frac{i/12}{1-(1+i/12)^{12T}}\right) = 450,000\left(\frac{0.07/12}{1-(1+0.07/12)^{36}}\right) = 3,488.85$$

7. The effective rate for the bond is 10.25%.

8. The current price of the bond is $1,136,442.

9. The banker's acceptance has the form of a discount loan. As a result, the company needs a loan amount of $166,666.67. The settlement amount is $150,000.

10. The company should purchase 100 futures contracts.

Glossary

Accounts payable	Money that the company owes to suppliers and vendors for products and services purchased on credit.
Accounts receivable	Money that is owed to the company for the supply of products and services provided on credit.
Accrual accounting	The system of accounting in which the income is reported when earned and expenses when incurred. In contrast to cash accounting, in which income is reported when cash is received and expenses when cash is paid.
Accrued expenses	Expenses that have not been paid by the end of the accounting period.
Amortization	The reduction in value of a liability in regular payments. A mortgage is amortized through the payments of the instalments. The writing of the value of an intangible asset over the lifetime of the asset, is also referred to as amortization.
Annual operating charge	The equivalent annual charge of the operating costs.
Arbitrage	The action of exploiting the price difference between identical assets or financial instruments that are sold in different markets or in different forms.
Arbitrageur	Someone who practices arbitrage.
Asset	An item of value owned by a person or company.
Audit	The examination of the company's financial and accounting records.
Balance sheet	The statement of the financial state of the business, particularly the assets of the business and the claims against them, that is the liabilities and owners' equity.
Bank overdraft	A loan arrangement with a bank that allows the account holder to have a negative balance.
Bankers' acceptance	A short-term loan issued by a firm and guaranteed by a bank.
Basic engineering	A phase in engineering in which the design is specified sufficiently for the price to be estimated to within 15%.
Bill of exchange	A note or order by one person for a second person to pay a third person.
Board of Directors	People elected by the shareholders to oversee the management of the company. They assume legal responsibility for the company.

Bond	A financial instrument used to raise debt for more than one year. It is a promise to repay the principal and the interest by a specified date according to a specified schedule.
Bond market	The market for the trading of bonds. It can be on an exchange or over-the-counter.
Book value	The value of an asset on the balance sheet. It is equal to the cost of acquisition of the asset less the accumulated depreciation.
Bottom line	Net income, or net profit.
Call option	An option that gives the owner the right but not the obligation to acquire an asset at a point in the future for a price that is decided now.
Capital	The cash and goods that are used to generate income. Also refers to the owners' claim on a business, that is, the assets less the liabilities.
Capital allowances	The allowances for the cost of an asset for the purposes of tax. Depreciation for the purposes of tax.
Capital budgeting	The procedures used to determine whether a long-term project is worth undertaking.
Capital gain	The difference in an asset's selling price and its original purchase price (if positive).
Capital project	
Capital recovery	The equivalent annual charge of the capital cost of an investment.
Capital structure	The proportion of permanent long-term financing of the company that is made up of long-term debt, shareholder's capital and retained earnings. See financial structure.
Cash flow statement	The statement of the company's cash flow over a reporting period. It shows the sources and uses of cash for the period.
CEO	Chief Executive Officer
Collateral	Assets that are pledged by a borrower to secure a loan. They are seized in the event of default.
Common stock	Equity ownership in a company. Also called shares.
Compound interest	Interest that is paid on the accumulated interest.
Concept	The earliest phase of an engineering or business study.
Contingent	Dependent on events
Controller	
Convertible bond	A bond that can be converted at the choice of the bondholder into shares of the company. Also called convertible loan stock.
Corporate venture capital	A division or subsidiary of a corporation that makes venture capital investments.
Cost of capital	The cost of raising funds by a company for its activities.
Cost of goods sold (COGS)	The cost of purchasing raw materials and manufacturing the final product. Can also be called the variable costs.
Cost-benefit analysis	The analysis of the benefits and costs to all stakeholders in a possible development.
Credit management	
Creditors	Same as accounts payable.
Current assets	The assets that are readily converted into cash.
Current liabilities	The liabilities that are due for repayment within a year.
Days payable	The average time that it takes a company to pay its suppliers and vendors.
Days receivable	The average time that it takes the company's vendors and suppliers to pay for purchases.
Debt	A liability in which one person owes to another and which must be repaid by a certain date. Forms of debt are loans, bonds, mortgages, *etc.*

Debt-holders	The lenders of permanent or long-term debt to a company.
Debtors	A person or company that owes another person or company a debt.
Depreciation	The allocation of the cost of an asset over time.
Derivative	A financial instrument whose value depends on the value of another asset or financial instrument, called the underlying asset.
Detailed engineering	A phase in the development of an engineering project.
Discounted cash flow	The discounting of anticipated future cash flows to their present values as a result of the time value of money.
Dividend yield	The dividend per share divided by the price per share.
Dividends	A payment made by a company to its shareholders. The board of directors declares the dividend amount, which must be paid from current or retained earnings.
Due diligence	The examination of the details and a verification of the facts of a potential investment.
Earnings	The profit, or net income, of a company. Equal to the revenue less costs, depreciation, interest, and taxes.
Earnings before interest and taxation	The same as operating income, which is revenues less cost of goods sold and SGA expenses.
Economic service life	The life of a machine at which the equivalent annual charge is a minimum.
Equity	Ownership, particularly the ownership interest in a company. Also the value of the owner's interest in a company.
Equity capital	Capital raised from owners.
Equivalent annual charge	The annuity amount that is equivalent to the net present value of an investment calculated over the life of the investment.
Feasibility	A stage in engineering design.
Finance	The raising, investment and management of funds.
Financial accounting	The reporting of accounting information to external parties.
Financial structure	The balance sheet items that details how assets are financed. Differs from capital structure in that it includes short-term items such as working capital.
Financial year	The twelve-month period for the preparation of the financial statements.
Fiscal year	Same as financial year.
Fixed assets	Long-term, tangible assets held for the generation of income, such as manufacturing equipment or plant.
Fixed costs	Costs that do not vary with the rate of production.
Fixed tangible assets	Same as fixed assets.
Free cash flow	Operating cash flow less capital investment (less dividends).
Futures	A contract to purchase or sell an asset for a fixed price at some specified time in the future. These are standardized contracts that are traded on exchanges.
GAAP	Generally Accepted Accounting Principles.
Gearing	The ratio of long-term debt to the total capital of the business. The use of long-term debt in the capital structure.
General, selling and administrative	The costs not directly associated with the manufacture of the product. Also called the overheads.
Goodwill	An intangible asset, such as brand name. In the even of an acquisition of a company, goodwill is the difference between the purchase price and the net asset value of the acquired company.
Historical cost	An accounting convention in which all costs for a company are recorded based on original price.
Hurdle rate	The required rate of return in the discounted cash flow analysis, above which the investment is viable and below which it is not.

IFRS	International Financial Reporting Standards
Income	Same as earnings.
Income statement	A statement of the sales, expenses and profit for a period.
Inflation	The loss in purchasing power of money, or the increase in prices of goods; measured by the consumer price index.
Interest	The fee charged for the use of money.
Internal rate of return	The discount rate at which the net present value is zero.
Inventory	The company's raw materials, and finished and unfinished products that have not been sold yet.
Junior debt	Same as subordinated debt.
Leverage	The degree to which a company is using debt in the financing of the company's activities.
Liabilities	A financial obligation or debt. A claim against the value of the company from supplying goods or lending money to the company.
Loan capital	Capital that is provided in the form of debt.
Long term	Greater than one year.
Long-term debt	Debt that is not due within a year.
Net asset value	The value of the assets less the value of the liabilities.
Net income	Sales less costs, interest, depreciation, taxes and other expenses. Same as net profit. Also the bottom line.
Net present value	The present value of an investment's anticipated future cash flows less the initial investment. If it is positive, the investment is recommended; otherwise it is not.
Net profit	Net income.
NPV	Net present value.
Option	A derivative instrument in which the hold has the right but not the obligation to purchase or sell an asset, called the underlying asset.
Order-of-magnitude	A stage in the engineering design process.
Ordinary shares, capital stock	A unit of ownership in a company.
Overheads	The fixed costs of a business, such as rentals, insurance and administration expenses.
PBIT	See Earnings before interest and taxation.
Pre-feasibility	A stage in the engineering design process.
Private placement	The direct selling of equity and bonds of a company, that is, not to the public at large or through an exchange.
Profit	The gain from an investment or business that is attributable to owners.
Profit and loss account	A statement of the sales, expenses and profit for a period. Similar to an income statement.
Property, plant and equipment	Same as fixed assets.
Public issue	The selling of the equity in a company to the public through a stock exchange.
Reserves	Same as retained earnings. May be non-distributable reserves in the event that the directors have re-valued property.
Revenues	Sales
Royalty	Amount paid for the use of property. Usually intellectual property such as patents. The royalty amount is usually calculated as percentage of revenues.
Salvage value	
Security	A document containing a right, such as notes, agreements, contracts.
Selling, General and Administration (SGA)	The costs not directly associated with the manufacture of the product. Also called the overheads.

Share	A portion of ownership in a company or partnership.
Share capital	The portion of capital that has been obtained directly from shareholders from the sale of shares.
Share premium	An amount above the nominal value of the share.
Shareholders	An owner of shares in a company.
Shares in issue	The total nominal value of the shares that have been issued to shareholders.
Shares outstanding	Same as shares in issue.
Stockholder's equity	Same as reserves.
Stocks	Same as shares.
Swap	A derivative in which one set of cash flows is exchanged for another set.
Tax depreciation	Depreciation for the purposes of tax.
Treasurer	The person in a company who is responsible for the collection and investment of funds.
Turnkey	A product or service that can be implemented without any additional expenditure.
Turnover	The same as revenue or sales.
Underlying asset	The asset on which the value of a derivative instrument depends.
Value at Risk	A method for determining the statistical chances of making a loss on an investment.
VaR	See Value at Risk.
Variable costs	Costs dependent on the rate of production.
Venture capital	The funding of start-up and small businesses that have growth potential.
Volatility	The relative rate at which the price moves up or down.
Withdraw	Take money out of an account.
Working capital	Current assets less current liabilities.

Bibliography

Books

1. Advancement of Cost Engineers International (1997) Cost Estimate Classification System: As Applied in Engineering, Procurement and Construction for the Process Industries. AACE International Recommended Practice No. 18R-97
2. Aggarwal R (1993) Capital budgeting under uncertainty. Prentice-Hall, Englewood Cliffs, NJ
3. Allen DH (1991) Economic Evaluation of Projects: A Guide, 3rd edn. Institution of Chemical Engineers, Rugby, England
4. Amis D, Stevenson H (2001) Winning Angels. Pearson Education, London
5. Atrill P, McLaney E (2004) Accounting and Finance for Non-Specialists, 4th edn. Pearson Education, Harlow, England
6. Bernstein PL (1992) Capital Ideas: The Improbable Origins of Modern Wall Street. The Free Press, New York
7. Bierman H, Smidt S (1975) The Capital Budgeting Decision: Economic Analysis and Financing of Investment Projects. MacMillan, New York
8. Blank L, Tarquin A (2002) Engineering Economy, 5th edn. McGraw-Hill, New York
9. Brigham EF, Ehrhardt MC (2002) Financial Management Theory and Practice, 10th edn. Harcourt College, Fort Worth, TX
10. Clark JJ, Hindelang TJ, Pritchard RE (1984) Capital Budgeting Planning and Control of Expenditures, 2nd edn. Prentice-Hall, Englewood Cliffs, NJ
11. Clemen, RT (1996) Making Hard Decisions: An Introduction to Decision Analysis, 2nd edn. Duxbury, Belmont CA
12. Copeland T, Antikarov V (2001) Real Options: A Practitioner's Guide. Texere, New York
13. Copeland T, Koller T, Murrin J (2000) Valuation: Measuring and Managing the Value of Companies, 3rd edn. Wiley, New York
14. Correia C, Flynn D, Uliana E, Wormald M (1993) Financial Management, 3rd edn. Juta and Co, Johannesburg, South Africa
15. Dixit AK, Pindyck RS (1994) Investment under uncertainty. Princeton, University Press, Princeton, NJ
16. Elton J, Gruber MJ (1995) Modern Portfolio Theory, 5th edn. Wiley, New York
17. Finnerty JD (1996) Project Financing: Asset-Based Financial Engineering. Wiley, New York
18. Fogler HS, LeBlanc SE (1995) Strategies for Creative Problem Solving. Prentice-Hall, Englewood Cliffs, NJ
19. Glantz M (2000) Scientific Management: Advances In Intelligence Capabilities For Corporate Valuation And Risk Assessment. American Management Association, New York
20. Groppelli AA, Nikbakht E (1995) Finance, 3rd edn. Barron's Educational Series, New York

21. Haug EG (1998) The Complete Guide to Option Pricing Formulas. McGraw-Hill, New York
22. Higgins RC (1999) Analysis for Financial Management, 5th edn. McGraw-Hill, New York
23. Hull JC (1997) Options, Futures and Other Derivatives, 3rd edn. Prentice-Hall, Upper Saddle River, NJ
24. Hull JC (1998) Introduction to Futures and Options Markets, 3rd edn. Prentice-Hall, Upper Saddle River, NJ
25. Institution of Chemical Engineers and the Association of Cost Engineers (1977) A New Guide to Capital Cost Estimating. The Institution of Chemical Engineers, Rugby, England
26. King, TA (2006) More than a Numbers Game: a Brief History of Accounting. Wiley & Sons, Hoboken, NJ
27. McConnell CR, Brue SL (2005) Economics: Principles, Problems, and Policies, 16th edn. McGraw-Hill, New York
28. Monahan GE (2000) Management Decision Making: Spreadsheet Modeling, Analysis, and Application. Cambridge University Press, Cambridge, UK
29. Newnan DG (1996) Engineering Economic Analysis, 6th edn. Engineering, San Jose, CA
30. Northcott D (1992) Capital Investment Decision-Making. Academic, London
31. Parker RH (1999) Understanding company financial statements, 5th edn. Penguin Books, London
32. Peters MS, Timmerhaus KD (1981) Plant Design and Economics for Chemical Engineers, 3rd edn. McGraw-Hill, New York
33. Rice A (2003) Accounts Demystified, 4th edn. Pearson Education, London
34. Ross S, Westerfield RW, Jordan BD, Firer C (2001) Fundamentals of Corporate Finance, 2nd South African edn. McGraw-Hill, Roseville, NSW, Australia
35. Silbiger S (1999) The 10-Day MBA, 2nd edn. Piatkus, London
36. Stermole FJ, Stermole JM (2000) Economic Evaluation and Investment Decision Methods, 10th edn. Investment Evaluations Corporation, Lakewood, CO
37. Timmons JA (1999) New Venture Creation: Entrepreneurship for the 21st Century. McGraw-Hill, Singapore
38. Trigeorgis L (2000) Real Options: Managerial Flexibility and Strategy in Resource Allocation. The MIT Press, Cambridge, MA

Journal Articles

39. Abel A, Eberly JC (1994) A unified theory of investment under uncertainty. Am Econ Rev 84:1369–1384
40. Abel AB, Dixit AK, Eberly JC, Pindyck RS (1996) Options, the value of capital, and investment. Q J Econ 111:753–777
41. Ang JS, Lewellen WG (1982) Risk adjustment in capital project evaluations. Finan Manage 11(2):5–14
42. Barnett M (1987) Role of merchant banker in projects. Project Manage 5(4):197–203
43. Ben-Horim M, Sivakumar N (1988) Evaluating Capital Investment Projects. Managerial Decis Econ 9:263–268
44. Beranek W (1975) The cost of capital, capital budgeting and the maximization of shareholder value. J Finan Quant Anal 10(1):1–20
45. Bernanke B (1983) Irreversibility, uncertainty and cyclical investment. Q J Econ 98(1):85–106
46. Bhappu RR, Guzman J (1995) Mineral investment decision making. A study of mining company practices. Eng Mining J 70: 36–38
47. Black F, Scholes M (1973) The pricing of options and corporate liabilities. J Polit Econ 81:637–569
48. Bradley PG (1998) On the use of modern asset pricing for comparing alternative royalty systems for petroleum development projects. Energy J 19:47–81

49. Brennan MJ, Schwartz ES (1985) Evaluating Natural Resources Investments. J Business 58:135–157
50. Brookfield D (1995) Risk and capital budgeting: avoiding pitfalls in using NPV when risk arises. Manage Decis 33(8):56–59
51. Bruner RF, Eades KM, Harris RS, Higgins RC (1998) Best practices in estimating the cost of capital: Survey and synthesis. Finan Pract Educ 8(1):13–28
52. Butler JS, Schachter B (1989) The investment decision: estimation risk and risk-adjusted discount rate. Finan Manage (Winter 1989):13–22
53. Cannadi J, Dollery B (2005) An evaluation of private sector provision of public infrastructure in Australian local government. Australian J Public Admin 64(3):112–118
54. Chambers DR, Harris RS, Pringle JJ (1982) Treatment of financing mix in analysing investment opportunities. Finan Manage 11(2):24–41
55. Chaney AR (1987) Financial guarantees. Project Manage 5(4):231–236
56. Coles S, Rowley J (1995) Revisiting decision trees. Manage Decis 33(8):46–50
57. Constantinides GM (1978) Market risk adjustment in project evaluation. J Finan 33(2):603–616
58. Copelan TE, Keenan PT (1998) Making real options real. McKinsey Q 1998(3)128–141
59. Copeland T (2002) The real-options approach to capital allocation. IEEE Eng Manage Rev First Quarter 2002:82–85
60. Corner JL, Kirkwood CW (1991) Decision analysis applications in the operations research literature, 1970–1989. Oper Res 39(2):206–219
61. Dillon RL, John R, von Winterfeldt D (2002) Assessment of cost uncertainties for large technology projects: a methodology and an application. Interfaces 32(4):52–66
62. Dorfman R (1981) The meaning of internal rates of return. J Finan 36(5):1011–1021
63. Dudley CL (1972) A note on reinvestment assumption in choosing between net present value and internal rate of return. J Finan 27(4) 907–915
64. Dulman SP (1989) The development of discounted cash flow techniques in US industry. Bus History Rev 63(3):555–587
65. Eschenbach TG, Lavelle JE (2001) MACRS depreciation with a spreadsheet function: a teaching and practice note. Eng Econ 46(2):153–161
66. Fama EF (1977) Risk-adjusted discount rates and capital budgeting under uncertainty. J Finan Econ 5: 3–24
67. Findlay MC, Gooding AE, Weaver WQ (1976) On the relevant risk for determining capital expenditure hurdle rates. Finan Manage 5(4):9–16
68. Graham J, Harvey C (2002) How do CFOs make capital budgeting and capital structure decisions? J Appl Corp Finan 15(1):8–23
69. Grant M (1997) Financing Eurotunnel. Japan Railway Transport Rev 11:46–52
70. Grinyer JR, Sinclair CD, Ibrahim DN (1999) Management objectives in capital budgeting. Finan Pract Educ 9(2):12–22
71. Gup BE, Norwood SW (1982) Divisional cost of capital: a practical approach. Finan Manage 11(1):20–24
72. Hacura A, Jadamus-Hacura M, Kocot A (2001) Risk analysis in investment appraisal based on the Monte Carlo simulation technique. Eur Phys J B 20:551–553
73. Hamilton D (2004) Economic analysis of pumping alternatives for the A.D. Edmonston pumping plant. Cost Eng 46(3):16–25
74. Hax AC, Wiig KM (1975) The use of decision analysis in capital investment problems. Sloan Manage Rev 17(2):19–48
75. Hertz DB (1964) Risk analysis in capital investment. Harvard Bus Rev (Jan–Feb 1964):95–106
76. Hillier FS (1963) The derivation of probabilistic information for the evaluation of risky investments. Manage Sci 9(3):443–57
77. Howard RA (1988) Decision analysis: practice and promise. Manage Sci 34(6):679–695
78. Ingesoll JE, Ross SA (1992) Waiting to invest: investment and uncertainty. J Bus 65:1–29
79. Jacoby HD, Laughton DG (1992) Project evaluation: a practical asset pricing method. Energy J 13:19–47

80. Jagannathan R, Meier I (2002) Do we need CAPM for capital budgeting? Finan Manage 31(4):55–77
81. Keef SP, Roush ML (2001) Discounted cash flow methods and the fallacious reinvestment assumption: a review of recent texts. Account Educ 10(1):105–116
82. Keeney RL (1982) Decision analysis: an overview. Oper Res 30(5):803–838
83. King P (1975) Is the emphasis of capital budgeting theory misplaced? J Bus Finan Account 2(1):66–82
84. Kira DS, Kusy MI (1990) A stochastic capital rationing model. J Oper Res Soc 41(9):853–863
85. Kumar S, Benusa JM (2004) Economic justification of a data acquisition system: a case study. Cost Eng 46(1):31–37
86. Laughton D (1998) The management of flexibility in the upstream petroleum industry. Energy J 19:83–114
87. Lintner J (1965) Security prices, risk, and maximal gains from diversification. J Finan 20(4):587–615
88. Lockett AG, Gear AE (1975) Multistage capital budgeting under uncertainty. J Finan Quant Anal 10(1):21–36
89. Love PED, Skitmore M, Earl G (1998) Selecting a suitable procurement method for a building project. Constr Manage Econ 16:221–233
90. Luehrman TA (1997) Investment opportunities as real options. Harvard Bus Rev July–August 1998 51–67
91. Luehrman TA (1998) Strategy as a portfolio of options. Harvard Bus Rev Sept–Oct 1998:89–99
92. Majd S, Pindyck RS (1987) Time to build, option value, and investment decisions. J Finan Econ 18:7–27
93. Markowitz H (1952) Portfolio selection. J Finan 7(1):77–91
94. McDonald R, Siegel D (1984) Option pricing when the underlying asset earns a below equilibrium rate of return. J Finan 39:261–265
95. McDonald R, Siegel D (1985) Investment and the valuation of firms when there is an option to shut down. Int Econ Rev 26:331–349
96. McDonald R, Siegel D (1986) The value of waiting to invest. Q J Econ 101:707–727
97. McDonald RL (1998) Real options and rules of thumb in capital budgeting. In: Brennan MJ, Trigeorgis L (eds) Innovation, Infrastructure, and Strategic Options. Oxford University Press, London
98. McNulty JJ, Yeh TD, Schulze WS, Lubatkin MH (2003) What's your real cost of capital? IEEE Eng Manage Rev First Quarter 2003:76–82
99. Merton RC (1973) The theory of rational option pricing. Bell J Econ Manage Sci 4:141–183
100. Meyer RL (1979) A note of capital budgeting techniques and the reinvestment assumption. J Finan 34(5):1251–1254
101. Miles JA, Ezzell JR (1980) The weighted average cost of capital, perfect capital markets and project life: a clarification. J Finan Quant Anal 15(3):719–730
102. Mustafa A (1999) Public–private partnership: an alternative institutional model for implementing the Private Finance Initiative in the provision of transport infrastructure. J Project Finan 5(1):56–71
103. Myers SC, Turnbull SM (1977) Capital budgeting and the asset pricing model: good news and bad news. J Finan 32(2):321–333
104. Nahapiet H, Nahapiet J (1985) A comparison of contractual arrangements for building projects. Constr Manage Econ 3:217–231
105. Obermaier B (2002) Comment on: analysis in investment appraisal based on the Monte Carlo simulation technique by A. Hacua, M. Jadamus-Hacura and A. Kocot. Eur Phys J B 30:407–409
106. OECD (2006) http://www.oecd.org/dataoecd/26/56/33717459.xls. Accessed 29 March 2006
107. Paddock JL, Siegel DR, Smith JL (1988) Option valuation of claims on real assets: the case of offshore petroleum leases. Q J Econ 103:479–508

108. Perrakis S (1975) Certainty equivalents and timing uncertainty. J Finan Quant Anal 10(1):109–118
109. Pindyck RS (1991) Irreversibility, uncertainty, and investment. J Econ Lit 24:1110–1148
110. Pindyck RS (1993) Investments of uncertain cost. J Finan Econ 34:53–76
111. Pindyck RS (1993) The present value model of rational commodity pricing. Econ J 103:511–530
112. Pohlman RA, Santiago ES, Markel FL (1988) Cash flow practices of large firms. Finan Manage 14:71–79
113. Prest AR, Turvey R (1965) Cost-benefit analysis: a survey. Econ J 75(300):683–735
114. Randolph GC, Schrantz A (1997) The use of capital markets to fund the Ras Gas project. J Project Finan 3(2):5–15
115. Robichek AA (1975) Interpreting the results of risk analysis. J Finan 30(5):1384–1386
116. Ross M (1986) Capital budgeting practices at twelve large manufacturers. Finan Manage 15(4):15–22
117. Rubenstein M (1973) A mean-variance synthesis of corporate financial theory. J Finan 28(1):167–181
118. Samis M, Poulin R (1998) Valuing management flexibility: a basis to compare the standard DCF and MAP valuation frameworks. CIM Bull 91:59–74
119. Senet LW, Thompson HE (1978) The equivalence of alternative mean-variance capital budgeting models. J Finan 33(2):395–401
120. Sharpe W (1964) Capital asset prices: a theory of market equilibrium under conditions of risk. J Finan 19(3):425–442
121. Shrieves RE, Wachowicz JM (2001) Free cash flow (FCF), economic value added (EVA™) and net present value (NPV): A reconciliation of variations of discount cash flow (DCF) valuation. Eng Econ 46(1):33–52
122. Sick GA (1986) A certainty equivalent approach to capital budgeting. Finan Manage 15(4):23–32
123. Stulz RM (1999) What's wrong with modern capital budgeting? Finan Pract Educ 9(2):7–11
124. Vriend NJ (1996) Rational behavior and economic theory. J Econ Behav Organ 29:263–285
125. Walker J (1994) A graphical analysis for machine replacement a case study. International J Oper 14:54–63
126. Wallis MR, Morahan GT, Dyer JS (1995) Decision analysis of exploration opportunities in the onshore US at Phillips Petroleum. Interfaces 25(6)39–56
127. Weaver SC, Clemmens PJ, Gunn JA, Dannenburg BD (1989) Divisional hurdle rates and the cost of capital. Finan Manage 18(1):18–25

Index